H. G. (Harry Govier) Seeley

**The Fresh-Water Fishes of Europe**

A History of Their Genera, Species, Structure, Habits, and Distribution

H. G. (Harry Govier) Seeley

**The Fresh-Water Fishes of Europe**

*A History of Their Genera, Species, Structure, Habits, and Distribution*

ISBN/EAN: 9783337143855

Printed in Europe, USA, Canada, Australia, Japan

Cover: Foto ©berggeist007 / pixelio.de

More available books at **www.hansebooks.com**

# THE
# FRESH-WATER FISHES

OF

# EUROPE.

A History of their Genera, Species, Structure, Habits,
and Distribution.

BY

## H. G. SEELEY, F.R.S.,

F.G.S., F.Z.S., F.L.S., F.R.G.S.,

PROFESSOR OF GEOGRAPHY IN KING'S COLLEGE, LONDON; FOREIGN CORRESPONDENT OF THE
ACADEMY OF SCIENCES OF PHILADELPHIA; FOREIGN CORRESPONDENT OF THE
K. K. GEOLOGISCHE REICHSANSTALT OF VIENNA.

*WITH 214 ILLUSTRATIONS.*

CASSELL & COMPANY, LIMITED:
*LONDON, PARIS, NEW YORK & MELBOURNE.*
1886.

[ALL RIGHTS RESERVED.]

# PREFACE.

In this Volume the Fresh-water Fishes of Europe are systematically described for the first time. Such an undertaking has been rendered comparatively easy by the valuable special memoirs which have been published upon the Fishes of the several European countries.

The classification of Dr. Günther has been generally followed; and, although we have often ventured to differ from this distinguished writer in minor matters, it would have been impossible to have brought the Fishes of Europe into intelligible order without the aid of his Catalogue of the Fishes in the British Museum. To that great work readers may refer who would investigate the nomenclature of the species described.

For the Fishes of Great Britain and Ireland we have consulted, among other works, Pennant's "British Zoology," Yarrell's "British Fishes," Couch's "Fishes of the British Isles," Houghton's "British Fresh-water Fishes," Day's "British Fishes," and memoirs in the various publications of the Linnæan and Zoological Societies. Buckland's "Natural History of British Fishes," and his Fishery Reports, have been frequently referred to, while among works of a more special character we may mention Mr. St. John's "Tour in Sutherlandshire," Mr. Russel's work on the Salmon, and the "Prose Halieutics" of the Rev. Dr. Badham.

The method of description has been in general that used by Heckel and Kner in their "Süsswasserfische der Östreichischen Monarchie," upon which we have moulded the descriptions of external characters of the Fishes described. Among other German authors we have been indebted only in a less degree to Von Siebold's "Süsswasserfische von Mitteleuropa," and Günther's "Fische des Neckars;" the former invaluable for its critical discrimination of species, the latter rich in anatomical observations. Benecke's "Fische, Fischerei, und Fischzucht, in Ost- und Westpreussen," has also been consulted.

For the Fishes of France Blanchard's "Poissons des Eaux Douces de la France," and Moreau's "Poissons de la France," have furnished information of the greatest value. For the Fishes of Switzerland the first volume of Fatio's work has been our chief guide. The Fishes of Spain, though described in some special catalogues, like that of Machado, dealing with the Fishes of the Guadalquivir, are best known from the writings of Steindachner, originally published by the Academy of Sciences at Vienna. Italian Fishes have been well catalogued and described by Bonaparte, in his "Fauna Italica;" by Canestrini, in his "Prospetto critico dei Pesci d'Acqua dolce d'Italia," and other memoirs and works; and in the catalogue of Giglioli. On Scandinavian Fishes the more important modern writers are Collett, who has given us "Norges Fiske;" and Lilljeborg, who has written "Sveriges och Norges Fiskar." Many valuable notes will be found in Lloyd's "Scandinavian Adventures."

The chief authorities on Russian Fishes are Nordmann, who described the Fishes in Demidoff's "Voyage dans la Russie Meridionale;" Grimm's "Fishing and Hunting in Russian Waters;" and Kessler's memoirs in the "Bulletin de la Société Imperiale des Naturalistes de Moscou." Besides these works, many general treatises like the "Histoire Naturelle des Poissons" by Cuvier and Valenciennes, and many volumes of travel, like those of Russigger and Filippi, have yielded information which has helped our history.

Shortcomings of omission of some species, almost inevitable in a work of this kind, may be remedied hereafter. But we trust that the fabric of the work will give a new interest to the Fishes of our own country, and may influence British peoples to a thrifty cultivation of the roving wealth which swims, little heeded, in our forms of fresh-water fish life—a cultivation like that which rewards the folk who breed and care for, and feed upon the Fresh-water Fishes of Continental Europe.

H. G. SEELEY.

*The Vine, Sevenoaks.*
*18th February, 1886.*

# CONTENTS.

### CHAPTER I.
#### INTRODUCTION.

The Orders of Fresh-water Fishes—Characters used in Classification of Fishes—The Forms of Fresh-water Fishes—The Structure of the Fins and the Fin Formulae—The Scales—Organs of the Senses—The Mouth—The Operculum and Gill-arches—The Skeleton—Air bladder - Digestive Organs—Reproductive Organs—Growth—Geographical Distribution—Table showing Range of Species—Table showing Classification of Fishes described ............... 1

### CHAPTER II.
#### FRESH-WATER FISHES OF THE ORDER ACANTHOPTERYGII.

FAMILY PERCIDÆ GENUS PERCA : The River Perch — The Bass — GENUS PERCARINA : Percarina demidoffii — GENUS ACERINA: The Ruff or Pope Acerina schraetzer—Acerina rossica GENUS LUCIOPERCA: The Pike-Perch — The Pike-Perch of the Volga GENUS ASPRO : The Zingel The Apron—The Streber—FAMILY COTTIDÆ—GENUS COTTUS : The Miller's Thumb - Cottus poecilopus—Father Lasher—FAMILY GOBIIDÆ—GENUS GOBIUS : Species of Goby—The Polewig—FAMILY BLENNIIDÆ—GENUS BLENNIUS FAMILY ATHERINIDÆ—GENUS ATHERINA : The Sand Smelt—FAMILY MUGILIDÆ—GENUS MUGIL : The Grey Mullet—FAMILY GASTEROSTEIDÆ—GENUS GASTEROSTEUS : The Sticklebacks ........... 23

### CHAPTER III.
#### FRESH-WATER FISHES OF THE ORDER ANACANTHINI.

FAMILY GADIDÆ - GENUS LOTA — Fresh-water Representatives of the Cod Family — The Burbot—FAMILY PLEURONECTIDÆ—The Flounder—The Sole ......... 8

### CHAPTER IV.
#### FRESH-WATER FISHES OF THE ORDER PHYSOSTOMI.

FAMILY SILURIDÆ—GENUS SILURUS—FAMILY CYPRINIDÆ GENUS CYPRINUS : The Carp: Its Varieties and Hybrids—GENUS CARASSIUS : The Crucian Carp—Goldfish— The Carassius bucephalus—GENUS BARBUS : The Barbels of Central Europe and Spain—GENUS AULOPYGE : Aulopyge hügelii—GENUS GOBIO : The Gudgeon—Gobio uranoscopus ... 90

### CHAPTER V.
#### FRESH-WATER FISHES OF THE ORDER PHYSOSTOMI (continued).

GROUP LEUCISCINA—GENUS LEUCISCUS : The Rudd, and its varieties—The Ide Leuciscus aula L. adspersus—The Roach—L. pigus—L. friesii—The Chub—The Dace—L. illyricus— L. pictus—L. svallize—L. ukliva- L. turskyi—L. microlepis—L. tenellus—L. muticellus The Minnow —L. hispanicus - L. arcasii—L. alburnoides— L. macrolepidotus— L. arrigonis—L. lemmingii - L. hoekelii L. pyrenaicus — L. polylepis GENUS PARAPHOXINUS—GENUS TINCA : The Tench GENUS CHONDROSTOMA : C. nasus C. phoxinus—C. genei—C. knerii—C. rysela C. polylepis - C. willkommii ............ 133

### CHAPTER VI.
#### FRESH-WATER FISHES OF THE ORDER PHYSOSTOMI (continued).

GROUP RHODEINA —GENUS RHODEUS : The Bitterling—GROUP ABRAMIDINA GENUS ABRAMIS : The Bream—A. vimba—A. elongatus—A. ballerus A. sapa—A. leuckartii - A.

bjorkna Hybrids A. bipunctatus—A. fasciatus—GENUS ASPIUS: A. rapax—GENUS ALBURNUS: A. lucidus—A. alburnellus A. mento A. chalcoides—GENUS LEUCASPIUS: L. delineatus—GENUS PELECUS: P. cultratus—GROUP COBITIDINA—GENUS MISGURNUS: M. fossilis—GENUS NEMACHILUS: The Loach—GENUS COBITIS: The Spinous Loach . . . 207

## CHAPTER VII.

### FRESH-WATER FISHES OF THE ORDER PHYSOSTOMI (continued).

FAMILY CLUPEIDÆ—GENUS CLUPEA: Shad—Twaite Shad—Whitebait—Black Sea Herring—Caspian Herring . . . . . . . . . 256

## CHAPTER VIII.

### FRESH-WATER FISHES OF THE ORDER PHYSOSTOMI (continued).

FAMILY SALMONIDÆ—GENUS SALMO: Salmon—Salmon Fisheries—Salmon Disease. TROUT: Gillaroo—Salmo hardinii—Sewin—S. argenteus—S. venernensis—S. mistops—S. nigripinnis S. obtusirostris—S. microlepis—River Trout—S. ausonii—Galway Sea Trout S. autumnalis S. lemanus—S. carpio—S. rappii—S. dentex S. spectabilis - S. marsiglii—S. lacustris S. schiffermülleri—Loch Leven Trout—Orkney Trout S. polyosteus—Grey Trout Salmon Trout—Great Lake Trout S. bailloni—S. genivittatus—S. labrax . . . . . 265

## CHAPTER IX.

### FRESH-WATER FISHES OF THE ORDER PHYSOSTOMI (continued).

FAMILY SALMONIDÆ (concluded): - THE CHARR TRIBE: Salmo salvelinus S. umbla — S. alpinus — S. nivalis S. killinensis—S. perisii—S. colii S. willughbii S. rutilus—S. carbonarius—S. grayi — S. hucho — S. lossos — GENUS LUCIOTRUTTA: L. leucichthys — GENUS OSMERUS: Smelt—GENUS COREGONUS: Species with the upper jaw prolonged—Species with the snout obliquely truncated—Species with the snout vertically truncated—Powan—Species with the mandible longer than the snout—Pollan—Vendace—GENUS THYMALLUS: Grayling—Thymallus microlepis . . . . . . 317

## CHAPTER X.

### FRESH-WATER FISHES OF THE ORDER PHYSOSTOMI (concluded).

FAMILY ESOCIDÆ—GENUS ESOX: Pike—FAMILY UMBRIDÆ—GENUS UMBRA—Umbra krameri—FAMILY CYPRINODONTIDÆ—GENUS CYPRINODON: Cyprinodon calaritanus—C. iberieus—GENUS FUNDULUS: Fundulus hispanicus—FAMILY MURÆNIDÆ—GENUS ANGUILLA: Common Eel—Broad-nosed Eel—Dalmatian Eel . . . . 360

## CHAPTER XI.

### FRESH-WATER FISHES OF THE ORDER GANOIDEI.

FAMILY ACIPENSERIDÆ—GENUS ACIPENSER: Acipenser glaber—Sterlet—A. guelini A. stellatus A. schypa A. güldenstädtii A. naccarii A. nardoi A. heckelii—A. nasus—The Sturgeon—A. huso . . . . . . . . . 381

## CHAPTER XII.

### FRESH-WATER FISHES OF THE SUB-CLASS CYCLOSTOMATA.

GENUS PETROMYZON: Lamprey—Lampern—Mud Lamprey, or Pride . . . 420

GENERAL INDEX . . . . . . . . 431
INDEX TO SPECIFIC NAMES . . . . . . . 440
INDEX TO COMMON NAMES . . . . . . . 443

# LIST OF ILLUSTRATIONS.

LEUCISCUS AULA. VAR. BASAK—LEUCISCUS AULA, VAR. RUBELLA—LEUCISCUS VULGARIS, VAR. CHALYBÆUS *Frontispiece.*

| | PAGE |
|---|---|
| Scale of Gobius | 6 |
| Scale of Leuciscus | 6 |
| Teeth of Salmon | 7 |
| Skeleton of Perch | 8 |
| Skull of Perch with the Operculum removed, to show the Bones of the Head, the Branchial Arches, and the Branchi-ostegal Rays | 9 |
| Vertebra of Herring | 10 |
| Air-bladder of Carp | 11 |
| Internal Anatomy of Carp | 11 |
| Internal Anatomy of Carp | 12 |
| Perca fluviatilis | 24 |
| Percarina demidoffii | 31 |
| Acerina cernua | 34 |
| Acerina schrœtzer | 37 |
| Lucioperca sandra | 40 |
| Lucioperca volgensis | 43 |
| Aspro zingel | 45 |
| Aspro vulgaris | 47 |
| Cottus gobio | 50 |
| Head of Cottus gobio, seen from the front | 50 |
| Cottus gobio, var. microstomus | 53 |
| Head of Cottus microstomus, seen from the front | 54 |
| Cottus gobio, var. ferrugineus | 54 |
| Head of Cottus ferrugineus, seen from the front | 55 |
| Cottus pœcilopus | 55 |
| Gobius martensii | 59 |
| Scale of Gobius martensii | 59 |
| Blennius vulgaris | 65 |
| Gasterosteus aculeatus | 74 |
| Gasterosteus aculeatus, var. brachycentrus | 79 |
| Lota vulgaris | 83 |
| Silurus glanis | 91 |
| Cyprinus carpio | 95 |
| Cyprinus acuminatus | 100 |
| Cyprinus hungaricus | 100 |
| Cyprinus hungaricus | 101 |
| Cyprinus carpio (var. hungaricus). Pharyngeal Teeth | 101 |
| Cyprinus regina | 102 |
| Carpio kollarii | 103 |
| Pharyngeal Teeth of Carpio kollarii | 103 |
| Carassius vulgaris | 105 |
| Carassius moles | 107 |
| Carassius gibelio | 108 |
| Pharyngeal Teeth of Carassius gibelio | 108 |
| Carassius oblongus | 109 |
| Pharyngeal Teeth of Barbus vulgaris | 114 |
| Barbus vulgaris | 115 |
| Barbus plebejus | 118 |
| Barbus plebejus | 118 |
| Barbus caninus | 119 |
| Barbus petenyi | 120 |
| Aulopyge hügelii (female) | 126 |
| Under side of the Head of Aulopyge hügelii | 126 |

| | PAGE |
|---|---|
| Aulopyge hügelii ; Lateral line and skin, showing pigment spots | 126 |
| Aulopyge hügelii (male) | 127 |
| Pharyngeal Teeth of Aulopyge hügelii | 128 |
| Gobio fluviatilis | 129 |
| Head of Gobio fluviatilis, seen from above | 129 |
| Pharyngeal Teeth of Gobio fluviatilis | 130 |
| Gobio uranoscopus | 131 |
| Head of Gobio uranoscopus, seen from above | 131 |
| Leuciscus erythrophthalmus | 134 |
| Pharyngeal Teeth of Leuciscus erythrophthalmus | 135 |
| Leuciscus erythrophthalmus, var. dergle | 136 |
| Leuciscus erythrophthalmus, var. plotizza | 137 |
| Leuciscus erythrophthalmus, var. scardafa | 137 |
| Leuciscus scardafa | 138 |
| Leuciscus erythrophthalmus, var. macrophthalmus | 138 |
| Leuciscus idus | 139 |
| Pharyngeal Teeth of Leuciscus idus | 139 |
| Leuciscus aula | 142 |
| Pharyngeal Teeth of Leuciscus aula | 143 |
| Leuciscus adspersus | 145 |
| Leuciscus rutilus | 146 |
| Leuciscus rutilus, var. pausingeri | 149 |
| Leuciscus pigus | 150 |
| Pharyngeal Teeth of Leuciscus pigus | 150 |
| Head of Leuciscus pigus, seen from below | 151 |
| Leuciscus pigus, var. virgo | 151 |
| Leuciscus friesii | 154 |
| Head of Leuciscus friesii, seen from above | 154 |
| Pharyngeal Teeth of Leuciscus cephalus | 157 |
| Leuciscus cephalus | 157 |
| Leuciscus cephalus, var. cavedanus | 159 |
| Leuciscus cephalus, var. albus | 160 |
| Leuciscus vulgaris | 161 |
| Heads of Leuciscus vulgaris | 163 |
| Leuciscus vulgaris, var. rostratus | 163 |
| Leuciscus illyricus | 164 |
| Large scale of Leuciscus illyricus | 165 |
| Leuciscus svallize | 166 |
| Leuciscus ukliva | 167 |
| Leuciscus turskyi | 169 |
| Leuciscus microlepis | 170 |
| Leuciscus tenellus | 171 |
| Leuciscus muticellus | 172 |
| Pharyngeal Teeth of Leuciscus muticellus | 172 |
| Leuciscus muticellus, var. savignyi | 174 |
| Leuciscus phoxinus | 176 |
| Pharyngeal Teeth of Leuciscus phoxinus | 176 |
| Paraphoxinus alepidotus | 186 |
| Pharyngeal Teeth of Paraphoxinus alepidotus | 187 |
| Tinca vulgaris | 189 |
| Pharyngeal Teeth of Tinca vulgaris | 190 |
| Chondrostoma nasus | 194 |
| Head of Chondrostoma nasus, seen from below | 194 |

|  | PAGE |
|---|---|
| Pharyngeal Teeth of Chondrostoma nasus | 197 |
| Chondrostoma phoxinus | 198 |
| Head of Chondrostoma phoxinus, from below | 198 |
| Chondrostoma genei | 199 |
| Head of Chondrostoma genei, seen from below | 199 |
| Chondrostoma soetta | 201 |
| Chondrostoma knerii | 202 |
| Head of Chondrostoma knerii, seen from below | 203 |
| Rhodeus amarus | 208 |
| Pharyngeal Teeth of Rhodeus amarus | 208 |
| Abramis brama | 210 |
| Pharyngeal Teeth of Abramis brama | 211 |
| Abramis brama, var. vetula | 214 |
| Abramis vimba | 215 |
| Abramis elongatus | 217 |
| Abramis ballerus | 218 |
| Abramis sapa | 220 |
| Abramis leuckartii, hybrid between Leuciscus rutilus and Abramis brama | 222 |
| Abramis (blicca) bjorkna | 223 |
| Pharyngeal Teeth of Abramis (blicca) bjorkna | 225 |
| Abramis bjorkna, var. laskyr | 225 |
| Abramis bipunctatus | 227 |
| Aspius rapax | 230 |
| Pharyngeal Teeth of Aspius rapax | 231 |
| Pharyngeal Teeth of Alburnus lucidus | 232 |
| Alburnus lucidus | 233 |
| Alburnus lucidus, var. breviceps | 235 |
| Alburnus alburnellus | 236 |
| Alburnus alburnellus, var. fracchia | 237 |
| Alburnus mento | 238 |
| Leucaspius delineatus | 241 |
| Leucaspius delineatus | 241 |
| Pelecus cultratus | 243 |
| Pharyngeal Teeth of Pelecus cultratus | 244 |
| Misgurnus fossilis | 246 |
| Nemachilus barbatulus | 249 |
| Cobitis taenia | 252 |
| Cobitis taenia, var. elongata | 254 |
| Clupea alosa, with caudal scales | 257 |
| Salmo salar | 267 |
| Head of Salmo salar | 268 |
| Salmo obtusirostris | 286 |
| Vomer of Salmo obtusirostris | 287 |
| Vomer palatines and maxillary arch of Salmo obtusirostris | 287 |
| Salmo fario, var. ausonii | 291 |
| Salmo carpio | 296 |
| Salmo dentex | 298 |
| Salmo spectabilis | 300 |
| Salmo marsiglii | 302 |
| Front view of Vomer of Salmo marsiglii | 302 |
| Salmo lacustris | 304 |
| Front view of Vomer of Salmo lacustris | 304 |
| Salmo lacustris, var. schiffermülleri | 305 |
| Front view of Vomer of Salmo schiffermülleri | 305 |
| Salmo genivittatus | 315 |
| Salmo salvelinus | 318 |

|  | PAGE |
|---|---|
| Salmo umbla | 322 |
| Salmo hucho | 329 |
| Coregonus lavaretus | 330 |
| Coregonus wartmanni | 342 |
| Coregonus hiemalis | 346 |
| Thymallus vulgaris | 351 |
| Esox lucius | 361 |
| Umbra krameri | 366 |
| Head of Umbra krameri, seen from above | 367 |
| Scale of Umbra krameri | 367 |
| Anguilla vulgaris | 374 |
| Anguilla curystoma | 380 |
| Acipenser glaber | 385 |
| Upper surface of Head of Acipenser glaber, showing Dermal Scutes | 385 |
| Under side of Head of Acipenser glaber, showing Mouth and Barbels, and Pores on the under side of the Rostrum | 385 |
| Acipenser ruthenus | 388 |
| Acipenser ruthenus, showing the Cranial Scutes on the upper side of the Head | 389 |
| Acipenser ruthenus, showing the under side of the Head, Mouth, Barbels, Pores, etc. | 389 |
| Acipenser ruthenus, var. gmelini | 391 |
| Acipenser gmelini, Head seen from above | 392 |
| Acipenser gmelini, Head seen from below | 392 |
| Acipenser stellatus | 393 |
| Acipenser stellatus, Head seen from above | 394 |
| Acipenser stellatus, Head seen from beneath | 394 |
| Acipenser güldenstädtii, var. schypa | 396 |
| Acipenser schypa, Head seen from above | 397 |
| Acipenser schypa, Head seen from below | 397 |
| Acipenser güldenstädtii | 399 |
| Acipenser güldenstädtii, Head seen from above | 400 |
| Acipenser güldenstädtii, Head seen from below | 400 |
| Upper surface of Head of Acipenser naccarii | 403 |
| Upper surface of Head of Acipenser naccarii, var nardoi | 403 |
| Acipenser heckelii | 405 |
| Head of Acipenser heckelii, seen from above | 406 |
| Head of Acipenser heckelii, seen from below | 406 |
| Head of Acipenser nasus, seen from above | 408 |
| Head of Acipenser nasus, seen from below | 408 |
| Acipenser sturio | 410 |
| Head of Acipenser sturio, seen from above | 411 |
| Head of Acipenser sturio, seen from below | 411 |
| Acipenser huso | 414 |
| Head of Acipenser huso, seen from above | 416 |
| Head of Acipenser huso, seen from below | 416 |
| Petromyzon marinus | 422 |
| Head of Petromyzon marinus, seen from below | 422 |
| Petromyzon fluviatilis | 424 |
| Petromyzon fluviatilis, Head seen from below | 425 |
| Petromyzon branchialis | 428 |
| Petromyzon branchialis, Head seen from below | 428 |
| Larval form of Petromyzon branchialis, formerly known as Ammocoetes | 429 |

THE
# FRESH-WATER FISHES OF EUROPE.

## CHAPTER I.

### INTRODUCTION.

The Orders of Fresh-water Fishes—Characters used in Classification of Fishes—The Forms of Fresh-water Fishes—The Structure of the Fins and the Fin Formulæ—The Scales—Organs of the Senses—The Mouth—The Operculum and Gill-arches—The Skeleton—Air-bladder—Digestive Organs—Reproductive Organs—Growth—Geographical Distribution—Table showing Range of Species—Table showing Classification of Fishes described.

WE commonly understand by Fresh-water Fishes such as live in rivers, ponds, and lakes. The majority of these species are incapable of existing in salt water, and rarely travel far in the stream or water which is their home. Other species are hardier, and not only endure sea-water without difficulty, but a migration to the sea, like an autumn holiday, is a more or less necessary condition of existence. Among such migratory fishes the European seas offer familiar examples in the Sturgeon, Salmon, migratory Trout, and Eels. Yet the circumstance that many Trout pass the whole of their lives in fresh water leads us to believe that the constitution of a fish is somewhat elastic, and capable of accommodating itself to more restricted conditions of existence. While, on the other hand, the presence in brackish waters of the Baltic Sea of Perch, Pike, Roach, and other familiar fresh-water types, makes it probable that after a fish has once become a denizen of fresh waters, it may again return to a salt-water life. About the mouths of rivers there are always marine fishes, which come into the shallow and sheltered places to spawn; and as some of them make their way far up the waters, they occasionally become acclimatised, and in any case may be mentioned as fishes found in fresh waters during their wanderings. Among such are certain of the flat fishes, like the Flounder, and, in some localities, the Plaice and Sole; the Whitebait; and, in some countries, species of Atherina and Grey Mullet. So numerous are these types that it would be impossible to describe them all without appearing anxious to augment the province of Fresh-water Fishes at the expense of their marine relations. Nevertheless, it is impossible to ignore the existence of such fishes in our rivers, because they exemplify one of the ways in which fresh-water species of marine genera originate.

## GROUPS OF FRESH-WATER FISHES.

Fresh-water Fishes may be referred to five of the great groups into which the Fish class is divided.

The spiny-finned order, named Acanthopterygii, is represented by the Perch family, the Cottidæ or Bull-heads, the Goby family, the Blennies, and the Sticklebacks, while the Grey Mullets and Atherines may be associated with them.

The second order, termed Pharyngognathi, in which the fins are spiny and the lower pharyngeal bones united together, has no representative in the fresh waters of Europe.

The third order, Anacanthini, in which there are no bony rays in the vertical fins, has but one representative in the Cod family, known as the Burbot; while the Pleuronectidæ or Flat Fishes are represented by the Flounder, Plaice, and Sole, which are otherwise marine.

The fourth order, Physostomi, has the air-bladder opening into the throat by a duct, so that the fishes are easily able to adapt themselves to varying pressure in the water; and this persistence of the pneumatic duct probably accounts for the circumstance that by far the larger number of fresh-water fishes are referable to this order. Among them are representatives in the fresh waters of Europe of eight families, the Siluridæ, represented by the Wels, the Cyprinidæ, which includes the Carp tribe, Barbels, Gudgeons, Roach, Dace, Rudd, Minnow, Tench, Bream, and a multitude of others whose Continental names have yet to become familiar. The Clupeidæ or Herrings are represented by species of Shad and the Whitebait. The Salmonidæ are known to us from Salmon, Trout and Charr, not to mention the various species of Coregonus, the Smelt, and the Grayling. The Pike is the only representative of the Esocidæ. Umbra represents the Umbridæ; the Cyprinodontidæ have some Continental representatives; and there are in Europe two or three species of Eels in the family Anguillidæ.

The Ganoid order of fishes is represented by the Sturgeons, which form the family Acipenseridæ.

And the remarkable group of Lampreys, which are comprised in the order Cyclostomata, constitute the family Petromyzontidæ.

## CLASSIFICATION.

The differences of structure which characterise the types of fishes here enumerated exemplify the variability of fish organisation. The nature of the fins, as consisting of bony spines or soft-jointed branched rays, is a difference sufficient to constitute an ordinal group. The obliteration or persistence of the pneumatic duct to the air-bladder is an important character in classification. The persistence of a cartilaginous condition of the skeleton, with bony plates on the body, similarly characterises the Sturgeons. And the lower condition in the Lampreys, in which the vertebral column remains cartilaginous, and without sub-divisions into distinct vertebral parts, may be taken to represent another grade of structure. But these Orders also vary in minor details of structure, so as to present a multitude of types, which are known as genera. And since every fish must be described in its genus and species, it becomes necessary to dwell upon the nature of the characters on which classification is made. For there are rarely broad lines of demarcation in Nature; and if such exist, we may reasonably believe that intermediate types have formerly lived upon the earth, and become fossilised during its geological history. Hence, whenever an assemblage of species can be united by common characters, and similarly separated from other fishes, they constitute a genus, which may comprise one species or a hundred. If the characters which distinguish genera are compared together in the following descriptions, they will be seen to be of a varying nature. Thus the Perch genus has about thirteen or fourteen spines in the first dorsal fin, and has the head free from scales on the upper part. These characters separate it from some South American Perches, which form the genus Percichthys, in which the spines in the first dorsal fin are reduced to nine or ten, and the head is covered with scales. When we turn to the Marine Perches, which form the genus Labrax, no difference in the number of spines is found, but the tongue, which was smooth in Perca and Percichthys, has become toothed, like the palatine bones.

The species of a genus are distinguished by characters of another kind. It is often difficult to say where a variety ends and a species begins. There are local races of many fishes which, under the changed conditions of physical geography which from time to time affect the distribution of life on the earth, have become isolated from the rest of the race, so as to live on table-lands or low plains, in cold mountain lakes or in shallow swamps, in sluggish waters or rapid torrents, and thus, differently circumstanced, have developed into varieties distinguished by size, form, colour, and certain internal and external differences in the organs and proportions of the body. A genus like Salmo or Coregonus shows a remarkable vitality under varied conditions in the evolution of specific characters; but these variations are, for the most part, so small, and so slightly differentiate from each other the forms they characterise, that it is open to us either to recognise that there is no clear distinction between such varieties and species, or to make a definite demarcation and group the varieties together, by taking some character which they have in common as the basis for a species. Thus there are many kinds of Charr, but we might easily group them together, and, in the same way, the gradations between the Trout which do not migrate, would absorb a large number of types which have usually been regarded as species. Fish characters being so variable, and changing with age, it becomes necessary for us briefly to examine them, so that their nature and value may be observed in living fishes.

### Fish Form.

There is less variety in the aspect of fresh-water fishes than in the larger assemblage which inhabits the sea. There are none of the marvellous modifications of snout, such as distinguish the Sword-fish and the Saw-fish; none of the singular conditions of fin which are seen in crawling Gurnards, or the fins of Flying Fishes. Nor are the fins ever modified into a sucker, as in several marine types. There are no such singularly-expanded fresh-water fishes in Europe as the Rays, no globular expansible spiny globe-fish like the Diodons, or armoured box-fish like Ostracion, or Ribbon Fish, or Sea-horses, or Sun-fish.* And even the most singular inhabitants of rivers are fishes which come up from the sea, like Lampreys, Eels, Sturgeon, and Flounders. If we exclude these genera, fresh-water fishes may be said all to possess the typical fish form, which is compressed from side to side, widest behind the head, tapering towards the tail, with the caudal fin evenly lobed, the other fins well developed, and the body, in most cases, covered with scales. The parts of the body are described as the dorsal region or back, sides, and abdomen. The proportions of the body furnish important characters in distinguishing genera and species. These are stated in the relation of the length of the head to the length of the fish, and in the proportions of depth and thickness of the body to its length. The abdomen may be rounded or form a sharp edge. The relative positions of the fins also furnish distinctive characters.

### The Fins.

Fresh-water fishes nearly all have two pairs of fins, which may be regarded as corresponding to the limbs of terrestrial animals (Fig. 4). The pair at the hinder margins of the head are termed pectoral fins. The pair beneath them, and usually placed farther back, are the ventral fins. Besides these, a fish has at least three other limbs, which are vertical. There are one or two dorsal fins on the median line of the back, but when there are two, one is placed behind the other. More or less opposite to the dorsal fin, but always behind the vent, is

---

\* See Article "Fishes," in "Cassell's Natural History."

the anal fin, which is in the inferior median line of the body. And, finally, the fish terminates posteriorly in the caudal fin. These fins vary greatly in size, composition of their skeletons, and form, being sometimes rounded, sometimes pointed, sometimes truncate. In the embryo fish these vertical fins are not differentiated from each other, but extend as a continuous fringe round the hinder part of the body, and in a few fresh-water fishes, like eels, this condition is retained throughout life.

The rays which form the fins are of three types: first, simple bony spines; secondly, softer spines, which are jointed in the upper part; thirdly, soft rays, which are not only jointed in the upper part, but sub-divide so as to terminate in a fringe. The rays which project above the body are jointed at their bases and articulated to internal portions of the fins, which extend between the extremities of the spinous processes of the vertebræ, especially on the dorsal side. In the Acanthopterygian group the first dorsal fin consists of spiny rays, and the second dorsal usually consists of soft-jointed rays, with one or two more or less developed spines in front. The number of rays in a fin is counted, and written in a formula. Thus, in the Perch, the formula of the dorsal fins is written 1 D. 13—15, 2 D. 1—2/13—14: signifying that there may be thirteen to fifteen spines in the first dorsal fin, and in the second dorsal fin, one or two spines followed by thirteen to fourteen soft rays. The other fins have their rays counted and expressed in a similar way, so that, as the anal fin has two spinous rays and eight or nine soft rays in the Perch, its formula is written—A. 2, 8—9. In this fish the ventral fin has one spiny ray and five soft rays, so that its formula is V. 1, 5. The formula of the pectoral fin is P. 14, and of the caudal fin C. 17. In each genus the species commonly vary in the number of rays in some or all of the fins, but the most variable are the dorsal and anal fins. In describing a fish, it is necessary to state the form, positions, and frequently the length of the fins, when the length is in any way unusual. In the Salmon tribe there is a second dorsal fin which contains no rays, but consisting of fatty substances, is known as the adipose fin.

When a fish is deprived of its fins it floats with the abdomen upward. They therefore maintain the animal in its vertical position, and are organs of motion.

## SCALES.

There are two principal types of scales among fresh-water fishes: first, the *ctenoid* scale, in which the free margin of the scale, and sometimes the whole exposed surface, is serrated or spiny; and, secondly, the *cycloid* scale, in which the margin shows no serrations, though it may be marked with delicate rays (Fig. 1). Ctenoid scales are well seen in the Perch and Pope, while the cycloid type characterises Cyprinoid fishes, Salmon, and the majority of fresh-water forms. The scales exhibit their simplest development in the Eel, in which they are remarkably thin, and consist of a single layer of cells, which preserve their forms distinct. The scales overlap each other in the majority of fishes, so that only the posterior margins are exposed; but the size and number vary in the different species, so that sometimes the scales are embedded in skin without overlapping. They often vary in size on different parts of the body. At times they are delicate and easily detached, while in other fishes their substance is strong, and firmly adherent to the skin. Usually the fins are free from scales, but in some types scales extend on to the bases of the fins, and may more or less completely cover the head. There is generally a row of perforated scales extending along the middle of the side of the fish, from the shoulder to the tail. This is known as the lateral line. Its position is variable, and in many fishes runs parallel to the dorsal contour, while in others it is parallel to the abdomen. As a rule, the scales of the lateral line differ somewhat in form from the others, and may be either smaller or larger. The lateral line is sometimes imperfectly developed, and its development may be different in individuals of the same species. The perforations on the scales are the outlet for the mucus canal, the secretion from which lubricates the body of the fish (Fig. 2). This canal is prolonged on to the head, and opens by pores on the lower jaw, and above and below the eye. A few fishes have no scales at all. The Sturgeon is covered with bony scutes, which, however, do not overlap, and differ in size.

Fig. 1.—SCALE OF GOBIUS.

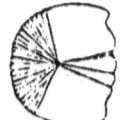 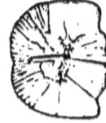

Fig. 2.—SCALE OF LEUCISCUS.

The scales in fishes with a bony skeleton are arranged in transverse series. The number of scales in the lateral line gives the number of these transverse rows in most cases. The number of scales in a transverse series is counted from the beginning of the dorsal fin downward and backward, so as to follow the free borders of a line of scales. The numbers above and below the lateral line are written in the form of a fraction. Some fishes are stated to shed their scales periodically.

## SENSORY ORGANS.

The size of the eye is distinctive of many species. Its size is stated as a third, fifth, seventh, or some other fraction of the length of the head. Its position in the head is defined by measuring the number of times its own diameter is distant from the snout and from the other eye. There is something resembling an eyelid in the Salmon tribe, and in the Shad there are two eyelids, which are anterior and posterior, like the third eyelid of birds rather than the eyelids of mammals.

The nostrils are remarkable for having a single opening in the middle of the upper part of the head in the Lamprey; but in most fishes they are double, and have two openings on each side. They are placed between the eye and the snout, and sometimes the two openings are different in size, and one may have a slight tubular prolongation.

On the under side of the snout sensitive tentacles are developed in the Sturgeon, and similar organs surround the mouth in Cobitis, Barbels, Silurus, or may occur at its angles, as in the Tench and some other fishes.

### The Mouth.

The mouth is often terminal. Sometimes the jaws are unequal, and either the upper or lower jaw may be the longer. When the upper jaw is very short the cleft of the mouth may be oblique, or even vertical, and this, with the distance to which its cleft extends back, is an important character in distinguishing many species. Sometimes the snout is long enough to make the mouth inferior, as in some species of Coregonus; though the Sturgeon is most singular in the inferior and backward position of its mouth, which is beneath the eyes. Lampreys are exceptional in opening the mouth from side to side, as well as in the mouth being suctorial.

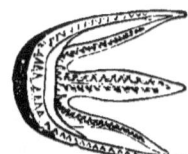

Fig. 3.—TEETH OF SALMON.

The lips may be thick and fleshy, or absent. The teeth are absent in Sturgeons, horny in Lampreys, but usually bony when they exist. There are no teeth in the mouth in fishes of the Carp family, but the pharyngeal bones are well developed and carry teeth, which are arranged in parallel series, of which there may be one, two, or three. This character is so important that it is always stated numerically in a formula—thus, 5·4—4·5, which means five teeth in the long row, and four in the short inner row on each side.

The teeth are often remarkably developed (Fig. 3), and may be present not only upon the pre-maxillary and maxillary bones, and mandible, but also upon the vomer, palatine, and pterygoid bones, and upon the tongue. There are often teeth of different sizes in the same jaw, as in the lower jaw of the Pike. They are frequently fine, dense, and villiform, as in the Perch. Occasionally the teeth may be in a double row, as in the upper jaw of the Flounder. The jaws are sometimes protractile.

### The Operculum and Gill-Arches.

The bones which form the gill-covers, or are connected with the branchial apparatus, furnish important distinctive characters for genera and species.

The gill-cover consists typically of four thin bony plates. First, there is an expanded hindermost plate, which is named the operculum; secondly, a long plate beneath the operculum, named the sub-operculum; thirdly, the inter-operculum, which prolongs the line of the sub-operculum forward; and, fourth, the pre-operculum, which is elongated vertically, and extends in front of the other opercular elements. The gill-cover is united with the suspensory arch, which connects the lower jaw with the skull, the opercular bone being united to the hyo-mandibular bone, and the inter-operculum has a ligamentous union with the suspensory arch, as well as with the lower jaw. A modified gill, which possessed respiratory functions in the embryo, but has lost those functions in the adult, is attached to the operculum. These gills are known

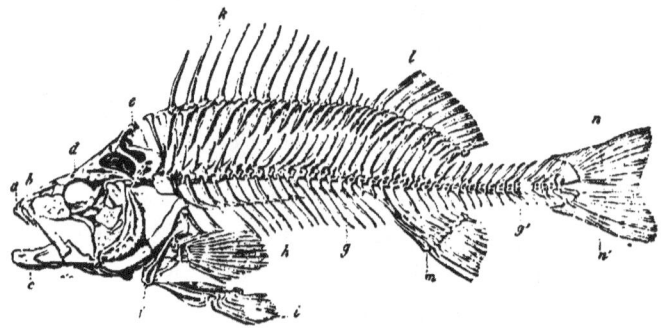

Fig. 4.—SKELETON OF PERCH.
*a*, Pre-maxillary bone; *b*, maxillary bone; *c*, mandible; *d*, palatine arch; *e*, cranium; *f*, inter-operculum; *g, g'*, vertebral column; *h*, pectoral fin; *i*, ventral fin; *k, l*, dorsal fins; *m*, anal fin; *n, n'*, caudal fin.

as pseudobranchiæ. They do not exist in all fresh-water fishes, and when present, carry aërated blood to the eye and other parts of the head. The gills are supported by the branchial arches; most bony fishes have four complete gills. The branchial arches commonly develop protuberances, which are called gill-rakers, and form a sort of net to filter the water from solid substances, which might otherwise be carrried into the gill-cavity. The number of branchi-ostegal rays, and the degree of development of the gill-rakers on the branchial arches furnish important characters in defining the species of fish. In the Sturgeon there is an external opening to a canal placed behind the orbit, which is known as a spiracle. These spiracles lead into the pharynx. In Lampreys there are a number of successive respiratory sacs, each opening externally by a short duct, and internally into a canal, which communicates in front with the back of the throat, termed the pharynx. In most fishes the water is taken in through the mouth, and passed out by the gill-openings.

SKELETON (Fig. 4).

Bony fishes, or TELEOSTEI, have the elements of the skeleton well ossified. They comprise all the fishes here described, except Sturgeons and Lampreys. The skull (Fig. 5) is formed upon the same general plan as the skull in any other vertebrate animal, the essential difference consisting in the development of the suspensory arch for the lower jaw, by the hyo-mandibular bone united to the skull, and the quadrate bone which articulates with the lower jaw; and these bones, with some other less important elements, support the oper-

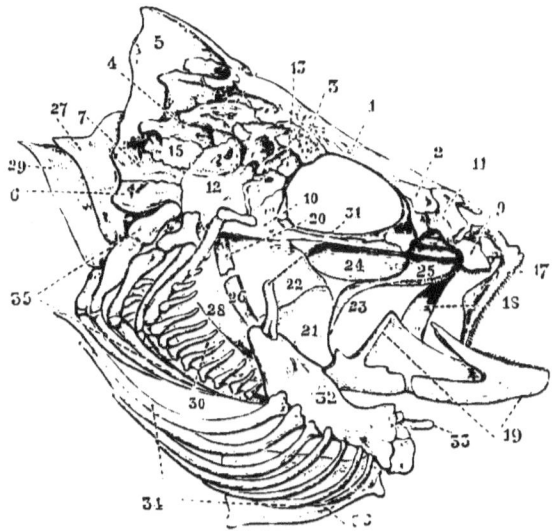

Fig. 5. SKULL OF PERCH WITH THE OPERCULUM REMOVED, TO SHOW THE BONES OF THE HEAD, THE BRANCHIAL ARCHES, AND THE BRANCHI-OSTEGAL RAYS.

1, Frontal; 2, pre-frontal; 3, post-frontal; 4, epiotic; 5, supra-occipital; 6, basi-occipital; 7, ex-occipital; 9, vomer; 10, para-sphenoid; 11, nasal; 12, ali-sphenoid; 13, orbito-sphenoid; 15, petrosal; 17, pre-maxillary; 18, maxillary; 19, dentary and articular elements of mandible; 20, hyo-mandibular; 21, quadrate; 22, meta-pterygoid; 23, pterygoid; 24, ento-pterygoid; 25, palatine; 26, symplectic; 27, operculum; 28, pre-oper-culum; 29, sub-operculum; 30, inter-operculum behind the gill rakers; 31, stylo-hyal; 32, epi-hyal, cerato-hyal, and basi-hyal; 33, glosso-hyal; 34, branchi-osteg-1 rays; 35, bones of the branchial arches; 36, uro-hyal.

culum, which is peculiar to fishes. The branchial apparatus is also a distinctive fish character; and the hyoid, or tongue-bone, attains a remarkable development in bony fishes, consisting of median pieces, from which a jointed arch is given off on each side.

There are five branchial arches (Fig. 5), but the fifth is modified, and instead of supporting a gill, becomes in many genera armed with teeth, and is known as the pharyngeal bone. The other branchial arches consist of three or four pieces. The outermost is the epi-branchial, the middle piece is the cerato-branchial, and, below this is the hypo-branchial. The uppermost segments of the fourth branchial arch are known as the upper pharyngeal bones.

It is remarkable that among the Sturgeons, in which the head consists of an enormously thick mass of cartilage, of similar texture to that which makes the rest of the Sturgeon's skeleton, the superficial dermal scutes upon the head are arranged so as to correspond in position with the bones which roof in the brain-case of bony fishes. And in describing the Sturgeon group we shall find it necessary to pre-suppose a familiarity with the structure of the skull, so far as is implied in knowing the nomenclature of the upper bones of the head, which is essentially the same for all animals. Thus, at the back of the head there is a single bone in the median line, named supra-occipital, and at its outer margins a bone on each side, called the epiotic. In front of the supra-occipital there is a pair of bones, known as the parietals, and, external to these, another pair, known as the squamosals. In front of the parietals is the pair of frontal bones, more or less separated in front by the ethmoid bone or bones. External to the frontal bones, so as to form the margins for the orbit of the eye, are post-frontal bones behind and pre-frontal bones in front, which fishes have in common with other cold-blooded vertebrata. All these elements are well developed in the dermal head-plates of the Sturgeon, and there are other ossifications, some of which may possibly represent the nasal bones, though the nasal apertures have no pre-maxillary bones bounding them in front, as in the majority of vertebrata.

The vertebral column is divided into an anterior, or thoracic portion, and a posterior, or caudal portion. The number of vertebræ varies in genera and species, and is, therefore, usually counted. Each vertebra (Fig. 6), has a bi-concave body, with the remains of the primitive cartilage, called the notochord, filling the spaces between the cups. An arch above protects the spinal cord, and in fishes of deep body is prolonged to a great height.

Ribs are attached to the vertebræ in the abdominal region, and these often possess complex accessory processes (Fig. 6).

The caudal region has a lower arch to each vertebræ, corresponding to the upper neural arch. The termination of the tail is commonly directed upward, as in Sharks and the Sturgeon, but the bony processes attached to it, as may be seen in the Salmon, attain a symmetrical development, so as to make an evenly-lobed, or homocercal fin.

Fig. 6.—VERTEBRA OF HERRING.
pd, rib; pb, appendage of rib

The structure of the bones which support the pectoral and ventral fins is full of interest in their correspondence with the similar arches of land

INTERNAL ANATOMY. 11

vertebrates, but we shall have no occasion to discuss these points in European fresh-water fishes.

AIR-BLADDER.

There is no air-bladder in the Lampreys, and the organ cannot be considered to show a respiratory function in any of the fresh-water fishes of Europe. It is commonly placed below the vertebral column in the abdominal cavity, and in most cases tapers in front and behind. In an early stage of existence the air-bladder communicates with the intestine by a duct, which is obliterated in many genera as the fish attains maturity.

Fig. 7.—AIR-BLADDER OF CARP.

In the Carp tribe the air-bladder is constructed so as to form anterior and posterior divisions (Fig. 7), and in the Loach tribe it consists of two parallel portions, which are contained in bony capsules at the sides of the anterior vertebræ. In the Sturgeon it is thick, and consists of several layers.

DIGESTIVE ORGANS (Figs. 8, 9).

Sometimes the intestine is simple, as in the Lamprey, in which it is difficult to separate the parts which are usually named œsophagus, stomach,

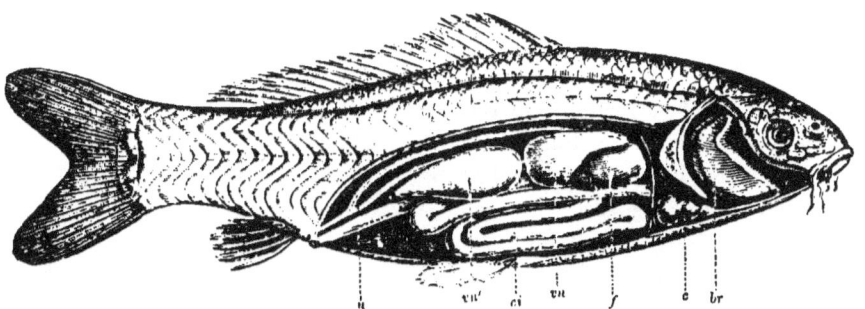

Fig. 8.- INTERNAL ANATOMY OF CARP.
*br*, branchiæ or gills; *c*, heart; *f*, liver; *vn*, *vn'*, air-bladder; *ci*, intestinal canal; *u*, urethra.

and intestine. In the Sturgeons the intestine is complicated, and receives the secretion from the liver and the pancreas, and terminates in a spiral valve, like that seen in Sharks. The form of the stomach varies, sometimes making a blind sac, and sometimes being an expanded part of the intestine, situate at an angular bend in its course. Behind the stomach, the intestine frequently carries a fringe of appendages, known as pyloric appendages, which are greatly developed in the Salmon tribe, and vary in number in the different species. The liver may be simple or lobed; it usually includes a gall-bladder, and in some fishes contains a good deal of fat. The spleen is a dark red organ placed near to the stomach. The intestine terminates in a pore, which frequently includes the outlets for the urinary bladder and the reproductive organs (Fig. 8).

### REPRODUCTIVE ORGANS.

The sexes of fishes are always distinct; though hermaphrodites are occasionally met with in some groups. All European fresh-water fishes are oviparous. The ovaries and milt are frequently mature long before the fish is full-grown, and they usually occupy a large portion of the abdominal cavity. In the Smelt there is an internal oviduct, but commonly the eggs are discharged into the abdominal cavity (Fig. 9). Some fishes have an external oviduct, which is of great length in the Bitterling, *Rhodeus amarus*. The eggs are

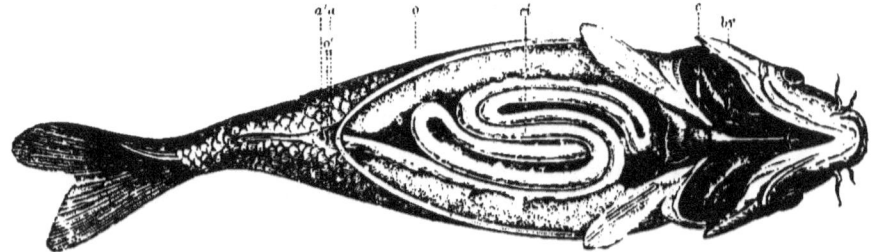

Fig. 9.—INTERNAL ANATOMY OF CARP.
*br*, branchiæ or gills; *c*, heart; *ci*, intestinal canal; *o*, ovaries; *a*, vent; *o'*, oviduct; *a'*, urinary outlet.

commonly globular, but oval in the Lampreys. They vary much in size in the same individual in different periods of life and in different species. In the Eels they are microscopic; the Sticklebacks have large eggs for their size, but among the largest eggs of our fresh-water fishes are those of the Salmon. Sometimes the eggs are deposited on plants; the Bitterling is said to deposit them in the shells of living bivalve Mollusca, while other fishes, like the Salmon, cover the eggs with gravel, and Sticklebacks construct nests for their better protection. Among the Cyprinoid fishes hybrids are constantly met with, and they are probably not uncommon in the Salmon tribe.

At the breeding season many fishes put on brilliant colours, which are lost when the spawn is deposited, and some, like most Cyprinoids, develop small tubercles upon the scales in both sexes, or in the male only. The Eels, though commonly accounted fresh-water fishes, spawn in the sea.

### GROWTH.

The proportions of the body change with growth, the head usually becoming relatively smaller, and the fins also become modified, so that the caudal, when forked in the young, may become truncated or rounded in the old fish. The Lampreys pass through a sort of metamorphosis, by which changes equivalent to a generic modification take place when the fish is three or four years old. The size of a fresh-water fish varies with its food, and many varieties which are small in one locality become large where more suitable food abounds. It is impossible at present to estimate the importance of the changes which varieties of food may induce, for when fishes which usually feed on small animals take to eating mollusca, the stomach may become thickened, as in some species of Trout, and the number of pyloric appendages appears also to vary with the food. The odour of fish would seem to depend on the food, and some insects possess a power of stimulating growth in the fishes which eat them, which is out of proportion to the weight of the insects eaten.

## Geographical Distribution.

Numerical statements of the distribution of species necessarily vary with the size of the species, and, therefore, species are scarcely so instructive in illustrating geographical distribution as genera. Thus, excluding types like the Bass, Grey Mullets, Atherine, Flounders, Plaice, and Soles, there are about forty-one genera of fresh-water fishes in Europe, and of these, twenty-five, Perca, Accrina, Cottus, Gasterosteus, Lota, Cyprinus, Carassius, Barbus, Gobio, Leuciscus, Tinca, Abramis, Alburnus, Misgurnus, Nemachilus, Cobitis, Clupea, Salmo, Osmerus, Coregonus, Thymallus, Esox, Anguilla, Acipenser, and Petromyzon, are represented in the British Islands, to which, however, no genus is limited.

The Acanthopterygian fishes are not so numerous in Great Britain as on the Continent, and are best represented in Austria, Prussia, and Germany. Austria and Russia have largest fish fauna. Both countries are rich in Sturgeons, Salmon, Herrings, and Cyprinoid fishes. By means of the annexed table, which gives the geographical range of the principal species, with their popular names in the chief European countries, a better idea of the distribution of the fishes will be gained than can possibly be conveyed in any analytical statement. For further details we must refer to the body of the work, where about two hundred and fifty fishes are described.

Weights and measurements of fishes are frequently given in the metric system. A kilogramme is 2·2 pounds avoirdupois. A mètre is 39·37 inches. The centimètre is about four-tenths of an inch.

## TABLE SHOWING THE GEOGRAPHICAL DISTRIBUTION IN EUROPEAN COUNTRIES OF SOME OF THE MORE IMPORTANT SPECIES OF FRESH-WATER FISHES.

*When the popular name is known or generally used it is given in each country; otherwise the distribution is indicated by an asterisk (\*).*

| Genus. | Species. | British Islands. | Scandinavia. | France. | Germany. | Russia. | Austria. | Italy. | Spanish Peninsula |
|---|---|---|---|---|---|---|---|---|---|
| PERCA | FLUVIATILIS | Perch | Aborren | Perche | Barsch | Okuny | Flussbarsch | Persico. | \* |
| LABRAX | LUPUS | Bass | Hafsaborren | | | | | Lapo. | \* |
| PERCARINA | DEMIDOFFII. | | | | | | | | |
| ACERINA | CERNUA | {Pope} {Ruffe} | Snorgers | Gremille | Kaulbarsch | Ersie | Kaulbarsch. Schrason. | | |
| | SCHRETZER ROSSICA | | | | | Nosary. | | | |
| LUCIOPERCA | VOLGENSIS SANDRA | | Gosen | | Zander | Sokreta Soolake | Sander. | \* | |
| ASPRO | ZINGEL VULGARIS STREBER | | | | Zingel Streber | Chope. | Strebr. | | |
| COTTUS | GOBIO Var. MICROSTOMUS Var. FERRUGINEUS POECILOPUS SCORPIUS | {River Bull-head, or Miller's Thumb.} | Stensimpan Bergsimpan. Rotsimpan. | {Chabot de la Rivière.} | Kaulkopf. | Bieloke | \* \* \* | | |
| GOBIUS BLENNIUS | MARTENSII VULGARIS Var. ALPESTRIS | | | Caguette. | | Babka | \* | {Ghiozzo. {Bottola. {Cagnetto. | |
| GASTEROSTEUS | ACULEATUS PUNGITIUS | Stickleback | Storspiggen Smaspiggen | {Epinoche {Epinochette {Piquante} | Stachelfisch | Kolushka | {Stichling. {Kleiner. {Stichling. | Spinarello. | |
| LOTA | VULGARIS | Burbot | | Lote | {Rutte {Quappe {Waller {Wels} | Meny | Aal-rutte. | | |
| SILURUS | GLANIS | | \* | | | Some | {Wels. {Schaiden. | | |
| CYPRINUS | CARPIO Var. ACUMINATUS. Var. HUNGARICUS. Var. REGINA Hybrid variety, KOLLARII | Carp | \* | Carpe | Karpfen Karpf-gareisl. | Karope | {Genuiner. {Donaukarpfe. | Carpa | Carpa. |

# TABLE OF GEOGRAPHICAL DISTRIBUTION.

|  |  | | | | | | | | | | |
|---|---|---|---|---|---|---|---|---|---|---|---|
| CARASSIUS | VULGARIS | Crucian Carp | Caraseiu | Dama-Ruda | | | Karausche. Giebel. | Karassy | { Karausche { Gareisol | Barbo. |
| | Var. GIBELIO | | Gibele | | | | | | | |
| | Var. OBLONGUS AURATUS BUCEPHALUS HUMILIS | Goldfish | | | | | (In Turkey). | | | Comba. |
| BARBUS | VULGARIS PLEBEJUS CANINUS COMIZA BOCAGEI SCLATERI CHALYBEATUS | Barbel | Barbeau | | | Barbe | Barbe Mreni Zeembling | Mareua | Barbe Semling | Barbo Barbio. |
| AULOPYGE | HUGELII | | | | | | Ukliva. | | | { Golione. { Bertone. |
| GOBIO | FLUVIATILIS | Gudgeon | Goujon | | | { Gressling { Gründling Steingressling | Weber Slipize. | | { Gründling { Gressling | Scardola. Triotto. |
| | URANOSCOPUS | | | | | | Gangling. | | | |
| LEUCISCUS | IDUS | Ide | Ide | Id | | { Aland { Norting | | Yase | | |
| | Var. MINIATUS ERYTHROPHTHALMUS | { Red-eye { Rudd | Rotengle | Sarf | | Rothauge | Rothange Pigo | Krasnoperka | Rothauge | { Encobin. { Pigo. |
| | AULA ASPERSUS RUTILUS | Roach | Gardon | | | Rothauge | Rothange | Beclia | Rothauge | |
| | PIGUS | | | | | | Pertösed | | | |
| | FRIESII | | | | | | { Attel. { Döbel. | | | |
| | CEPHALUS VULGARIS ILLYRICUS SVALLIZE UKLIVA TURSKYI Var. MICROLEPIS TENELLUS | Chub Dace | Chevaine. Vandoise. | Stamm | | Döbel Haslinr | Hazel. Kleni Svallize. | Vyirczbe. Goloveny Eletse | Döbel Haslinr | Kleni |
| | MUTICELLUS | | | | | | | | | |
| | PHOXINUS | Minnow | Vairon | Elrizor | | Elritze | Langen Pfrille | | Elritze | { Mozzetta. Vairone. { Sanguinerola Fregarolo. |
| TINCA | VULGARIS | Tench | Tanche | | | Schleihe | Schleihe | Leeny | Schleihe | Tinca. Tenca. |
| CHONDROSTOMA | NASUS GENEI SOETTA KNERII PHOXINUS FOJILEPIS | | Nez | | | { Nase { Nasling | Nasling. Lau. | Podcoste | { Nase { Nasling | Lasca. Savetta. |

## TABLE SHOWING THE GEOGRAPHICAL DISTRIBUTION IN EUROPEAN COUNTRIES OF SOME OF THE MORE IMPORTANT SPECIES OF FRESH-WATER FISHES (continued).

| Genus. | Species. | British Islands. | Scandinavia. | France. | Germany. | Russia. | Austria. | Italy. | Spanish Peninsula. |
|---|---|---|---|---|---|---|---|---|---|
| PARAPHOXINUS | ALEPIDOTUS, CROATICUS | | | | | | | | |
| RHODEUS | AMARUS | | | | Betterling | Gorchake | Grundel. | | |
| ABRAMIS | BRAMA | Bream | | Brême | Blei | Leshy | Bitterling. | | |
| | VIMBA | | | | Zarthe | Rybetze | Brachsen. Russmase. Blaunase. | | |
| | ELONGATUS | | | | | | | | |
| | BALLERUS | | | | Zope. | Seenetse. Kiepetse. | Pleinzen. | | |
| | SAPA | | | | Leiter | Leshytaramike | | | |
| | LEUCKARTII | | | | Güster | | | | |
| | BJOERKNA | White Bream. | Bjelk | Bordelière | Bliche | Goostera. | Zobel-pleinze. | | |
| | BIFURCATUS | | | Ablette Sperlin | | | Laube. | | |
| | FASCIATUS | | | | | | | | |
| ASPIUS | RAPAX | | Asp. | | Rapfen | Belezna | Rapfen. | | Alaur. |
| ALBURNUS | LUCIDUS | Blenk | Löja. | Ablette | Schiesl | Verkhovodka. | Lauhe. Alborella. Laube. | | |
| | ARBURNELLUS | | | | Uckelei | | | | |
| | MENTO | | | | | | | | |
| | CHALCOIDES | | | | | | | | |
| LEUCASPIUS | DELINEATUS | | | | Mottke | | | | |
| PELECUS | CULTRATUS | | Säärluxen | | Ziege | Chokony | Sichling | | |
| MISGURNUS | FOSSILIS | Pond Loach | | Loche d'Étang. | Schlammpeitzger Bissgure. | Vyune | Schlammbeisser. | | |
| NEMACHILUS | BARBATULUS | Loach | | Loche Franche. | Schmerle. Bart-grundel. | Avlushka | Bart-grundel. Schmerl. | Cobite Barbotte lo. | |
| COBITIS | TÆNIA | Groundling. Spinous Loach | | Loche de Rivière. | Dorn-grundel. Steinpeitzger. | Sheepovku | Steinbeisser | Cobite Fluviale. | |
| CLUPEA | ALOSA | Allis-shad | | Alose Commune. | Maifisch | | Maifisch | Ceppino Agone | Sabalo. |
| | FINTA | Twaite-shad Whitebait | | Alose Feinte | | | | Cappa | Saboezo. |
| | HARENGUS | | | | | | | | |
| SALMO | SALAR | Salmon Gillaroo. | Lax | Saumon | Lachs | Petrouga. | | | |
| | STOMACHICUS | | | | | | | | Salmo. |
| | HARDINII | | | | | | | | Lachs. |
| | CAMBRICUS | Sewin Bull-Trout | | Truite de Mer. | | | | | |
| | ARGENTEUS | | | | | | | | |
| | VEXENSENSIS. | | | | | | | | |
| | HISTOPS. | | | | | | | | |

# TABLE OF GEOGRAPHICAL DISTRIBUTION.

| | | | | | | | | |
|---|---|---|---|---|---|---|---|---|
| SALMO (continued) | NIGRIPINNIS. | | | | | | | |
| | OBTUSIROSTRIS. | | | | | | | |
| | MUCHOLETH. | | | | | | | |
| | FARIO | . River Trout | . Truite de Rivière | Bach-forelle | . | . Lachsforelle | |
| | GALLIVENSIS | . Galway Sea-trout | | | | | |
| | AUTUMNALIS. | | | | | | | |
| | LEMANUS. | . Fjalloret | | | | . Pastrova. | |
| | CARPIO | | | | | . Lachsforelle. Illanken. Mailacha. | . {Carpoine. Trutta del Lago. |
| | RAPHII | | | Grundforelle | | | |
| | DENTEX | | | | | | |
| | SPECTABILIS. | | | | | | |
| | MARSIGLII | | | | | | |
| | LACUSTRIS | | | Schwebforelle. | | | |
| | SCHIFFERMÜLLERI. | | | | | | |
| | LEVENENSIS | . LochlevenTrout | | | | | |
| | ORCADENSIS. | | | | | | |
| | FOLIOSTEUS. | | | | | | |
| | TRUTTA | . Salmon-trout | . {Laxoring Orlax} | {Maiforelle Lachsforelle} | | | |
| | | . GreatLakeTrout | | | | | |
| | FEROX | | | | | | |
| | BAILLONI | | | | | | |
| | GENIVITTATUS. | | | | | | |
| | LABRAX. | | | | | | |
| | SALVELINUS | . Charr | . {Oubre Chevalier} | {Stibling Rödnel Roth-forelle} | . Sälbling. | | |
| | UMBLA | | | | . Röttsforelle. | | |
| | ALPINUS. | | | | | | |
| | NIVALIS | . [In Iceland]. | | | | | |
| | KILLINENSIS. | | | | | | |
| | PERISII. | | | | | | |
| | COLII. | | | | | | |
| | WILLUGHBYI. | | | | | | |
| | RUTILUS. | . Gautesfisk. | | | | | |
| | CARBONARIUS. | | | | | | |
| | GRAYI | | | | | | |
| | HUCHO | | | | | | . {Rothfisch. Huch. |
| | LOSSOS | | | | | | |
| OSMERUS | EPERLANUS | . Smelt | . Slom | Eperlan | | . Schnapel | |
| COREGOSUS | OXYRHYNCHUS | . Näbb-sik | Outil | | | . {Schnapel Maräne} | Sig |
| | LLOYDII | | | | | | |
| | LAVERETUS | . Knubb-sik | | | | | . {Kropfling. Riedling. |
| | LAPROXICUS. | | | Schnapel. | | . {Kilch. Kropfelchen. Renke Felchen} | |
| | GRACILIS. | | | | | | |
| | WIDEGRENI. | | | | | | |
| | HIEMALIS | | | | | | |
| | WARTMANNI | . Powan. | | Gravenche | | | |
| | CLUPEOIDES | . Loi-sik. | | | | | |
| | MAXILLARIS | | | | | | . Rheinanken. |

TABLE SHOWING THE GEOGRAPHICAL DISTRIBUTION IN EUROPEAN COUNTRIES OF SOME OF THE MORE IMPORTANT SPECIES OF FRESH-WATER FISHES (*continued*).

| Genus. | Species. | British Islands. | Scandinavia. | France. | Germany. | Russia. | Austria. | Italy. | Spanish Peninsula. |
|---|---|---|---|---|---|---|---|---|---|
| COREGONUS (*cont*). | HUMILIS. MEGALOPS. NILSSONI. ALBULA. NORWEGICA. FINSICA. VIMBA. POLLAN. VANDESIUS. | Pollan. Vendace. | Martinsmess-sik. Blasik. Sik-loja. Sik-wimma. | | Kleine Mirane. | | | | |
| THYMALLUS. | VULGARIS. ELIANI. | Grayling. | Harr. | Ombre. | Aesche. | | | Temolo. | |
| ESOX. | LUCIUS. | Pike. | Gädda. | Brochet. | Hecht. | Shcoka. | Hecht. | Luccio. | |
| UMBRA. | CRAMERI. | | | | | | Hundsfisch. | | |
| CYPRINODON. | CHALARITAS. FASCIATUS. IBERICUS. | | | | | | | Nono. | • |
| FASPITUS. | HISPANICUS. | | | | | | | | • |
| ANGUILLA. | VULGARIS. LATIROSTRIS. EURYSTOMA. | Eel. | Al. | Anguille. | Aal. | | Aal. | Anguilla. | Anguilas. |
| ACIPENSER. | GLABER. RUTHENUS. GMELINI. STELLATUS. SCHYPA. GULDENSTAEDTII. NARDOI. HECKELI. NASUS. STURIO. HUSO. | Sturgeon. | Störge. | Esturgeon. | Sternhausen. Stor. | Sterled. Sevruga. Verze. Osetre. Beluga. | Glattdick. Sterlet. Dick. Waxdick. Hausen. | Storione. Copene. | Sollo. |
| PETROMYZON. | MARINUS. FLUVIATILIS. BRANCHIALIS. | Lamprey. Lampern. Mud-lamprey. Pride. | Hafsnejnoga. Nejnoga. | Lamproie. Lamprillon. PetiteLamproie. | Meer-lamprete. Neunauge. Kleines Neunauge. | | Pricke. Neunauge. | Lamproia. Lamprelone. | |

## CLASSIFICATION OF THE FRESH-WATER FISHES OF EUROPE.

*Giving the Orders, Families, Genera, Species, and principal Varieties.*

### ORDER—ACANTHOPTERYGII.

*PERCIDÆ:*

PERCA (Artedi)
   *fluviatilis* (Linnæus)

LABRAX (Cuvier)
   *lupus* (Lacépède)

PERCARINA (Nordmann)
   *demidoffii* (Nordmann)

ACERINA (Cuvier)
   *cernua* (Linnæus)
   *schrœtzer* (Linnæus)
   *rossica* (Cuvier & V.)

LUCIOPERCA (Cuvier)
   *sandra* (Cuvier)
   *volgensis* (Pallas)

ASPRO (Cuvier)
   *zingel* (Linnæus)
   *vulgaris* (C. & V.)
   *streber* (Siebold)

*COTTIDÆ:*

COTTUS (Artedi)
   *gobio* (Linnæus)
   Var. *microstomus* (Heckel)
   Var. *ferrugineus* (Heckel & Kner)
   *pœcilopus* (Heckel)
   *scorpius* (Linnæus)

*GOBIIDÆ:*

GOBIUS (Artedi)
   *martensii* (Günther)
   Var. *accensensis* (Canestrini)
   Var. *panizzæ* (Canestrini)
   Var. *punctatissimus* (Canestrini)
   *semilunaris* (Heckel)
   *minutus* (Linnæus)

*BLENNIIDÆ:*

BLENNIUS (Artedi)
   *vulgaris* (Pollini)
   Var. *alpestris* (Blanchard)

*ATHERINIDÆ:*

ATHERINA (Artedi)
   *lacustris* (Bonaparte)

*MUGILIDÆ:*

MUGIL (Artedi)
   *capito* (Cuvier)
   *cephalus* (Cuvier)
   *chelo* (Cuvier)
   *septentrionalis* (Günther)

*GASTEROSTEIDÆ:*

GASTEROSTEUS (Artedi)
   *aculeatus* (Linnæus)
   *pungitius* (Linnæus)

### ORDER—ANACANTHINI.

*GADIDÆ:*

LOTA (Cuvier)
   *vulgaris* (Cuvier)

*PLEURONECTIDÆ:*

PLEURONECTES
   *flesus* (Linnæus)
   *italicus* (Günther)
   *platessa* (Linnæus)

SOLEA (Cuvier)
   *vulgaris* (Cuvier)

### ORDER PHYSOSTOMI.

*SILURIDÆ:*

SILURUS (Linnæus)
   *glanis* (Linnæus)

CYPRINIDÆ:
CYPRINUS (Artedi)
  carpio (Linnæus)
  Var. acuminatus (Heckel)
  Var. hungaricus (Heckel)
  Var. regina (Bonaparte)
  Var. kollarii, hybrid

CARASSIUS (Nilsson)
  vulgaris (Nilsson)
  Var. moles (Agassiz)
  Var. gibelio (Nilsson)
  Var. humilis (Heckel)
  Var. oblongus (Heckel & Kner)
  auratus (Linnæus)
  bucephalus (Heckel)

BARBUS (Cuvier)
  vulgaris (Fleming)
  plebejus (Valenciennes)
  caninus (Cuvier)
  petenyi (Heckel)
  comiza (Steindachner)
  bocagei (Steindachner)
  sclateri (Günther)
  chalybeatus (Pallas)
  albanicus (Steindachner)
  Hybrids

AULOPYGE (Heckel)
  hügelii (Heckel)

GOBIO (Cuvier)
  fluviatilis (Rondeletius)
  uranoscopus (Agassiz)

LEUCISCUS (Cuvier)
  erythrophthalmus (Linnæus)
  Var. deryle (Heckel & Kner)
  Var. scardafa (Bonaparte)
  Var. plotizza (Heckel & Kner)
  Var. macrophthalmus (H. & K.)
  idus (Linnæus)
  Var. orfus (Linnæus)
  Var. miniatus (Heckel & Kner)
  Var. lapponicus (Günther)
  borysthenicus (Kessler)
  aula (Bonaparte)

LEUCISCUS (continued)
  Var. rubella (Bonaparte)
  Var. basak (Heckel)
  adspersus (Heckel)
  rutilus (Linnæus)
  Var. pausingeri (Heckel)
  pigus (Lacépède)
  Var. virgo (Heckel)
  Var. roseus (Bonaparte)
  friesii (Nordmann)
  cephalus (Linnæus)
  Var. caredanus (Bonaparte)
  Var. albus (Bonaparte)
  vulgaris (Fleming)
  Var. lancastriensis (Yarrell)
  Var. bearnensis (Blanchard)
  Var. burdigalensis
            (Valenciennes)
  Var. chalybæus (Heckel & Kner)
  Var. rodens (Agassiz)
  Var. rostratus (Heckel)
  illyricus (Heckel & Kner)
  pictus (Heckel & Kner)
  svallize (Heckel & Kner)
  ukliva (Heckel)
  turskyi (Heckel)
  microlepis (Heckel)
  tenellus (Heckel)
  muticellus (Bonaparte)
  Var. savignyi (Bonaparte)
  phoxinus (Linnæus)
  hispanicus (Steindachner)
  arcasii (Steindachner)
  alburnoides (Steindachner)
  macrolepidotus (Steindachner)
  arragonis (Steindachner)
  lemmingii (Steindachner)
  heckelii (Nordmann)
  pyrenaicus (Günther)
  polylepis (Steindachner)

PARAPHOXINUS (Bleeker)
  alepidotus (Heckel)
  croaticus (Steindachner)

TINCA (Cuvier)
  vulgaris (Cuvier)

## CLASSIFICATION.

CHONDROSTOMA (Agassiz)
  nasus (Linnæus)
  Var. cœrulescens (Blanchard)
  Var. auratus (Schaefer)
  Var. dremæi (Blanchard)
  Var. rhodanensis (Blanchard)
  phoxinus (Heckel)
  genei (Bonaparte)
  soëtta (Bonaparte)
  knerii (Heckel)
  Var. meigii (Steindachner)
  rysela (Agassiz) hybrid
  polylepis (Steindachner)
  willkommii (Steindachner)
  Hybrid

RHODEUS (Agassiz)
  amarus (Bloch)

ABRAMIS (Cuvier)
  brama (Cuvier)
  Var. vetula (Heckel)
  Var. gehini (Blanchard)
  vimba (Linnæus)
  elongatus (Agassiz)
  ballerus (Linnæus)
  sapa (Pallas)
  leuckartii (Heckel)
  björkna (Linnæus)
  Var. laskyr (Pallas)
  Hybrid abramorutilus
    (Hollandre)
  bipunctatus (Heckel & Kner)
  fasciatus (Nordmann)

ASPIUS (Agassiz)
  rapax (Agassiz)

ALBURNUS (Heckel)
  lucidus (Heckel & Kner)
  Var. lacustris (Heckel & Kner)
  Var. breviceps (Heckel & Kner)
  Var. fabrææi (Blanchard)
  Var. mirandella (Blanchard)
  arburnellus (Martens)
  Var. fracchia (Heckel & Kner)
  Hybrid
  mento (Agassiz)
  chalcoides (Güldenstadt)

LEUCASPIUS (Heckel)
  delineatus (Heckel)

PELECUS (Agassiz)
  cultratus (Linnæus)

MISGURNUS (Lacépède)
  fossilis (Linnæus)

NEMACHILUS (Van Hasselt)
  barbatulus (Linnæus)

COBITIS (Artedi)
  tænia (Linnæus)
  Var. elongata (Heckel & Kner)

*CLUPEIDÆ:*

CLUPEA (Cuvier)
  alosa (Linnæus)
  finta (Cuvier)
  pontica (Eichwald)
  Whitebait

*SALMONIDÆ:*

SALMO (Artedi)
  salar (Linnæus)
  Trout
  stomachicus (Günther)
  hardinii (Günther)
  cambricus (Donovan)
  argenteus (Cuvier & Val.)
  venernensis (Günther)
  mistops (Günther)
  nigripinnis (Günther)
  obtusirostris (Heckel)
  microlepis (Günther)
  fario (Linnæus)
  ausonii (Linnæus)
  galliveusis (Günther)
  autumnalis (Pallas)
  lemanus (Cuvier)
  carpio (Linnæus)
  rappii (Günther)
  dentex (Heckel)
  spectabilis (Cuv. & Val.)
  marsiglii (Heckel)
  lacustris (Willughby)
  schiffermülleri (Val.)
  levenensis (Walker)
  orcadensis (Günther)

SALMO (*continued*).
    *polyosteus* (Günther)
    *brachypoma* (Günther)
    *trutta* (Linnæus)
    *ferox* (Cuvier & Val.)
    *bailloni* (Cuvier & Val.)
    *genivittatus* (Heckel & Kner)
    *labrax* (Pallas)
    salvelini or Charr
    *salvelinus* (Linnæus)
    *umbla* (Linnæus)
    *alpinus* (Linnæus)
    *nivalis* (Günther)
    *killinensis* (Günther)
    *perisii* (Günther)
    *colii* (Günther)
    *willughbyi* (Günther)
    *rutilus* (Nilsson)
    *carbonarius* (Ström)
    *grayi* (Günther)
    *hucho* (Linnæus)
    *lossos* (Günther)
LUCIOTRUTTA (Günther)
    *leucichthys* (Güldenstadt)
OSMERUS (Artedi)
    *eperlanus* (Linnæus)
COREGONUS (Artedi)
    *oxyrhynchus* (Linnæus)
    *lloydii* (Günther)
    *laveretus* (Linnæus)
    *lapponicus* (Günther)
    *gracilis* (Günther)
    *widegreni* (Malmgren)
    *hiemalis* (Jurine)
    *wartmanni* (Bloch)
    *clupeoides* (Lacépède)
    *maxillaris* (Günther)
    Var. *humilis* (Günther)
    Var. *megalops* (Günther)
    *albula* (Linnæus)
    Var. *norvegica* (Günther)
    Var. *maræanula* (Günther)
    Var. *fiunica* (Günther)
    *vimba* (Linnæus)
    *pollan* (Thompson)
    *randesius* (Richardson)

THYMALLUS (Cuvier)
    *vulgaris* (Linnæus)
    Var. *æliani* (C. & V.)
    *microlepis* (Steindachner)

ESOCIDÆ:
ESOX (Artedi)
    *lucius* (Linnæus)

UMBRIDÆ:
UMBRA (Kramer)
    *crameri* (Müller)

CYPRINODONTIDÆ:
CYPRINODON (Lacépède)
    *chalaritanus* (Bonelli)
    Var. *fasciatus*
    *ibericus* (Valenciennes)
FUNDULUS (Lacépède)
    *hispanicus* (Valenciennes)

MURÆNIDÆ:
ANGUILLA (Cuvier)
    *vulgaris* (Turton)
    Var. *latirostris* (Risso)
    *eurystoma* (Heckel & Kner)

ORDER—GANOIDEI.
ACIPENSER (Artedi)
    *glaber* (Marsil)
    *ruthenus* (Linnæus)
    Var. *gmelini* (Fitzinger & Heckel)
    *stellatus* (Pallas)
    *schypa* (Güldenstädt)
    *güldinstadtii* (Brandt & Ratz)
    *naccarii* (Bonaparte)
    Var. *nardoi* (Heckel)
    *heckeli* (Fitzinger)
    *nasus* (Heckel)
    *sturio* (Linnæus)
    *huso* (Linnæus)

ORDER—CYCLOSTOMATA.
PETROMYZONTIDÆ:
PETROMYZON (Artedi)
    *marinus* (Linnæus)
    *fluviatilis* (Linnæus)
    *branchialis* (Linnæus)

## CHAPTER II.

### FRESH-WATER FISHES OF THE ORDER ACANTHOPTERYGII.

FAMILY PERCIDÆ —Genus Perca: The River Perch—The Bass—Genus Percarina: Percarina Demidoffii—Genus Acerina: The Ruff or Pope—Acerina schraetzer—Acerina rossica—Genus Lucioperca: The Pike-Perch—The Pike-Perch of the Volga—Genus Aspro: The Zingel—The Apron—The Streber—FAMILY COTTIDÆ Genus Cottus: The Miller's Thumb—Cottus pœcilopus—Father Lasher—FAMILY GOBIIDÆ— Genus Gobius: Species of Goby—The Polewig—FAMILY BLENNIIDÆ Genus Blennius— FAMILY ATHERINIDÆ—Genus Atherina: The Sand Smelt—FAMILY MUGILIDÆ —Genus Mugil: The Grey Mullet—FAMILY GASTEROSTEIDÆ—Genus Gasterosteus: The Sticklebacks.

## FAMILY: PERCIDÆ.

### Genus: Perca (Artedi).

THE Perch is the type of the group Percina, a tribe of about eighteen genera of fishes, which agree in having an oblong and elevated form of body, denticulated opercular bones, and ctenoid scales, with two dorsal fins, distinguished by a number of spines, which is constant in each species. The Perch is found in rivers and lakes. The variety *Perca flavescens* ranges over the United States and Canada, and the Canadian variety, *Perca gracilis*, is probably only a variation of that type. The number of spines in the first dorsal fin is greater in Perca than in the marine Bass (*Labrax*), which is its nearest ally, a genus which lives alike in mouths of rivers and in the seas of the northern parts of the world. The Perch family have the teeth simple and conical, have no barbels to the mouth, and the lateral line of mucus pores extends from the head to the caudal fin.

### Perca fluviatilis (Linnæus).—The River Perch.

1 D. 13—16, 2 D. 1—2/13—15, A. 2/8—9, V. 1/5—6, P. 14—17, C. 17,

Scales: lateral line 54—72, transverse $\frac{7-9}{13-18}$.

The Perch (Fig. 10) is sub-ovate in body, three to four times as long as high, and moderately compressed from side to side. Its greatest height occurs above the ventral fins. The length of the head, to the extremity of the operculum, is nearly equal to the height of the body, but may be a little longer or a little shorter. The eye varies in relative size with age, and is placed from one to one and a third times its own diameter behind and above the extremity of the snout, which is occupied by the jaws. Both jaws are evenly covered with broad bands of teeth, like short bristles, and the vomer and palatine bones both carry similar teeth. The tongue is large and fleshy, and is smooth, like the middle of the palate. There are seven branchiostegal rays on each side; these are attached to the hyoid bones, termed cerato-hyals and epi-hyals, and are seen below the operculum. Four branchial arches, which support the gills, are attached to the skull. These arches consist of an upper part, termed epi-branchial, and a lower cerato-branchial portion. Both bones of the branchial arch carry bony processes, termed gill-rakers, which are directed towards the mouth, and defend the entrance to the branchial clefts. The gills are attached to grooves on the

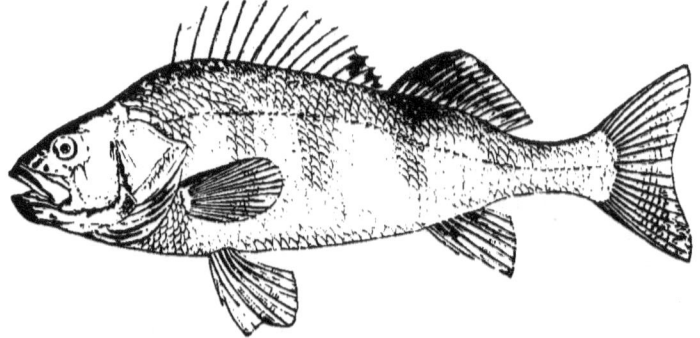

Fig. 10. —PERCA FLUVIATILIS (LINNÆUS).

outer side of the epi-branchial and cerato-branchial bones. The gills are comb-shaped, and greatly developed in the young animal.

The opercular bones give characters which define the species. The margins of the pre-opercular bone form a right angle or an obtuse angle, the upper or vertical part of which is very finely denticulated, and the lower edge more coarsely serrated, with the serrations directed forward; the first process is generally strongest and largest. The opercular bone bears scales only on its upper half, and ends posteriorly in a long spike, under which are seen one or two other serrations. The blunter point of the sub-operculum extends farther backward, and the lower third of its margin is finely and evenly serrated, though the serrations are scarcely distinguishable in the fresh fish. There are minute serrations on the inferior border of the inter-operculum. Scales extend over the cheeks and the sub-operculum; but the fins, the entire upper part of the head, the sub-orbital ring, and the inter-operculum are naked.

The first dorsal fin is placed at a distance behind the snout equal to the length of its own base, and begins just over the spinous extremity of the opercular bone. The fin rays are strong, pointed, thirteen to fifteen in number. The last ray is very short, and the fifth is highest, and equal to two-fifths the height of the body. The second dorsal fin is only a little less elevated; its base measures two-thirds the length of that of the first dorsal fin. Its first spinous ray is very short, and is united to the first dorsal by membrane.

The anal fin commences beneath the anterior half of the second dorsal. Its two strong spinous rays are less elongated than the succeeding soft rays.

The pectoral fins are moderately developed, and have a rounded outline. The first and last rays are about half the length of the middle and longest rays; there are fourteen rays in the fin.

The ventral fins commence below the middle of the pectorals, and are about as long as those fins, but the rays are broader. The fin extends backward half-way to the vent.

The caudal fin is evenly lobed, and only moderately concave in the outline of its hinder margin. This fin forms one-seventh of the length of the Perch.

The lateral line is nearly parallel to the back; its length includes from sixty to sixty-eight scales. At the base of the ventral fin there are from thirteen to fifteen rows of scales below the lateral line, and seven to nine rows of scales above it. The scales are finely serrated at the free edge, are generally broader than long, are largest on the abdomen and the sides, and smallest on the throat. The caudal fin is the only fin with scales at its base. There are three pyloric appendages to the stomach, of moderate size. The air-bladder adheres to the vertebræ and ribs.

The shade of colour varies from a brassy yellow to a bluish or greenish tinge, becoming whitish on the belly, golden yellow at the sides, and blackish-green on the back. From five to nine brownish-black vertical bands extend down the sides of the Perch between head and tail. These bands vary in length, breadth, and intensity of colour, and are sometimes represented by blackish-clouded spots, or spots and bands may both be absent. The large black spot between the last two or three rays of the first dorsal fin is, however, always present. The first dorsal fin is violet-grey; the second dorsal has a yellowish base, becoming reddish and green at the borders; but the colours of the fins vary. The ventral, anal, and caudal may be orange or vermilion, while the pectoral is described by Yarrell as pale brown, by Heckel and Kner as yellowish-red, while Fatio finds it to be yellow. These differences confirm Blanchard's statements that the colours of the fins and of the fish vary with the locality and season at which it is taken, and, we might also add, with age. Various Continental writers mention a yellow variety, which Siebold considers to resemble the Yellow Perch (*Perca flavescens*) of North America. Cuvier and

Valenciennes distinguished a *Perca italica*, characterised by being free from bands, and by having the head slightly larger than the common type; but this is to be regarded as one of the many varieties of the species. The Perch of the Danube, known as *Perca vulgaris*, has no better claim to a distinctive name.

Another variety, found in the Lakes Longemer and Gerardmer, in the Vosges, is distinguished by the more elongated body, less elevated back, associated with a number of minor differences in the form of the snout, sub-orbital pits, and serrations of the pre-operculum, with a scaliness of the cheek. It is usually smaller, and is probably an ill-nourished variety of the Common Perch.

The usual size of the Perch in England is from nine inches to a foot. Isaac Walton mentions one measuring nearly two feet, and the longest recorded measured twenty-nine inches. But the head of a reputed Perch is said to be preserved in the Church of Lulea, in Lapland, which is nearly a foot in length, so that the fish, presumably, must have been between three and four feet in length. The weight of large fishes varies from two or three pounds to eight or nine pounds. In Russia they reach eight pounds in Lake Seligher. In Austria the Perch is rarely more than a foot long, and a pound and a half in weight. In the Zeller See it reaches a weight of three to four pounds. In the Swiss lakes the average weight is three pounds, increasing exceptionally to six. In the higher Alpine lakes the Perch is always smaller. In North Germany it attains a larger size in the Haffs on the Prussian coast than in the smaller lakes. It is a common fish in the Baltic; but in that sea rarely exceeds from half a pound to a pound in weight.

There are no marked differences to distinguish the sexes; but the males have the body slightly higher, and relatively shorter. At Salzburg only one-tenth of the fishes caught proved to be males.

The Perch frequents clear water, and commonly rests among plants at a depth of two or three feet. In Scandinavian lakes it passes the winter in deep water, but on the breaking up of the frost makes for the shore in large shoals. Siebold describes it as living in Lake Constance at a depth of 210 feet, under pressure so great that when it is brought to the surface the fish may be found torn open or with the viscera forced into the mouth. In Swiss lakes it is found as high as 4,400 feet above the sea. In such positions it never exceeds one pound in weight, though high up in the Jura Mountains it reaches a larger size. The warmer the water the larger the fish.

It is gregarious when young, and is often found in great shoals. It feeds on small fishes, worms, larvæ of insects, fresh-water crustacea, and amphibians. Perch will even attack and kill water-rats. Like all voracious fishes, it is easily taken by the hook, especially when baited with minnow or lob-worm; but in Germany, where it is more valued as food than in England, it is generally captured with a net. In a Swedish lake two men with rod and line may often take about midsummer from three hundred to three hundred and sixty pounds weight of Perch in three or four hours.

Sometimes a decoy is used in attracting Perch in shallow streams, a number of live minnows being placed in a glass bottle, the mouth of which is closed with perforated metal to allow of the water circulating. The angler then drops his line in the shoal of Perch which gathers round the bottle.

The Perch grows slowly. According to Kröyer, the young fish at the commencement of the first winter is only one inch long; in the third year it is six inches long, and weighs three ounces; in the sixth year it weighs a pound and a half, and is sixteen inches long. It is believed to spawn for the first time in its third year.

A Perch of half a pound weight may contain 280,000 eggs. It spawns in March, April, and May. The female disencumbers herself of the eggs by rubbing her body against stones, so that they become enveloped in mucus; and the eggs then hang to the stones in strings, which are frequently connected, not unlike the meshes of a net, the mass often being five or six feet long. The eggs are preyed upon by birds and various fishes, such as the Trout and Eel, and by fishes of the same species; and are sometimes cast ashore in storms.

Specimens have been taken in the Rhine with the milt well developed as late as September.

The flesh of the Perch is white, firm, and well flavoured when taken from clear water, but liable in rivers to acquire what may be described as a muddy flavour. In the brackish water of the Norfolk Broads its flavour improves with a diet of shrimps. If it is baked, all the rough scales are carefully removed; if it is boiled, the skin and scales are taken off together after it is cooked. It is often served with orange-juice or vinegar.

In Russia it is an important article of food, and goes to market fresh, frozen, partly salted, or dried. Dried Perch is known as "soosh."

From the skin an excellent isinglass is prepared similar to that of the Sturgeon; and in Scandinavia this substance is used as a glue. Lloyd states that the skins are dried, and subsequently steeped in cold water. The scales are then scraped off. The skins are next placed in a bullock's bladder, which is tied securely to keep out the water. The bladder is then placed in a cauldron, and boiled till the skins are dissolved. The scales were formerly used in Scandinavia for embroidery on ribbons, reticules, &c. In North Germany artificial flowers are made from the scales.

The fish is not migratory, though it is capable of living out of water for some hours, and is often carried long distances in wet grass. In Catholic countries individuals remaining unsold in the market are again returned to the ponds, as is the case in some parts of the middle of England, as at Peterborough.

Many deformities of the Perch have been described, the back becoming greatly elevated, and the tail distorted. With old age Perch become dark in colour, and blind. The density of the cornea has been attributed to inflammation, produced by parasites in the aqueous humour of the eye. Remarkable epidemics have destroyed the Perch in some of the Swiss lakes in great numbers. One of these has been attributed by Dr. Forel and Dr. Du Plessis to the presence of bacteria in the blood, and is regarded as a form of typhus fever. The Perch suffers from many small parasites; the mouth is infested by a crustacean (*Achtheres percæ*); the body is attacked by leeches (*Ichthyobdella percæ*); the skin, muscles, cephalic cavities, branchiæ, liver, digestive canal, are all subject to the attacks of entozoa and other parasites. The Perch is introduced into Trout ponds in North Germany as food for the Trout.

In Britain the Perch is absent from the extreme north; and Yarrell speaks of it as only sparingly met with in the lochs north of the Forth. In Scandinavia it occurs as far north as Lapland, where it reaches a large size. It is spread nearly all over Europe, and found generally in the Alps to the height of upwards of 4,000 feet.

It occurs in the Sea of Azov, in the brackish water of the Caspian, and over a large part of Northern Asia.

The name Perch was first used by Aristotle, and is, therefore, of Greek origin. It is a term applied to the dusky colour of ripening grapes, and is supposed to indicate the spotted or banded marking so characteristic of this fish. With the Romans it was *Perca*, with the Germans it is *Barsch*, in which the "*P*" has become changed into "*B*." From this the Dutch *Baars* is readily derived, a name which appears to have been adopted into the east of England with slight variation of spelling. All these Teutonic names signify barred or banded fish. In France it is *la Perche*, in Italy *Persico*. In Bavaria it is *Bürstel*, in Lake Constance it is the *Egli*, in Sweden the popular name is *Aborren*. Dr. Day mentions that the Saxons represented one of their gods as standing with naked feet upon the back of a Perch, as an emblem of patience in adversity.

## Labrax lupus (Cuvier).—The Bass.

1 D. 8—9, 2 D. 1 12—13, P. 16, V. 1/5, A. 3/10—11, C. 17.

The Bass is essentially a sea Perch, with seven branchiostegal rays and well-developed pseudo-branchiæ. The body is elongated. The pre-operculum is serrate, the operculum spinous, and the pre-orbital bone entire. The teeth are villiform, are found in both jaws, on the vomer, the palatine bones, and the tongue. The scales are ctenoid, but with the denticulations rather less conspicuous than in the Perch, and the scales are rather smaller. There are only nine spines in the first dorsal fin, and usually three spines in the anal fin, so that the essential difference between Labrax, the sea Perch, and Perca, the river Perch, consists in the former having teeth on the tongue, fewer spines in the first dorsal fin, and usually one spine more in the anal fin, though this character is less constant.

In North America several species of Labrax are found in the rivers and along the eastern shores of the United States, but in Europe the Labrax is more typically marine, though the type species, *Labrax lupus*, is found often enough in European fresh waters to justify a notice of it among fresh-water fishes.

The *Labrax lupus* was known to the Greeks under its generic name, but the Rev. Dr. Badham remarks that it is uncertain whether the name refers to the fish's sin of gluttony or of violence, though the Latin name of *lupus*, now become specific, was always recognised as an appropriate designation.

It was esteemed far more by ancient epicures than by those of modern times. Aristophanes terms this the wisest of fishes, on account of its cleverness in escaping from dangers; but, says Dr. Badham, every one has a weak point to lead him astray, and the Bass's foible is inordinate greediness. He enjoys prawns exceedingly, and on meeting a shoal, opens his mouth, and fills it at a gulp with hundreds of these nimble and prickly crustaceans. The prawns no sooner find themselves on the wrong side of the barrier, and going down quick into the pit of their enemy's stomach, than they hold on hard, and run the sharp serrated rostrum of their heads into his palate and throat, and so stick the greedy fish, who, unable to cough them up or otherwise detach them, dies either of spasmodic croup or of an ulcerated sore throat. Heliogabalus, it is said, would only eat the brains and milt of the Labrax; Rondeletius pronounces its liver finer than that of goose or turkey; and the dried roe is still highly esteemed by the modern Greeks as a substitute for caviare. Ancient and modern writers agree that the fish is in excellent condition when it has been fattened for a month or six weeks in fresh water. They do not breed in fresh water, but at the spawning time retire to the salt marshes, and commonly run up streams all round the Mediterranean. It is a foul-feeding fish, and was especially famous when caught in the Tiber, between the bridges where the great drain emptied itself. And although the taste for fishes so fed is past, they have not ceased to frequent the feeding-ground; and Dr. Badham narrates how he had often stopped to watch the net used in this fishery perpetually revolving with the current, while some pallid waterman has stood up feebly in the tethered old boat, wan as the Stygian ferryman, with ague in his veins and no quinine in his pocket, eyeing the meshes, and putting forth a spectral arm to secure the small Bass or other prey, and then letting down the net again into the floating feculence of the river.

Yarrell mentions that it had been successfully retained in Mr. Arnold's fresh-water lake in Guernsey, and Steindachner mentions it as characteristic of the embouchures of the rivers in the west of the Spanish peninsula. Its ordinary length is from twelve to eighteen inches, though Couch measured one that was thirty-one inches long. The weight is rarely in proportion to the length, and does not commonly exceed twenty pounds in large fishes. Its colour is grey on the back, becoming silvery at the sides, and as bright as new silver beneath. There is a dark spot on the upper half of the operculum. The young have small dark spots on the body. The fins are grey, with a yellowish tinge in the pectoral and ventral fins.

## Genus: **Percarina** (Nordmann).

This genus is essentially a Perch, in which all the fins are strongly developed except the first dorsal, and in which the palatine bone carries no teeth. The form of the fish is compressed and elongated, resembling Acerina. The mucus-cavities in the bones of the skull are greatly developed. The first dorsal contains ten spines, and is united to the second dorsal by a narrow membrane. The operculum has one spine; the pre-operculum is denticulated. The scales are small. The genus is known only from the south of Russia and the adjacent borders of Austria.

## Percarina demidoffii (Nordmann).

1 D. 10, 2 D. 3 9—10, A. 2 9—10.

This species (Fig. 11) is the only example of the genus; it is characterised by a line of round black spots along the back, and a lunate brown spot on the neck. All the fins are transparent, and without spots. It differs from Acerina, and resembles Lucioperca in possessing two closely-connected dorsal fins. The body is more compressed from side to side than in *Acerina cernua*. Its greatest height is at the first dorsal fin, and is twice the thickness of the fish; the height is also equal to the distance from the extremity of the snout to the margin of the pre-operculum. The entire length of the

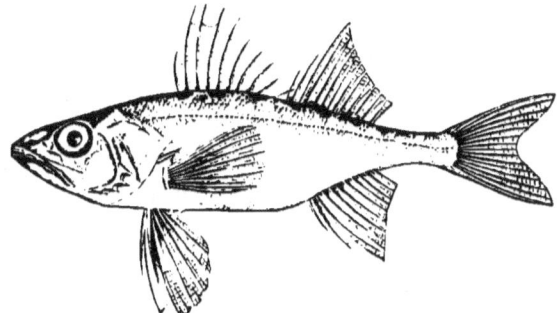

Fig. 11. PERCARINA DEMIDOFFII NORDMANN.

animal is three and three-quarter times the length of the head. The eye is large, separated from the snout by its own diameter, and the inter-orbital space is two-thirds of the orbital diameter. The mouth is wide; it reaches back to beneath the middle of the eye. The pre-maxillaries are prolonged to form the extremity of the jaw, and carry fine densely-placed teeth throughout their length. The lower jaw has not only a small band of villiform teeth along its upper edges, but the teeth are produced on both sides, so that the lower jaw is covered with villiform teeth, like a spinous cylinder—a form of dentition which is not known in any other fish. The vomer is provided with only a small group of teeth. The operculum is less developed than in *Acerina cernua*, but possesses a similar spine. The pre-operculum along its entire length has the aspect of being double, there being two parallel spinous borders, between which the branch of the cephalic canal is contained which is supplied to the lower jaw. Owing to the wide gape of the mouth, its backward extension, and the breadth of the upper jaw behind, there is but little space for the front portion of the sub-orbital ring, which is therefore small, and the cavities of the cephalic canal in this region are, therefore, less obvious than in Acerina; the other portions, however, of this vascular mucus system are more strongly developed, and the head appears to be surrounded by mucus vesicles. The external apertures of the nares, which are anterior and posterior to each other, are separated by a broad process of membrane.

The first portion of the dorsal fin commences over the base of the pectoral fin. Its rays radiate, but the last is the shortest, and is joined by membrane to the succeeding longer ray which Nordmann included in the second dorsal, but which is not very distinctly separated from the first dorsal, so that it is almost a matter of taste whether the fish is accredited with one or two dorsal fins. The second dorsal includes three spines, and nine to eleven soft rays. The termination of the soft part of the fin is exactly opposite to, and coincides with, the termination of the anal fin, though the anal commences somewhat farther back, immediately behind the conspicuous ovarian papilla. The anal aperture is nearly in the middle of the length of the body. The ventral fins are close under the bases of the pectoral fins, than which they are somewhat shorter, for the pectoral fins, when turned back, reach nearly to the anal aperture, and exceed in length the deeply-forked, evenly-lobed caudal fin.

The scales are all ctenoid, and rather delicate. There are thirty-four to thirty-six scales along the lateral line, which have tubular perforations. The number of scales is greater in the row which is next above the lateral line. There are no scales on the head and throat.

The colour is yellowish-white, shading into violet on the back, with the sides, operculum, and belly like burnished silver. The base of the dorsal fin is marked by numerous round brownish-black spots, which extend to the caudal fin. The lateral line is margined by spots of black pigment.

Nordmann, who first found this interesting species at Ackermann, in Bessarabia, states its size to be rather smaller than *Acerina cernua*.

Heckel and Kner describe it from the Dniester; Dr. Grimm states that it is found in the Black Sea, at the mouths of the Dniester, Bug, and Dnieper.

GENUS: **Acerina** (CUVIER AND VALENCIENNES).

This is a small group of Perches in which there is but one dorsal fin, and that always has the upper border concave. Its anterior part is formed of spinous rays and its hinder part of soft rays, so that the single dorsal is formed by the blending of the two dorsals of the Perches, owing to the obliteration of the interval between them. The teeth are similarly villiform, but there are no teeth on the palatine bones. The skull has the system of mucus channels greatly developed in the bones. The operculum and pre-operculum are both spiny. The scales are small. The throat and abdomen have few scales. There are few pyloric appendages to the stomach.

This genus is limited to the Old World and to its Palæarctic portion, being found distributed over Siberia, and in most, if not all, European countries.

## Acerina cernua (LINNÆUS).—The Ruff, or Pope.

D. 13—15/12, A. 2,5—6.

This species (Fig. 12), popularly known in England as the Ruff, or the Pope, has a close general resemblance to the Perch in form. The snout is blunt, but rounded; the dorsal fin may have twelve or fifteen spinous rays. The fish is about four and a half times as long as high, and about a third higher than thick. The greatest height is over the base of the ventral fin, and is nearly equal to the length of the head. The height of the head is two-thirds of its length; it is more than half as broad as long. The eye is large, sub-circular, or slightly oval, and placed well to the side, but so as to look a little forward. Its diameter is rather less than one-fourth of the length of the head. The mouth is surrounded by fleshy lips. It descends obliquely, is capable of a slight protraction, and the gape extends at least as far back as the nasal apertures. The tongue is moderately developed. The pre-maxillary bones and lower jaws possess rows of teeth of even size. The vomer carries only a small group of teeth. The nasal apertures are double, and occupied chiefly with a remarkable expansion of the nerve of smell. The anterior aperture is small, round, margined by a valve, and midway between the orbital border and the extremity of the snout. The posterior aperture is large, sub-triangular, and near to the eye.

The head is characterised by the broad and deep pits which carry the system of mucus-canals of the head, which are covered over only by the external skin. Many pits lie between and in front of the eyes; one row is contained in the curve of the sub-orbital bones. A bow-shaped line extends along the cheek, but the most remarkable branch extends from the pre-operculum to the lower jaw. This branch of the cephalic canal is covered by the spines from the operculum. The pre-operculum is rounded posteriorly and inferiorly, and is armed at its border with ten or a dozen spines, which diverge. The lower spines are larger and farther apart than the upper ones, which are smaller and serrated. The operculum ends in a short, sharp spine, which is covered by a small flap of skin. The sub-operculum is large, and slightly

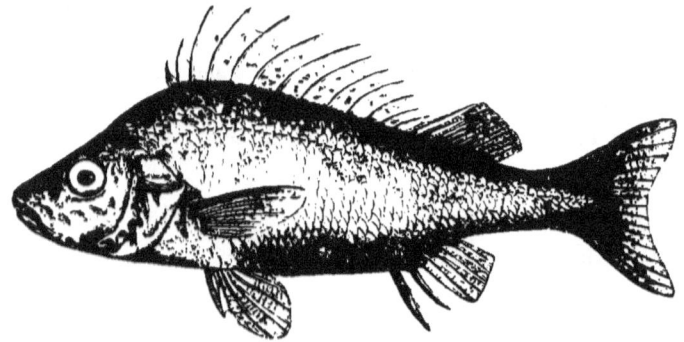

Fig. 12.—ACERINA CERNUA (LINNÆUS).

denticulated on its anterior border. Villiform teeth occur on the pre-maxillary, the maxillary, and pharyngeal bones. Small teeth are also irregularly distributed on the tubercles of the branchial arches. Some small teeth have been observed on the palatine bones in specimens from the River March, in Austria.

The dorsal fin occupies nearly the entire length of the back, and commences above the opercular spine and pectoral fin. It usually contains fourteen spinous rays, of which the fourth to sixth are the longest, and the succeeding rays decrease in length to the last spinous ray, so that the outline of this part of the fin is a curve. There are eleven to fourteen soft rays in the hinder part of the fin.

The anal fin is opposite to the membranous part of the dorsal, but does not reach quite so far back. The pectoral and ventral fins are of similar length—the pectorals are oval, and contain thirteen rays; the ventral fin has one thick spiny ray and five soft rays. The caudal fin is moderately notched, and is formed of seventeen rays. The anal aperture is behind the middle of the length of the fish.

The ctenoid scales are moderately large; fifty-five to sixty are found in the lateral line, with six to seven rows above, and ten to twelve rows below the lateral line. A part of the breast in front of the ventral fin is naked. The scales resemble those of the Perch and the Apron, but they are more oval, and the spines are sharp and conical, longer than those of the Apron, and more numerous than in the Perch. The lateral line is nearly straight, and follows the outline of the back; it is very conspicuous, on account of the large size of the mucus-canals, which open through the scales.

The internal organs have much in common with those of the Apron and the Perch. The eggs are of large size and yellowish in colour; the ovaries form closed sacs; the uro-genital opening is a perforation in a thick short papilla behind the anal aperture. The stomach has three pyloric appendages, and the intestine has three folds. The vertebræ usually number thirty-five or thirty-six, according to Fatio, though Valenciennes records thirty-seven, of which fifteen are abdominal and twenty to twenty-two caudal.

The colour is greenish-olive, mottled and spotted with brown. The sides are brassy yellow, the belly an opalescent white, with the breast and throat pale red. Upon the spinous part of the dorsal fin the brownish-black spots generally form four or five rows between the spines. On the caudal fin the spots often cover the rays. The pectoral fin is sometimes free from spots or irregularly marked. The anal and ventral fins are white, with red tinting. The iris is black, with a golden-yellow colour below.

In England this little fish is rarely more than three or four inches long, but sometimes reaches a length of six or seven inches. An example from the Elbe is said to have measured nine inches, and Bloch mentions one from a Prussian lake which was a foot long. In Siberia it is taken seventeen to eighteen inches long, and weighing a pound and a half.

This is essentially a northern species, being common in the rivers of Russia, Northern and Central Scandinavia, and Siberia, and spread through England, France, Switzerland, and Central Europe. According to Cuvier, it is especially common at the confluence of rivers. It has not been recorded from Spain, Italy, or Greece. In Scandinavia it is known, from its slimy covering, as Snorgers, *suor* being nasal mucus, and *gers* the popular name of the fish.

This fish, the *Kaulbarsch* of the Germans and *Gremille* of the French, is found indifferently in rapid streams and sluggish waters, where it frequents sandy bottoms, and takes to deep water in the winter. It does not thrive on a clay bottom, perhaps because it lives at the bottom of the water. It passes most of the year in solitude, is sluggish, moves by short shoots, and waits for its prey; but when alarmed it is active enough to have originated the Swedish saying, "agile as a Ruff."

*Acerina cernua* feeds on the eggs of fishes, on insects, worms, and aquatic organisms, and will also eat grass and earth. It does not attempt to capture large fishes or those that move rapidly.

It spawns in April and May. The yellow eggs are deposited on the roots of water-plants in connected strings, like those of the Perch. Dr. Day found the roe, weighing four and a half ounces, to contain upwards of 205,000 well-developed eggs.

In some parts of France there is an absurd popular idea that this fish is a cross between the Perch and the Gudgeon.

It is easily caught with fine meshed nets in the spring of the year, or may be taken with the rod and line. When storms drive away other fishes, the Ruff remains unaffected. It is easily tamed. Wherever it is abundant, its flesh is highly prized; it is firm, white, and palatable. But, owing to its small size, it is neglected in districts where it is rare. It is sufficiently tenacious of life to be easily taken from place to place, and, Lloyd says, may be kept alive a long time if frozen as soon as captured, and afterwards thawed in cold water.

Frank Buckland mentions that in the Thames, and in other parts of England, a custom called "plugging a Pope"—probably connecting its name of Pope with Roman Catholic persecution—consists in pressing a wine-cork on the spines of the dorsal fin. The people of Sheffield, Leeds, and York, to the number of hundreds, go to Crowel Bridge, in Lincolnshire, for fishing matches provided with corks, which they fix on the dorsal spines of this fish, and then return them to the water. He states that it is a funny sight to see the surface of the canal for so many miles covered by these unfortunate Popes. In Russia the Pope is dried in ovens to make "soosh."

## Acerina schrætzer (LINNÆUS).

The *Acerina schrætzer* (Fig. 13), which is peculiar to the Danube, is regarded by some as a geographical variety of the *Kaulbarsch*. It has an elongated body, with the snout prolonged forward, and a dorsal fin, which contains eighteen or nineteen spinous rays, and runs nearly the whole length of the back. As compared with the *Kaulbarsch* (*Acerina cernua*), it is more elongated; the greatest height at the pectoral fin is only one-fifth to one-sixth of the total length, the breadth is barely two-thirds of the height, and is equal to half the length of the head. The diameter of the eye is one-fourth of the length of the head. The snout is more elongated, and the profile nearly straight. The mouth reaches back to the anterior aperture of the nares. The teeth are like those of *Acerina cernua*. The tongue is round, smooth, and rather free. The openings of the cephalic canal, and the opercular spines resemble those of *Acerina cernua*. At the upper edge of the preoperculum are seven strong sharp spines, one at the angle, two on the under edge, and four or five covered by skin. The operculum terminates in a very pointed spine; the sub-operculum only shows fine denticulations.

The dorsal fin commences over the pectoral fin; its bony rays rapidly increase in height to the fourth, and then decrease gradually to the last, which is about half the length of the succeeding soft rays, which are much closer together than are the spinous rays. The anal fin is placed beneath the soft portion of the dorsal fin, though its base does not reach back so far. The anal aperture is in about the middle of the length of the body.

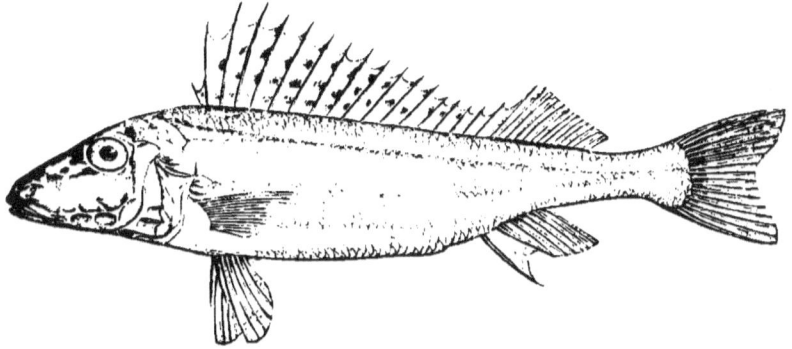

Fig. 13. ACERINA SCHRAETZER (LINNÆUS).

The pectoral fin is of irregular form; the longest rays are the earliest. In the ventral fin the strong spinous ray is only half the length of the succeeding soft rays. The hinder margin of the caudal fin is only moderately concave; the fin measures one-seventh of the total length of the fish.

Scales are generally absent from the head and the regions of the paired fins, though a few small scales are sometimes found in front of the ventral fin on the breast.

Thus it is seen that the chief distinctions in the fins of this species, other than those of form, are in the dorsal containing nineteen spinous rays instead of twelve or fourteen, while there is occasionally an additional ray in the anal and the pectoral fin.

The lateral line is parallel to the back, is on the upper fourth of the side, and contains sixty to seventy scales. There are seven or eight rows of scales above it, at the commencement of the dorsal fin, and thirteen or fourteen rows of scales below it.

The general colour is yellowish, becoming a brownish-olive or olive-green towards the back, and silvery-white on the belly. It is marked on the sides with three or four black bands, which are longitudinal; the first is near to the back, the second is near to the lateral line, the third commences above the pectoral fin, and joins the second near the tail. The fourth band, which is sometimes made up of spots, runs between the pectoral and ventral fins. The membrane of the whitish dorsal fin is marked with three or four rows of blackish spots, which do not extend on to the portion with soft rays. At the end of the caudal fin there are small black spots, and the other fins are more or less yellow. This fish weighs about half a pound; it grows slowly, and lives for fifteen or twenty years. Its habits are similar to those of *Acerina cernua*.

It is not uncommon in the Danube and some of its tributaries. In Bohemia it is known as the *Gesdik*, at Budweis it is termed the *Schrazl*; but in general it is known on the Danube as the *Schrasen*, or *Schratzel*.

## Acerina rossica (CUVIER).

D. 17/12, P. 15, V. 1/5, A. 2/6, C. 17.

This fish is not uncommon in the south of Russia, where it is known as *Biritchok*, or *Babyr*, though it has other local names. It is found in the Dniester, Dnieper, and Bug, and is especially abundant in the Don and Donetz.

The anterior part of the head, between the eyes and the snout, is greatly prolonged, so that the eyes, which are very large, are placed far back. The mouth is small. The round spots on the sides of the body are arranged quincuncially, or in series of five. Their colour is blue-black. The dorsal fins are spotted with black.

There are fifty-five scales in the lateral line.

## GENUS: Lucioperca (CUVIER).

The Pike-Perches have an elongated form of body, which suggests the Pike, as do the large teeth, which occur among the villiform teeth in the jaws. In general character Lucioperca resembles the Perch, having two dorsal fins, with twelve to fourteen spines in the first dorsal. The anal fin has two spines, the pre-operculum is serrated, the palatine bones are toothed, the head is elongated with a straight upper outline, and the cleft of the jaws extending as far back as the middle of the eye. The species of this genus are mostly found in rivers and lakes of the northern parts of the Old World and America. *Lucioperca marina*, which is found in the Black and Caspian Seas, is the only marine species. The North American species are distinct from those of Europe. Two species occur in the fresh waters of the European Continent.

## Lucioperca sandra (Cuvier).—The Pike-Perch.

Fins: 1 D. 13—14, 2 D. $\frac{1-2}{21-22}$, A. $\frac{2}{11-12}$, P. 15, V. 1/5, C. 17.

The form of the body of the Pike-Perch (Fig. 14), compared with that of the Perch, is much more elongated and less compressed. The body is highest in front of the dorsal fin, but that height is scarcely more than one-sixth to one-fifth of the length of the fish. The head, measuring to the point of the operculum, is about one-quarter of the length of the fish. The eye is placed in the anterior third of the head, is separated from the eye of the opposite side by its own diameter, and is distant from the extremity of the snout one and a half times the orbital diameter. The external nostrils are small, and near to the eye; the upper and lower jaws are equal. The maxillary bone often extends behind the eye when the mouth is closed. The tongue is smooth. Between the fine villiform teeth arranged in bands, which are seen in both the upper and lower jaws, are tusk-like teeth, in the position of canines at the corners of the mouth. The entire edge of the lower jaw carries an even row of rather shorter pointed teeth, which extend behind the villiform teeth. The villiform teeth are absent on the pre-maxillary bones, where only a few large pointed incisors extend between the canines. The anterior teeth on the palatine bone are stronger, and include longer teeth than those on its hinder part. The vomerine teeth are small, and occur only in bands. As the stronger teeth are often lost, owing to the predatory habits of the animal, the dentition varies in different individuals. The vertical part of the pre-operculum is finely serrated at its edge; the serrations on the horizontal part are larger and uneven. The edge of the long bone, termed the supra-scapula, also carries firm denticulations. The first gill-arch carries long notched spines, termed gill-rakers, while the later arches have only prominences, upon which denticulations are collected. There are seven branchiostegal rays, which become successively smaller.

The first dorsal fin is about as long as the head. It is conspicuously spiny; the spines are pointed and powerful, and increase in length to the fourth, fifth, and sixth. The later spines decrease in length, so that the last ray is the smallest. The first dorsal fin is often connected by a low membrane with the second dorsal, but three to four rows of scales sometimes separate them. The first spinous ray of the second dorsal fin is slender, and is lower than the succeeding jointed rays, which at first are as high as the first dorsal, and gradually diminish. Its base is as long as the base of the first dorsal. The anal aperture is placed beneath the commencement of the second dorsal fin. The anal fin is placed behind it, and has some of the rays longer than those of the second dorsal. The base of the ventral fin

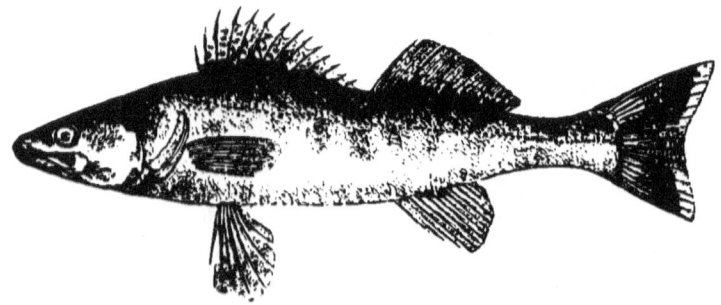

Fig. 14. LUCIOPERCA SANDRA (CUVIER).

is but little behind the base of the pectoral fin; both are of about equal length.

The caudal fin is equally lobed, and moderately concave, with its terminal rays measuring from one-half to two-thirds of the length of the head.

The scales on the body are relatively smaller than those of the Perch; but they are equally rough and denticulated, and vary in size and number. The lateral line is parallel to the back; it includes from seventy-five to ninety scales. Above the lateral line are twelve to fourteen rows of smaller scales, and beneath it sixteen to twenty rows of larger scales under the first dorsal; but at the extremity of the tail there are only four to six rows of scales on each side of the lateral line. The scales are smallest on the throat, abdomen, and fore-part of the back. Usually they become larger towards the tail; but their size varies with the stream, the Danube yielding large-scaled specimens, while in the streams of Galicia, Northern Russia, and Sweden small-scaled varieties are found. The scales vary a little with age; in the young the whole of the upper part of the head, the opercular bones, and base of the pectoral fin are, for the most part, without scales; but with age the operculum, parts of the cheeks, and crown of the head between the eyes, often carry scales; while, with the exception of the first dorsal, all the fins have scales at their bases. The sub-orbital ring is small; it reaches back behind the broad jaws.

The colour of the back is greenish-grey, passing into silvery-white towards the belly; brown clouded spots extend from the back down the sides, but are only occasionally united into regular transverse bands, though in the young eight or nine such bands are commonly seen. The sides of the head are marbled brown. Both dorsal fins have a grey ground, marked along their length with blackish spots between the rays, which often form in both fins five or more longitudinal bands, which are interrupted by the pale colour of the rays. Similar spots ornament the tail-fin; the other fins have a more or less pale yellow colour.

The specimens of this fish from the Platten See, in Hungary, and from Galicia, show, according to Heckel and Kner, six pretty long pyloric appendages. This number had been observed previously by Bloch, but Cuvier found only four. Dr. Benecke, in North Germany, finds the pyloric appendages vary from four to eight. The Pike-Perch grows to very nearly the size of the Pike. It is often from three to four feet long, and weighs from twenty-five to thirty pounds. In Russia its size is becoming smaller, owing to the demand being greater than the supply. About 26,000,000 of these fishes are exported from Astrakhan every year. This species is distributed over a large part of Northern, Eastern, and Central Europe, both in lakes and rivers. It is known, for example, from the Danube below Ulm, and several of its tributaries, such as the Leitha, Salzach, Save, &c.; it extends southward into Lombardy. It is found in the Oder and Vistula, and along the Baltic coast of Prussia, and is met with in the Haffs; it also occurs in the Elbe. It is spread over Sweden and Russia, but is absent from the valleys of the Weser and Rhine, and from Switzerland and France. It is comparatively rare in the lakes of South Germany, though found in the Platten See, Traun See, Atter See, Ammer See, and some others; but in North Germany it is more frequently found in the lakes.

This fish, commonly known in Sweden as *Gös*, in Germany as the *Zander*, and in Austria as the *Sander*, or *Sandel*, prefers clear, deep-flowing water, and generally remains at some depth. Its movements are heavy and not graceful, and its inactive disposition has originated the Swedish saying, "stupid as a Pike-Perch." It lives chiefly on defenceless fishes, and prefers young Smelts, but is extremely voracious, and will eat small Pike, and even its own young, besides various invertebrata. It is not common in turbid water or on clayey bottoms. It spawns between the end of April and beginning of June, when it leaves the deep waters for shallow spots overgrown with water-plants, termed marshes in Central Europe, upon which to deposit its eggs; but in the Swedish lakes it is said to spawn in deep water, and only at night. A single animal will deposit between 200,000 and 300,000 eggs, which have a pale yellow colour, and are each about one millimètre in diameter. This roe, with that of the Bream and Roach, is made by the Russians into scaled-fish caviare, which is termed *Ichastikori*, and is exported from Astrakhan to Turkey and Greece. With abundant food the fish grows rapidly, especially if it remains in the marshy districts, attaining in the first year a weight of one pound and a half; in the second year it weighs two pounds and a half, and in the third year from five to six pounds. In the lower waters of the Danube, however, its weight in the first year is only three-quarters of a pound, and in the second year two pounds. It is not tenacious of life, and is, therefore, transported with difficulty. It lives only from eight to ten years. It is most readily captured when it comes up from the deep water to spawn, being then bold and incautious. At other times only single individuals are caught with the line. When taken out of the water it discharges the air from the swim-bladder with some noise. In the Platten See it is captured in winter with a large drag-net pushed under the ice, and the larger individuals reach the market in Vienna. Swedish fishermen, according to Eckström, pierce the tail near the caudal fin, so that the blood runs freely, when the flesh is whiter. In summer the fish is in less demand, and finds but little sale. The flesh is then divided down the length, salted, and dried in the air, and in many places in Austria and Russia great heaps of the dried fishes may be seen along the banks of lakes and rivers, built up like stacks of wood. Lloyd says its fat is used by Scandinavian peasants as an embrocation for the cure of rheumatism and sprains. Its flesh is white, firm, well-flavoured, and esteemed, especially when the fishes have been well fed. It should be eaten boiled as soon as caught. The price varies, according to Brehm, from one shilling a pound to five pounds for eighteen-pence.

It is not suited as a breeding-fish for culture. It flourishes well in ponds which have a current of water flowing through them, a depth of at least eight or ten feet, a sandy or hard bottom, and banks covered with water-plants. Under these favourable conditions its growth is as rapid as that of the Pike, and in a few years it will attain a weight of twenty pounds. Its voracity can only be satisfied by a large supply of small fishes. It needs more than ordinary care in confinement, for though easily captured in June, it is not often successfully transferred to other waters.

Its gills are sometimes affected with disease, becoming covered with small bladders of gelatinous fluid, in which state the flesh is regarded as unwholesome.

## Lucioperca volgensis (Pallas).

This species (Fig. 15) has the body less compressed, the head shorter and higher, the mouth smaller than in the previous species, and the angle of the pre-operculum inclined more forward. It occurs in the extreme east of the Austrian Empire, is characteristic of the south of Russia, is well known in the Dniester and Volga; it is also found in the rivers of Asiatic Russia. It is of about the same size as *Lucioperca sandra*, but differs from it in the proportions of the body, as well as in the markings. The greatest height of the head is about equal to its length, and its length, which reaches back to the fleshy point of the operculum, is contained about four and a

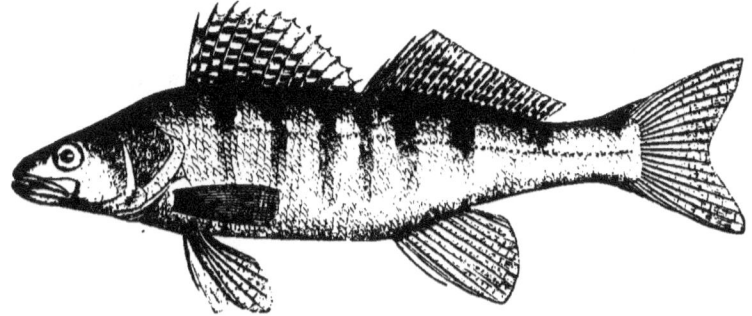

Fig. 15.—LUCIOPERCA VOLGENSIS (PALLAS).

quarter times in the entire length of the fish. The breadth of the body is nearly equal to half its height, so that in general shape it is higher than *L. sandra*, and has a rounder aspect. The eye is large, its diameter being one-fifth of the length of the head, and it is more than its own diameter behind the extremity of the snout, and separated from the other eye by less than its own diameter. The profile of the head rises more quickly than in the Sander, as it extends backward, so that the anterior part of the back is arched in front of the first dorsal fin. The upper jaw terminates backward under the middle of the eye. The pre-operculum is evenly notched, and its upper branch is inclined forward, so that the angle of the bone extends conspicuously backward.

The first dorsal spine is a little shorter than the second, which is as long as the third and fourth. The spines afterwards diminish in length, becoming more inclined, and the last and shortest, which lies horizontally, has a short membranous connection with the second dorsal fin.

The longest rays in the second dorsal fin are quite as long as those of the first dorsal fin, and are nearly half the height of the body. The rays of the anal fin are more elongated than those of the dorsal fins, but still longer rays are found in the ventral fin. The pectoral fin has the shortest fin-rays, though they are longer than the shorter rays of the dorsal fins. Another character, which aids in discriminating this species, is furnished by the scales. They are larger than those of the Sander; less than eighty occur along the lateral line, and below the first dorsal fin there are two rows above the lateral line, and seventeen or eighteen rows beneath it. The scales extend high up the bases of all the fins except the first dorsal. The anal aperture is in the middle of the length of the fish. Specimens preserved in spirits are marked by eight blackish transverse bands, which are more regular than those of the Sander, and the colour of the dorsal and caudal fins is deeper.

This species has constantly a black ocellate spot on the operculum, such as is often seen in the Sander. The iris is a brassy-yellow, darkly spotted. There are three pyloric appendages to the stomach.

## Genus: Aspro (Cuvier).

The genus Aspro is formed for two European species, characterised by having the body elongated and spindle-shaped, becoming almost cylindrical. The snout is thick and projects beyond the mouth. The eyes look upward as well as forward and outward. All the teeth are villiform; they extend in bands in both jaws, and are present on the vomer and the palatine bones. There are two dorsal fins; the scales are small; the operculum is spinous, the pre-operculum serrated; and the pre-orbital contour has an unbroken outline. There are seven branchiostegal rays. The genus is hence distinguished from *Lucioperca* rather by the flattened form of the head, elongation of the snout, and fusiform-shaped body, than by important details of structure.

## Aspro zingel (Linnæus).

This species (Fig. 16) was well known to Linnæus, Bloch, Cuvier, Valenciennes, and the other older writers on fishes. It is distinguished by having thirteen to fourteen spinous rays in the first dorsal fin.

In the opercular region the breadth and depth of the body are equal, and either of these measurements is about one-seventh of the total length. Towards the tail the body becomes more compressed than in the other species. The head, which is rather long, is flattened, and has the inflated snout projecting over the mouth. The upper and lower lips are thickly set with papillæ, which resemble teeth. The front row of dense teeth of the pre-maxillary bone is stronger, and the bands of teeth on the vomer and palatine bones are much broader than in *Aspro vulgaris*. The mouth, which is horseshoe-shaped, reaches back behind the hinder narine. The forward narine is the smaller, and farthest from the eye, being placed midway between the anterior orbital border and the extremity of the snout. The eyes are separated by something less than the orbital diameter. This diameter is one-fifth of the length of the head, midway in which length the eyes are placed. The pre-operculum is

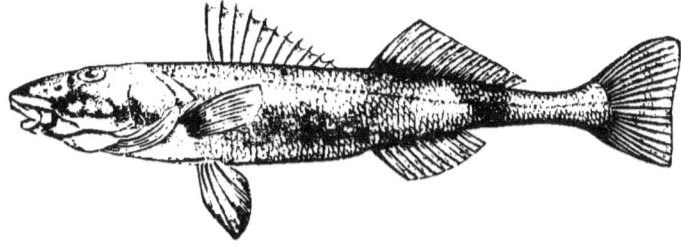

Fig. 16.—ASPRO ZINGEL (LINNÆUS).

finely-toothed along its length, but its angles have larger denticulations. The operculum terminates in a strong spine which has a smaller spine beneath it. Scales extend over the snout to near its end; they cover the top of the head, the temporal region, and opercular bones, but there are no scales below the eyes, on the jaws, cheeks, or throat. The upper scapular bone is finely denticulated.

The commencement of the first dorsal fin is opposite to the base of the ventral fin; its first ray is the shortest, the third to fifth are the longest, and then the rays gradually decrease in length as they become more inclined towards the back of the fin. Between the two dorsal fins is a small interspace occupied by from three to five rows of scales. The second dorsal fin has a longer base than the first dorsal. Its first spinous ray is its shortest ray. Behind that are about twenty soft rays, of which the last ten are nearly equal in length. The anal fin has a shorter base by two-fifths than the second dorsal; it commences opposite to the anterior limit of that fin; its first ray is the shortest, and the longest rays are nearly equal to those of the second dorsal. The rays of the pectoral fin correspond in length with the terminal rays of the caudal; both are somewhat shorter than the rays of the ventral fins. The anal aperture is half-way down the length of the body.

The scales are small, hard, and have their margins finely denticulated: the lateral line contains ninety scales, and at the commencement of the first dorsal fin there are seven rows of scales above the lateral line, and thirteen to fourteen rows below it. Towards the tail the lateral line descends, so as to occupy a lower position, and beneath the second dorsal fin it runs along the middle of the side of the body, and has eight or nine rows of scales above it. The throat is for the most part naked, but scattered groups of scales are found here and there. Small scales cover the base of the pectoral fin, and behind the ventral fin the belly is entirely covered with scales. The system of cephalic canals is conspicuously developed, and forms, with the upward branch of the pre-operculum, a wide, deep channel, which is covered by six thin bony plates, which are indicated by as many depressions in the skin which extends over them. The accessory gills are small and comb-shaped; and the branchiostegal rays are seven in number, as in the other species of the genus.

The back is greenish-brown, the sides are yellowish or grey-yellow, and the belly whitish. Irregular spots and dark-brown marbling extend across the four cloudy brownish-black bands, which are more or less conspicuous, and run obliquely from front to back down the sides. The first band descends in front of the first dorsal, and extends below its base, the second band is under the hinder half of the same fin, the third and fourth bands similarly descend from the beginning and the end of the second dorsal fin. The snout and operculum are brownish, and irregular oblique blackish streaks occur on the cheeks.

This fish often reaches a length of one foot, and a weight of two pounds, and thus grows to a much larger size than the *Aspro vulgaris*. Its flesh is agreeable to the palate, and is easily digested.

The species belongs to the basin of the Danube, and only visits the larger tributary streams. It has been found in the Salzach, and Alt, in the Sieben-bürgen. Like the other species, it spawns in May. It feeds chiefly on worms, larvæ, and small fishes. It lives for seven or eight years, and is well defended from all enemies, except the Pike, by its rough scales and sharp spines.

### Aspro vulgaris (Cuvier and Val.).—The Apron.

This species (Fig. 17), originally defined by Cuvier and Valenciennes, and commonly known as the Apron, was confounded with the Streber until Von Siebold separated them. It is characterised by having nine spinous rays in the first dorsal fin, with thirteen rays in the second dorsal, of which the first is spinous. The tail is short and compressed. The colour is golden-brown, with four broad oblique blackish bands.

Von Siebold gives the usual lengths of Rhône specimens at five inches, and it so closely resembles the young Zingel as to be distinguished only by counting the rays in the dorsal fins. In France it is nowhere abundant, and is found only in the north-eastern provinces, but owing to its small size is not taken by the fishermen.

The body is elongated, nearly cylindrical in the middle, and tapering towards the tail. The head forms about one-fifth of the entire length of the fish, is large, rather depressed above, and covered with scales, even as far forward as the interspace between the eyes and nostril. The snout is smooth and remarkable for its blunt termination. The diameter of the eyes is one-fifth of

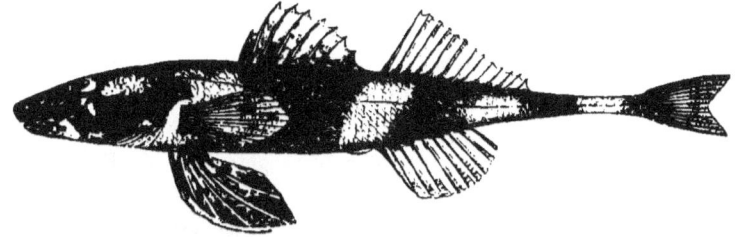

Fig. 17.—ASPRO VULGARIS (CUVIER AND VALENCIENNES).

the length of the head; they are placed in the middle of its length, and resemble the eyes of the Perch. The nasal apertures are double, close together, with the larger one close in front of the eye. The mouth is narrow. The tongue is smooth. The pre-operculum is covered with skin in the fresh condition; it is serrated, but the denticulations are so fine as to be scarcely visible. The operculum is rounded on its inferior border, and terminates posteriorly in a long and sharp spine.

The fins of this species differ remarkably from those of its allies. The dorsal fins are of moderate height and extent. The first dorsal, which has a convex curved outline, contains only nine spinous rays. The second dorsal fin has the first spinous ray short and weak, followed by twelve soft rays, of which the first is simple and the others branched in their upper portion. The pectoral fin is ovate, and includes fourteen rays. The ventral fins are remarkably long; they have only six rays, of which the first is spiny, and the others stout. The anal fin is below the second dorsal, has a shorter base, and is made up of ten rays, of which the first is spinous. The caudal fin terminates with a notched or crescent-shaped outline, and includes twenty-one rays.

The body is covered with scales except in the pectoral region. The scales are large, rough, and finely-denticulated at the margin. The scales resemble those of the Perch, but differ in being longer, in wanting the basal expansion, and in being rougher. The lateral line runs parallel to the back, and near to it. The mucus-canal in these scales is very large. In this line seventy scales may be counted. There are seven rows of scales above the line and fourteen rows below it.

There are two ovaries equally developed. The eggs are stated by Blanchard to be larger than those of the Perch, although the animal is so small. The stomach is oval, and the intestine has only two folds. There are forty-two vertebræ, of which seventeen are abdominal, and twenty-five are caudal.

The Apron is found chiefly between Lyons and Vienne, in the Upper Rhône; it also occurs in the lower part of the river, but has not been noticed below Avignon. It has also been found in the Saône, Doubs, Ouche, Ognon, &c. Like other Perches, it lives on small fishes and insects. It spawns in March and April. It lives at the bottom, and comes to the surface only in bad weather with a north or west wind, when other fishes take refuge at the bottom. This habit has given it a bad reputation with French fishermen, who term it the "sorcerer." In the Côte d'Or the fishermen for a long time threw these fishes away whenever captured, believing that they brought bad luck, but having discovered that they are excellent eating, and flavoured like the Perch, they believe now that the capture brings bad luck to the fish. In the Isère their capture is regarded as a bad omen.

Although the body is widest in the opercular region, the width in the Apron is less than that in the Streber. In the former the tail is shorter and more compressed; it has fewer scales about the head than has the latter. The banding on the body is different; there are five bands in the Streber and four in the Apron, according to Siebold, or three according to Blanchard. In the Streber the principal bands descend from the bases of the first and second dorsal fins, and are proportionately wide; while in the Apron the narrow bands descend forward from between and behind the dorsal fins.

## Aspro streber (VON SIEBOLD).

This somewhat rare fish, formerly identified as *Aspro vulgaris*, is, according to Von Siebold, limited to the Danube and its tributaries, and he regards the identifications as erroneous which have recorded it from the Rhine and the Lake of Geneva. Being small and rare it has not attracted much attention. It spawns in March and April.

The colour is olive, shading into yellow-brown or a reddish tint above, and becoming whitish below. There are four to five transversely oblique blackish bands on the side. The first is generally on the nape of the neck. The fins are of a yellowish-grey colour. The body is slender, with a long, very thin tail. The ventral fins are greatly developed. The length does not exceed sixteen centimètres.

There are only seventeen rays in the caudal fin, instead of twenty-one, as in the Apron; while the anal fin has twelve soft rays to compare with nine in the Apron. Its habits are similar to those of the Apron, living in clear, deep water, and feeding on worms and other small aquatic animals. Its flesh is well flavoured, but the fish is not sought for with the line, and is caught only by accident when fishing with the large net. The Streber is tenacious of life. Many points of difference from the Apron are enumerated in the description of that species (*see* p. 48).

## Family: COTTIDÆ.
### Genus: Cottus (Linnæus).

The family of the Cottidæ includes a large number of carnivorous fishes, which are bad swimmers. They mostly live at the bottom. Frequently there is no air-bladder, though it is present in the Gurnards. The Cottidæ are mostly marine, and distinguished by having the sub-orbital ring articulated with the pre-operculum.

Some of the fishes enter rivers, and others live exclusively in fresh water. The latter belong to the division of the Cottidæ having the genus Cottus for its type. These fishes have the spinous portion of the dorsal fin less developed than the soft portion, and less developed than the anal fin, while the body may be naked, scaled, or variously armoured.

The fishes forming this genus have the head rounded in front, broad, and depressed; the body is almost cylindrical in the middle, becoming laterally compressed posteriorly. The skin is soft and scaleless, but the lateral line is marked. The jaws and vomer have villiform teeth; the pyloric appendages are few, and there is no air-bladder. The ventral fins are under the throat; the pectoral fin is always rounded, and has some of its rays simple. The two dorsal fins are moderately developed. This genus is characteristic of northern seas, especially in North America and Europe, extending to the coast of France. In includes a large number of species, which differ in the characters of the pre-opercular spine, and in the presence or absence of teeth on the vomer. Some of the species are met with in the fresh waters of Northern Asia, Europe, and North America.

## Cottus gobio (Cuvier).—The Miller's Thumb.

1 D. 8—9, 2 D. 15—18, A. 12—13, P. 13—14, V. 1 4, C. 13.

The head in the River Bullhead, or Miller's Thumb (Fig. 18), is the broadest part of the animal, and is widest in the region of the gill-opening. It is nearly always as broad, or broader, than long, though its breadth differs with the form of the operculum. The length of the head is usually equal to one-fourth the entire length of the fish. The deepest part of the body below the first dorsal fin is equal to one-fifth of the total length, but at

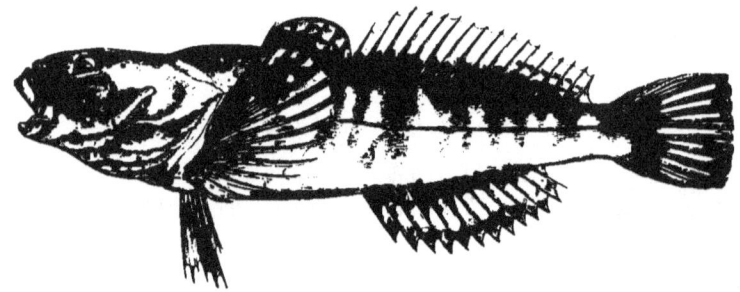

Fig. 18.- -COTTUS GOBIO (CUVIER).

the caudal fin the height of the body is only one-twelfth or one-thirteenth of its length. Behind the head the body is compressed from side to side in a wedge shape to the tail. The eyes are superior (Fig. 19), but directed obliquely outward, are nearer to the snout than the gill-opening, and are small, having a diameter of less than one-fifth of the length of the head.

Fig. 19. - HEAD OF COTTUS GOBIO, SEEN FROM THE FRONT.

The profile of the fore part of the head is a segment of a circle; both jaws are equal, and the pre-maxillary and mandible, as well as the vomer, possess broad bands of villiform teeth. The cleft of the mouth generally reaches back beneath the middle of the eye. The pre-operculum is armed with a spine directed upward, and a much smaller spine beneath it hidden in the skin. The operculum ends behind in a rounded point; the sub-operculum has a short sharp spine directed forward, but generally covered with skin. The gill-opening is moderately large, the number of branchiostegal rays is six, the accessory gills are large and pectinate, while the pharyngeal bones possess villiform teeth. The gill-arches, with the exception of the first, possess rake-like teeth.

The first dorsal fin begins over the base of the pectoral, and possesses in most individuals eight undivided rays, of which the middle and longest rays are one-third the height of the body. A narrow membrane unites the first dorsal with the second dorsal, the first ray of which is very short, while the fourth to the sixth rays are longer than any in the first dorsal. Its rays are all undivided. The anal fin begins behind the second dorsal, and does not reach so far back. One specimen from the Tyrol has only nine rays in the anal fin. Its longest rays equal those of the dorsal fins. All are simple spines with the exception of the last, which is usually divided to the base. The ventral fin commences a little behind the pectoral, its spinous ray is joined to the first of the four simple soft rays. They never reach back beyond the last ray of the first dorsal. In Scandinavia the rays of this fin are branched. The pectoral fins are very much developed; their length is nearly one-fourth of the length of the fish, and they reach back to the beginning of the second dorsal fin. Their aspect is that of a broad, rounded, expanded fan. Young individuals have the rays unjointed, but with age jointing begins with the first and extends gradually to five or six rays, but the succeeding rays always remain simple. The rounded caudal fin generally measures one-sixth of the total length; its eight middle rays are all jointed. The anal aperture lies in the fore part of the body, and behind it is the long genital papilla, which is of different form in the two sexes.

The head and body are without scales. The lateral line is in the upper third of the body in its anterior part, but afterwards descends to the middle, and its mucus-canals open in from twenty-six to twenty-eight simple pores. The pores of the cephalic canal extend along the lower jaw.

The colour varies, but always has a ground tint of grey; the back is often some shade of brown, spotted with dark brown, or flecked or clouded with brown colour, which sometimes forms transverse bands. The belly is whitish-grey, occasionally streaked with brown. The dorsal, pectoral, and caudal fins are covered with brown stripes, which usually extend along the rays, though the skin is sometimes banded.

The anal fin is not banded with colour, and the ventral fins are mostly unstreaked, or possess only a few scattered spots.

The first dorsal fin is reddish. The iris is red.

The stomach is a broad, round sac with three or four rather wide pyloric appendages. The abdominal walls are covered internally with a black pigment. There is no air-bladder. The milt of the male is double, and, like the ovaries in the female, has a wide external aperture near to the pear-shaped urinary bladder, opening through the uro-genital papilla.

When young, individuals are characterised by having very conspicuous pores above and below the lateral line, as well as by the undivided rays of the pectoral fin already mentioned. The male is usually the darker fish, has a rather broader head, and longer uro-genital papilla. In the female the ovaries become excessively enlarged at spawning-time. The fish attains a length of five or six inches.

This species is found in the smallest streams wherever the water is clear. It is common in all parts of Scotland, but so rare in Ireland that its existence in Derry has been more than once questioned. Blanchard enumerates localities for it in all parts of France, Fatio finds it in Switzerland up to a height of between 6,000 and 7,000 feet, Heckel and Kner meet with it not only in the Danube and its tributaries, but in the small Austrian lakes; and some varieties of it are found in the Hartz. It is found in Scandinavia, and may sometimes descend the streams into the Baltic. Professor Canestrini enumerates many localities in the north of Italy, where it is found, but it has not been recorded from Spain or Greece. It is common in Siberia.

The River Bullhead, or Miller's Thumb, is an active, voracious fish of solitary habit, which lies concealed under stones, and feeds on insects, insect larvæ, and the fry of young fishes; and females have been found with their own eggs and fry in the stomach.

When attacked by other fishes, or by carnivorous birds, it defends itself by expanding the pectoral fins and the operculum, so that the opercular spine becomes a formidable weapon, which has often proved fatal to birds which have attempted this armoured morsel.

It spawns in March and April, when the females deposit the eggs under stones, or excavate special depressions in which they may be placed.

Experienced fishermen narrate how the male, having discovered a suitable depression between stones, valiantly defends it against other males who would dispossess him of his holding; and a Cottus is often captured holding in its mouth the head of an antagonist which he is unable to swallow. If a female comes by she is welcomed to the retreat which the male Goby has secured, and after being installed she deposits her eggs. These are about two millimètres in diameter, of a pinkish colour, are commonly connected like bunches of grapes, and occasionally adorn the nuptial chamber by being suspended from the roof. The mass of ova may cover an area of an inch and a half of ground, and they have been compared in aspect to frog's spawn. The number of eggs varies with the age of the individual. In the nest of one which measured four inches in length Fatio counted a deposit of 761 eggs. After the eggs are laid the female usually retires to a respectful distance, though some of the older writers stated that she cemented the eggs to her breast. Then her place in the nest is taken by the male, who not only defends them against the curiosity or gluttony of idle fishes in the stream, but with a mother-like devotion watches over them for four or five weeks without intermission, never leaving his self-imposed duties except to obtain food. He is not easily disturbed, and if attacked by the angler with a rod or stick he will seize the intruding object, and is often killed in defending the nest. Some writers, however, have questioned the existence of this remarkable instinct.

The Miller's Thumb is caught in Germany with a small net known as the "Koppensegge," or it may be taken in baskets at weirs, or with a rod

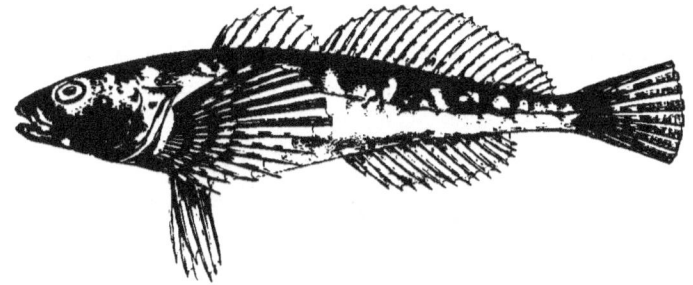

Fig. 20.—COTTUS GOBIO, VARIETY MICROSTOMUS (HECKEL).

and line, and is valued as a bait for eels. Its flesh is white and well flavoured. In Pallas's time it was often worn round the neck by Russian peasants as an amulet against intermittent fever.

In many parts of North Germany it furnishes a favourite sport to children, who pursue it bare-legged in the shallow streams, armed with an ordinary table-fork, with which the fish is speared.

In Britain it is not eaten. Yarrell mentions an instance in which its flesh became red when boiled. It is tenacious of life, and lives out of water for an hour or two if kept wet.

The young individuals are at first gregarious, and keep to the nest; but as its limited space becomes inconveniently crowded, they seek solitary habitations. They begin to breed when two years of age.

Yarrell remarks that the popular English name of Miller's Thumb has reference to the smooth, broad, rounded form of the head, which is shaped like the thumb of a miller moulded by testing the grinding of his mill.

His hand is constantly at the mill-spout to ascertain the quality of the meal produced, so that his thumb becomes the gauge of value of the produce, and this has led to the adage "worth a miller's thumb," and the proverb "an honest miller hath a golden thumb." It is remarkable how the thumb of the miller comes by incessant use to resemble the head of the fish which is usually found in his mill-stream. In Sweden its popular name is *Simpa*.

There are two varieties of this species, which were regarded as of specific rank by Heckel and Kner, but their true nature was indicated by Dr. Günther. One of these is the variety *microstomus*, the other the variety *ferrugineus*. There is considerable difference in the form of the head in these varieties, as may be seen in the profile and in the front view of the face (Figs. 20, 21, 22, 23). *Cottus microstomus* has the head smaller in front, as seen from above; it is less blunt, has a smaller mouth; while the eyes are larger, with a greater diameter than the distance between them, and the spine of the pre-operculum is more developed, and the

Fig. 21.—HEAD OF COTTUS MICROSTOMUS, SEEN FROM THE FRONT.

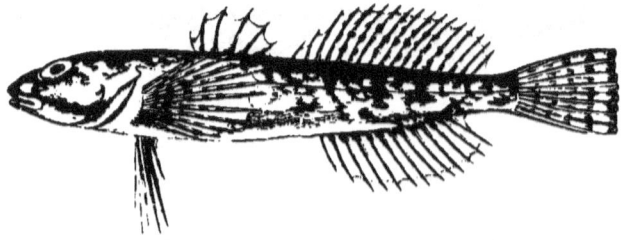

Fig. 22.—COTTUS GOBIO, VARIETY FERRUGINEUS (HECKEL AND KNER).

bands of teeth on the pre-maxillary and vomer are broader. The caudal fin is less rounded than in *Cottus gobio*, its six middle rays subdivide twice, while the rays above and below are simply forked, and are succeeded at the outer parts of the fin by three or four undivided rays. The lateral canal opens by thirty-four or thirty-five pores, a larger number than is found in *Cottus gobio*. The skin is smooth, but has a rough texture on the upper part of the head arising from a multitude of small ossifications like the minute dermal bones which are embedded in the skin above and below the lateral line. Its length is about four inches. It is found principally in Southern Russia, but extends into Austria by way of Cracow, and is found in the Poprad, which is a tributary of the Vistula.

The second variety, *Cottus ferrugineus* (Figs. 22, 23), has a more slender, rounder body. Its pectoral fin is smaller, and the anal fin has longer rays: in both fins the rays are undivided. The head is relatively small, but is contained four times in the entire length. The least height of the tail is one-fourteenth of the length. The snout is more pointed, the eyes are separated from each other by their own diameter, and are the same distance from the snout. They are rather smaller than in the other variety.

The villiform teeth in both jaws are longer and thinner than in *Cottus. gobio*. The skin nowhere shows pores or traces of rough, bony depressions. There are usually one or two fewer rays in the anal fin. The males have a thicker head and broader mouth than the females. Their bi-lobed milt is not covered with a dark peritoneum such as wraps round the simple heart-shaped ovary of the female. The largest examples are three inches long, but frequently the fish does not exceed half this length.

Fig. 23. — HEAD OF COTTUS FERRUGINEUS, SEEN FROM THE FRONT.

Its distribution is more southern than that of the other varieties, being known from the Lake Garda, from Xegar in Dalmatia, and from Servia.

## Cottus pœcilopus (HECKEL).

1 D. 8—9, 2 D. 16—18, P. 14, V. 1/4, A. 13—14, C. 13.

This species (Fig. 24) closely resembles the preceding, so that it is chiefly necessary to indicate such differences as distinguish it. The eyes approximate

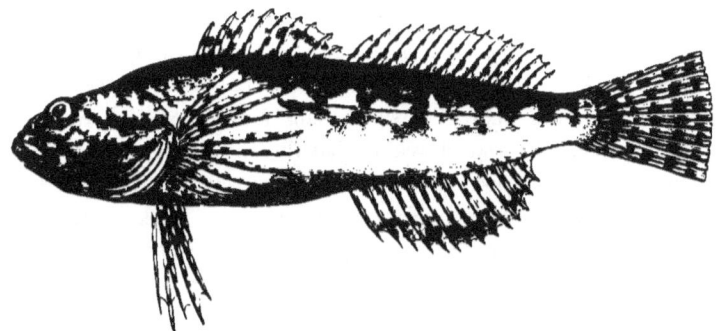

Fig. 24.—COTTUS PŒCILOPUS (HECKEL).

more closely together, being separated by less than their own diameter. The hatchet-shaped free opercular spine, which curves forward, is beneath the thick skin. The ventral fins are somewhat longer; they commence just under the pectoral fins, and reach back to the vent. The pectoral fins are somewhat shorter, and reach as far back as the beginning of the second dorsal.

All its rays are simple spines. The caudal fin is less than one-tenth of the entire length of the body. It possesses fourteen rays of nearly equal length, of which nine in the middle are subdivided into bifid forks.

The number of rays in the fins is very similar to that in *Cottus gobio*. The colour, too, is very similar. All the fins are dotted with brown; the spots form vertical bands in the pectoral and caudal fins, and horizontal bands in the other fins. These bands number six or seven in the ventral fins. The number of branchiostegal rays is six, occasionally five, or they may be unsymmetrical, as in Salmon, six on one side, and five on the other. The number of pyloric appendages also varies; Heckel and Kner record five in the male and four long ones in the female.

This species is met with in the Pyrenees, Carpathians, in the Vistula of Galicia, and many mountainous localities in Hungary and Bukowina. Von Siebold remarks that he is disposed to attach less importance than do Heckel and Kner to the jointed rays of the pectoral fin, and states that in *Cottus gobio*, the jointing of the rays is variable, the same ray being as often simple as jointed.

## Cottus scorpius (Bloch).—The Sea Bullhead, or Father Lasher.

D. 10/14, A. 11—12, P. 17, V. 3, C. 18, Vertebræ 12—13/22.

The body is club-shaped, strongly compressed towards the tail, with a large head, which is somewhat depressed. The mouth aperture is wide, and reaches back to behind the eyes; it has a projecting upper lip. There are two small spines above the snout, and four obtuse tubercles on the crown of the head. The pre-operculum has three spines, of which the upper one is the largest; the operculum has one large spine. There are teeth in the jaws, and on the vomer. The tongue is short, thick, hard, and free from teeth. The eyes are large, and elevated in position on the side of the head; the iris is yellowish. The nasal apertures are simple, tubular, and are placed midway between the eyes and the labial margin. The skin is naked, thick, and soft. It sometimes contains bony granules, which in Prussian specimens are arranged in a few irregular rows. The two dorsal fins are united by a low membrane. The very large pectoral fins are placed at the sides of the head, and when expanded are quite as high as the body, and are larger in the male than in the female. The small ventral fins are placed below the pectorals, on the throat. All the fins except the caudal have unjointed rays.

The colour of the upper side of the body is dark brown-blackish, or dark olive-green, spotted or marbled with grey. The sides are white, with blackish-grey marbling; the belly is yellow in the males, and white in the females. The dorsal and anal fins have broad black and grey bands; the other fins are banded with black, grey, and orange. The liver is very large, and flesh-red. The stomach has four pyloric appendages.

There are thirty-five vertebræ. The upper surface of the skull is formed chiefly by the frontal bones, which have concave excavations to receive the eyes. The crown of the head is flat, but the space between the eyes is concave, and an obtuse ridge running from the back of the eye defines the crown from the side of the head. The vomer is anchor-shaped, tapering posteriorly. The maxillary bone is more elongated than the pre-maxillary. The dentary bone of the mandible sends off a free superior fork. Below the orbit are three oblong flat bones, with several mucus pits. Both the skull and dentary bone are well supplied with mucus channels.

This species spawns in December and January, and deposits its eggs on sea-plants. They are one millimètre in diameter, of an orange-red colour, and are contained in a thick envelope. The fish usually lives in deep water, but frequents the coast in summer, when it occasionally enters rivers. It occurs in the north of the Gulf of Bothnia; but it is nowhere frequent in fresh water. The liver is well flavoured, and is occasionally used for the manufacture of oil, when the fish is taken in large shoals. The length of this species varies from twenty or thirty centimètres to one mètre.

The Horned Bullhead, *Cottus quadricornis* (Linnæus), is found in Lake Ladoga, in Russia. It occurs in Lake Wetter, in Sweden, and in other large Scandinavian lakes; is found in the Gulf of Bothnia and the Baltic; and is not rare on our own coasts.

## Family : GOBIIDÆ.

### Genus : Gobius (Artedi).

Under the name *Acanthopterygii gobiiformes* Dr. Günther comprises a division of fishes containing the *Discoboli* and the *Gobiidæ*. In the former the ventral fins have the spine and five rays rudimentary, and they form the bony support for a round disc, margined by a membranous fringe, which forms a sucker, very well seen in the marine Lump-sucker.

The *Gobiidæ* comprise fishes which have the body elongated, naked in some species, and scaly in others; rarely tuberculate, as in the *Discoboli*.

The spinous part of the dorsal fin is feebly developed, and the spines are flexible. The ventral fins sometimes unite into a sucking disc. This family includes about thirty genera of small carnivorous fishes, which frequent shores in temperate and tropical regions, and sometimes live in fresh waters. Among these fishes are some interesting types, such as the *Periophthalmus* of the tropics, which at the ebb of the tide leaves the waters, and hunts for crustacea and small animals in the mud. The several genera of *Gobiidæ* are distinguished chiefly by the condition of the ventral fins, which may be united into one fin, or remain separated, as well as by the presence or absence of scales, and the condition of the teeth, and gill-openings.

*Latrunculus pellucidus* is a small Goby, which lives for one year only, and in Europe is the only instance known among the vertebrata of life thus limited.

*Gobius* is the only genus in the family which has fresh-water representatives. The genus is distinguished by having a scaly and somewhat elongate body. Teeth are in several series in the upper jaw, conical and fixed. The anterior dorsal fin usually has six flexible spines; the posterior dorsal fin is often more developed than the anterior dorsal. The anal fin is beneath, and corresponds to the second dorsal, and both terminate considerably in advance of the caudal fin. The ventral fins unite to form a disc which is not attached to the abdomen. The gill-openings are vertical. There are no pyloric appendages to the intestine. According to Günther there are nearly three hundred species of Gobies.

## Gobius martensii (Günther).

1 D. 6, 2 D. 1/10, A. 1/7—8, V. 10 = 5/5, P. 13—14, C. 3—4/13/3—4.

In this fish (Fig. 25) the anterior extremity of the head is convexly rounded. The head measures about one-quarter of the total length; it is longer than broad. The greatest breadth is about equal to the greatest height of the body of the fish. The eye is placed near to the frontal outline, and has a diameter of one-fifth the length of the head, is separated by its own diameter from the other eye, and is in the anterior half of the head. The well-developed muscles give a swollen aspect to the occipital region and to the cheeks. The gape of the mouth is widened, and the mandible is sometimes longer than the pre-maxillary. In the maxillary and pre-maxillary bones the teeth are arranged in parallel rows, but those of the first row in both jaws are somewhat longer than the others. There are teeth on the pharyngeal bones. The nares are unusually small, near to the orbit of the eye; the posterior narine has no valve.

The head and operculum are naked. The gill-opening does not ascend above the base of the pectoral fin; the pseudo-branchiæ are large. There are twenty-nine to thirty vertebræ. The outline of the large pectoral fin is rounded, and its longest middle rays reach back as far as the end of the first dorsal. All the rays are jointed except the two or three lowest, and sometimes the uppermost rays are unjointed; but, as a rule, these rays are divided. The ventral fins, with the fleshy funnel at their base, scarcely attain half the length of the head. They unite laterally in front and behind, forming an oval disc, which is margined by the fin-rays. The disc is free behind. It is slightly in front of the base of the pectoral fins; and the ventral rays, which are all jointed, except the first on each side, form a median transverse fan-like fin or funnel, the middle of which has the longest rays.

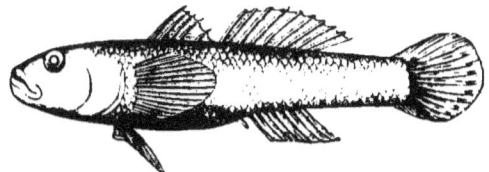

Fig. 25.—GOBIUS MARTENSII (GÜNTHER).

There is sometimes a slightly greater interval between the rays of the first and second dorsal fins in the female than in the male, but in the young fish the interval is less than in the adult. The second dorsal has higher rays than the first dorsal, and it is a larger fin than the anal, which, however, terminates opposite to the termination of the second dorsal fin. The caudal fin is rounded, and is about one-fifth the length of the fish. It somewhat resembles the pectoral fin in form. About thirteen rays in the middle of the fin are jointed, and there are three or four short rays, external to these above and below, which are unjointed.

The skin is naked in the anterior part of the back as far as the first dorsal, and along the throat and abdomen. Heckel and Kner count thirty-six to forty scales along the lateral line. The scales are limited to the sides of the fish, and become larger towards the tail; they are remarkably pectinate at the anterior edge, while the rest of the scale is marked with furrows, which radiate like a fan from a small centre in the middle of the anterior margin (Fig. 26). Behind the second dorsal six or seven scales cover the height of the fish, and at the end of the tail the number is four or five.

Fig. 26.—SCALE OF GOBIUS MARTENSII.

The colour varies in different individuals, and with the season, and may be affected by varying external circumstances, and by internal impressions. In the breeding season the upper parts of the body are yellowish-grey or green, marbled with brown. The face is generally darker than the body. The under side of the body is paler, and free from spots. The pectoral and ventral fins are pale-yellow or greenish, often without spots. The anal fin is sometimes colourless, sometimes bluish, with variable black spots, often becoming a black band. The first dorsal fin is the most variable in colour; it often has a broad black band at the base, and the upper part of the fin shows brilliant metallic tints margined externally with white. The second dorsal fin is transparent grey, with horizontal lines of brown spots. The iris is yellow.

In the breeding period the body and fins often become covered with little tubercles. The males are usually the thinner fish. The females are most easily recognised by the blunter snout and broader urogenital papilla. They have two large ovaries, and a smaller air-bladder. The male sexual organ varies considerably among the Gobies. The common size of the fish does not exceed a length of three inches, though larger and smaller specimens have been captured. This species commonly hides under stones, to which the female attaches her eggs. They are stated to be spindle-shaped, connected in a row, and to float about on the surface of the waters, and develop in June.

The fish is eaten, and regarded as well flavoured. It is abundant in all the canals and small streams of Lombardy, in the Rivers Isonzo and Treviso, and it is found in the Lakes of Garda and Maggiore. The Lombards term it *Bottola* and *Bottina*, and the Tuscans name it *Ghizzo*.

This fish was named by Cuvier *Gobius fluviatilis*.

The species includes several varieties. Three of these have been defined by Professor Canestrini as *Gobius avernensis*, 1 D. 6, 2 D. 1 12—13, A. 1 8—9, from the Arno; *Gobius panizzae*, 1 D. 5, 2 D. 1 8, A. 1 7, from Lago di Garda; *Gobius punctatissimus*, 1 D. 6—8, 2 D. 1 7—8, A. 1/7—8, V. 1/5, P. 1 16, C. 13.

The first of these varieties is distinguished from the type chiefly by the more elongated form of body, the more pointed form of snout, and especially by having about forty-seven scales in a longitudinal row on the side. But Fatio regards this species as so variable that it would be easy to select individuals to which distinctive names might be as well applied as those which have been distinguished as species by the Italian ichthyologist; and although we have no personal knowledge of the variation of this Goby in Italy, the characters assigned to Canestrini's types seem better to accord with varieties than species, though in this matter our judgment is necessarily governed by the relative persistence of characters.

*Gobius punctatissimus* (Canestrini) is found near Bologna, at Mantua, Modena, and Venice. The head is one-quarter of the length, and is equal to the height. The diameter of the eye is one-third of the length of the head. The body is marbled with long brown spots, extending transversely. There is a triangular black spot at the base of the tail. The female when full of eggs has the body remarkably deep. The ventral and anal fins are black. There are one or two blue spots on the last rays of the dorsal fin. The second dorsal and anal fins are less high in the male than in the female. There are thirty vertebræ.

Among the species which are found in Russia may be mentioned *G. marmoratus* (Pallas), coming up the rivers which flow into the Black Sea and the Caspian; *G. lugens* (Nordmann), in the River Kodar, which falls into the Black Sea; *G. mucropus* (Filippi), in Lake Paleostom, near the Black Sea; *G. constructor* (Nordmann), in small rivers which flow into the Caspian; *G. melanostomus* (Pallas), which goes far up the Dniester, Dnieper, and Volga; *G. lactens* is a Dniester species; *G. fluviatilis* (Pallas) similarly ascends the rivers which fall into the Black Sea, Sea of Azov, and Caspian; *Gobius kessleri* (Günther) enters the mouths of the Black Sea and Caspian rivers; *G. gymnotrachelus* (Kessler) is found in the Dniester, Bug, and Dnieper; *G. burmeisteri* at the mouth of the River Rion, in Georgia.

## Gobius semilunaris (HECKEL).

1 D. 6, 2 D. 18, A. 14. Scales: lateral line 34—37.

This species has no characters of form to distinguish it from the other Gobies. The body is six times as long as high at the first dorsal fin, and ten times as long as high in front of the caudal fin.

The head has a conical shape, and is two-ninths the length of the fish. The mouth is small, and its cleft extends back only to the nasal apertures. The jaws are of equal length, and armed with a small band of short teeth. The nares are midway between the eyes and the end of the nose. The eyes are near together, high up in the anterior half of the head, and separated by an orbital diameter from the snout. The operculum and preoperculum are rounded, and covered by a thick naked skin. There are five branchiostegal rays, which are embedded in the edge of the skin of the throat.

The pectoral fin is broad, and its point is rounded; it extends back to the beginning of the second dorsal. It consists of thirteen or fourteen rays, of which the fourth and fifth are jointed at their extremities. The ventral fins are below the pectorals, and in contact with each other, forming a transverse band on the throat; they reach back to the star-shaped vent, which is in the middle of the length of the body. Each fin consists of six rays. The two dorsal fins are rather distant from each other. The first dorsal commences behind the base of the ventral, and consists of six unjointed rays, forming an arch which is half as high as the body is deep. The second dorsal is formed by eighteen unjointed rays, commencing over the vent and terminating near the caudal fin. The first ray is very short; the succeeding rays are nearly equal, and as high as the body below them is deep. The anal fin is beneath the second dorsal, but does not extend so far forward or so far back; it has fourteen unjointed rays, which are two-thirds the height of the rays in the second dorsal. The caudal fin is one-sixth the length of the fish, and its free border is slightly convex, as in other Gobies.

This species does not exceed a length of two inches.

The scales are small on the head, along the base of the first dorsal fin, and between the throat and the vent. They are largest on the sides of the body between the dorsal and anal fins. The free border of the scale is convex, sometimes festooned with about fifteen indentations, which correspond to as many sub-parallel diverging rays. In the middle of the base there is a thickened point, from which the excentric lines of growth extend backward; but along the attached margin is a row of thorn-like spines embedded in the skin. There are from thirty-four to thirty-seven scales in the lateral line, which is not very distinct, but runs in the middle of the side. There are seventeen to eighteen scales in a transverse series below the first dorsal fin.

The fish is of a pale yellowish-brown colour, brighter on the belly and cheeks. There is a dark brown lunate spot on each side of the first dorsal fin, with its convexity directed upward. There are three other spots on each side of the second dorsal fin. Small irregular spots extend along the base of the anal fin. There is a dark brown band at the base of the caudal fin, behind which are many imperfect bands. They are seen on all the fins, but are almost invisible on the ventrals.

It is found in the River Maritza, near Philippopolis, in Eastern Roumelia.

Among other species of Gobius is the spotted Goby, *G. minutus*, known in the Thames as the Polewig. It has sixty scales in the lateral line. The body is about seven times as long as high, and the head is one-fourth of the total length. The eye is a fifth of the length of the head, and the eyes approximate towards each other, so as to look upward. The inter-orbital space is naked. The pectoral fin is broad; the ventral fin, which is below it, extends back nearly to the vent. Yarrell mentions that this species is constantly taken with shrimps. When full grown it is only about three inches long. Its wide mouth is armed with several rows of small teeth, which are directed inward.

The first dorsal fin is higher than the second. It is above the pectoral fin when that fin is laid back. It contains six rays. The second dorsal fin, which is opposite to the anal fin, commences behind the vent. The anal fin is rather shorter than the second dorsal. The colour of the body is yellowish-white, varied with minute yellowish or black spots, with occasional large spots along the lateral line. The caudal fin is vertically barred with rows of spots.

Another variety, named *G. minutus* by Eckström, is found in the River Gotha, in Sweden, where it was taken from the stomach of *Cottus scorpius*. It is under three inches in length. Günther proposes to name it *G. eckströmii*, since it differs from *G. minutus* in having one spiny ray, and thirteen soft rays in the second dorsal fin, instead of nine to eleven rays, as in *G. minutus*, and one spiny ray and eleven soft rays in the anal fin, instead of the nine or ten soft rays of the spotted Goby.

## FAMILY: BLENNIIDÆ.

### GENUS: **Blennius** (ARTEDI).

The Blenniidæ are a group of carnivorous fishes most frequently found as bottom fishes along shores, but with some representatives in fresh waters. The fishes are long with somewhat depressed or sub-cylindrical bodies, naked in some genera and covered with small scales in others. They are characterised by having no articulation between the sub-orbital ring and the pre-operculum. The dorsal fin may be in one, two, or three parts; sometimes the whole fin is composed of spines, sometimes the spinous part and the soft part may be equally developed, but in any case the dorsal fin usually extends along nearly the whole of the back. The ventral fins are always jugular, except in the genus *Pseudo-blennius* of Japan, in which they are thoracic. They are formed by fewer than five rays. In the elongated genus *Nemophis* both caudal and ventral fins are absent.

Usually there is no air-bladder; and there are no pyloric appendages. The genera are defined according to the presence or absence of molar teeth, and the condition of the dorsal, ventral, and caudal fins. Some are known to be viviparous, such as the marine *Zoarces*, and others build a nest and care for the young in a manner that resembles the habit of the Stickleback.

Blennies are abundant in all temperate and tropical seas. They are mostly small fishes, and include some of the smallest, though the Cat-fish, *Anarrhichas lupus*, is a gigantic Blenny growing to a length of more than six feet.

The characters of the genus Blennius, so named by the Greeks from the slimy surface of the skin, are, first, the ventral fins are jugular; second, the spinous and soft portions of the dorsal fin are nearly equal in extent; third, the jaws are armed with a single series of immovable teeth, and sometimes with a small curved posterior canine tooth, but never with flattened crushing teeth. The gill-opening is wide, body naked, and the caudal fin distinct.

## Blennius vulgaris (Pollini).

D. 12/17—18, A. 19—20, V. 2—3, P. 13, C. 11.

In this species (Fig. 27), as in most Blennies, the head is short with the face inclined in front so as to make the snout somewhat prominent. The height of the body is about the same as the length of the head, which is one-fifth of the length of the fish. There is a strong curved tooth which is functionally canine, in both jaws, on each side. There are twenty or more teeth in the premaxillary bone, and from ten to twelve teeth in the mandible. The eye is near to the angle which the face makes with the upper part of the head; its diameter is one-fifth of the length of the head. The male has a tentacle over the eye, but it is often insignificant except during spawning, and is wanting in the female. There is a low, fleshy crest which runs along the middle line of the head in males when spawning, which at other times subsides to a low ridge or may be entirely wanting. Fleshy lips margin the mouth and extend back to below the anterior orbital border. The jaws are equal in length.

The gill-opening does not ascend so high as the level of the orbit, but descends to the middle of the throat. The operculum is covered with skin, and has neither serrations nor spines. The pseudo-branchiae are small and fringe-like. There are short and sharp gill-rakers on the gill-arches.

The dorsal fin commences over the gill aperture, at the origin of the lateral line. Though it is undivided, the anterior part is distinguished by having shorter rays than the posterior part. The anterior spinous rays are of nearly equal length, not half the height of the body, but the posterior soft part of the fin has the rays higher, and they are longer than those of the opposite anal fin. The rays of the ventral fin, which are used as a locomotor organ, are a little shorter than those of the caudal fin. The pectoral fin has a somewhat elongated oval outline, and is about as long as the head.

Neither the dorsal nor anal fin unites with the caudal. All the fins except the caudal have undivided rays, and only the last rays of the dorsal and anal fins are jointed. The vent lies within the anterior half of the fish; behind it is the urogenital papilla, which differs in form with the sex. The lateral line at first lies above the pectoral fin, but behind that fin it bends down with a strong curve to the middle of the side, but is not continuous to the tail; and in the variety *Blennius pollinii* it terminates behind the pectoral fin. The course of the cephalic mucus-canal lies over the gill-opening, round the eye, and along the lower jaw; and its direction, like that of the lateral canal of the body, is made manifest by white pores.

No matter how much the colour may vary, there are always black spots on the head, and a row of fine black spots along the base of the dorsal fin, and often there is a row of five or six black clouded spots along the back. A broad

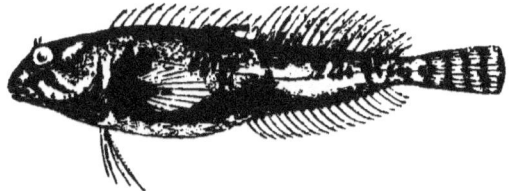

Fig. 27. BLENNIUS VULGARIS (POLLINII).

black longitudinal band is commonly seen on the anal fin. The ground colour of the fish in life is greenish-grey or green.

The males are distinguished by having a tufted papilla in connection with the first two short rays of the anal fin. In the female there is no trace of a tufted papilla, but in the corresponding region there are three openings, of which that nearest to the first spine of the anal fin is urinary. The species is not viviparous. The ovaries are large and symmetrical.

The largest specimens are four inches long. The species is gregarious, living in small shoals on a stony bottom, with habits like those of the Loach. Its movements are very rapid. It spawns in the summer months. It is valued for food, and has a white, well-flavoured flesh.

It is met with in the Lago di Vrana (in Dalmatia), in the River Isonzo, in the Lago di Garda, and other lakes of Northern Italy, in the Lake Bracciano, in the Tiber, and in the River Oreto, near Palermo, in Sicily. In France it is found in the department of Var, and is reported from other parts of the south of France, but it does not occur in any other part of Europe. Its common name in Italy is *Cagnetto*, in France *Cagnette*.

A variety of this fish, which Professor Blanchard names *Blennius alpestris*, occurs in a little river which falls into the Lake of Bourget, in Savoy. The longest specimen measures about two and a half inches. Its form, as compared with the foregoing type, is more graceful, owing to a greater lateral compression of the body. The head is shorter and more compressed, and it is distinguished by its colouring. The shining skin is a bright chestnut, finely sanded with black, and marked with large black spots, which are diffused over the head and all the body except the ventral region, which is of a uniform yellowish-white. On the sides of the head and on the back, large irregular spots of a blackish-brown form short transverse bands down the fish, and are close together.

The fins also have brown and black spots on them, especially the caudal and hinder rays of the dorsal.

The head is more abruptly truncated than in the type of the species, and the occipital crest is less marked. The dentition shows no variation. The dorsal fin never contains more than twenty-six to twenty-nine rays; the pectorals have eleven rays, and the anal fin seventeen or eighteen rays. These differences seem to us to mark a variety rather than a species.

## Family: ATHERINIDÆ.

### Genus: **Atherina** (Artedi).

This genus is widely distributed in temperate and tropical seas, and is familiarly known from its type, the Sand Smelt, *Atherina presbyter*.

The body is thick sub-cylindrical, or slightly compressed. The snout is blunt. The cleft of the mouth extends at least as far back as the front of the eye. The teeth are small. The scales are of the cycloid type. A silvery band extends along the side of the body. The ventral fins are placed behind the pectorals. The intestine has no pyloric appendages. An air-bladder is present.

About twenty-four species are known from Australia, Tasmania, the East Indian Archipelago, the seas of China and Japan, the east coast of Africa from the Cape of Good Hope to the Red Sea, the West Indies, the Mediterranean, the Black Sea, and the British shores.

*Atherina lacustris* is the only European species limited to fresh water, though *A. mocho*, *A. boyeri*, *A. hepsetus*, enter fresh waters in Spain and the south of Europe.

## Atherina lacustris (Bonaparte).

1 D. 5—9, 2 D. 1/10—13, A. 1/12—15.  Scales: lat. line 46—60, trans. 10—11.

This is a small species, with an elongated body, eight times as long as high, and one and a half times as high as thick. The dorsal contour is nearly horizontal, except towards the head, where it becomes convex. The abdominal outline is similar, but rather more curved, so that the tail is attenuated.

The head is one-fifth of the total length. Its upper surface is flat. A median longitudinal ridge extends between the eyes, and is prolonged forward, as are the lateral ridges from the superciliary region. The eye is two-fifths the length of the head. The inter-orbital space, which is equal to the length of the snout, is two-thirds of the orbital diameter in width. The cleft of the mouth is directed obliquely downward. The maxillary bone extends below the front margin of the eye. There is a simple row of small pointed teeth on the pre-maxillary bone, and on the mandible; they are longest in front. There are no teeth on the vomer or palate. The gill-aperture is large. The gill-rakers are small and tooth-like. The pre-operculum forms a right angle, with the angle rounded. The head is covered with scales as far forward as the eyes.

Von Martens counted the rays in the dorsal and anal fins in sixty specimens, with the result that they exhibited twenty-three variations in the fin formula. Nearly two-thirds had seven rays in the first dorsal, three-fourths had twelve rays in the second dorsal, while in the anal fin nearly half had thirteen rays, and in more than a third there were twelve rays. There is no relation between the size of the fish and the number of fin-rays.

The pectoral fin is pointed; it is four-fifths of the length of the head. The ventral fins are rather shorter, and their insertion is in front of the extremity of the pectoral fin when laid back. The small first dorsal fin is two or three scales farther back than the base of the ventral; it is the length of the head behind the operculum, and a like distance separates it from the second dorsal fin. The base of the anal fin is a little in advance of the base of the second dorsal; the fin is rather larger than the second dorsal. In both fins the second ray is twice as long as the first; the rays diminish in length rapidly; none of them are branched. In both fins the posterior border is concave. The caudal fin is deeply forked, and the rays are subdivided.

The lateral line is nearly horizontal. It usually contains sixty scales; but Von Martens records instances in which the number is only forty-five. They are silvery and deciduous, like the other scales in the longitudinal lateral silvery band; but the pale olive-green scales on the back adhere better. Black pigment is scattered over the scales of the back, forming a sort of network. There are four rows of scales above the lateral line, and six or seven below. The scales are broadly pentagonal, with rounded corners. The middle scales are about one-fourth of the orbital diameter. There are forty-three or forty-four vertebræ. The extremity of the swim-bladder is enclosed by processes from the nineteenth and twentieth vertebræ.

The fish is found in the Lake of Albano, in Italy, where it may be three inches long; in the Lake of Nemi it may exceed four inches.

## Family: MUGILIDÆ.

### Genus: Mugil (Artedi).

The Grey Mullets are migratory fish, found in all temperate and tropical regions. They enter fresh waters, but always pass a portion of the year in the sea. They have an oblong compressed body, on which there is no lateral line. The cleft of the mouth is transverse, and short. The jaws are without teeth. The mandible sometimes has the anterior margin ciliated, and it is always sharp. The eyes are placed laterally. The gill-openings are wide. There are five or six branchiostegal rays. There are two dorsal fins; the anterior dorsal is longer than the second dorsal, and contains four stiff spines. The ventral fins are abdominal in position, and attached to the elongated coracoid bone.

Dr. Günther observes that fishes of this genus feed on organic substances which are found mixed with sand or mud, and that the mud is worked for some time between the pharyngeal bones, when the roughest materials are ejected from the mouth.

The pharyngeal bones have an arched irregular form, tapering in front. They are covered with a membrane, which is studded with horny cilia, and rest upon a mass of fat. Each branchial arch carries close-set gill-rakers, which fit into those of the adjacent arch, forming a perfect filter.

The stomach is remarkable for its muscular character. In structure it resembles the stomach of a bird. There is a first membranous portion, which is globular, and a second portion, which is formed by an exceedingly strong muscle, of equal thickness in the whole circumference. The internal cavity of the stomach is small. A small circular valve in the pyloric region divides the stomach from the intestine. There are about five short pyloric appendages. The intestine forms many convolutions, and in the species *Mugil chelo*, variety *septentrionalis*, was seven feet long in a fish measuring thirteen inches. The liver has a large free gall-bladder. The ovaries are large and long, and arranged in transverse laminae. There is a separate ovarian opening behind the vent. The air-bladder is large and simple, and connected with the abdominal muscles in front. The pubic bones are long and flat, and connected with the coracoid by ligament. There are about eleven vertebræ in the abdomen, and thirteen in the tail. The length of body in these regions is nearly the same.

The species number nearly seventy, besides others imperfectly defined. The characters on which they are formed are the presence or absence of a fatty eyelid on the posterior side of the iris, the thickness of the upper lip, the number of rays in the anal fin, the number of scales in the lateral line, the condition of the maxillary bone, the form of the snout, the proportions of the body, and the position of the fins.

The genera most nearly allied to *Mugil* are *Agonostoma*, found in the fresh waters of the West Indies, Central America, New Zealand, Australia, Celebes, Mauritius, and the Comoro Islands, and *Myxus*, found on the coast of Australia, the Pacific coast of Central and South America, and in the Island of Ascension.

## Mugil capito (Cuvier).—Grey mullet.

1 D. 4, 2 D. 1/8, A. 3/9. Scales: lat. line 45, trans. 14.

The Grey Mullet of England is widely distributed on the coasts of Europe and Africa, reaching as far south as the Cape of Good Hope. It is common in the Nile, and is known in Egypt as the *Bouri*. It has been recorded from some fresh-water lakes in Tunis. Günther mentions it from Rome, where it is known as the *Cefalo calamita* ; at Venice its common name is *Caustello*. In France it ascends the Gironde, Loire, Seine, and the Somme, and is known as the *Muge capiton*. Steindachner records it from the rivers of the Spanish peninsula. It is common on British shores.

In this species the total length is about four and a half times the length of the head; and the fish is fully five times as long as high. The dorsal and ventral contours are so moderately convex as to give the fish a long and flattened aspect. The snout is wide and depressed; the rather convex space between the eyes is about two-fifths the length of the head.

The eyes, which are rather large, are distinguished by wanting the fatty eyelid which is found in many species of the genus. The nostrils are close together. The lips are not covered by the nasal bones. The interspace between the mandible and the inter-opercular elements is broad in the anterior part, but tapers both in front and behind. It extends well behind the orbits. The scales are nearly as long as high, and extend as far forward as the snout.

The pectoral fin terminates at the eighth or ninth scale in the longitudinal series. The first dorsal fin commences at the twelfth or thirteenth scale, and the second fin begins at the twenty-fourth or twenty-fifth scale. On the sides of the body the scales have the aspect of being arranged in parallel longitudinal series. These scales have the free border rather angular, and the base truncate. Fine striæ are seen on the covered part of the scale, and minute asperities extend over the exposed part. On the summit of the head and snout the scales are much smaller, and are sometimes deformed.

The colour is a dark steel-blue on the back, the sides are paler, and the abdomen silvery-white.

There are seven to eight dark longitudinal stripes on the sides, which extend along the series of scales, and there is usually a blackish spot above the base of the pectoral, which is situate above the middle of the side of the body. The ventral fin is midway between the pectoral fin and the spinous dorsal; its base is usually orange-coloured.

The Grey Mullet feeds on soft and fat food, preferring that in which decomposition has commenced. The only living prey which Couch observed it to take is the common sand-worm. This observer states that it is most readily taken with bait formed of the fat entrails of a fish, or cabbage boiled in broth.

It spawns about Midsummer in the sea; and in August the young, which are then about an inch long, are seen entering fresh waters.

The fish is taken in rivers when the tide is coming in, as it returns with the ebb to salt water.

Grey Mullet have been confined in a fresh-water pond in Guernsey and in France, with the result that the fish grew fat, deep, and heavy, and improved more in fresh water than any other salt-water fish similarly treated.

There are twelve vertebræ in the abdomen, and the same number in the caudal region.

## Mugil cephalus (CUVIER).

1 D. 4, 2 D. 1 8, A. 3 8, V. 1 5, P. 2 16.

The distribution of this species is very similar to that of *Mugil capito*. It is common on the west coast of Africa, in the Mediterranean, and in the Nile. In Italy it is known as *Muggine cefalo*. It enters most of the rivers in the peninsula, and is especially common about Venice. In France it is characteristic of the Rhône, in which it is taken at Avignon as late as September; but as the cold weather comes on, it descends to the sea. Its French name is *le Muge céphale*. In Spain it is found in the Ebro and Guadalquivir.

In this species the body is five times as long as high, and four and a half times as long as the head.

The head is slightly convex from side to side, and the width of the interorbital space is three-sevenths the length of the head. The eyes are hidden behind a broad fatty membrane, which leaves only a vertical slit for the pupil. The upper lip is not conspicuous for its thickness; the mandibular bones form an obtuse angle in front. The cleft of the mouth is more than twice as wide as deep at its angles. The maxillary bone is covered by the pre-orbital bone. The posterior nasal opening is midway between the orbit and the anterior nasal aperture. The space on the chin between the mandibles is broadly lanceolate.

This species differs but little from *Mugil capito* in form and proportions, but the body is thicker. The scales are relatively large, and rather longer, with a very small median canal. They do not extend to the snout, and are not found upon the vertical fins.

In the first dorsal fin the first two spines are half as long as the head; the pectoral fin, as in *M. capito*, has its base above the middle of the body. It extends to about the eighth scale of the longitudinal series. There is a long carinate triangular scale at its base.

The colour is like that of the Grey Mullet, being dark bluish-grey on the back, paler on the sides, silvery-white below. About seven sub-parallel longitudinal lines of dark-blue colour, with golden iridescence, extend along the sides. There are black spots on the second dorsal fin. The pectoral fin is brown  The anal fin has a black border.

The fish weighs three to four kilogrammes, and may be half a mètre long. It spawns in May. There are two pyloric appendages in this species, and six or seven in *M. capito*.

Other species enter the rivers on the coasts of France, among which Blanchard mentions the *Mugil auratus* and *Mugil saliens*, which both range into the Black Sea.

## Mugil chelo (Cuvier).

1 D. 4, 2 D. 1 8, A. 3/9, V. 1 5, P. 2/16.  Scales: lat. line 45, trans. 16.

This is essentially a Mediterranean Grey Mullet, but is best known in Italy. Among other rivers it enters the Sile, and is caught at Treviso. The height of the body is one-fifth of the total length, or a little more, and exceeds the length of the head, which is three-elevenths of the length. The snout is blunter than in the variety *M. septentrionalis*, and is broad and depressed. The width of the inter-orbital space is about half the length of the head. The upper lip is remarkably thick, with three rows of broad white papillæ on its lower half. They are soft, but have the aspect of teeth. There is a slight ridge on the pre-orbital bone, which commonly has a rounded extremity, and does not cover the maxillary bone.

The two bones of the mandible meet at an obtuse angle. There is no fatty eyelid. The nostrils are close together, and the space on the chin between the mandibular bones is very narrow, when it exists at all, and not longer than the eye. The first two spines of the first dorsal fin are long. The root of the pectoral fin is somewhat above the middle of the body; the extremity of the fin reaches back to the thirteenth scale. The first dorsal fin commences at the fifteenth, and the second dorsal at the twenty-fifth scale.

The ventral fin is midway between the pectoral and the spinous dorsal. As in other Grey Mullets, six or seven dark stripes extend longitudinally along the scales. The scales are higher than long.

As in the other species, the back is bluish-grey, and the abdomen silvery. The pectoral fin is yellow, with a dark spot towards the upper part of the base. The species reaches a length of about a foot and a half.

## Mugil septentrionalis (Günther).

1 D. 4, 2 D. 1/8, A. 3/9.

This form is essentially a northern variety of *Mugil chelo*.

The thick-lipped Grey Mullet is a gregarious fish, first recognised in the mouths of British rivers by Mr. Jonathan Couch. It enters fresh waters in the winter months, when it is often left in pools by the retiring tide. Like all the Mullets which have been observed, it has the habit of escaping from nets by leaping over the head-line.

This is a northern representative of the *Mugil chelo*, and is common on the British and Scandinavian coasts, where it may reach a length of two feet. It is distinguished from *M. chelo*, by the shorter pectoral fin, longer tail, thinner upper lip, different form of pre-orbital bone, and by having five pyloric appendages, instead of seven. Dr. Günther states that the number of vertebræ is the same as in *M. chelo*—eleven in the abdomen, and thirteen in the tail. The head is about one-fifth of the total length, or a little less. The snout is blunt; the width of the inter-orbital space is half the length of the head. The lower third of the thick upper lip is whitish, and covered with two rows of flat tooth-like papillæ. The nostrils are near the orbits. There is no fatty eyelid. The mandibles nearly cover the chin. The dorsal fin is midway between the snout and the base of the caudal fin. Pointed scales extend along the base of the first spiny dorsal fin. There is no pointed scale at the base of the pectoral fin. The back is greenish, with paler sides and silvery abdomen. Dark greenish stripes run along the sides. This species, under the name of the Arundel Mullet, is one of the celebrated good things of the county of Sussex, sometimes found in the Arun, nearly twenty miles from the sea.

Moreau mentions that the roe of the Mullet is salted and dried in Provence, and made into a kind of caviare, which is sold under the name Boutargue.

## Family : GASTEROSTEIDÆ.
### Genus : **Gasterosteus** (Artedi).

The Sticklebacks belong to a great division of the spiny-rayed fishes in which the spinous dorsal fin, when present, is represented by isolated spines, and the ventral fins are either in the throat, or, if abdominal, have the pubic bones attached to the arch which supports the pectoral fin. The mouth is small, and at the extremity of the snout. In this group of fishes Dr. Günther includes two families, the *Gasterosteidæ*, which comprise the single genus Gasterosteus, and the *Fistulariidæ*, or Tobacco-pipe fishes, which are all marine.

These fishes have an elongated and compressed body; the opening of the mouth is oblique; the teeth are villiform, and placed on the jaws. The infra-orbital bones cover the cheeks. There are no spines to the opercular bones. There are no scales, but large scutes form an incomplete armour along the sides. Formidable isolated spines occur in front of the soft dorsal fin. The ventral fins are joined to the pubic bone, and each is composed of a spine and a small ray. There are three branchiostegal rays. The air-bladder is simple and oblong, and there are few pyloric appendages to the stomach.

### Gasterosteus aculeatus (Linnæus).—Stickleback.

D. 3, 10—12, A. 1/8—9, V. 1/1, P. 10, C. 4/12/5.

This Stickleback (Fig. 28) is characterised by its three isolated dorsal spines, which are separated from each other; the middle and longest of the spines is about three-fourths of the greatest height of the body. The form of the body is moderately compressed, highest under the second dorsal spine, with the height exceeding the length of the head, and between one-third and one-fourth of the length of the fish. The head is about three and a half times as long as the eye. The eyes are separated by two-thirds of their diameter. The cleft of the mouth reaches under the nares. The teeth are relatively large in both jaws, and pointed. The longest axis of the fish runs through the middle of the cleft of the mouth, and touches the edge of the pupil of the eye. The very broad sub-orbital bone extends back to the angle of the pre-operculum, with which it is closely united, and, like the anterior smaller elements in the sub-orbital ring and the operculum, is finely striped.

The profile rises in a regular arch from the margin of the mouth to the second dorsal spine. The body is four times as high under the dorsal spine as at the tail.

The pectoral fins are truncated, and extend back as far as the second dorsal

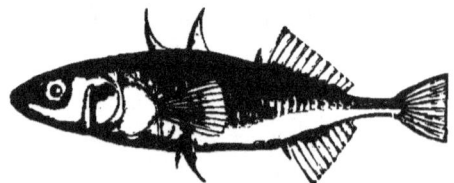

Fig. 28.—GASTEROSTEUS ACULEATUS (LINNÆUS).

spine, which is nearly opposite to the ventral fin. The third, or short, dorsal spine is farther removed from the first two than are these from each other. The soft dorsal fin has its upper edge straight and inclined backward, because the anterior rays are nearly three times as long as the short posterior rays. The vent is opposite to the third dorsal spine, and therefore nearer to the tail than to the snout. The anal fin resembles the soft dorsal, and reaches equally far back, but has a shorter base, and is preceded by a short, broad compressed spine. The spine of the ventral fin is rough, and furrowed down its length; the fin also possesses a soft curved ray, which often reaches as far back as the three-cornered pelvic shield, and therefore below the third dorsal spine. In front of its base the pelvis forms an upward forked shield, which lies between the lateral shields, when they are present, and reaches to half the height of the body. The caudal fin is vertically truncated, and has a length of about one-eighth of the length of the fish. In front of the first dorsal spine, and close behind the back of the head, there is a small spine with its point turned backward, followed by two others with their points directed forward, but concealed under the skin. All the fin rays, except those of the pectoral fin, are bifurcate at the end.

The shining silvery pigment of the skin of the trunk is naked, except where covered by the median dorsal shields and lateral armour that defends the ventral side of the fish.

The armature of the skin is sufficiently variable to almost suggest specific differences, and on this evidence the species was divided by Cuvier into *Gasterosteus trachurus* and *G. leiurus*, or *gymnurus*. But when individuals from many districts are compared, when the young are compared with the old, and those caught in summer are contrasted with fish taken in the winter, the two extremes are found to be united by intermediate types, in which the dermal shields show every possible gradation of form, as was pointed out by the Swedish ichthyologists, Fries and Eckström. These naturalists regarded both colour and armature of the skin as varying with summer and winter, so that the Stickleback in its brightly-coloured bridal dress, when it is naked, is *Gasterosteus leiurus*, while *G. trachurus* is the same species in its winter dress.

The form named *Gasterosteus leiurus* is not met with except at the spawning-time, though in the very young state, when only eight lines long, examples of it have been described with slightly developed plates on the pectoral and pelvic girdles. Dermal plates shield the back. In front of the dorsal spine are two broad plates developed one behind the other, which are in contact with the median shield of the occipital region; and shields of varying size extend on each side, along the base of the dorsal fin, and often reach the lateral shields.

The lateral shields are best developed in the region of the pectoral girdle. They are elongated vertically, in contact with each other, partly imbricated, and vary in size and form, and in number from three to twenty-eight. When most developed the bucklers descend to the abdomen, and have their asperities finely furrowed. Their number is unequal on the two sides of the body. There is a remarkable system of ventral armour which covers the throat below the inferior angle of the operculum, and extends back to the pelvis. The pelvis develops a shield which is sub-triangular behind, and sends a branch up the side of the fish for about half the height of the body; but it varies in form and ornament. This plate is the fulcrum for the ventral fin.

The course of the lateral canal along the upper third of the side of the fish is very marked. The accessory gills are thin fimbricated filaments. The ovary is double, but is stated by Fatio to be sometimes only deeply bi-lobed. The air-bladder is simple, and ends in two short fine air-canals, which open above the dorsal side of the stomach. The digestive canal is shorter than the animal, and almost straight.

The stomach is large, and has two small pyloric appendages. The vertebræ vary in number from thirty-one to thirty-three.

The colour of the back is bluish-black or greenish-brown, with the sides and belly silvery. The throat and breast vary from pale rose to blood-red; the fins are transparent and greenish; the pupil is oblique, and the iris like silver. These colours reach their greatest intensity in the spawning season. The colours are less variable in the female and the young than in the male.

This species is rarely more than three inches long. It is somewhat tenacious of life, and will often throw itself out of the water and live in damp grass for a time. It is very voracious, and its movements are rapid. In an aquarium a small Stickleback in five hours devoured seventy-four young dace about a quarter of an inch long. They are sometimes found in such quantities as to be used for manure. The fish is frequently afflicted with a kind of tape-worm (*Bothriocephalus solidus*), which often fills the entire body cavity and kills the Stickleback, when the worm is liberated, often to find a resting-place and development in the duck, or some other water bird.

The Stickleback is widely distributed, and is found in shallow streams in almost all parts of Europe, and in the Baltic and Frische-Haff, though the varieties are differently distributed. The variety a *Gasterosteus gymnurus*, which has only four or five plates above the pectoral fin, and the rest of the body naked, is found in England, France, Southern Germany, and the Baltic. The variety β *Gasterosteus semiarmatus*, which has ten or fifteen plates on the front part of the sides, is found in England, Belgium, and France. The variety γ *Gasterosteus semiloricatus*, which has scaly plates reaching to the caudal region, is limited to France and Ireland. The variety δ *Gasterosteus trachurus* has scaly plates covering the tail as well as the body, and is found in the northern parts of Europe, England, France, and Germany. The variety ε *Gasterosteus noreboracensis* differs from *G. trachurus* only in having a longer ventral spine. It is found in Greenland and in North America.

Many writers attach specific importance to these varieties, and Yarrell adopted most of the types which had been recognised by Cuvier and Valenciennes, and are still preserved by Blanchard.

The species appears to be wanting in the Danube.

In Pennant's time the Stickleback was remarkably abundant in the Fens of Lincolnshire, and every seven or eight years amazing shoals appeared in the Welland, at Spalding, and came up the river in a vast column. They were used to manure the land, and occasionally oil was extracted from them by boiling, as in Sweden and France. As an illustration of the abundance in which they were found, Pennant mentions that a man employed by the farmer to capture them for manure earned for a considerable time four shillings a day by selling them at the rate of a halfpenny a bushel.

On the Continent the refuse is used for manure after the oil is extracted.

The fighting instinct is limited to the males, and may be as well observed in captivity as in nature. It is then that the use of the spines and the armour becomes manifest, for on their skilful use the life or death of the combatant depends. Sometimes the cause of battle is the desire to possess a cave, or recess, or corner already occupied, but the sense of honour is so delicate in these little fishes, that even a longing glance thrown towards a spot which might become a happy home is sufficient to provoke an assault on the passer-by. Or, if one fish has a position of vantage, and feels the presence of less fortunate companions irksome, his spines and spears are at once erected to drive the gazers away, saying, no doubt, with Horace, "Odi profanum vulgus, et arceo." The battle begins; the combatants swim round each other rapidly, biting and thrusting, retreating and advancing, to avoid or deal a blow. It is not often that the weaker seeks safety in flight, for he knows that no quarter will be shown to the vanquished if captured. He rather tries to act on the defensive until such time as the stronger fish shall forget his discretion, and sooner or later the abdomen of the enemy will be ripped up. When opponents are thus disposed of or driven away, the valiant Stickleback, no longer worn with care, grows fat and increases in length. And as the season advances he begins to construct a nest. According to Professor Blanchard, the first step is the excavation of a depression on the bed of the stream by a rotatory movement of the body, and then, as Monsieur Costa had previously recorded, stalks of grass, rootlets of trees, and other delicate fibres of vegetable matter, are laid in the hollow, and cemented together with mucus from the skin. Walls are then raised, and ultimately roofed over, a small aperture only being left as an entrance. The nest is nearly an inch in diameter.

The male now puts on his most brilliant colours, and, conspicuous in his scarlet breastplate, and sides shining with a glossy lustre, leaves his labours and goes in search of a mate, by no means contenting himself with the first female he sees; but selecting, with many caresses, one ready to spawn, he allures her to the nest, and, should there be any indisposition to enter his nest, something more than gentle pressure, it has been observed, is used to force her into it. She enters, and in a few minutes deposits a few eggs, and forces a way out on the opposite side and escapes. The male follows her through the nest, and fertilises the eggs. Next day he again goes in search of his mate, but the conjugal instinct is not very strong among Sticklebacks; and, failing to find her, another female is brought to the nest. The male in the same way rubs his side against the female, and passes through the nest over the eggs. This process may be continued day by day until, according to Von Siebold, sixty or eighty eggs are deposited, before the male closes the exit. The male now watches over his treasure for the period of incubation, which may vary according to the temperature of the water from ten or fifteen to twenty-five or thirty days. During this time he defends the nest against all invaders, and especially against the curiosity of his wives. He enters the nest from time to time, and by movements of the fins circulates the water; and, as the young become hatched, goes in search of food, which is first prepared in his own mouth for their wants. And only when the young are sufficiently developed to eat and swim does the parent allow them to leave their first home. Five or six days after hatching, according to Fatio and other observers, the male opens the nest and spreads its materials, upon which the young fishes assemble. They are then three to four millimètres long, and have a natatory membrane which extends along and around the body. This disappears in ten or twelve days, when the fins begin to develop. The fish is now five or six millimètres long. The number of young varies with the nest; some observers record twenty to thirty, while others mention sixty to one hundred.

If the male is captured, Houghton records that the nest is attacked by a crowd of hungry, marauding Sticklebacks, of all ages and both sexes. In the ovary Dr. Günther has counted ninety eggs; Blanchard counted one hundred to one hundred and twenty; but Fatio, and other writers, have found a smaller number. The ripe eggs are about one and a quarter millimètre in diameter, and of a yellow colour. The breeding season varies according to the locality; and a male may have three nests between the 1st of April and the end of June. They begin to breed at the age of one year, and are reputed to live for three or four years. Where powerful carnivorous fishes abound, such as Perch, Pike, Trout, Salmon, their numbers may be kept down, but the Stickleback fights for life to such purpose that it is not often molested; indeed, Blanchard suggests that they might adopt for motto the French proverb, "Qui s'y frotte s'y pique."

The Pike has been known to eject them from the mouth when captured.

In France it is known as *Epinoche*; in Germany it is *Stachelfisch*; in Italy, *Spinarello*; in Poland, *Ciernik*; in Sweden, *Hundslagg* or *Stor Spigg*.

The variability of a Stickleback, though not so marked as that of a pigeon or a dog, is so considerable that the best ichthyologists differ as to the forms which shall be regarded as species, and those which they class as varieties. There is no doubt that if the characters of the varieties were constant, the conclusions of Cuvier would still be maintained, and numerous species of Sticklebacks of the three-spined type be recognised, which would show remarkable modifications with geographical distribution. But since the characters are very variable there has been a tendency to push the doctrine of specific variation as far as possible, and group under the name *Gasterosteus aculeatus* all the three-spined Sticklebacks of Europe and America. If we contrast the extreme terms of this series as represented by the *Gasterosteus aculeatus* with lateral armour extending from the head to the tail, with a form like *Gasterosteus argyropomus*, in which the sides of the body are entirely free from plates, then no one could hesitate, especially keeping in view the short first dorsal spine of the latter form, to regard it as a separate species, which, being limited to Italy, might be looked upon as a geographical representative. But when we find in the fresh waters of France every intermediate variation in lateral armature, the idea of separating the Italian type as a species fails. There are only two varieties known in Italy, that to which we have just referred, and the partly-armoured *Gasterosteus brachycentrus*; and both these are united by Professor Canestrini as furnishing the Italian type of *Gasterosteus aculeatus*. And Dr. Steindachner, after study

Fig. 29.—GASTEROSTEUS ACULEATUS, VARIETY BRACHYCENTRUS.

of the Spanish types, adopted the conclusion that the *Gasterosteus brachycentrus* must also be merged in the *G. aculeatus*.

As a type, however, which has been generally regarded as a species by the writers of Great Britain and the Continent, we have thought it desirable to point out its more distinctive variations and characters (Fig. 29). All the fin-spines are shorter, and the first dorsal spine begins over or somewhat behind the base of the pectoral fin, instead of in front of it. The head is longer than in the Common Stickleback, and longer in the male than the female. The greatest height is never equal to the length of the head, even in the female at the time of spawning. The diameter of the eye is a quarter of the length of the head; it is separated from the other eye by its own diameter. The mouth is more oblique, and the mandible projects over the pre-maxillary, especially in the males. The dentition is scarcely visible. The dorsal spine, like that of the ventral fin, is always shorter than in the Common Stickleback, but always broader at the base, and is more strongly notched on both sides. The base of the soft-rayed dorsal fin is longer, and extends just before the vent. The anal fin begins farther back, and its rays are somewhat shorter. The caudal fin is about one-half the length of the head; the three-cornered plate between the ventral fins is smaller and more pointed in males than in the females, in which it is considerably broader than in *G. aculeatus*. There are four lateral shields. They touch the base of the shield below the spinous ray, and reach down to the pelvic girdle. This variety is found at Görz, in Istria, and the Lake of Garda, and other localities in Lombardy, where it is known as *Spinarola*.

A four-spined Stickleback has been named *Gasterosteus spinulosus* by Jenyns and Yarrell. It is recorded from near Edinburgh, Arran, and Berwick. A five-spined Stickleback, taken near Warrington, has been recorded by Mr. J. Peers. These can only be regarded as another series of variations of *Gasterosteus aculeatus*.

## Gasterosteus pungitius (LINNÆUS).—The Ten-spined Stickleback.

D. 9—11/11, P. 9—10, V. 1/1, A. 1/9—11, C. 5/12/6.

The body in this fish is compressed, elongated, with the sides naked, except for about a dozen keeled scales, which terminate the lateral line on the tail, though they are not present in all varieties. The lateral line is indicated only by a furrow. The operculum is rather larger than in *G. aculeatus*. There are from nine to eleven, more frequently ten, low, free, dorsal spines, which are similar and nearly equal in size, each with a slight fin-membrane. They are erectile at the will of the fish. There is no interval between the last dorsal spine and the dorsal fin. The anal fin, which has a recurved spine in front, is opposite to the dorsal. The ventral fin consists of one spine and one soft ray. The spine is attached to a triangular pelvic shield, which is keeled, and has something of the form of a sternum. The fins are transparent.

The colour is more uniform than in the Common Stickleback. The upper side of the body is green or bluish-black, with darker transverse bands. The belly and sides are silvery, but with change of season become silver-blue, orange, or reddish. The male in summer is often black on the under side in Germany, and in Britain its wedding dress is an intense velvety black, without the slightest trace of red. The black colour develops first on the belly in March, and gradually extends over the body. In Scandinavia it is olive-green, with a coppery-silvery tone below, according to Lloyd. Its habits are similar to those of the other species, being equally fond of society, and often found in large shoals; but it is even fiercer and more pugnacious.

Lloyd states that the fishing season for Sticklebacks in Sweden begins in November, and continues till the ice begins to form. At nightfall the spot where the fishes are congregated is seen on a clear evening from the ruffled state of the water. Then two men go to the spot in a boat, with a torch of wood burning at one end. One man steadies the boat with a pole driven into the ground, while the other scoops up the fish with a landing-net; and sometimes they obtain several boatloads in this way in a single night.

The eggs are orange. When abundant the fish is boiled for oil, four bushels of fish yielding two or three gallons of oil, which is used for lamps in Scandinavia. It is common in the Baltic and North Sea, is met with in the estuaries of rivers, but is found chiefly in small tributary streams far up their course, and occurs in almost every lake and river in Sweden, where it is called the *Sma Spigg*. Pigs are sometimes fed upon it. Blanchard distinguishes varieties in France, which he names *G. burgundianus*, *G. loris*, *G. lotharingus*, and *G. breviceps*, but they are less easily characterised than the varieties of *G. aculeatus*. Günther includes the *Gasterosteus occidentalis* (C. and V.) as a North American variety of this species.

The fifteen-spined Stickleback, *Gasterosteus spinachia*, is a marine species, common on the coasts of all the northern countries of Europe. It is particularly abundant in some parts of the Baltic Sea, where it is used for manure and as food for pigs; but it is rare on the Prussian coast. It rarely enters rivers and fresh waters in Scandinavia, and is almost unknown in the fresh waters of Britain.

# CHAPTER III.

### FRESH-WATER FISHES OF THE ORDER ANACANTHINI.

FAMILY GADIDÆ.—GENUS LOTA— Fresh-water Representatives of the Cod Family—The Burbot—FAMILY PLEURONECTIDÆ—The Flounder—The Sole.

## FAMILY: GADIDÆ.

### GENUS: **Lota** (CUVIER).

THE *Gadidæ* of Dr. Günther is a family in the order *Anacanthini*, characterised by having small smooth scales on a more or less elongated body. The genera may have three or two dorsal fins, or one dorsal fin. These fins occupy nearly the whole of the back, and the hindermost dorsal has the rays well developed. There may be one or two successive anal fins, and the different combination of the dorsals and anals, and the degree of separation of these fins from the caudal, furnish an important series of generic characters. The ventral fins are placed in the throat. The development of the ventral fin helps to define many genera. The gill-openings are wide, and the gill-membranes are not attached to the isthmus. There is an air-bladder.

The family is most characteristic of Arctic and temperate seas, but is common on coasts in Europe and America.

The only genus met with in the fresh waters of Europe is Lota, of which but one species is known—the Common Burbot—of which several varieties, differing but slightly from each other, are met with in the United States and Canada.

The genus Lota is characterised by having two dorsal fins, one anal fin, a separate caudal fin, narrow ventral fins, composed of six rays, with ten to thirteen rays in the first dorsal. Villiform teeth occur on the jaws and on the vomer. The chin has a barbel.

## Lota vulgaris (CUVIER).—The Burbot.

This fish (Fig. 30) has a wide flattened head, an elongated body, and compressed tail. The mandible is shorter than the upper jaw. The upper surface of the head is always slightly arched. The length of the head is one-fifth the total length of the fish. The width of the fish is nearly two-thirds the length

of the head ; and the height about one-half the head length. Under the first
dorsal fin the height and thickness are about equal. The diameter of the eye,
which is placed laterally, is one-seventh of the length of the head. The
iris is golden. The nares are small ; the anterior nasal aperture has a raised
lobe of skin, hardly amounting to a barbel. It is placed midway between
the eye and the end of the snout. The gape of the mouth is wide, and
reaches back to the eye. Its outline is semicircular. The fleshy upper
lip projects over the lower lip, which has a long barbel under the middle
of the chin, and near it is sometimes a shorter barbel. There are small
pointed teeth on the pre-maxillary bone, and in the mandible, where

Fig. 30. LOTA VULGARIS.

they are arranged in dense rows; and there is a broader row on the pre-
maxillary, parallel to the ligament between the bones. The teeth on the
vomer are stronger.

The gill-aperture is wide ; the rough skin of the branchiostegal rays forms
a sort of curtain in front of them. The branchiostegal rays are seven in
number. The rake-teeth on the four gill-arches are short, thick, and finely
serrated ; the accessory gill-rays are very small, and concealed under a deep
fold in the skin.

The first dorsal fin begins in the second third, and ends in the middle of
the length of the fish, opposite to the anal aperture. Its middle jointed rays
are the longest, and about a third of the height of the body depth beneath
them. The second dorsal immediately succeeds the first dorsal, and extends to
near the caudal fin, from which it is sometimes scarcely separated; and in some
localities, as in Lemberg, the two dorsal fins have a membranous connection.
The base of the dorsal fin is necessarily nearly equal to half the length of
the fish. Its jointed rays are forked at the extremity, and become somewhat longer towards the end of the fin. The anal fin resembles the second

dorsal, but has a shorter base and shorter rays. The pectoral fin is somewhat ovate; its longest rays are more than half the length of the head. The ventral fins are about equally long; their first and second rays generally terminate in fine points. The caudal fin is pointed in the middle, and the rays radiate like a fan. All the fins are extremely soft, and terminate in membrane, so that the rays are counted with difficulty.

Fine scales are found behind the eyes and on the sides of the head and throat, but the body is covered with thick ovate scales, with concentric growth. Very delicate scales are found on the bases of the fins, and the second dorsal has scales upon its hinder half. The scales are embedded in the skin, which surrounds them, and they give it the aspect of being covered with pits. The smallest scales on the body are found along the lateral line, which is near to the back at its commencement, descends to the middle at the side, and runs horizontally to the tail.

The stomach has an elongated form like a wide sack. The pyloric region is surrounded by about thirty appendages, which are of unequal size, grouped in bundles, and connected with each other by a common canal, which opens into the intestine. The liver is tri-lobed. The air-bladder extends the entire length of the abdominal cavity. Its anterior extremity is constricted or heart-shaped, and is supplied with a large blood-vessel. The principal mass of the kidney is situate at the extremity of the abdominal cavity, but sends small lobes forward. The large, long urinary bladder terminates with the ovary or milt duct in a genital papilla, which is situate behind the vent. The ovaries are contained in a thin peritoneal sac, which is scarcely half as long as the abdominal cavity. The milt is shorter still. The colour of the back, sides, and fins is a more or less clear olive-green, marbled with cloudy spots of blackish-brown. The belly, ventral fin, and throat are whitish.

The size varies in different parts of Europe. In the Danube it is from one foot and a half to two feet long; but the weight is only three or four pounds, though in the Austrian lakes, such as the Fuschler See and Atter See, it is usually from eight to twelve pounds, and occasionally sixteen pounds. In North Germany it rarely exceeds a length of about nineteen inches, and a weight of two pounds, though specimens weighing thirty pounds have been captured in the Rhine. Blanchard relates that one weighing forty-two pounds is said to have been purchased by the city of Strassburg for 600 francs for a breakfast given to Charles X., but does not pledge himself to the accuracy of the tradition.

This fish is hardy, and found indifferently in large rivers, small streams, elevated lakes, and low-lying ponds, though it usually lives in deep water, and in lakes is often found at a depth of thirty or forty fathoms; but under such circumstances its colour is paler than when living in shallow water. It commonly lives under stones and in holes, waiting for its prey, and hence has come, from a rabbit-like habit, to be sometimes named the Coney-fish. Its instincts, however, are those of robber and pirate. It waylays the female and the young brood, especially of the Perch, and is a terror to all small fishes.

The Burbot spawns at different times in different localities. Some deposit their eggs in November and December, others not till March. At spawning-time they become gregarious, and in the rivers of Germany a hundred may be found together interlaced in a compact mass like Eels. The eggs are hatched in about six weeks. The young grow slowly, and at four years of age reproduce for the first time. They are in best condition for the table at spawning-time, when they are taken with a ground-net or line. In many places the liver is considered a great delicacy, while the oil which it yields was formerly esteemed in medicine, under the name *Liquor hepaticus mustelæ fluviatilis*. The air-bladder is made into an inferior kind of isinglass, but the hard roe, in Austria, at least, has the reputation of being poisonous, and is used in Germany as a purgative.

Von Siebold mentions that Dr. Steinbuch in 1802 speared two examples of this fish which were united into one mass, head to head and belly to belly, by a constricting band round their bodies, which had the same colour and slimy character as the skin.

The Burbot is essentially a fresh-water Ling, from which it differs in having a few more rays in the second dorsal fin, and fewer in the first dorsal, and a few more rays in the anal fin.

It is widely distributed in Europe, and is referred to by Yarrell as occurring in Siberia and India. In Britain it is found along the eastern side of England. It is met with in Sweden, France, Switzerland, Germany, Austria, the north of Italy, and Russia.

The North American specimens reach a length of three feet. They have been described under various specific names, but are associated by Dr. Günther with the European type.

## Family: PLEURONECTIDÆ.
### Pleuronectes flesus (Linnæus).—The Flounder.

D. 62—60, P. 10, V. 6, A. 39—45, C. 14. Lat. line 85.

Though best known as a sea-fish, the Flounder everywhere frequents fresh waters, often ascending rivers for long distances. It is known in Sweden as the *Flundra*, from which the English name *Flounder* may be derived, though it is also known in North Germany and Central Europe as *Flunder*. The Dutch term it *Bot*, which is probably the origin of the term *Butt*, by which it is known in the backwaters behind Yarmouth and other parts of the east coast, where it abounds. In France it is indifferently known as *Picaud*, *Flondre*, or more commonly as *le Flet*. In Scotland and the north of England it shares with Plaice the name of *Fluke*.

Yarrell says it is taken as high up the Thames as Teddington and Sunbury. Before the construction of the Thames Embankment it was frequently taken in London by boys fishing from the river-banks. It formerly ascended the Avon to near Bath, and Mr. Day, in former years, took it at Shrewsbury. Pennant remarks that though they never grow large in our rivers, the fishes are sweeter in flavour than those which live in the sea. In Belgium, according to Selys-Longchamps, it goes up the Scheldt, and passes up the Nethe beyond Brussels to Waterloo.

It ascends the Rhine, and has been taken in the Moselle at Trier and Metz, and has been recorded in the Rhine at Mainz. It is common in the Baltic, and there enters the rivers, not, however, with the regularity of Salmon, or other migratory fishes like the Eel, but probably in search of food and love of quiet. And it similarly ascends the rivers of France, where it has been taken in the Dordogne and other inland streams. It ascends all the rivers of Russia from the Black Sea, the Sea of Azov, the Baltic, and the White Sea, and reaches many of the lakes.

It possesses the power of adapting its colour to the locality in which it lives. On black mud it is often indistinguishable from the bottom on which it rests, while in clear water, with white sands, it may be white on both sides. Sometimes, as Pennant and others have remarked, the colour is pink or red, and specimens have been found with orange spots in some localities, while other specimens occasionally have very dark brown spots. It frequently occurs reversed with the colour and eyes on the left side. The Flounder is not very discriminating in food, but prefers worms, insects, and small fishes, and will also take small mollusca and most animal substances.

It spawns in February, March, and April, and the young are frequently seen before the beginning of May. In some localities the spawning does not take place till late in April, or in North Germany in May. It rarely reaches a large size, and a length of nine inches and a weight of four pounds are exceptional, though Pennant mentions specimens which weighed six pounds.

The number of eggs progressively increases with the size of the fish, and more than one million and a quarter have been found in one weighing a pound and a half. It is often used as bait, but is generally eaten wherever it occurs, and formerly much more commonly than now, in a salted and dried condition, was imported from the Baltic. In Germany swine are sometimes fed on it.

In form the Flounder has a strong family resemblance to the Plaice. The head is about one-quarter of the length of the fish, and the greatest breadth inside the fins is one-third of the length, or a little more. The caudal fin is long, as are the dorsal and ventral fins. The eyes are elevated above the level of the head, and are close together, side by side, hardly separated by an orbital diameter. The mouth is small; the lower jaw is the longer; the jaws open obliquely. The teeth are blunt and conical, small, and form two rows in the upper jaw, and one row in the lower. The anterior nostril is tubular; the posterior is oval, and has no distinct tube. The dorsal fin extends from the eye to the narrow part of the tail in front of the caudal fin. Its margin is serrated. Its highest part is behind the middle of the body, so that the anterior edge is longer than the truncated posterior edge. The anal fin commences behind the pectoral, and reaches as far back as the dorsal. It is quite as high as the dorsal, but being shorter, the anterior and posterior portions are nearly equal. A row of rounded or star-like spiny tubercles margins much of the length of the bases of these fins, one tubercle being placed in each interspace between the fin-rays. The pectoral fins are alike on both sides of the body. They are short, but rather longer than the ventral fins, which are placed below and slightly in advance of them. The caudal fin is truncated. The scales are small, with concentric lines and delicate radiating branched rays, which, however, are developed only on the posterior two-thirds of the scale, and make its margin irregular. The scales on the head are rudimentary. The lateral line ascends a little over the pectoral fin, and its course is marked with star-shaped tubercles, and similar tubercles margin it above and below. The cephalic canals are very distinct, running forward in a curved course between the eyes, and giving off a sub-orbital branch and a longer branch to the lower jaw. The common colour of the upper side of the body varies from brown to olive-green. The fins are usually paler. The skin is white on the under side.

There are twelve vertebræ in the body and twenty-three to twenty-four in the tail.

## Pleuronectes italicus (Günther).

D. 60—64, A. 41—48, P. 11—12. 10—11, V. 6, C. 12.

This species, the *Platessa passer* of Bonaparte, is known in Italy, according to Canestrini, as *Pianuzza passera*. It is common in the Adriatic Sea, and goes up the Adige and other streams. Many Continental writers regard it as identical with the Flounder.

A bony keel divides the head into two parts. The space between the eyes is naked and small. At the bases of the dorsal and anal fins are rows of small accessory spines. The anal fin has one spine directed forward. The small scales have a semicircular form. The teeth are strong, blunt, and arranged in a single row.

The head is one-fourth the length of the fish. The height of the body is to the total length as ten to twenty-seven or twenty-nine.

The vertical fins have large, brown, irregular spots.

The Plaice, *Pleuronectes platessa* (Linnæus), has been found by Steindachner in the rivers of the south of Spain, and is common in many estuaries of the south of France and England. In East Friesland it has been placed in freshwater ponds, where it thrives well.

## Solea vulgaris (Cuvier).—The Sole.

D. 73—86, P. 7, V. 5—6, A. 61—73, C. 16. Lat. line 160.

The Sole is essentially a marine fish, yet it is not only capable of being naturalised in fresh water, but in some rivers is said to have developed marked characteristics. It is remarkably hardy, and commonly migratory. It retires to deep water in winter, and comes into the shallow sea with the warmer weather of April or May. It lives on a sandy bottom, above which it does not rise far.

It varies in size and colour with the fishing-ground, being small on the east coast of England, and large on the south and south-west coast. The largest Sole mentioned by Yarrell was twenty-six inches long, and weighed nine pounds.

McCulloch was the first to draw attention to the circumstance that Soles have been kept in confinement in fresh-water ponds in Guernsey, where they became twice as thick as fishes of the same length taken in the sea. And Yarrell tells us that Soles frequent the River Arun, in Sussex, nearly up to the town of Arundel, remaining in the river the whole year, breeding in it and burying themselves in the sand in the winter. They are well known to the Sussex fishermen, having long been taken with the trawl, sometimes weighing two pounds, but usually smaller, and, like the Guernsey fishes, they are relatively thicker than those caught at sea.

Soles seldom take bait; but, at sea, feed at night on small mollusca, small echinoderms, marine worms, and fish spawn. Nothing is known of their food in rivers. The eggs of the Sole are very minute, and Buckland estimated that there are 134,000 eggs in a fish of one pound weight. So long as the young fishes have the eyes on opposite sides of the head the swimming position is vertical, and locomotion performed by means of the fins; but as the eye migrates across the head, so that both eyes are on one side of the head, the body assumes a horizontal position, and locomotion is performed by serpentine undulations of the body.

The body has a long oval form, narrowing a little to the tail. The length of the head is between one-fifth and one-sixth of the entire length. The height of the body, excluding the dorsal and anal fins, is one-third of the length. The mouth is not terminal; its aperture is curved, and extends under the hinder eye. The upper jaw is slightly the longer. Both jaws carry minute teeth, which are best developed on the under side. The anterior narine is tubular. The eyes are small, rather distant, the upper one being in advance of the lower.

The dorsal fin, which is low, commences in front of the anterior eye, and extends almost to the caudal fin. The ventral fin is very small, in advance of the base of the pectoral, distinct from the anal but not separated from it by an interval. The anal fin is similar to the dorsal, but shorter. The pectoral fins are similar on both sides of the body, are less than half the length of the head, and often black at the extremity. The caudal fin is short, but longer than the pectoral, with its extremity rounded.

The lateral line is straight, and runs between the upper border of the eye and the middle of the tail. The scales are small, and fimbriated at the free margin. They cover the body and extend on to the fins, but are absent from the larger part of the under side of the head, where papillæ replace them about the mouth.

The colour of the under side is white, though Soles have occasionally been taken with both sides coloured. The upper side is commonly brown, but the shade varies, sometimes having a bluish tinge. Not infrequently there are blotches of darker brown or black.

It is characteristic of the North Sea, from which it goes into the Baltic, and migrates into the English Channel. It is known on all the French coast, the coasts of Spain and Portugal, and it enters the Mediterranean.

## CHAPTER IV.

### FRESH-WATER FISHES OF THE ORDER PHYSOSTOMI.

FAMILY SILURIDÆ—Genus Silurus— FAMILY CYPRINIDÆ Genus Cyprinus: The Carp: Its Varieties and Hybrids—Genus Carassius: The Crucian Carp—Goldfish—The Carassius Bucephalus Genus Barbus: The Barbels of Central Europe and Spain—Genus Aulopyge: Aulopyge hügelii—Genus Gobio: The Gudgeon—Gobio uranoscopus.

## Family: SILURIDÆ.

### Genus: **Silurus** (Linnæus).

The great family of Siluroid fishes is characterised by having the skin naked, or armoured with bony scutes, but scales are not developed. The margin of the upper jaw is formed by the pre-maxillary bone, and the maxillary bone is usually a rudiment, which forms the base for a maxillary barbel. Barbels are always present. The sub-opercular bone is always wanting. The air-bladder generally communicates with the organ of hearing by auditory ossicles.

Dr. Günther remarks that in the typical Siluroids the cranial cavity is closed at the sides, as in the *Cyprinidæ*, by the orbito-sphenoid and ethmoid bones, which unite with the pre-frontal bones. The supra-occipital bone is greatly developed, and many Siluroids have the skull enlarged posteriorly by a kind of helmet formed of dermal ossifications, which spread over the region of the neck.

As a rule, there is a strong dorsal spine, which is articulated to a buckler formed of the second and third inter-neural bones.

The Siluroids have been separated by Günther into eight great groups, distinguished by the relative length, development, and position of the dorsal and anal fins, the gill-membranes, the barbels, the nostrils, and the vent. These sub-families are named *Homalopteræ*, *Heteropteræ*, *Anomalopteræ*, *Proteropteræ*, *Stenobranchiæ*, *Proteropodes*, *Opisthopteræ*, and *Branchicolæ*.

The genus Silurus is included in the *Heteropteræ*, which have the dorsal fin very short, the anal fin elongated, the ventral fins below or behind the dorsal, and the caudal fin rounded. Nearly all Siluroid fishes, which include fully one hundred and twenty genera, characterise fresh waters, though some enter the sea. They are most numerous in the Tropics.

The genus Silurus ranges from Europe, eastward through Afghanistan to Cochin China, China, and Japan.

It has no adipose fin, and the single dorsal fin, which includes more than eight rays, has no sharp spine. There is a barbel to each maxillary bone and to each mandible. The eye is lateral and above the angle of the mouth. Five or six species are known, but the European *Silurus glanis* is the type of the whole family.

## Silurus glanis (LINNÆUS).

D. 1 4, A. 90, V. 12—13, P. 1 17, C. 17.

The head is nearly as broad as long (Fig. 31). It has a long barbel on each side of the upper jaw, and four short barbels on the lower jaw. The head is somewhat compressed, flatly arched, rounded in front, and is nearly

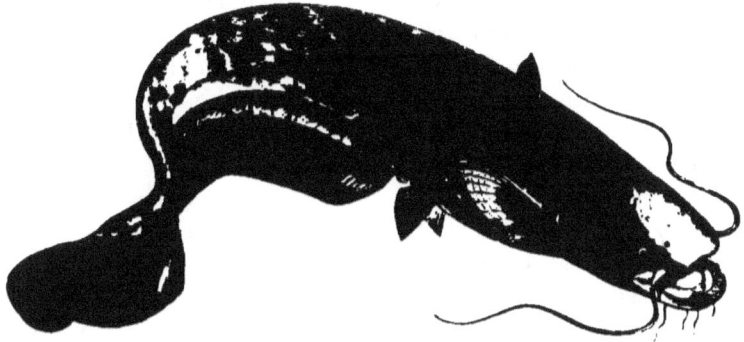

Fig. 31.—SILURUS GLANIS (LINNÆUS).

one-sixth longer than wide. Its height is two-thirds of its width, and its length is about one-sixth of the length of the fish.

The mouth is wide, with bow-shaped jaws, margined by thick, fleshy lips which cover the teeth. The barbels of the upper jaw reach back at least as far as the point of the pectoral fin, and behind the barbels the skin is folded so as to form a deep furrow parallel to the edge of the lower jaw. The eyes are very small, with a yellow iris, and a small pupil, which is elongated vertically. Between the eyes are the apertures of the small posterior nasal openings, but the anterior nasal apertures are prolonged forward as a tube, and placed near to the margin of the upper lip. The mandible is somewhat longer than the upper jaw. The teeth are compressed and conical, with their points recurved. The teeth in the lower jaw form a band of four or five rows, divided into lateral portions by the symphysis. On the pre-maxillary bone the teeth form two parallel rows placed behind each other. The hinder row is shorter and unbroken; but in the front row there are interspaces between the teeth. There are two small patches of smaller sickle-shaped teeth on the vomer. The tongue is short, thick, triangular, and toothless. The gill-rakers constitute four comb-like arches. Their teeth-like processes are short, curved, blunt, and distant. When the gill-aperture is closed, a rod-like process of skin is seen to overlap it on each side. There are sixteen branchiostegal rays.

The superior contour from the head to the tail is comparatively straight. Behind the dorsal fin the breadth of the back diminishes steadily to the tail.

The short and feebly-developed dorsal fin is intermediate between the pectoral and ventral fins. The pectoral fins are fan-shaped. The first ray is bony and strong, but not serrated. The ventral fins are similar, but shorter; they reach back over the vent to the anal fin. The anal fin occupies the entire space between the vent and the caudal fin, but remains distinct, and its rays maintain a nearly even length throughout. Close behind the pectoral fin, and over its root, is a remarkable narrow aperture, which penetrates apparently between the pectoral muscles; but Von Siebold observes that although these apertures have been detected in many Siluroids, no one has suggested any purpose in the economy of the fish which they serve, though Heckel and Kner describe the walls of the cavity as well supplied with nerves.

The lateral line runs near to the back, and can be distinguished down to the caudal fin. The branch of the cephalic mucus-canal running along the lower jaw is marked by a row of large pores. Behind the vent there is a long perforated papilla, on both anterior and posterior aspects of which there are apertures which the papilla divides. The tail of the fish is remarkably compressed and relatively long.

The colour of the skull, back, and the edges of the fins, especially the ventral and anal, is nearly blue-black. The sides of the body are greenish-black, becoming paler towards the belly, and spotted with olive-green. The belly may be of a reddish tinge or yellowish-white, with blue-black marbling. There is a clear yellow band in the middle of the ventral and pectoral fins. The upper jaw barbels are white on the under side; those of the lower jaw are reddish. In old age all the fins acquire reddish borders.

Next to the Sturgeon, the *Silurus glanis* is the largest and heaviest fresh-water fish in Europe. It is often from six to nine or ten feet long, and occasionally reaches a length of thirteen feet. In the Danube it often attains a weight of four hundred to five hundred pounds, and in South Russia may exceed six hundred pounds. With age it increases chiefly in circumference, and sometimes is as much as two men can span.

It is found in rivers and lakes, and prefers quiet depths with a muddy bottom, where it lies concealed under roots of trees, awaiting prey. The barbels are used to attract fishes; but if fishes are scarce it will eat frogs, crayfish, water-birds, or, in fact, anything that lives in the water or comes into it. It will seize on swimming ducks or wading geese; and Heckel and Kner mention that a poodle and the remains of a boy have been found in the stomachs of old fish.

The Wels has an evil reputation along the Danube, and on the Murtner See a superstitious belief prevailed that a fisherman always died when a Silurus was captured, a legend which may have arisen from the dangerous wounds which it is able to inflict with the sharp bony ray of the pectoral fin. In stormy weather it comes to the surface of the water, and is hence looked upon as a weather prophet.

It spawns in June and July, when it rises to the banks to lay its eggs on water plants. The eggs are pale yellow, about three millimètres in diameter, and, according to Dr. Benecke, 60,000 may occur in a fish of four pounds. The young are extruded from the eggs in eight to fourteen days. In the first year the weight may be as much as one pound and a half, and in the second year may reach three pounds, though it is usually less. According to Hungarian fishermen, it lives for ten to twelve years, but probably has a much longer life. The flesh is white, flaky, and well flavoured. When young, the fish is fat, and is then used for food; but in old age it becomes tough, and is not much prized. The fat is used for dressing leather. The swim-bladder is treated, like that of the Sturgeon, for the manufacture of isinglass and glue. It is not easily caught, for, living at the bottom, the nets often pass over it; and a large fish is powerful enough to break such nets as are likely to be used.

During the spawning-time old fishes may be taken at night, but young examples are captured with the rod and line. In Russia it is salted, and exported from Astrakhan.

It is distributed in Europe to the east of the Rhine, in which it is rare. In the Frische-Haff, specimens twenty centimètres long are abundant, but examples one mètre long are uncommon. The largest individuals are caught in lakes. There are some local varieties in Continental rivers, such as in the Maros and Szamos; and Heckel and Kner mention a beautifully-coloured, spotless, yellowish fish with a darker back, which is seen in the Vienna fish-market.

FAMILY: CYPRINIDÆ.   GROUP: CYPRININA.
GENUS: **Cyprinus** (ARTEDI).

The Cyprinoid family of fishes includes more than one hundred genera, which are widely distributed in the fresh waters of the Old World and of North America. Much as they vary in aspect and minor details of organisation, they have many important internal characteristics in common. The margin of the upper jaw is formed by the pre-maxillary bones, which in mammals carry the incisor teeth. But in these fishes the mouth is toothless; and though teeth exist, they are arranged in one, two, or three series upon the pharyngeal bones in the throat, which are parallel to the branchial arches. The stomach, too, has no blind sac, and the intestine is without pyloric appendages. The air-bladder is usually large, and divided by a constriction into anterior and posterior portions; though in the tribe called *Cobitidina* by Günther the air-bladder is small, commonly divided into right and left portions, more or less completely contained in bony capsules, which are connected with the head and early vertebræ. A small group of East Indian fishes of the genus Homaloptera has no air-bladder at all. As a rule, the head is naked, and the body covered with scales, though there are some exceptions, sufficient to show that the scales are not a very important family character.

The family is divided into large groups by the condition of the air-bladder, the number and forms of the pharyngeal teeth, the length of the anal fin, the characters and position of the dorsal fin, the sharpness or roundness of the abdominal margin, and the position of the lateral line. On these differences of structure Dr. Günther founds fourteen subdivisions of the family, most of which comprise many genera. The groups which occur in the fresh waters of Europe are the *Cyprinina*, which includes the genera Cyprinus, Carassius, Barbus, Aulopyge, and Gobio; the *Leuciscina*, including Leuciscus, Paraphoxinus, Tinca, and Chondrostoma; the *Rhodeina*, represented by Rhodeus; the *Abramidina* by Abramis, Aspius, Alburnus, Leucaspius, and Pelecus; and the *Cobitidina* by Misgurnus, Nemachilus, and Cobitis. Thus there are five tribes of Cyprinoid fishes in Europe, and eighteen genera.

The *Cyprinina*, as represented in Europe, comprise fishes with the pharyngeal teeth in three series. They have the dorsal fin opposite to the ventral fin: the anal fin is short, and always consists of from five to seven-branched rays. The lateral line runs in the middle of the tail, and the abdomen is rounded from side to side.

In this group the genus Cyprinus is conspicuous in having four barbels, with a rounded, rather blunt snout and narrow mouth, which opens anteriorly. The long dorsal fin has a serrated bony ray, and the anal fin is short. The pharyngeal teeth have a grinding character, and are arranged in the series 3· 1· 1·—1· 1· 3. The genus is limited to Europe and Asia, and is found only in temperate regions.

## Cyprinus carpio (LINNÆUS).—The Carp.

D. 3—4/17—22, P. 1/15—16, V. 2/8—9, A. 3/5, C. 17—19.
Scales 5—6/35—39, 5—6.

In this variable species (Fig. 32) typical forms often have the height of the body equal to nearly one-half the length, yet the shape is elongated, laterally compressed, and the body has its greatest breadth at the operculum, where the

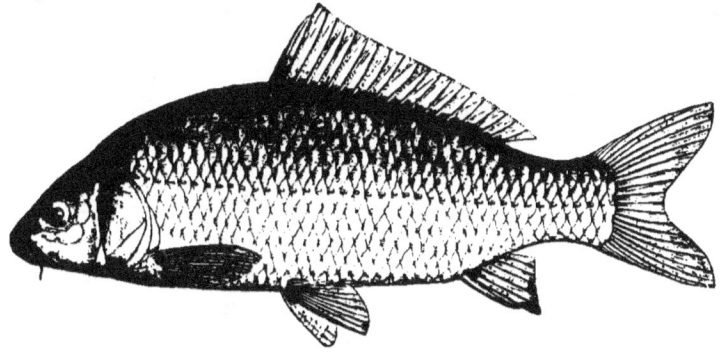

Fig. 32.—CYPRINUS CARPIO (LINNÆUS).

thickness is nearly half the height of the fish. The snout is blunt, the nose thick, and the forehead and profile of the back form a low bow, which rises uniformly to the dorsal fin. The length of the head varies with age; usually it is one-quarter of the length of the body, but in old individuals the head is nearly one-fifth the length of the fish. The diameter of the eye is about one-sixth the length of the head, and, as in so many allied fishes, is about twice its diameter from the snout; it may be three times its diameter from the other eye, and has an elevated position over the mouth. The angle of the mouth is so slightly oblique as to be almost horizontal, and the gape scarcely reaches back to the anterior nasal opening. The yellow or red barbels at the corners of the mouth are always longer than the olive barbels of the upper jaw; but both are variable, sometimes unsymmetrical, and frequently absent. The lips are thick and fleshy; the lower lip is somewhat the shorter; the tongue is smooth. The palate is covered with a white and very sensitive skin (Carp's tongue). The nostrils of each side are separated by a projecting flap of skin; the smaller hinder nostril is near to the eye.

The beginning of the long dorsal fin is immediately over the ventral fin. Both commence in front of the middle of the length. The base of the dorsal fin measures one-third of the length of the fish. Its greatest height is about one-third of its length; but the height diminishes posteriorly, so that the last rays are not half as long as the earliest. The first three or four rays, which are undivided, are very short, and may be only one-third of the height of the earliest-jointed ray. The third spinous ray, which is more flexible towards the summit, carries on its hinder border on each side of a median furrow a row of eleven to fourteen denticles, which increase in size from below upward, and have their points directed downward.

The anal fin is higher than long; its termination is opposite to the termination of the dorsal fin. Its bony ray (whose curved extremity is often broken away in old fishes) is like those of the dorsal fin. The longest rays of the ventral fin are equal to those of the anal, and are never long enough to reach so far back as the latter. The somewhat elongated pectoral almost reaches as far back as the ventral. The caudal fin has even lobes, with rounded points. The longest caudal rays never exceed the length of the head.

The scales are large, thick, and make an approach to a four or five-sided outline. The largest are in the anterior part of the sides; their diameter is equal to about one and a half times the diameter of the eye. Their free, thick margin is often marked with a fan-like ornament. The lateral line extends concavely, or nearly straight between the upper angle of the operculum and the middle of the tail, and varies a little in position with age, sex, and condition of the individual. In the lateral line an oblique or slightly curved tube traverses the middle portion of each scale, which is similar in form to the adjacent scales on the sides. The cephalic canals are marked on the sub-orbital ring, where they are furnished with numerous tubes and pores, which have thick walls, and extend along the branch of the pre-operculum. All the branchial arches are covered with rake-teeth of nearly equal length, and compressed sword shape, with their points turned inwards, and their edges serrated.

The back in its darkest parts is bluish-green; the sides are yellowish and greenish, and shade into the darker colour of the back. The abdomen is whitish. Blackish spots often extend down from the lateral line. The hinder edge of every scale is bordered with black, so as to form a dark net-work over the fish; or the pigment is sometimes more developed in the middle of each scale, and then gives the effect of sub-parallel longitudinal black lines. The lips are yellow. The dorsal fin is grey. The pectoral, ventral, and caudal fins are violet. The anal fin is reddish-brown, with orange rays. The iris is golden. All these colours vary with age, nutrition, state of the water, season of the year, and other conditions of existence. Occasionally varieties are found with unsymmetrical colouring, and a fish may sometimes show glittering golden stripes on one side of the body and pale steel-blue on the other. Sometimes typical Carp are black, bluish, green, red, golden, silvery, and even white. Dr. Fatio records that he has kept Carp in confinement which, originally green or golden, became colourless in an opaque vase.

The alimentary canal is twice as long as the body. It has no pyloric appendages. The air-bladder is divided into two parts by a transverse constriction. The hinder part, which is the smaller and longer, ends in a point. From the hinder portion, close to the constriction, the air-canal, or trachea, is given off, which at first is rather wide, and has its outlet on the hinder part of the gullet. The base of the anterior part of the air-bladder, which is somewhat bi-lobed, is connected on each side with the auditory bones.

The rate of growth of the Carp is rapid, but varies with the food, and when six years old the fish may vary in weight from four to ten pounds. After this age the growth is slower, and the animal increases more rapidly in height than in length, so that there is no ratio between the length of a fish and its weight.

In Switzerland it rarely attains a length of three feet, and a weight of twenty to twenty-five pounds. In France the Carp are usually smaller. Large fishes are bred in the department of Oise, where they reach a weight of eighteen pounds in twenty years, and sell for twenty-four francs each. Heckel and Kner mention Austrian Carp, three to four feet long, weighing thirty-five to forty pounds, and quote Bloch as authority for a Carp caught at Bischoffshausen, near Frankfort-on-the-Oder, in 1711, which weighed seventy pounds. A Carp was taken near Petersfield, in England, which weighed twenty-four and a half pounds, and had scales as large as florins.

The Carp prefers tranquil water, with a soft muddy bottom, in which it can dig with its head and find food. It is essentially a vegetable feeder, but subsists upon confervæ, young shoots, water-plants, decomposing plant remains, mud rich in organisms, larvæ, worms, and insects, and becomes fat wherever the droppings of animals, especially of sheep, occur. Like many other fishes, it feeds most frequently before the spawning season.

In winter, as soon as the water begins to freeze, the Carp explore the bottom of the pond or lake, and finding the deepest place, excavate holes, and here, often pressed tight together, they hybernate for the winter, necessarily taking no nourishment, and undergoing very little emaciation. In warm seasons activity returns to them in May, but more generally in June in Germany. When activity is restored the fishes seek places which abound in water-plants, or reeds, or flooded meadows, and soon begin to deposit their eggs. Von Siebold mentions that at spawning-time the males develop white spots, like warts, on the skull, cheeks, and inner side of the pectoral fin. Two or three males may attend upon one female. The males become greatly excited, the French expression, "saut de carpe," or somersault, defining the character of their movements, but the motion is a spring executed by folding the body, and sometimes the jump is high, sometimes broad. The female is meanwhile perfectly calm, and remains under the water-plants depositing the eggs in the stream. As soon as she quits the spot one of the acrobatic males darts to the eggs, and fecundates them, clouding the water with a milky tinge. The eggs are small—one third of a millimètre—and of a greenish tint. Their number varies with the age of the fish, and has been stated at 750,000 in an individual weighing ten pounds. The fish begins breeding at the age of three years, and at the age of five years will lay 300,000 eggs. The fecundated eggs hatch in about eight days. The growth is rapid, so that in the third year in the best ponds a Carp weighs from half a pound to a pound. In pond culture it is transferred when about six or eight inches long, and in a pond with a clay bottom may then in three years attain a weight of four or five pounds.

In a state of nature the Carp lives for twelve or fourteen years, but survives much longer in confinement, although subject to many sicknesses, deformities, and wonderful variations. The Carp of Fontainebleau, Chantilly, and other royal residences in France have been supposed to grow more than a hundred years old; but Blanchard disposes of these fables by observing that in 1789, 1830, and 1848 the fish-ponds fell under the control of the sovereign people, who loved the Carp too well to permit great size or age to be an obstacle to its ministering to their pleasure.

The hearing of the Carp is excellent, and there are many examples of their answering a call; and it moves by hearing even when it cannot see. It makes an audible sound in eating and in swallowing air.

It is tenacious of life, and can live for many hours out of water if kept in damp grass or moss.

The flesh of the Carp is esteemed, especially in Germany and in France, and in both countries a great industry is carried on in breeding it for the market; but in Switzerland, owing, perhaps, to the abundance of Trout, it is held in less favour. In England, the Carp of the Thames, when large, are remarkable for richness and delicacy of flavour. The flavour of the flesh is best from autumn to spring, and that of the river and lake Carp is firmer and more prized than that of the pond Carp, which, however, may be improved by being placed for a week in better water. It is best boiled, and eaten with melted butter and pickled walnut. It is usually taken with the net, the weir basket, and with the rod, when worms or cooked egg, or pieces of bread, cheese, or grain, may be used as bait. In Holland the Carp is often kept alive in cellars for months, and fattened for the table on bread and milk. Dr. Badham tells us that the Jews of Constantinople, being forbidden by the Levitical law to eat caviare, discovered that the large roes of the Carp made an excellent substitute; and, the fish being scaly, it was perfectly orthodox to partake of its roe. But the roe was removed from the living fish, and it is recorded that Samuel Full somewhat modified the Eastern practice. He cut open male and female Carp, entirely removed the milts and ovaries, for which he substituted pieces of felt, and then united the wounds by suture, and placed the patients in their pond. They not only recovered strength, but grew rapidly, became fat and heavy, and when they were cooked for table the flavour proved more delicate than that of Carp which had not been "felted." Occasionally sexual varieties are produced, some Carp being sterile, or neuters, and more rarely some are hermaphrodites.

Among the diseases to which this fish is subject, one of the most obvious is a mossy growth on the head. When the water in which it lives is too warm, vesicles like small-pox develop beneath the scales. It is liable to ulcerations of the liver; to visceral obstructions, produced by feeding too freely on chickweed; it is affected by carbuncles, internal and external parasites, a skin disease ending in blindness, and various epidemic diseases.

The varieties are numerous. One of them is known as the Mirror Carp, or Carp-king. It differs in the arrangement of the scales, which are three or four times the ordinary size, and do not cover the body, but form one to three rows along the sides, with naked skin between. This variety is known only in ponds, where it breeds freely, and it is recorded that a female King-Carp produced more than one thousand young when fertilised by an ordinary Carp. When a King-Carp has a row of large scales down the back, as well as scales in the lateral line, it is termed a Saddle-Carp. These scales readily fall off with age, and when scaleless the skin becomes dark and leathery, and the fish is known in Germany as the Leather-Carp.

Three varieties have been thought sufficiently important in Germany to receive distinctive names. They are the *Cyprinus acuminatus* of Heckel,

*C. hungaricus* of Hecke', and *C. regina* of Bonaparte. As these varieties differ conspicuously in form of the body, we offer a brief description of each.

The variety *Cyprinus acuminatus* (Fig. 33) is especially characteristic of the Valley of the Danube, where it is more common than *C. carpio*. It

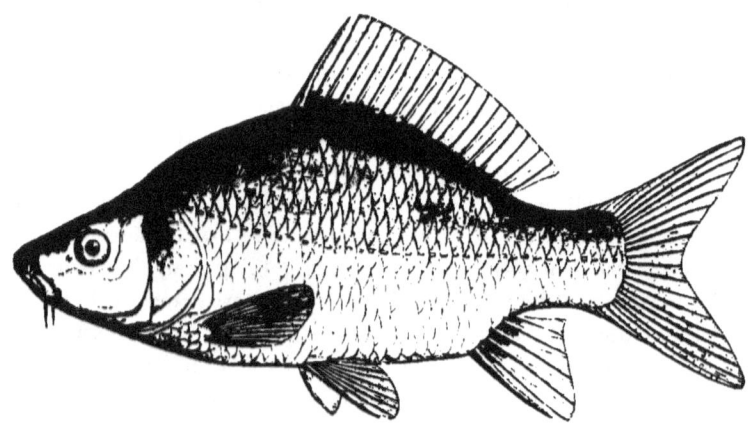

Fig. 33. CYPRINUS ACUMINATUS (HECKEL).

is distinguished by having a more pointed head, with the profile rising more sharply to the back. The body is two and a half to two and three-quarter times as long as high, and the breadth in the opercular region is

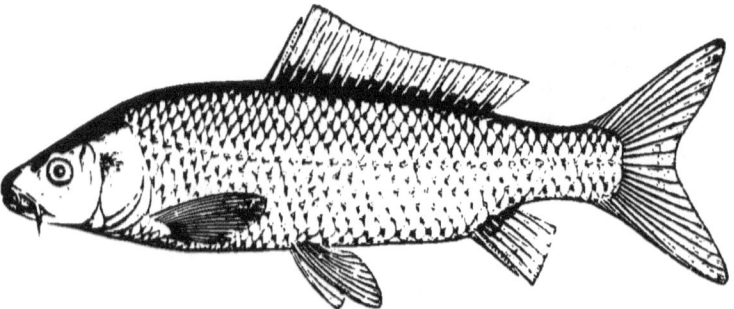

Fig. 34. CYPRINUS HUNGARICUS (HECKEL). (YOUNG INDIVIDUAL.)

more than half the length of the head, but much less than half the height of the body. The forehead is smaller than in the type already described.

The gape of the mouth is more inclined, and rather wider; the upper fleshy lip projects somewhat beyond the lower lip. The barbels of the upper jaw are only half as long as the true barbels, which reach as far as the orbit of the eye when laid back.

# CYPRINUS HUNGARICUS.

The dorsal fin begins half-way down the body, and is two and a third to two and a half times as long as high. Its base is nearly always less than one-third of the length of the fish. The anal fin may be twice as deep as long, though commonly less; its bony ray is relatively strong. The caudal fin is more developed than in the *Cyprinus carpio*, and has the lobes more pointed. This variety is also found in the Platten See and the Neusiedler See.

The *Cyprinus hungaricus* (Fig 34) is a variety with a more elongated and more cylindrical body. The height of the body in old specimens is equal to the length of the head, but usually it is a quarter of the length of the

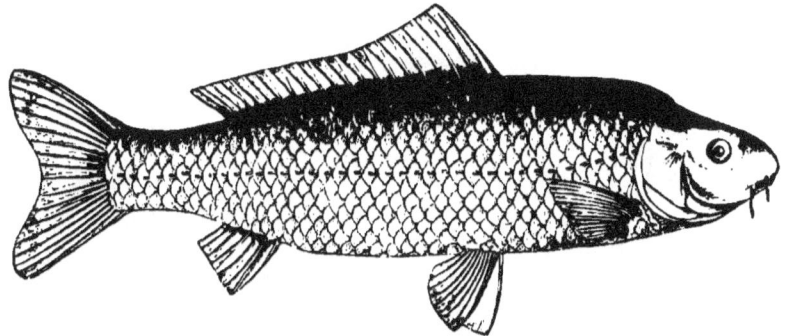

Fig. 35.—CYPRINUS HUNGARICUS (HECKEL). (OLD EXAMPLE.)

body, or rather less. The thickness is never less than half the height of the body, and may increase to four-fifths in old age (Fig. 35). The eye is larger than in *C. carpio*, its diameter being one-fifth of the length of the head, while in *C. carpio* it is one-sixth of the length. The position of the mouth is relatively lower, and the gape only curves slightly downward. Both pairs of barbels are longer than in *C. carpio*. The ventral outline is nearly horizontal between the pectoral and anal fins, while the dorsal outline is a very flat arch on the back, becoming in old age almost horizontal to the end of the dorsal fin. The caudal fin is more deeply forked, and has more pointed lobes than the typical forms of the species. The pharyngeal teeth (Fig. 36) are relatively stronger than in the type.

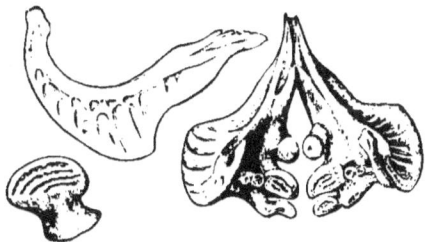

Fig. 36.—CYPRINUS CARPIO (VAR. HUNGARICUS).
PHARYNGEAL TEETH.

In the Neusiedler See it reaches a weight of from twenty to thirty pounds.

It fattens with age, and the body then becomes nearly cylindrical. It is brought to the Vienna market under the name of Lake Carp, but its flesh is less prized than that of the Danube Carp.

The *Cyprinus regina* was described by Bonaparte from Italy, and afterwards identified by Heckel as occurring in the Treviso; but Canestrini follows Günther in regarding this, like the other types we have named, as only a variety of *C. carpio*. The body is less elongated (Fig. 37) than in *C. hungaricus*, and is more compressed, so that while the back is angular in *C.*

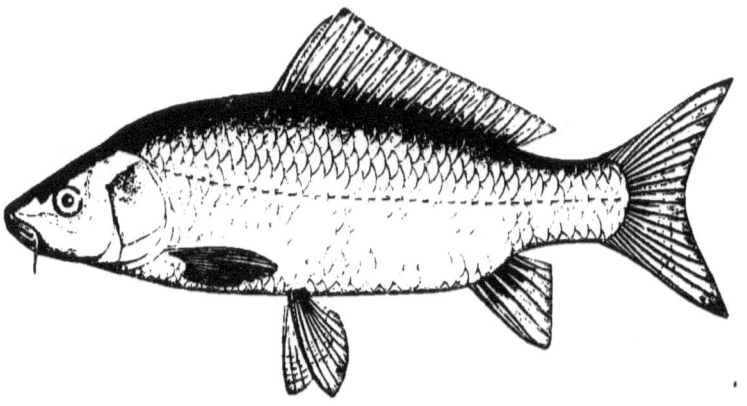

Fig. 37.—CYPRINUS REGINA (BONAPARTE).

*regina*, in *C. hungaricus* it is broad. The scales are larger than in that variety—the largest above the lateral line exceed in size the diameter of the eye.

A broad variety of *Cyprinus carpio*, found in the muddy rivers and ponds of Southern Russia, is known as the *C. nordmanni* (C. and V.). It reaches a weight of two and a half pounds.

## Hybrid between the Common Carp and the Crucian Carp.

D. 4 18—20, A. 3/5—6, V. 2 8, P. 1 17, C. 19.

Wherever the *Cyprinus carpio* and *Carassius vulgaris* are kept in a domesticated state this hybrid is met with; but its hybrid nature was not at first recognised, and is stated to have been first proved by Dybowski. Indeed, the celebrated ichthyologists Heckel and Kner regarded it as the type of a distinct genus, and described the form as a species, under the name *Carpio kollarii*; and on this account it may be interesting, from its bearing upon evolution of species, to give some detailed description of its characteristics (Fig. 38).

CARPIO KOLLARII.                                    103

The body is compressed and high, two and a half times as long as high, with a greatly arched back, so that its form closely resembles that of the Crucian Carp. The head is about one-fifth the total length, and is longer than high. The diameter of the eye is one-fifth of the length of the head. The angle of the oblique mouth does not reach quite as far back as the nares, and the centre of the mouth lies in a plane that would pass through the orbit and the middle of the body. It resembles the Carp in having two pairs of barbels. They may be short, and are sometimes only one line long. The

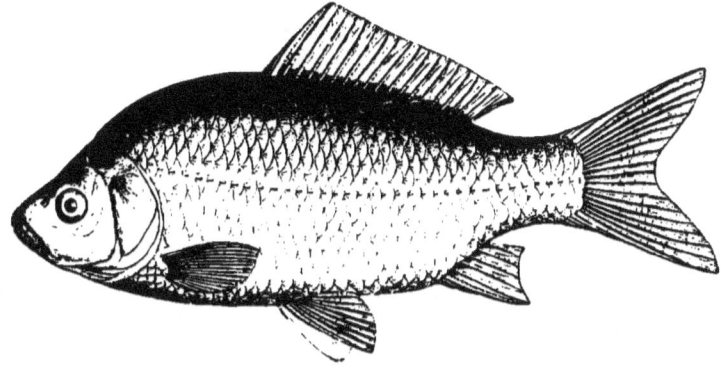

Fig. 38.   CARPIO KOLLARII.

contour of the forehead ascends from the rounded snout in a nearly straight or slightly concave line to the back, where the highest curve is passed before the commencement of the dorsal fin. The large dorsal fin is opposite to the ventral fin; it is more than twice as long as high, and its base is rather more than one-third of the length of the fish. The fourth bony ray ends in a curved point, and is of the same height as the succeeding jointed rays, though towards the end of the fin the height of the jointed rays diminishes one half. The anal fin is higher than long, and reaches back towards the caudal fin rather farther than does the dorsal. As in *Cyprinus carpio*, its bony ray is stronger than that of the dorsal.

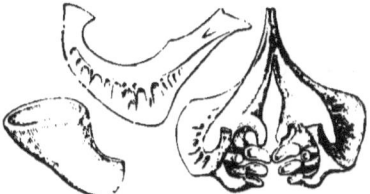

Fig. 39.— PHARYNGEAL TEETH OF CARPIO KOLLARII.

The ventral fins do not reach back to the vent, and the pectoral fins do not reach back to the ventrals. The tail is evenly lobed.

The scales are like those of *Cyprinus carpio*; the largest only slightly exceed the diameter of the eye.

The crowns of the middle pharyngeal teeth (Fig. 39) are less thick than in *Cyprinus carpio*, but are quite as flat, and have only one furrow. The fifth tooth, which forms a second row, is very small.

The colour is dark green on the back, yellowish-green on the sides, yellow on the belly. All the fins are black, though the dorsal fin is sometimes brownish-grey, with black dots. This variety rarely exceeds a length of eight inches; when young its form is relatively more elongated, and the height of the arch of the back increases with age.

It is remarkable that while Heckel constituted a genus for this fish, he mentions that the fishermen of the Neusiedler See have always regarded it as a bastard of the Crucian Carp and the Lake Carp.

## Genus: **Carassius** (Nilsson).

The genus Carassius differs from Cyprinus in having no barbels, and in having the pharyngeal teeth arranged in a single series of four on each side.

Three species have been described, which are known as the Crucian Carp, the Carp of Salonica, and the Gold Fish. The first and last named are both extremely variable, and the domesticated varieties of Gold Fish are remarkable enough to furnish characters for a multitude of genera if they could become constant.

## Carassius vulgaris (Nilsson).—The Crucian Carp.

D. 3/15—18, A. 3/5—6, P. 13—14, C. 19, V. 2/7.   Scales 7—8/5—6.

The Crucian Carp (Fig. 40) is a variable species, so that it is necessary, first of all, to describe the type, in which the back ascends in a sharp arch to the dorsal fin. The head is usually less than one-third of the length of the body, and is shorter and blunter than in the Carp. The height of the body is half the length of the fish, and the greatest breadth is nearly one-third of the height, so that the aspect of the fish is a form more than usually high and compressed. The head is only a little less high than long. The eye is rather small, not more than its own diameter from the snout, and twice its diameter from the other eye; the iris is silvery, with a coppery or golden border.

The mouth is inclined obliquely downward, and its corners reach as far back as the nares.

The dorsal fin commences opposite to the ventral, rather less than half-way down the body, and is commonly rather less than twice as long as high, but its height never exceeds the length of the head, and its base measures about one-third of the total length of the fish. The outline is more or less truncated, and the last ray is half the length of the longest. The anal fin commences opposite to the termination of the dorsal fin, and the extremities of its rays extend back to the beginning of the tail. It is deeper than long. The first jointed ray, which is the longest, is equal to the measurement from the point of the operculum to the pupil of the eye, but the base of the fin is scarcely longer than

Fig. 40.— CARASSIUS VULGARIS (NILSSON).

half the length of the head. The ventral fins are well shaped, somewhat pointed, and reach back to the vent, and the relatively small pectoral fins extend back to the base of the ventrals.

The caudal fin is commonly weak, evenly-lobed, and its margin is often concave. Its terminal rays are about as long as the head.

The scales have an irregular radiate pattern, and are more than half overlapped, so that their free edges appear higher than wide, though the height and breadth are equal. They are arranged in fourteen or fifteen longitudinal rows, usually with thirty or thirty-five scales in each row. The scales are marked with simple lines, which are furnished with regular elevated points. This, according to Blanchard, is one of the most distinctive characters of the fish. The attached part of the scale is notched and marked with canals, which converge to a sub-central point.

The lateral line is nearly straight, and nearly in the middle of the side, with cylindrical mucus-canals of moderate size. The mucus-canals of the head are well developed, and unite with a branch from the lower orbital border, which extends towards the upper lip and has numerous invisible pores. The pseudo-branchiæ are invisible. The first gill-arch has long, close-set rake-teeth, which are toothed on their inner border. The other gill-arches are smaller, but similarly toothed. The vertebræ number nineteen in the abdominal region, and thirteen in the tail. The alimentary canal is similar to that of the Carp, but is relatively shorter, being from one and a half to one and three-quarter times the length of the body.

The colour varies with habitat and with the season of the year. But typically the head is olive-green above, with brassy-yellow sides. The back is a darker greenish-brown, with the brassy-yellow of the cheeks prolonged down the sides of the body. The belly is reddish-white. The pectoral, ventral, and anal fins are reddish, and the other fins yellow, with grey borders.

The Crucian Carp is seldom larger than six inches long, and one pound and a half in weight. A fish of two pounds is exceptionally large.

For the greater part of the year it frequents the bottom of the water, preferring ponds and lakes, and never moving far from its native place. It burrows in the mud for worms and larvæ, and swallows some mud with its food. According to Von Siebold, it sometimes spawns as early as the end of May, but most observers record the spawning about midsummer, when it comes often to the surface of the water.

At spawning-time the fishes collect in shallow places where the bottom is covered with plants, and assemble in great numbers, and then, like the Carp, are seen continually smacking their lips on the surface of the water. The eggs, which are attached to the water-plants, number, according to Benecke, from 100,000 to 300,000.

This species is hardy and tenacious of life, and increases rapidly; but under ordinary conditions of nutrition grows slowly. In ponds where animal excreta abound, or where the fishes are fed with burnt malt, they become fatter and better flavoured; and in two years may reach a weight of three-quarters of a pound. They live for six or seven years. Their flesh is not usually valued for food, though it does not acquire a muddy flavour. It is said to disagree with some people, but in certain parts of Sweden is looked upon rather as a dainty. Crucian Carp are kept principally to feed predatory fishes, which are preserved in ponds. Notwithstanding the wide gill-aperture, they will live for hours out of water, and can be easily packed or transported in snow or damp leaves.

They are captured with nets of all kinds.

In England this Carp is found in the Thames and neighbouring ponds to the west of London, where it spawns in April or May. It is widely distributed over the Continent from Scandinavia, where it spawns in June, southward to Palermo, in Sicily; and from France and Belgium eastward to Siberia, where in some districts, according to Blanchard, it is an important food product. In Austria it is found in the Theiss and other tributaries of the Danube, as well as in the Platten See and Neusiedler See.

The synonymy of this species is somewhat difficult.

The Crucian Carp which we have described is recognised as such by all writers, but the varieties have been variously named. The Prussian Carp is

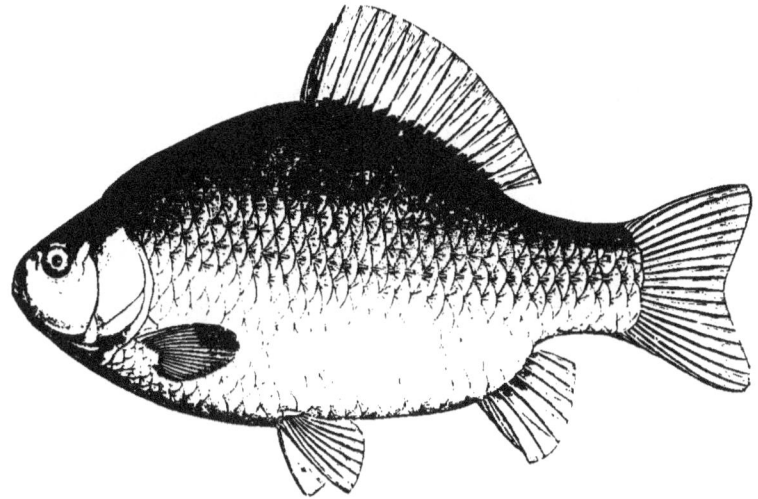

Fig. 41—CARASSIUS MOLES (AGASSIZ).

regarded as a distinct form, and by most writers is referred to the *Cyprinus gibelio* of Bloch (Fig. 42). Dr. Günther regards this as a mere variety of the *Carassius vulgaris*, following the views of Eckström, and he includes with this type the *Cyprinus moles* of Agassiz (Fig. 41), which the Austrian naturalists, Heckel and Kner, rank as a species, distinct from the *Cyprinus gibelio*. The *Carassius oblongus* of Heckel and Kner (Fig. 44) is also classed by Von Siebold and Günther as a variety of the Prussian Carp. Hence naturalists are divided as to the importance to be attached to the characters which distinguish these fishes, though there is very little difference of opinion as to the propriety of separating them.

*Carassius vulgaris, var. gibelio* (Fig. 42).—In this fish the lateral compression and height of the body are less than in the type, the height of the body being less than half the length. The head is more massive, with a blunter snout. The aperture of the jaws is nearly vertical, and its angle scarcely reaches so far back as the eye. The cheek is marked with asperities, and is free from the pits and depressions in front of the eye in the type. The operculum has a rough, irregularly furrowed surface. Fig. 43 shows the pharyngeal teeth.

According to Blanchard, the scales are larger than in the Crucian Carp, and though they have the same form, the concentric striæ are farther apart,

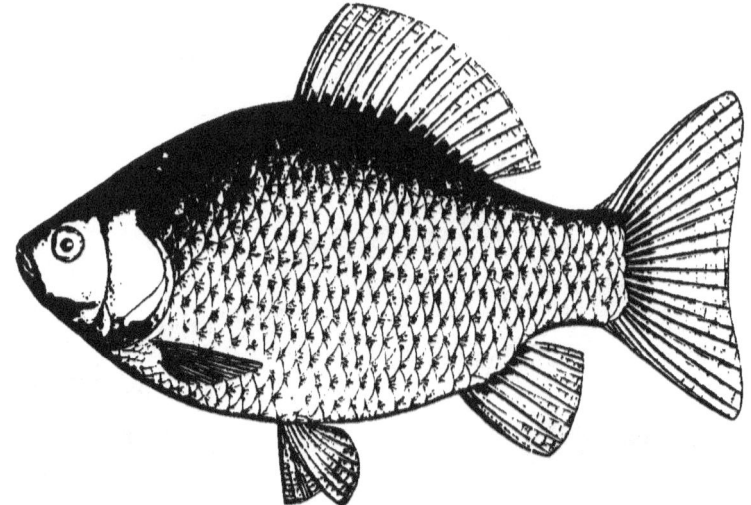

Fig. 42.—CARASSIUS GIBELIO (NILSSON).

and the canals are less numerous. The basal border of the scale is marked with very slight curvatures, and has not the festooned character of scale seen in the Crucian Carp. In Alsace and Lorraine these differences appear to be constant. The variation in the fins is less conspicuous. The dorsal fin is less elevated, and is stated by Blanchard to have twenty-one branched rays, as in some examples from Central Europe, though Austrian specimens have only fourteen to sixteen. The anal fin has seven branched rays in the specimens described by Blanchard, and five to six in the German and Austrian examples. The colour is more uniform and less bright than in the Crucian Carp, the back being blackish-green, changing to bluish, and the belly brownish-yellow, with a golden tint. The pectoral and ventral fins are reddish-brown at their bases; the other fins are black. The iris is greenish-black.

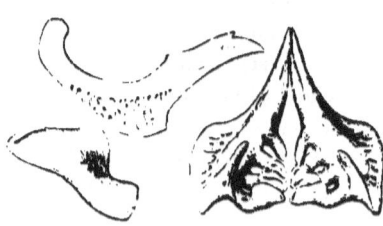

Fig. 43.—PHARYNGEAL TEETH OF CARASSIUS GIBELIO.

That all these differences are due to food and habitat, so that the same stock in a lake develops into the Crucian Carp, and in a pond into the Prussian Carp, would seem highly probable when we notice the close similarity of fin structure and scales. Its distribution is like that of the *Carassius vulgaris*. The Swedes term it *Damm-Ruda*, or Pond Crucian, to distinguish it from the Common Crucian, or Lake Crucian (*Sjö-Ruda*).

*Carassius vulgaris, var. oblongus* (Fig. 44).—This was regarded as a distinct species by Heckel and Kner, on account of its more elongated form, flatter back, large thick head, larger eye, and longer and more pointed tail. The length of the body is two and-two-third times as much as its height. Its

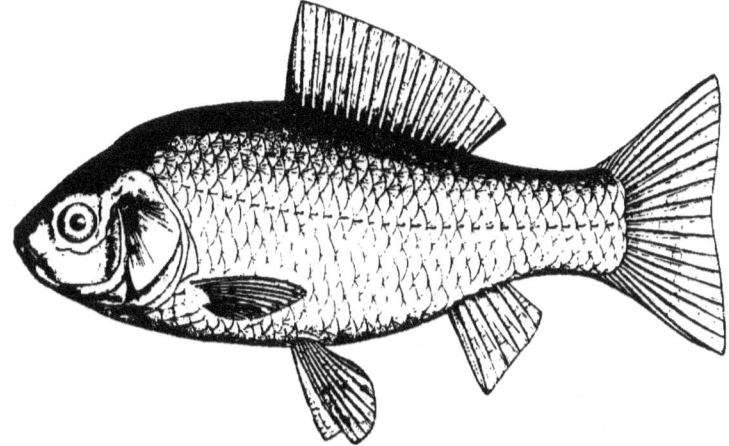

Fig. 44.—CARASSIUS OBLONGUS (HECKEL AND KNER).

greatest breadth at the operculum is one-half of the height of the body. The eye is larger than in any other variety, its diameter being one-fourth of the length of the head; its own diameter separates it from the extremity of the snout, and there is an interspace of only one and a half times the orbital diameter between the eyes. The snout is more pointed than in other varieties, and the contour from the mouth to the back is very slightly arched. The dorsal fin commences in the anterior half of the length, and extends for fully one-third of the length of the body. Its bony ray is conspicuously serrated on its hinder border, rather feeble, and as long as the jointed rays which succeed it. The anal fin is deeper than long, and its commencement is anterior to the termination of the dorsal. The ventral fins are placed half-way down the body behind the commencement of the dorsal fin; they are somewhat elongated, and reach the vent. The pectoral fins have also an elongated aspect, and they reach back to the ventrals.

The extremities of the caudal fin are more pointed than in other varieties, and the terminal caudal rays are quite as long as the head.

The longest specimen of this variety found in Austria measures rather less than six inches. It is known in that part of Europe from Galicia, Lemberg, the Styrian rivers, and the Carpathians. The British Museum possesses specimens obtained by Professor Von Siebold from Munich and Eastern Prussia; and Dr. Günther records it from Norway and the Baltic.

There is a variety of the Crucian Carp, which Heckel regarded as a distinct species, and named *Carassius humilis*. It is about three inches long, and known only from a lake at Palermo.

The back is less elevated and the head is thicker; the scales are larger, the caudal fin is longer, and the abdomen silver-white. The pectoral fins are more elongated, and when laid back, cover the bases of the ventral fins. The caudal fin is one-quarter of the total length of the fish, and is deeply notched. There are four compressed pharyngeal teeth in a row; the hindermost have their crowns broader than the bases.

The back of the fish is black, becoming silver-white on sides and abdomen, like that variety of the Golden Carp which is known as the Silver-fish.

## Carassius auratus (Linnæus).—The Gold Fish.

D. 19, A. 8. Scales: lat. 26, transverse 12.

The home of the Golden Carp, popularly known as Gold Fish, is in China and its islands, and Japan. There is no record of its introduction into Europe, but it is commonly stated to have reached England in the seventeenth century. It is now domesticated in nearly all civilised parts of the world, and under varied conditions of existence has developed varieties even more numerous and remarkable than those which are met with in its original home. Dr. Günther states that in Europe the form of the body is ordinarily more elevated than in Asia, and the fish usually has a larger number of longitudinal rows of scales above the lateral line.

It is the only fish which is truly domesticated. The form of the body is of moderate length, thick, so as sometimes to be well rounded, with a less elevation than is seen in the other species of the genus. The head is massive and obtuse, with the jaws equal and the operculum striated. There are twenty-five to twenty-eight rows of large round scales in the length of the body. There are usually four rows of scales above the lateral line, and seven rows below. The dorsal fin has the rays branched, and varying in number from fifteen to eighteen, defended in front by a large, curved, denticulated spine. The anal fin has three bony rays, followed by five soft branched rays. The pectoral fin is rather large. The caudal is moderately emarginate. The pharyngeal teeth form a row of four on each side, as in *Carassius vulgaris*.

Dr. Günther finds the chief variations in the domestic races produced in the vertebral column, which may be variously deformed so as to shorten or elongate the back; secondly, in the fins; thirdly, in the eyes. The fins vary greatly in development. Sometimes the dorsal fin is reduced to a serrated ray and a few soft rays; or it may be absent, without the other fins being in any way modified. Occasionally the loss of the dorsal fin is associated with the development of a double anal spine, and Fatio mentions that the anal fin itself may be double. The caudal fin, on the other hand, is liable to increase, not always in compensation for loss of the dorsal fin, for that fin is sometimes present with a double tail, or three-lobed, or even four-lobed, caudal fin; while in one variety, known as the Telescope Fish, the development of a lobed tail is associated with large eyes which protrude on pedicles, like the eyes of stalk-eyed crustacea.

The legend runs that the Gold Fish was introduced into France as a present to Madame de Pompadour. On this circumstance Badham moralises in the following terms:—"The Pompadour's reign of beauty was soon over, but her lubric rivals have maintained the breed, spread their conquests into distant lands, and secured to themselves hosts of admirers in every part of the civilised world." He goes on to say: "They are not, however, perfect beauties, and in symmetry of form yield the palm to the silvery Bleak, and darting Dace, while not a few have personal defects, such as lame fins and goggled eyes, or the mouth and sometimes the whole body screwed to one side." The colours of the fish vary with its age. In the words of Dr. Badham: "It is at first of a dark sooty colour, and the first change is indicated by the appearance of small silvery points, which are dispersed over the scales. These points spread and deepen in tint till the whole body is encased in a spangled robe of gold. In old age its brilliancy fades, and it at last dies with the body bleached and cheeks colourless." In the Seine, Blanchard observes that the colour is soon lost, so that the fish resembles ordinary brown or greenish Carp, and Fatio states that the red colour is regained when the green fish is placed in a glass bowl. In ponds the colours may be silvery, violet, reddish-brown, bright red, and piebald with these colours. Gold Fish are raised for export in large numbers in Portugal. And Buckland refers to a German farm near Oldenburg which consists of 120 ponds, bordering on the river Hunte. They cover seven and three-quarter acres of peat land. In some of the ponds the temperature is 100° Fahr., the water being obtained from the engines of a hemp factory. The stock consists of about three thousand fish. Here they breed in the spring and summer, and sometimes there is a third brood in the autumn. The fish hatched in the spring grow to a length of one or two inches in the autumn, when they are known as Globe Fish. In a year they may be five inches long. Pennant remarks that " in China every person of fashion keeps them for amusement, either in porcelain vessels or in the small basins that decorate the courts of Chinese houses. The beauty of their colours and their lively motions give great entertainment, especially to the ladies, whose pleasures, by reason of the cruel policy of that country, are extremely limited." In Europe they oftener pass their lives in perpetual circumnavigation of small glass globes. Buckland says that they should be preserved from the heat of the sun and the cold of winter; that their water must be frequently changed, have a few sprigs of *Vallisneria* in it, and a layer of duckweed on the top; and that they should be fed on scalded vermicelli and small red worms. But in Germany their dietary includes finely-chopped meat, ants' eggs, and insect larvæ, with clotted blood and barley. They seldom grow in globes, for want, apparently, of natural food; but in ponds like those at Hampton Court attain a fine development and show remarkable varieties of colour. Dr. Badham states that he has watched the evolutions of the large shoal kept in the basin in the park at Brussels, and has seen the whole corps in close pursuit after some delinquent member, whose tail first one and then another seized upon, and bit with boneless gums, all equally eager to hunt him down and secure the *brush*. Evidences of affection between Gold Fish are recorded by Jesse. Their sense of sound is sufficiently acute to obey a familiar call. The Chinese are said to assemble them in ponds at feeding-time in this way; but in ponds where visitors feed them in Europe they presumably detect the new comer by sight; for we have noticed that a gathering never fails to greet visitors on their appearance at public gardens in which these fishes are exhibited.

In thunder-storms they sometimes die, but this is not a failing peculiar to Gold Fish. They are eaten in Switzerland; and have been introduced into the Mauritius, where they are regarded with favour as an acceptable article of diet. In the mills and manufactories of the north of England, where the steam from the engines heats the water in the engine dams, Gold Fish thrive and multiply rapidly. Yarrell regards them as by no means useless inhabitants of the dams, since they consume the refuse grease, which would otherwise accumulate on the surface of the water, and retard its cooling.

## Carassius bucephalus (Heckel).

D. 3/16, A. 3/5, P. 14, V. 9, C. 6/17/6. Scales: lat. line, 32, trans. 8/5.

This species is known only from warm springs near Salonica, in Macedonia. It closely resembles *Carassius vulgaris*, but is characterised by its thicker, blunter head, large projecting eyes, and smooth scales; and it differs from *C. humilis* in height of body, number of horizontal rows of scales, and shortness of the pectoral and caudal fins.

The profile of the head is flatter than in the Crucian Carp. The body is much thicker, and less elevated. Over the pectoral fins the thickness is equal to nearly half the height in front of the dorsal, which is about a third of the entire length.

The head is one-fourth of the entire length; it is thicker than the body, and noticeable for its broad forehead. The large wide projecting eyes are nearly one-third the length of the head. Their posterior border is behind the middle of the head, and the width of the inter-orbital space is equal to one-half the length of the head.

The mucus-canals are well developed; their pores are conspicuous to the naked eye, and situate in small papillae, which, as in most Cyprinoids, extend above and below the eye, along the lower jaw and the margin of the pre-operculum. In form and position the mouth resembles that of the Crucian Carp. The fins are similar to those of *C. vulgaris*. The dorsal fin extends between the middle of the body and the first ray of the anal fin.

The scales are not quite so large as in the Crucian Carp. The lateral line is marked by small pores, but the scales are not perforate behind the ventral fin, so that there are only about twelve to sixteen scales, with mucus pores. The scales are smooth, and have the free margin marked with from two to four diverging rays. The colour is only known from specimens preserved in spirit, when the upper part of the sides and back are black, on a ground which has a silvery lustre. The under side is a dull yellow. There is a large annular spot on the tail adjacent to the caudal fin.

Specimens are about five inches long.

## Genus: **Barbus** (Cuvier).

The Barbels form a large genus, widely distributed, and presenting considerable diversity of character. Dr. Günther remarks that a further division of the 163 species which he enumerates must appear highly desirable; yet adds that nothing would be more contrary to the idea of natural genera, since the transition is perfect from one extreme species to the other. He believes that the size of the scales, the development of the third dorsal ray, and the form of the snout are useless as generic characters, because complete series of intermediate forms are met with. The lips, too, present in the same species variations which have been regarded as of generic value; and the number and size of the barbels is so variable that it is difficult to distinguish whether a species possesses two or four of these appendages. Hence it is not easy to define this genus; and the characters given by Dr. Günther himself are not readily grasped. Barbus is characterised as having the third longest simple ray of the dorsal fin ossified, enlarged, and often serrated, with never more than nine branched rays in the dorsal fin, and these branched rays commence opposite to the root of the ventral fin. The anal fin is short and high. The eye has no fatty eyelid; the mouth is arched; the barbels may be four, two, or absent. The lateral line runs in the middle of the body to the tail. The pharyngeal teeth are in three rows, in the formula 5—3—2, or 4—3—2, usually five on one side, and four on the other (Fig. 45).

This large genus may be divided into three sections:—first, fishes with four barbels; secondly, fishes with two barbels; third, fishes without barbels. And to facilitate identification of species these groups are further divided according to the number of scales in the lateral line, which may be more than forty or less than thirty. Other characters are found in the presence or absence of pores or tubercles on the snout, and the absence or presence and condition of the third dorsal ray, which, when it exists, varies in size, and in having or wanting serrations on its margin. The fringed condition of the barbels and size of the mouth furnish other characters. But in view of the facility with which external soft parts may vary, we should

Fig. 45.—PHARYNGEAL TEETH OF BARBUS VULGARIS.

have preferred to make the primary classification of the genus upon the characters of the third dorsal ray. The greater number of the species live in the fresh waters of India and the East Indian Archipelago; but the genus is widely represented in Asia and Africa, and is not rare in Europe, though the number of species decreases westward to two in France and one in Britain.

## Barbus vulgaris (Fleming).—The Barbel.

The fin formula of this species is differently stated by almost every observer, from which we infer a greater variability of characters than is commonly admitted, or the existence in some localities of types which make a transition towards other species—

D. 3—5 8—9, A. 3—5, V. 2 7—8, P. 1 15—17, C. 19.

The Barbel (Fig. 46) has the body fully five times as long as high, nearly cylindrical, with an elongated head, and the mouth somewhat under the projecting fleshy upper lip. These fishes have been compared to pigs, on account of their small eyes and the routing, burrowing movement of the snout in taking food. The tail part of the body is much compressed laterally. In old individuals the head is equal to about one-fourth of the length of the fish. The eye, which in the young is about one-fifth the length of the head, subsequently becomes one-sixth or one-seventh of its length, and the eyes are at first nearer together than in the adult fish, in which they are separated by twice their diameter. The angle of the aperture of the mouth extends back to the nares. The prolonged fleshy nose, which thickens with age, extends over the thick upper lip.

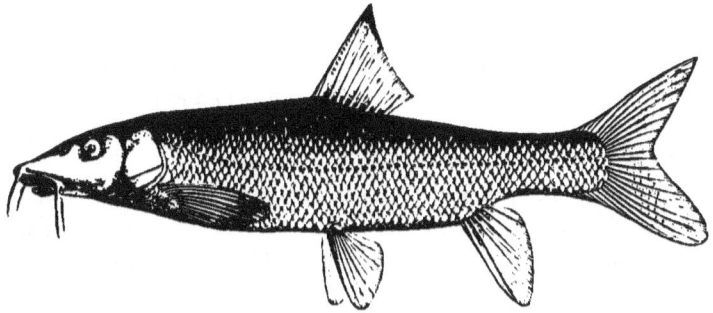

Fig. 46. BARBUS VULGARIS (FLEMING).

The barbels of the upper jaw elongate a little with age, but would always reach to the nares. The barbels of the angle of the mouth laid back would reach beneath the eye. The contour of the top of the head rises convexly to the occipital region, and there is a long flat arch from the occiput to the dorsal fin.

The short dorsal fin begins half-way down the body ; it is higher than long, and commences a little in front of the ventral fin; its bony ray is serrated posteriorly, and is longer than the other rays; the last ray is about one-third that length. This fin has been compared to the sail of a Thames barge.

The anal fin is placed far back in the middle of the hinder half of the body ; it is higher than long, but its longest ray is shorter than that of the dorsal. It is truncated behind, though its free margin is rounded. The ventral fins often have the rays as long as in the anal fin, but these fins do not reach back to the vent. The pectoral fins have rays of about the same length. The tail fin, according to Heckel and Kner, consists of seventeen jointed rays, which are supported by rather shorter pseudo-rays, to the number of ten in the upper lobe and eight in the lower lobe. The caudal fin is deeply emarginate, evenly lobed, and pointed at the extremities.

The scales cover the entire body except the pectoral region: they are numerous and small; they commonly increase in size from the fore part of the back towards the tail; the largest is scarcely as long as the eye. There may be from sixty to seventy in a line between the ear and the base of the caudal fin, according to Blanchard, though the number commonly counted in the lateral line is from fifty-five to sixty-two. There are from eleven to twelve rows of scales above the lateral line, fourteen to fifteen rows below it, though Blanchard counts thirty rows in the height of the body. The scales are longer than deep, lancet-shaped at their free margin. They are marked with radiating lines.

The lateral line is nearly straight, and in the middle of the body. The cephalic canals are not very distinct, except the branch in the sub-orbital ring, which forms many long depressions. The branch to the lower jaw has large pores. The pseudo-branchiae are free, rather large, and pectinate; the gill-rakers of the first gill-arch are spatulate and curved inward.

The colour of the back is olive-green, gradating into a paler tint on the flanks, which becomes greenish-white towards the belly. The throat and belly are pearly white. The cheeks glisten with golden tints, but the sides of the head and operculum are speckled with fine black spots, and the speckling is sometimes continued down the sides of the body. The dorsal fin is bluish; the tail has a similar tint, often with a bluish border; the other fins are red, the iris is brown.

The size of the Barbel varies with the locality. The fish is often two feet long, and weighs eight to ten pounds in the Danube, where, however, specimens have been caught weighing eighteen pounds. And Heckel and Kner mention an example taken in the Salzach in 1853 that weighed twenty-five and a half pounds. But Blanchard refers to individuals from the Volga that weighed from forty to fifty pounds.

The growth of the Barbel is rapid. The fish is tenacious of life, and lives for fifteen or twenty years. It is found throughout Europe in all rivulets, rivers, and lakes, both in low-lying districts and among mountains. It feeds on worms, small fishes, insects, mollusca, mud, and especially upon the excreta of animals; and the great length of the intestine may be taken to indicate that vegetable substances largely contribute to its sustenance. Barbels when young are reputed to live in association with Gudgeon, but as they grow larger leave their early associates, and become solitary.

The Barbel first spawns in its fourth year, in May or the beginning of June, and the fish then becomes banded, and often acquires a reddish or orange colour. The spawning period varies with atmospheric conditions. Bloch mentions that the eggs are of the size and colour of millet seed, and that the ova do not exceed eight thousand in a full-sized fish. The eggs are deposited on gravel, and at the breeding season the fish form troops of about a hundred, moving one behind the other. The old females swim in front, followed by the old males, with the young males bringing up the rear. The roe is in most places reputed poisonous, and is always carefully removed before the fish is cooked; yet Bloch records that it was eaten by himself and his family without unpleasant results. It is commonly reputed to produce effects like those of Asiatic cholera. According to Dr. Günther the males develop more rapidly than the females, and when only six or seven inches long are seen pursuing the females.

The flesh is white. It is valued in some countries, but not in others. In England Barbel were formerly eaten and valued. Mr. W. S. Mitchell records that at the installation of the Archbishop of York, in the sixth year of King Edward the Fourth, Barbel appeared in the dinner feast in the second course. Holinshed's "Chronicle" mentions Barbel as one of the fishes whose preservation is provided for by law; and in 1650 T. Venner writes that the Barbel is "of very pleasant taste, of good nourishment, but somewhat muddy; the greater excel the lesser for meat, because their superfluous moisture is amended by age."

At the present day there is a prejudice against the fish, and Dr. Badham declares that the Thames angler, having hauled a noble Barbel from the water, weighed, registered, and frescoed his full-length portrait on the walls of the inn, gives the carcase to the landlord's cat. But he also states that simply boiled in salt and water, and eaten cold with a squeeze of lemon-juice, the Barbel will be found by no means despicable fare. Blanchard, who excuses himself from pronouncing any personal opinion, as having no taste in the matter of fish, reminds us how often the sign "Aux trois Barbeaux" invites the traveller on the banks of the Loire. And in Italy the prejudice of the Milanese against Barbel may be balanced against the excellent reputation of the fish in Rome. Heckel and Kner recommend that Barbel should be kept for a few days in fresh spring water before being eaten, when their flesh will be found fairly well flavoured.

The species is stated by Graells to occur in Spain. In Russia it is found in the rivers flowing into the Sea of Azov and the Black Sea.

## Barbus plebejus (Valenciennes).

D. 3 8, A. 3 5, V. 9. Lateral line 66—75, transverse 14—17 14.

This species (Fig. 47) is limited in its distribution to Italy and Dalmatia, and said by Graells to occur in Spain. It has the body thicker and more rounded, with a shorter blunter snout, and the scales are smaller than in the preceding

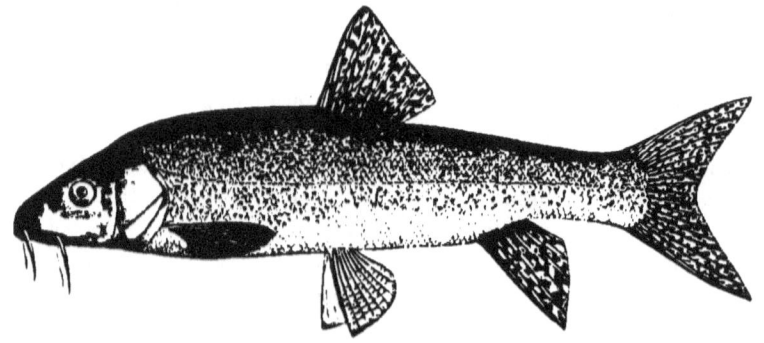

Fig. 47.—BARBUS PLEBEJUS (VALENC.).

species. The differences in the fin rays are sufficiently unimportant to justify Filippi in regarding it as but a variety of *B. vulgaris*. Dr. Günther, however, enters it as a good species, though he would unite with it the *Barbus eques*

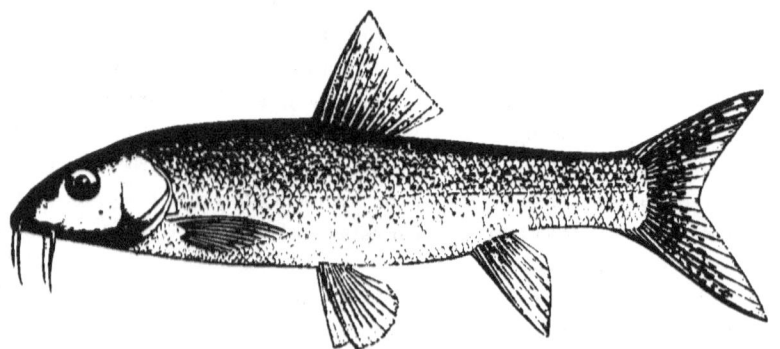

Fig. 48.—BARBUS PLEBEJUS (VALENC.), VAR. EQUES.

(Fig. 48) of the Austrian ichthyologists, while hesitating to follow Canestrini in uniting with it the *Barbus eques* of Cuvier and Valenciennes.

The head is less than a quarter the length of the body and about one-fifth the length of the fish. At the gill-apertures it is twice as long as broad.

In the young the head is five and a half times as long as the eye, in the old fish it is six and a half times as long as the eye, and with age the eye recedes a little from the snout. The frontal region is broader than in the preceding species; the mouth is horse-shoe-shaped. The barbels of the upper jaw reach back to the anterior margin of the eye, the barbels at the angle of the mouth reach back to the operculum. Both are longer than in *Barbus vulgaris*.

The dorsal fin is similar in position to that of the preceding species, and like it in size; its bony ray is thin, and finely serrated for one-third of its length. The dorsal and caudal fins are thickly covered with fine blackish-brown dots, but the sides of the tail, the anal and ventral fins are usually unspotted. The largest specimen is over one foot long. The variety which is found in Upper Italy and Dalmatia, and was regarded by Heckel and Kner as the *Barbus eques* (Fig. 48), is scarcely five inches long.

## Barbus caninus (CUVIER).

D. 3 7—8, A. 3,5, V. 9. Scales: lat. 48—52, transverse 10 13.

The body of this fish (Fig. 49) has much the same general form as in the preceding species. Its height is nearly one-fifth of its length. The tail is sub-cylindrical; the caudal fin is short, moderately forked, with the middle rays

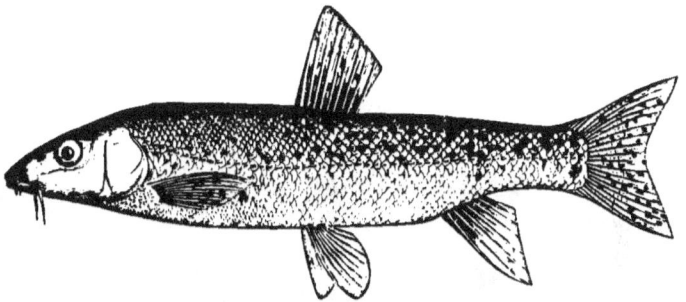

Fig. 49.--BARBUS CANINUS (CUVIER).

half as long as the outermost rays. The head is one-quarter the length of the body. The eye is less than one-fifth of the length of the head. The barbels of the upper jaw are short, and those at the angle of the mouth reach to the hinder border of the eye. Both are shorter than in *B. plebejus*. The snout is blunt and small. The back is very slightly arched.

The dorsal fin has the osseous ray no stronger than the others; it is flexible, and not serrated. This fin is opposite the ventrals, is higher than long, and rather truncated behind. The anal fin is higher than long; its longest rays are longer than those of the dorsal. The longest external rays of the caudal fin are shorter than those of the anal. The rays of the ventral fin are somewhat shorter than those of the pectoral.

The scales, especially along the lateral line, are larger than in the other species in harmony with their diminished number, but they are much smaller on the fore part of the back and belly.

This species is characterised by having the entire body covered with larger brownish-black spots, in addition to the small spots seen in the other species. The pectoral, dorsal, and caudal fins are all flecked with small spots, which are less developed on the anal fin. There are six rows of scales between the lateral line and the base of the ventral fin.

*Barbus caninus* is recorded from Idria, in Austria; from the Arno, and the northern parts of Italy; from Languedoc and Provence; from the Pyrenees and eastern parts of Spain. Examples from the River Xucar, which Steindachner named *Barbus guiraonis*, are referred by Günther to this species.

## Barbus petenyi (Heckel).

D. 3/8, A. 3/5. Scales: lat. 55—60, transverse 11—12, 8—9.

This species (Fig. 50) is known in Transylvania and Hungary, where it is termed the Semling. It has an elongated form; the anal and caudal fins

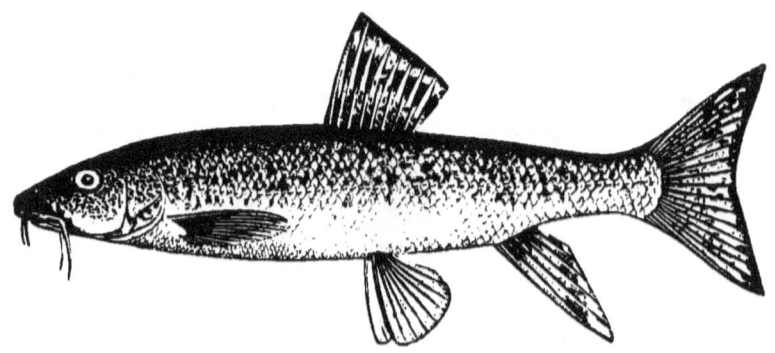

Fig. 50.—BARBUS PETENYI (HECKEL).

both have long rays; the dorsal is destitute of a notched bony ray; and the fore-part of the back and back of the head are broad.

This Barbel is one of the best known and most widely distributed of the Austrian species. It is distinguished from *B. vulgaris* by its blunt snout, head broader behind, less fleshy lips, absence of the serrated ray, and by the long rays of the anal fin, which is three times as high as broad. Similar characters, with its larger scales, separate it from the *B. plebejus*, which is less elongated. It is distinguished from *B. caninus*, with which Günther suggests it may, perhaps, prove to be identical, and with which it certainly agrees in colour, chiefly by the higher anal fin, longer caudal, the rounded furrowed back, and stronger form.

In front of the beginning of the dorsal the greatest height is between one-fifth and one-sixth of the length of the body. The head is one-quarter the length of the body. The anterior barbels are the shorter. There are large brown-black spots on the body, oftentimes touching each other. It varies in size from seven inches to one foot.

It is found in all the rivers and brooks of the Mittelgebirge in Bohemia, and in the plains of Transylvania and Hungary, but would appear to descend from the Carpathians. Von Siebold records it from the tributaries of the Oder; and Benecke mentions an example from Braunberg, which presented the characters of this species.

## Barbus comiza (Steindachner).

D. 3, 8—9, A. 3/5, V. 2/8, P. 1/16. Scales: $\frac{9 \text{ above}}{49-51 \text{ lat. line.}}$
$5-6$ below.

The body is moderately elongated and compressed. The head is Pike-like, with the snout compressed, much produced, inflated at its extremity; having the cleft of the mouth terminal, and bent somewhat upward. The lips are thin, the jaws are usually of equal length. The general shape of the head approaches to that of *Barbus vulgaris*, but differs in the forehead being more compressed, with a straight profile; and there is a lesser number of scales in the lateral line and above it. The body is four to four and two-fifth times the length of the head. The height of the body does not exceed one-fifth of its length. The eye is placed high on the side of the head; in the young fish it is one-sixth of the length of the head, while in the adult its relative length is one-ninth. The cheeks are rather compressed. The barbels are slender; the anterior barbel reaches back to the nostril, the posterior barbel extends back to the anterior border of the orbit.

The anterior half of the dorsal contour is conspicuously convex; the greatest height is attained in front of the dorsal, and is double the smallest height of the tail. The dorsal fin is opposite the ventral. The first two dorsal rays are very short and delicate, and nearly buried beneath the skin. The visible bony ray is remarkably broad and long, though rather shorter than the first membranous ray; its hinder side is toothed, with the denticles directed downward. The anal fin has a shorter base than the dorsal, is more pointed, and varies in length, but does not reach back to the caudal.

The caudal fin is deeply forked, and its lobes are usually equal. The lateral line is approximately parallel to the ventral outline; its last three scales are on the base of the caudal fin.

The cephalic canals are well developed behind the eye, and at the forward angle of the operculum the free edge of the pre-operculum is slightly pitted. The largest scales lie behind the shoulder-girdle, the smallest on the anterior part of the back. The free border of the scale is convex, and marked with many fine radiating lines.

The colour varies with conditions of life. In the muddy waters of the Tagus the fish is of a dirty yellowish-brown, while in the Guadiana, where the water is clear and the bed rocky, the colour is blue-grey with a metallic lustre, becoming silver-grey on the belly. The unpaired fins are usually spotted black, and a black band margins the base of the opercular region.

The species is not common in the Lisbon market, but is abundant at Toledo, where the local name for it is *Comba*. It is regarded by the fishermen of Mertola, on the Guadiana, as the male of *Chondrostoma willkommii*.

Dr. Steindachner is of opinion that crosses occur between *Barbus comiza* and *Barbus bocagei*; and since some fishes have the characters of *Barbus comiza*, with the jaws and mouths of *Chondrostoma polylepis*, he regards such individuals as hybrids. But Dr. Günther remarks that a similar combination of characters distinguishes certain Cyprinoid genera of Western Asia.

## Barbus bocagei (STEINDACHNER).

The body of this species is much compressed and elongated. The proportion of length of the head to the body is as one to four and five-sixths. The height varies with age, the fish becoming more slender as it gets older. The diameter of the eye is one-fifth to one-seventh of the length of the head, and the eye is more than twice its own diameter from the extremity of the snout. The gape of the mouth is small. The upper lip varies in thickness, and the snout extends over it. The barbels of the upper jaw do not reach back quite so far as the orbit of the eye. The barbels at the corners of the mouth reach across the orbits. The profile of the head is slightly arched, and thence extends in a convex curve to the dorsal fin, much as in *Barbus vulgaris*. The least height of the tail is equal to about one-half of the greatest height of the body.

The dorsal fin commences in front of the ventral, is in advance of the middle of the body, and is higher than long, though the height is very variable; the length of its base is about one-half of the length of the head. The last or fourth unjointed bony ray is more or less deeply toothed in the middle third of its length; but in old examples the serrations vanish, or are represented by slight irregularities. The superior hinder outline of the dorsal fin is slightly concave. The broad anal fin is placed far back, so that its rays sometimes reach to the base of the caudal fin, or it is sometimes much shorter. The short ventral fin is under the middle of the dorsal; it is always shorter than the anal fin. The pectoral fin may be five-sevenths of the length of the head, and is always longer than the dorsal fin. The caudal fin is deeply notched, with pointed lobes, of which the lower may be slightly the longer. This fin is seldom as long as the head.

The scales are longer than high, and have a fan-like ornament. The larger scales are above the pectoral fin, the smallest are on the neck and on the throat. The lateral line runs along the middle of the side, and cephalic canals extend from the inferior border of the orbit. The number of perforated scales in the lateral line varies from forty-six to fifty-one; the last three lie on the base of the caudal fin. There are eight or nine rows of scales between the lateral line and the dorsal fin, and five rows between the lateral line and the ventral fin.

The lowest and fifth pharyngeal tooth is a mere needle-point; but the two succeeding teeth are remarkably large.

The colour varies with the locality. In cold clear water the back is dark green or deep golden brown, with greenish spots; the sides are lighter, and the belly whitish. In muddy lakes and rivers, with warmer water, the back is dirty brown, and the belly yellow. The young fishes are sometimes spotted with brown.

This species is found in the ponds, rivers, and lakes of the interior of Spain, and reaches a length of one and a half to two feet. It is not met with in the east of Spain, but is common in the Tagus, Douro, Minho, and Xucar.

## Barbus sclateri (Günther).

D. 11, A. 8—9. Scales: lat. line 45—46. $\genfrac{}{}{0pt}{}{\text{above } 8}{\text{below } 8}$

This Barbel reaches a length of seven to twelve inches, and at present is known only from the Guadalquivir, in Spain. It has the bony ray of the dorsal fin very strong and sharply serrated. The body is somewhat compressed, nearly as high as the head is long, and the height is about one-fourth of the total length, exclusive of the caudal fin. The head is flat above, rather depressed, with a pointed and somewhat prolonged snout, so that the mouth is inferior in position. The lips are thick, the barbels are rather long, the upper barbel reaching beyond the front margin of the orbit, and the lower to the angle of the pre-operculum. The eye is in front of the middle of the head, its diameter being one-sixth of the length of the head in the full-grown fish, and two-fifths of the length of the snout. There are four or five longitudinal series of scales between the lateral line and the ventral fin.

The dorsal fin commences in the middle of the length of the fish, excluding the caudal. The anal fin is very narrow and pointed; the length of its base is fully two-fifths of its height, but when laid backwards it does not extend to the caudal. The caudal fin is as long as the head, but with the middle ray one-third of the length of the outer rays, so that it is deeply forked and has pointed lobes. The pectoral fin extends back to the twelfth scale of the lateral line. The ventral fin is somewhat shorter than the pectoral, and commences opposite to the spine of the dorsal.

The scales have numerous radiating rays. The colour is uniform greenish, with a silvery lustre, and a darker stripe along each series of scales. The dorsal and caudal fins are blackish.

## Barbus chalybeatus (PALLAS).

This Russian Barbel has the body and fins marbled and spotted with black. The head somewhat resembles that of *Misgurnus fossilis* in its elongated form and colouring. The snout is conical; the upper jaw extends beyond the mandible. The lips are fleshy. There are ten series of scales above, and eight below, the lateral line. Individuals reach a length of nine inches, but Nordmann describes no specimens above seven inches long in the south of Russia. It is found in the rivers between the Black Sea and the Caspian, and is common in the south of the Caspian Sea, where it enters the River Koora. It is also found in the Sea of Aral.

Among Russian Barbels Dr. Grimm enumerates *Barbus tauricus* (Kessler) in the small rivers of the Crimea; and many other species, chiefly found in the River Koora, and described by Kessler, Filippi, Güldenstadt, and Gmelin.

### HYBRID BETWEEN *BARBUS BOCAGEI* AND *CHONDROSTOMA POLYLEPIS*.

Professor Steindachner found a fish in the Rivers Tagus, Ebro, and Guadiana, by no means rare, which he regards as a hybrid between the species above named. It has the form and pharyngeal teeth of *Barbus bocagei*, with similar barbels to the jaws, and a serrated bony ray to the dorsal fin; but the jaws, especially the lower jaw, have the same broad, cutting, horny sheath which characterises species of the genus *Chondrostoma*, and the form and striping of the scales are such as occur in that genus. Dr. Günther remarks on the interest of this observation should it be confirmed, since a like combination of character is found in Cyprinoid genera of Western Asia, which are certainly not hybrids.

The fish is sometimes seen in the fish markets of Madrid.

Steindachner also records hybrids between *Barbus comiza* and *Chondrostoma willkommii* from the Guadiana, at Mertola. In these examples the snout is shorter, the mouth rounder, and less inclined downward than in *B. comiza*. It resembles that species in shape of body, in curve of the neck, height of the dorsal, and robustness of the four saw-like bony rays, and the barbels. In the scales and sheathing of the jaws it resembles *Chondrostoma willkommii*.

Another hybrid appears to be formed by *Barbus graellsii* and *Chondrostoma miegii*.

Spanish specimens, which have a long, small form of head, a feeble denticulation of a more slender bony ray in the dorsal, strongly-developed lips and barbels, and projecting snout, are thought by Steindachner to indicate a bastard between *Barbus comiza* and *Barbus bocagei*.

## Genus: **Aulopyge** (Heckel).

This remarkable fish has the body naked, intermediate in shape between Barbels and Minnows. It has four barbels; and four chisel-shaped pharyngeal teeth in a single row on each side. The ventral fin has a serrated bony ray; the anal fin in the female supports the urogenital tube.

The genus has been found only in Dalmatia and adjacent districts.

## Aulopyge hügelii (Heckel).

This species (Figs. 51, 54) has a very pointed, small, conically-elongated snout, and it is distinguished by its naked skin, which is silvery on the back and sides, or may be marked with blackish-brown spots. The male and female fishes differ so greatly in appearance that we will notice each separately. In the fully-grown female the body is much higher than in the males or young, and the dorsal outline rises from the back of the head to the dorsal fin in a conspicuous curve. The depth of the body is nearly one-quarter of its length.

The head is conical and pointed, it is half as broad as long (Fig. 52); the eye is rather small, nearly in the middle of the head, and separated from the opposite eye by twice the orbital diameter. The nasal opening is single, nearer the eye than the snout; it is defended at its front edge by a fold of skin, which can close it like a valve. The small mouth is nearly horizontal and horse-shoe-shaped; it does not reach as far back as the nasal valve. The rounded extremity of the snout projects over the mouth, which is covered by lips that are not fleshy. Both pairs of barbels are short, the longer barbels at the

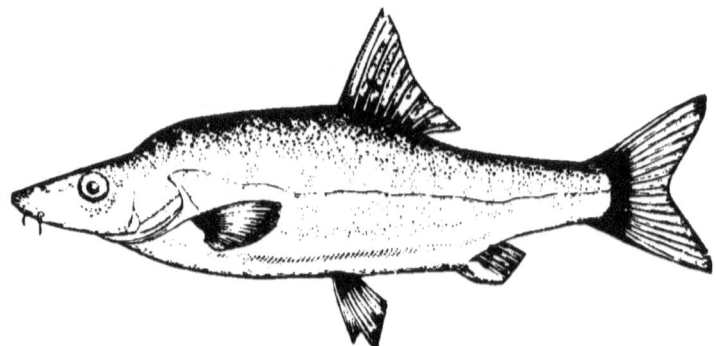

Fig. 51.—AULOPYGE HÜGELII (HECKEL).—FEMALE.

corner of the mouth being barely equal to the diameter of the eye. The depth of the body at the tail is one-third the greatest depth below the dorsal fin.

The dorsal fin commences nearer to the tail than the head. It is opposite the ventral fin, is higher than long, though it is fully half as long as the head; it is truncate behind; its second ray is remarkably broad and strong, and the third is deeply and strongly serrated down the entire length at its hinder edge. All the succeeding rays are jointed, and divide three times. The anal fin is posterior to the end of the dorsal, is as high as long, with a short base, and is truncate behind. This species is distinguished from all other known fishes by the remarkable condition of the anal aperture in the female.

Fig. 52.—UNDER SIDE OF THE HEAD OF AULOPYGE HÜGELII.—FEMALE.

The intestine and the urogenital system both leave the body in a thick, fleshy tube, which is united to the first ray of the anal fin in front, and near the extremity of that ray this tube has two openings, the urogenital being posterior. A somewhat similar condition has been described in the male of the American genus Anableps. The succeeding rays of the anal fin divide dichotomously.

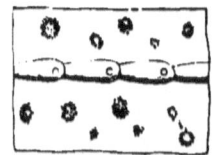

Fig. 53.—AULOPYGE HÜGELII: LATERAL LINE AND SKIN, SHOWING PIGMENT SPOTS.

The ventral and pectoral fins are moderately developed; the pectorals do not reach to the ventrals, nor the ventrals to the anal. The caudal fin is evenly lobed and deeply notched. The lateral line is in the middle of the body, but is irregularly waved and curved; it opens externally by simple pores, which are placed in close series. The orbital branches of the cephalic canal are very distinct, but the mandibular branch is obscure.

The male Aulopyge is always smaller than the female, and rather approaches to the shape of the Minnow, *Leuciscus phoxinus*. The height of body is less than in the female, the curve of the back is flatter, the dorsal fin is higher, the pectoral shorter, the anal fin commences farther forward, and the caudal often contains two more jointed rays than are found in the tail fin of

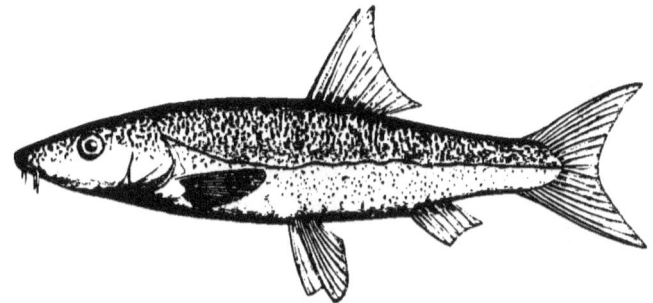

Fig. 51.—AULOPYGE HÜGELII (HECKEL).—MALE.

the female. Fig. 55 shows the pharyngeal teeth. There is no trace of the cloacal tube, the vent being in front of the anal fin, and the intestinal and urogenital openings being behind each other, in the usual way.

The young fish has a relatively longer head, the body being only three and a half times as long as the head. The height of the body is relatively less, the eyes are relatively larger, the frontal region is smaller, with a convex frontal outline, and the dorsal fin is as high as the body is deep.

In this fish the ground colour is silvery, becoming greenish towards the sides and on the back. The sides often show brownish or blackish spots and dots, which may be like scattered powder (Fig. 53) or gathered into large patches or broad bands. The iris is greenish-yellow. The fins are yellowish-grey, but the dorsal and caudal fins have their rays marked with more or less regular transverse rows of black spots.

The internal organs present some distinctive modifications. There is no distinct stomach, and there are no pyloric appendages to the stomach. The biliary duct opens, as in the genus Anableps, well above the stomach. The intestine makes three convolutions as it coils down the entire length of the ventral cavity. The air-bladder, as in other Cyprinoids, is divided transversely. There is a round and short anterior part, and a posterior part which ends in a blind point, and is provided with an air-canal. Under the termination of the air-bladder, the lobes of the ovary or milt unite in a common duct, by which their ripened elements are discharged. There is a small pear-shaped urinary bladder. The milt and ovaries are symmetrical and double; the ovaries are closed sacs, which run the whole length of the ventral cavity and get smaller behind as they join in the common urogenital canal. The milt lobes are much less developed, and do not extend far into the ventral cavity, but their small glandular lobes similarly unite in the outlet of the bladder to form a common urethro-genital opening. In the male the abdominal skin is coloured with a black pigment.

Fig. 55.—PHARYNGEAL TEETH OF AULOPYGE HÜGELII.

The size is variable; the largest known females are five inches long.

This fish was first discovered by Heckel in 1840, in the fish markets of Dalmatia. Its habits are entirely unknown, and Heckel records it only from the brooks and rivers of Dalmatia and Bosnia.

It is eaten in Bosnia cooked in oil, and is stated to be well flavoured.

GENUS: **Gobio** (CUVIER).

The Gudgeons are a small genus, closely allied to the Barbels, and limited to Europe. The dorsal fin has no spine. The jaws are equal in length. Both have simple lips, and there is a well-defined barbel at the angle of the mouth. The pharyngeal teeth are in two rows: five in the principal row, and two or three in the secondary. They are hooked at their extremities. The gill-rakers are very short.

## Gobio fluviatilis (RONDEL).—The Gudgeon.

D. 3/7, A. 3/6, V. 2/7—8, P. 1/14—15, C. 19. Scales: lat. line $\frac{\text{above } 6\frac{1}{2}}{40—44}$ below $6\frac{1}{2}$.

The body of the Gudgeon (Fig. 56) is four and a half times as long as high, and a third higher than wide. The head is more than one-quarter of the length of the body; the eye, which is near the top of the head, is about one-fifth of its length. The horizontal mouth is covered by the fleshy lips, and does not reach back so far as the nasal apertures, which are near the eyes. Fig. 58 shows the pharyngeal teeth. The short barbels are at the corners of the mouth. The form of the head (Fig. 57) varies so much, especially in length, that we might be inclined to suspect specific differences in diverse individuals were it not for the insensible gradations by which the extreme forms are connected. The frontal outline rises from the blunt nose to the hinder region of the head in a regular arch, which becomes more flattened towards the dorsal fin. The greatest depth of the body is two and a half or two and a third times the depth at the tail.

The dorsal fin is rather less than half-way down the body, and commences a little in front of the ventral fin. It is higher than long. Its height is almost

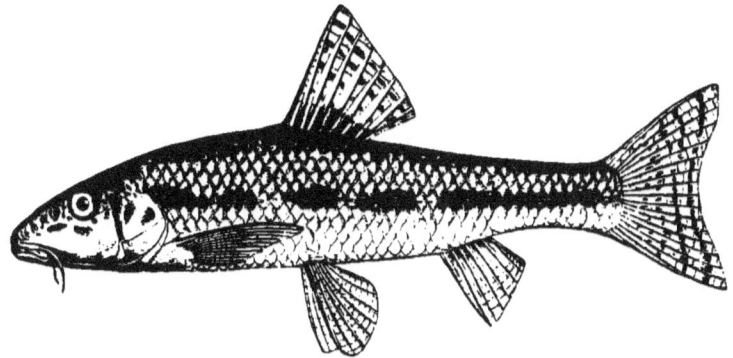

Fig. 56.—GOBIO FLUVIATILIS (RONDELETIUS).

equal to the depth of the body, but the last ray is only half as long as the first branched ray. The anal fin is placed well behind the termination of the dorsal. It is less developed than the dorsal, is similarly truncate behind, and higher than long. The ventral fins nearly reach to the anal, and always cover the vent. The pectoral fin, which is as long as the dorsal is high, reaches back as far as the beginning of the dorsal. The caudal fin is evenly lobed and forked, with its outer rays longer than those of the dorsal.

The back and sides are blackish-grey, sometimes brown or even fawn colour, spotted with dark green or black, especially

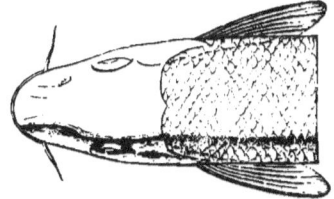

Fig. 57.—HEAD OF GOBIO FLUVIATILIS, SEEN FROM ABOVE.

along the lateral line and back. Sometimes there are about seven ocellate spots in the lateral line, which have often a tendency to pass into a band. The dorsal and caudal fins are usually spotted in Austrian specimens; the dorsal and anal fins are spotted in French. The spots are blackish-brown, on a reddish or yellowish ground. The iris is golden.

There is a variety of the fish, grey at the back, with four dark transverse bands on the silvery-white belly; while another variety, commonly darker in colour, has the body spotted, often with a yellowish band at the side of the tail.

The scales are deeper than long. The free edge of each is the segment of a circle, finely notched by the termination of the fifteen to eighteen radiating grooves which ornament the surface. The canal in the lateral line is cylindrical.

Fig. 58.—PHARYNGEAL TEETH OF GOBIO FLUVIATILIS.

The largest specimens found in France are five and a half inches long.

The Gudgeon is common in brooks, rivers, lakes, and is met with in marshes. It will live underground, and is found in the Adelsberger Cave, and in the warm springs of Teplitz, in Croatia. It prefers clear running water, with a sandy or pebbly bottom.

These fish live in large shoals, and feed on worms, fish-spawn, plants and decaying flesh. Those which inhabit lakes ascend the streams in spring. They begin to spawn in May or June, and lay their eggs, which are small and of a pale blue colour, under stones. The spawning is prolonged for about a month, and in the autumn the fishes return to the lakes. The period of incubation is about four weeks. The Gudgeon grows rapidly, and is tenacious of life, but has many enemies, among which Blanchard mentions the intestinal parasite, *Filaria ovata*.

It is caught both with the rod and net, chiefly in September and October; and in Paris it is estimated that about one million are captured in the year between the bridges of the Seine. Its flesh is well-flavoured, and is reputed to be easy of digestion. It is especially esteemed as a breakfast delicacy in Paris. In England it is rarely eaten, though its reputation was formerly as good in that country as it has always been on the Continent.

The Gudgeon is probably universally distributed throughout Europe; but, although the ancient Greek writers appear to have written warmly in its praise, it has not been recorded from Greece, and is at present unknown in Spain.

## Gobio uranoscopus (Agassiz).

In this species (Fig. 59) the body is more elongated than in *G. fluviatilis*, and the arch of the back is lower. The barbels are longer; the eyes are placed obliquely, so as to be directed upward, more towards the forehead. The dorsal and caudal fins are frequently free from spots, but sometimes marked with one, two, or three series of blackish dots, running transversely. The body of the fish is about six times as long as high; its thickness is more than two-thirds of the height; the body is about four and a half times the length of the head, or, taking in the total length of the caudal fin, the head is just under one-fifth of the length. The head (Fig. 60) is more elongated than

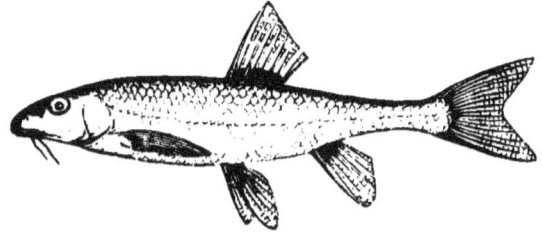

Fig. 59.— GOBIO URANOSCOPUS.

in *G. fluviatilis*; the eye is large, separated by its own diameter from the opposite eye, and it is twice its diameter from the snout. The mouth is similarly horizontal; its cleft does not reach to the nares; the barbels attached at its angles reach back to the operculum. The tail is less compressed from side to side than in the preceding species, so that it appears to be relatively cylindrical. The dorsal fin is only slightly higher than long; it commences opposite to the ventral. All the other fins possess relatively longer rays than those of *Gobio fluviatilis*, the pectoral reaching as far as or behind the beginning of the dorsal, the ventrals reaching to the anal, while the terminal rays of the more deeply-forked tail are as long as the head.

According to Günther, there are six and a half scales above the lateral line, seven and a half below; but Von Siebold, Heckel and Kner, and other writers, agree in reading five rows above the lateral line and four below; and there are differences in reading

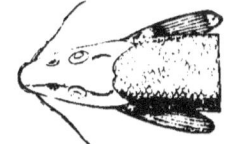

Fig. 60. — HEAD OF GOBIO URANOSCOPUS, SEEN FROM ABOVE.

the fin formulae, the Continental writers giving nine rays in the dorsal and seven in the anal, while Günther counts ten in the dorsal and eight in the anal.

The ground colour is silvery-white, becoming grey on the back, with diffused pigment. The trunk and fins may be free from spots, or a row of large brown spots may extend down the back and lateral line. The tail has a dark border, and the dorsal fin may have a dark transverse band in the middle. This species is usually from three and a half to four and a half inches long, occasionally reaching five inches. It is found in the Salzach, in Austria, and in the Save and Idria, but was first found by Agassiz in Bavaria, in the Iser. Its habits are similar to those of the Common Gudgeon.

## CHAPTER V.

#### FRESH-WATER FISHES OF THE ORDER PHYSOSTOMI (*continued*).

GROUP LEUCISCINA.—GENUS LEUCISCUS: The Rudd, and its Varieties—The Ide—Leuciscus aula—L. adspersus—The Roach—L. pigus—L. friesii—The Chub—The Dace—L. illyricus—L. pictus—L. svallize—L. ukliva—L. turskyi—L. microlepis—L. tenellus—L. muticellus—the Minnow—L. hispanicus—L. arcasii—L. alburnoides—L. macrolepidotus—L. arrigonis—L. lemmingii—L. heckelii—L. pyrenaicus—L. polylepis—GENUS PARAPHOXINUS—GENUS TINCA: The Tench—GENUS CHONDROSTOMA: C. nasus—C. phoxinus—C. genei—C. knerii—C. rysela—C. polylepis—C. willkommii.

## GROUP: LEUCISCINA.

### GENUS: **Leuciscus** (CUVIER).

LEUCISCUS is one of the most important genera of Cyprinoid fishes. Dr. Günther groups with it, in an assemblage termed Leuciscina, the genera *Ctenopharyngodon* of China, *Mylopharodon* of California, *Paraphoxinus* of Dalmatia, *Meda* of South America, *Tinca*, *Leucosomus* of North America, *Chondrostoma* of Europe and Western Asia, *Orthodon* of California, and *Acrochilus* of the Columbia River. Its European allies all belong to genera poorly represented in species; but Dr. Günther enumerates in his catalogue no fewer than thirty-five species of Leuciscus in the Old World, and forty-nine species in America. These species, however, have been divided by modern naturalists into a multitude of genera, founded on characters which have little value. The genus, as understood by Cuvier and by Günther, has the body covered with imbricated scales; the lateral line runs along the middle line of the tail, or a little below it; the dorsal fin is almost always opposite to the ventrals; the anal fin generally has from nine to eleven rays, and is rather short. The pharyngeal teeth may be conical or compressed; the intestine is short, with few convolutions.

The Old World species admit of being subdivided by the characters of the pharyngeal teeth, lateral line, and dorsal and anal fins. First, there are the species with the pharyngeal teeth in a single series; and some Continental writers limit the genus Leuciscus to such species of this type as have at least ten rays in the anal fin, and have the dorsal fin opposite to the ventrals; but when the anal fin has but nine rays, and the dorsal fin is behind the ventrals, they name the type *Pseudophoxinus*. Secondly, the species with the pharyngeal teeth in a double row are subdivided, so that the name *Phoxinus* has been used for those which have the lateral line incomplete; while types with the lateral line complete have been named *Squalius*, *Idus*, *Scardinius*, and *Telestes*. The American species are divided primarily by the number of scales in the lateral line, which is more than fifty or less than fifty, and minor divisions are based on the position of the dorsal fin and arrangement of the pharyngeal teeth.

## Leuciscus erythrophthalmus (LIN.).—The Rudd.

D. 11—12, A. 13—15, V. 9—10, P. 16, C. 7/17 6.

The fish known in England as the Red-eye is widely distributed on the Continent; it is the *Sarf* in Sweden, the *Scardola* of Italy, *le Rotengle* in France, and *das Rothauge* of Germany (Fig. 61).

The Rudd has an elevated body, with a narrow oblique terminal mouth, the angle of which reaches under the nares. The origin of the dorsal fin is con-

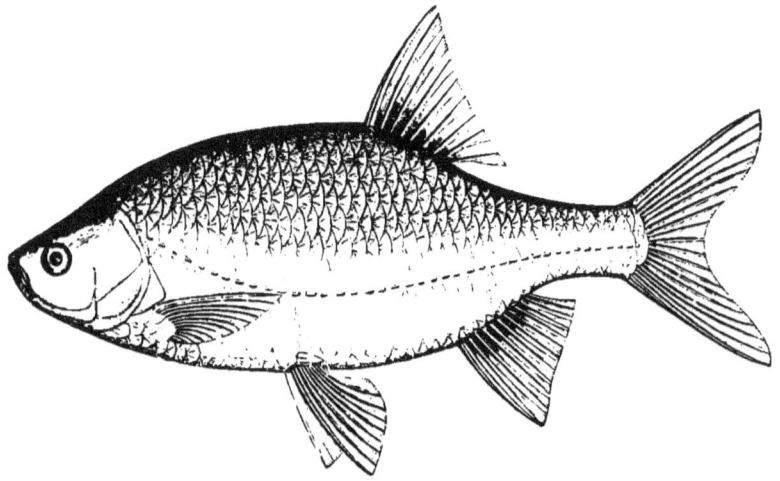

Fig. 61.—LEUCISCUS ERYTHROPHTHALMUS (LINNÆUS).

spicuously behind the ventral. There are three rows of scales between the lateral line and ventral fin, and the belly behind the ventrals forms a sharp edge covered with scales; but the species is extremely variable, and the proportions, measurements, and colour vary with sex, age, and habitat. The body is four times as long as high in the young fish, and three times as long as high in old females, in which the back rises in a conspicuous curve. At the tail the height diminishes to one-third. The body is compressed, so that the height is two and a half times as much as the thickness. The head is as high as long; the eye varies with age, being one-quarter of the length of the head in the young fish, or sometimes more, while it is less than one-fifth of the length of the head in the adult. The forehead widens with age; the dorsal fin is nearer to the caudal than to the snout, higher than wide, with its longest ray as long as the head. The vent is in a line behind the end of the dorsal and immediately in front of the anal fin. In the male fish the pectoral fins reach to the base of the ventrals; the ventrals are equally long, but do not reach to the vent. All the fins have the rays rather longer in the males than in the females. The latter, on the other hand, have the bases of the anal and dorsal fins relatively elongated. Fig. 62 shows the pharyngeal teeth.

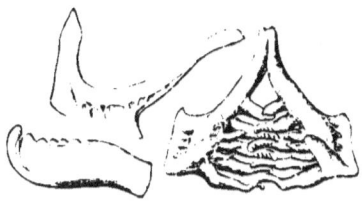

Fig. 62.—PHARYNGEAL TEETH OF LEUCISCUS ERYTHROPHTHALMUS.

The scales are strong, firmly adherent, and overlap each other considerably; they are longer than wide, rounded at their free border, and more truncate behind. There is a central knot, from which three or more furrows or canals radiate, and between these there are usually several indistinct furrows. Hence the concentric lines of growth are sinuous, and the free border of the scale is slightly undulating. The largest scales are in the lateral line, and exceed the diameter of the eye. The number of scales in the lateral line varies from thirty-nine to forty-two. According to Dr. Günther, there are seven and a half scales above the lateral line and five and a half below, and, according to Heckel and Kner, four rows above and three below.

In clear running water the back of the fish has the shining aspect of polished steel, with blackish smears; the sides are brassy-yellow, the belly silvery, the anal and ventral fins, and more rarely the dorsal, are blood-red; the bases of the caudal and dorsal are usually black; the caudal is red at its extremities, but the bases of the anal and ventral fins are white; the pectoral fins are nearly colourless, though sometimes tipped with red. The iris is orange-yellow, with red flecks. The young are paler-coloured; in the first year their fins become red at the margins, and it is not till the end of the second year that the full coloration of the mature fish is acquired. The colours, however, vary with the locality, and differences have been which characterise the fish in different parts of Europe. Dr. Day mentions that when taken from deep water the colours are similar to those of Roach or Dace.

A variety with black fins, especially the ventral and anal, is found in the lakes of Northern Italy, in the ditches of the rice-fields, and in the Engadine. It is known to the Italians as the Devil's Fish (*Pesce del Diavolo*); and as *Scardola* in some localities, and *Piotta* in others.

In size the Rudd does not exceed one foot—as in some of the Swiss and Hungarian lakes—but nine to ten inches is a good size; the weight is from one and a half to two pounds, though English specimens have been recorded which weighed from two to three pounds.

The Rudd is met with in company with the Crucian Carp and Tench, is cautious and rapid in its movements, prefers water which moves slowly, and hence is frequent in ponds and lakes. It feeds on insects, worms, and water plants, and Fatio has attributed the dark colour of the fins in some varieties to an exclusive diet of dark green vegetable growth. It is taken easily

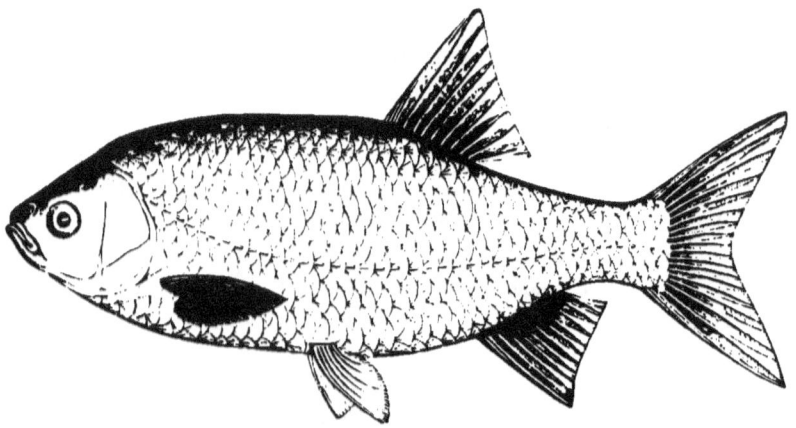

Fig. 63.—LEUCISCUS ERYTHROPHTHALMUS (LINNÆUS), VAR. DERGLE (HECKEL AND KNER).

in the Norfolk waters with the red worm, at sunrise and sunset. It is tenacious of life, and on that account is used as bait for the Pike.

It breeds freely, and furnishes food for Perch, Trout, and other predacious fishes. It spawns in April or May, and lays its eggs by the margins of streams where aquatic plants are plentiful. The eggs number about 100,000. The skin of the head and back at this time becomes covered in the male with little asperities.

It lives for about four or five years. The flesh is too full of bones to be much valued as food, but it is considered better eating than the Roach; and Yarrell mentions having seen it exposed for sale by the dozen in the old Hungerford Fish Market, which stood on the site of the Charing Cross Railway Station in London.

The variety which Heckel and Kner distinguished under the name *Dergle* presents minor differences of aspect (Fig. 63). It approaches closely to the dark-finned variety *hesperidicus*, which is found not only in the north of Italy, but in the Island Cherso, which lies to the east of Istria. But the snout is sharper, the mouth wider, the dorsal fin is higher than long, and similar to the anal; it is chiefly met with in Bosnia and Dalmatia, where it is common in

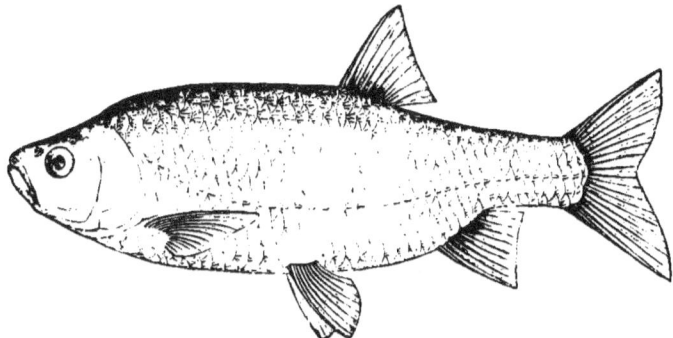

Fig. 64.—LEUCISCUS ERYTHROPHTHALMUS, VAR. PLOTIZZA (HECKEL AND KNER).

the Rivers Kerka and Zermagna, and is distinguished by the local fishermen as the Dergle.

A second variety, originally indicated by Bonaparte as *L. scardafa* (Fig. 65), is characteristic of Italian waters, and is known in Italy as the *Carezzal*; it is

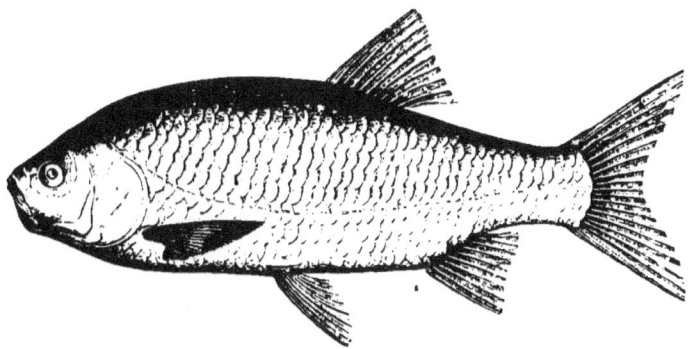

Fig. 65.—LEUCISCUS ERYTHROPHTHALMUS, VARIETY SCARDAFA (BONAPARTE).

also found in the Narenta, in Dalmatia, and is there called the *Pesquelei*, or *Peschkegl*. The only characters which distinguish it even as a variety are the more abrupt angle at which the lower jaw ascends, the slightly concave profile of the forehead, and the horizontal outline of the abdomen. The scales (Fig. 66) have a regular fan of rays, which indent the convex contour of each posterior free border. The sides of the head have a mother-of-pearl lustre, but the colours are not otherwise remarkable.

A third form, named by Heckel and Kner *Plotizza* (Fig. 64), is also Dalmatian, but has been met with at Livno, in Bosnia. The name *Plotizza* is locally given to White Fish, and those authors appear to have considered the verdict of fishermen as to the distinctness of fishes binding on scientific nomenclature. The unpaired fins are rather more feebly developed than in the typical Rudd. The profile of the frontal region is horizontal, and the dorsal outline less convex. The dorsal fin is higher than long, but the longest rays of the anal fin are shorter than those of the dorsal. It reaches a length of fourteen inches.

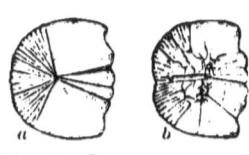

Fig. 66.—LEUCISCUS SCARDAFA—SCALES: *a*, ABOVE THE LATERAL LINE; *b*, IN THE LATERAL LINE.

A fourth variety was named by the Austrian authors *L. macrophthalmus* (Fig. 67). It is known only in the Tyrol, where it is distinguished as the Red Carp. Its common length is from three to seven inches. The body is covered with a net-work of black pigment; all the fins are grey, except the ventral. The head is one-fifth of the length

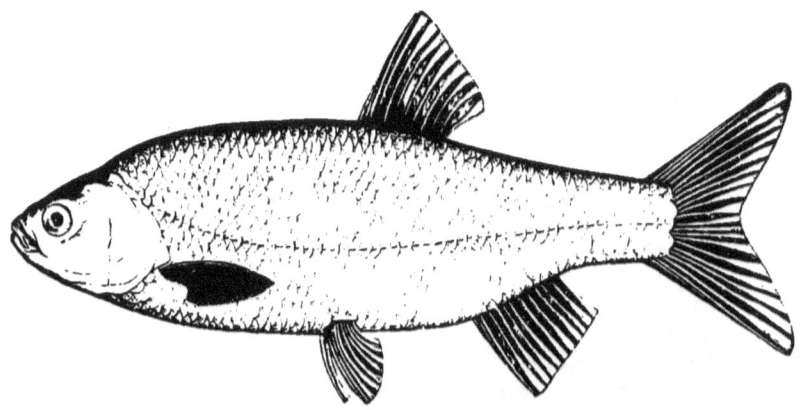

Fig. 67.—LEUCISCUS ERYTHROPHTHALMUS, VAR. MACROPHTHALMUS.

of the fish, and is three and a half times the length of the eye. Most of the scales have only two or three rays with concentric marking, and the largest scales are scarcely two-thirds of the diameter of the eye. It is distributed throughout Europe, and is recorded from Asia Minor. This habit of distinguishing local varieties by names, if followed out persistently by naturalists, may lead to a recognition of the steps by which species vary and merge in each other, but is ill-suited to giving a conception of the fish, which is substantially the same in all localities in which it is found.

## Leuciscus idus (Lin.).—The Ide.

The fish is known in Sweden as the *Id*; in Germany it is the *Nerfling*; in Prussia, the *Aland*. It is the *Göse* in Silesia, *Bratfisch* in some parts of Hungary, and the *Göngling* in Austria. It is placed by Heckel in a distinct genus, named Idus, and in this sub-division of Leuciscus he is followed by nearly all the Continental writers. The larger grouping of Dr. Günther,

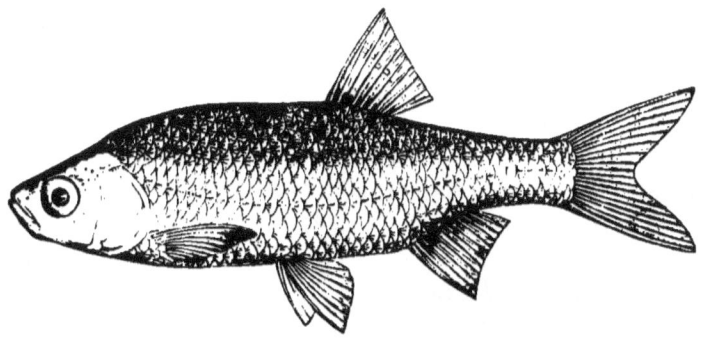

Fig. 68. LEUCISCUS IDUS (LINNÆUS).

however, preserves the species in association with its allies, the Roach, Chub, Dace, Rudd, and Minnow (Fig. 68).

The body is moderately elongated and compressed; its height is one-quarter of the length of the fish. The length of the head is only one-sixth of the total length; the greatest thickness is less than half the height.

The eye is less than one-quarter of the length of the head; it is separated from the other eye by twice its own diameter in the mature fish, and separated from the snout by a single diameter. The forehead widens with age. The mouth is small; the jaws are equal in length, and the angle of the mouth reaches only to the nares. The mucus membrane forms numerous folds on

Fig. 69.—PHARYNGEAL TEETH OF LEUCISCUS IDUS.

the throat, and they are covered with delicate papillæ, some of which put on the appearance of teeth. The pharyngeal teeth (Fig. 69) have the formula 5·3—3·5. The dorsal profile is moderately and evenly arched, with a depression at the hinder part of the head; the ventral outline is similar.

The dorsal fin is in about the middle, or rather behind the middle, of the length of the fish. It is nearly as high as the head is long. The anal fin is in the posterior third of the length of the body. It is nearly as high as long, but is not so deep as the dorsal. The ventral fins are placed rather in front of the dorsal, and reach to the vent. The pectoral fins are rather short, and do not extend to the ventrals. The caudal fin is deeply forked and evenly lobed, with its terminal rays longer than the head.

The lateral line is sub-parallel to the ventral outline, and reaches its lowest point between the ventral and anal fins.

The scales are rather smaller than in the Chub; they have strong concentric striae, with the rounded free border slightly festooned, and marked with a few radiating grooves. The attached border is truncate. The largest scales are smaller than the diameter of the eye. At spawning-time white tubercles appear upon the scales, head, operculum, and first ray of the pectoral fin; the colour varies with age, locality, and season of the year. During spring and spawning-time, in April and May, the colours are conspicuously bright. On the back the tint is blackish-blue, with a metallic lustre; the sides are whitish, the abdomen is silvery, and the head and operculum are golden. The iris is yellow, with dark spots in its upper part. The dorsal and caudal fins are grey-blue or violet, while the other fins are more or less red; though, according to Von Siebold, all the fins have a reddish tinge, with a tendency towards violet. In autumn the colours grow darker: the back becomes black or greenish-blue, but the brassy or golden lustre changes into yellowish-white; only the ventral and anal fins retain a dull red colour.

In the young the red colour of the fins is brightest, and is especially remarkable in the anal fin; and the back is then much paler, and has sometimes a brassy tone.

These fishes are usually about one foot long, and commonly weigh four or five pounds. They occasionally reach a length of eighteen or twenty inches, and then weigh upwards of six pounds in the Danube, but larger specimens are found in North Germany.

Its habits cause it to prefer pure cold water, so that it is not common in shallow streams, and it lives in deep places in winter. It is not confined to fresh water, being found in the Baltic and about its islands.

It is timid, but crafty, and comes to the quiet surface of the water only in the evening; it swims rapidly. It lives for eight or nine years. It is not a fish much sought after, but when fished for, whether with net or line, the bait used consists of grasshoppers, dung-fly, or small fishes. Its white flesh when boiled in salt water becomes yellowish or red, like Salmon Trout.

It is widely distributed in Eastern Europe. According to Dr. Grimm, it is found all over European Russia, as far north as Petchora, but is absent from the Caucasus. It is abundant in the middle and southern parts of Scandinavia, is found in all the rivers of Central Europe, and in the Danube and its tributaries. It is met with in Belgium and the eastern part of France.

There is a variety of this fish known as the Golden Ide, the Orfe, or Gold Nerfling; the *L. orfus* of Linnæus. This fish is, like the Golden Tench or Golden Carp, a sort of Albino variety of the common type, for the Orfe and Ide agree in all the essentials of structure. The proportions, the number of rays, and number of scales are the same.

In magnificence of colour it is scarcely inferior to the Gold Fish, while its vermilion tint is more durable; and specimens kept many years in spirit preserve their original colour perfectly. The back and sides are vermilion or orange-red. The belly is silvery. A broad indistinct band of violet tint runs longitudinally to the tail, and divides the deep red of the back from the pale tint of the abdomen. All fins are red at the base and white at the points. The terminal rays of the caudal fin and the first rays of all the other fins are white. The iris is golden-red, with a black pupil. In Austria this variety is chiefly found as a cultivated fish, fine examples being preserved in the Imperial ponds at Laxemburg Castle, near Vienna, and in preserves at Munich, and Dünkelsbühl. According to Bloch, the Orfe has forty-five vertebræ and twenty-two pairs of ribs, but the Gängling is said to have only forty-one vertebræ and fifteen pairs of ribs.

Dr. Günther gives the number of vertebræ for the species at forty-seven, twenty-six in the body, and twenty-one in the tail.

A variety of the Orfe, originally brought from the Tyrol, is kept in various ornamental waters in Austria, and was regarded by Heckel and Kner as a species, which they named *L. miniatus*. The head is larger and longer, the back of a paler red, the sides devoid of the median violet band, the abdomen reddish, while the sides are marked with blackish transverse bands, which extend over the trunk and head. It has about the same weight and size as the Orfe.

Another variety of *Leuciscus idus* is distinguished by Dr. Günther under the name of *L. lapponicus*. It is known from the river Muonio, a tributary of the Tornea, forming the frontier between Sweden and Lapland, where it reaches a length of seven or eight inches. The height of the body is about one-third of the total length of the fish, exclusive of the caudal fin, and the origin of the dorsal fin is nearly vertical over the region of the ventral fin.

*Leuciscus borysthenicus* (Kessler) has the fin formula D. 11, A. 11—12, V. 9, with thirty-seven or thirty-eight scales in the lateral line. There are seven rows of scales above the lateral line, and two and a half rows between the lateral line and the root of the ventral fin. The head is broad, with a very oblique mouth, and is one-fifth of the length. The height of the body is one-quarter of the total length without the caudal fin. It is found in the Dnieper.

## Leuciscus aula (Bonaparte).

D. 11—12, A. 11—12, V. 10, P. 15—17, C. 19. Scales: lat. line 37—46. $\frac{\text{above} \quad 8}{\text{below} \quad 4\frac{1}{2}}$

This species (Fig. 70) is a southern representative of *L. rutilus*, found only in Europe south of the Alps. The body is elevated, the head short and

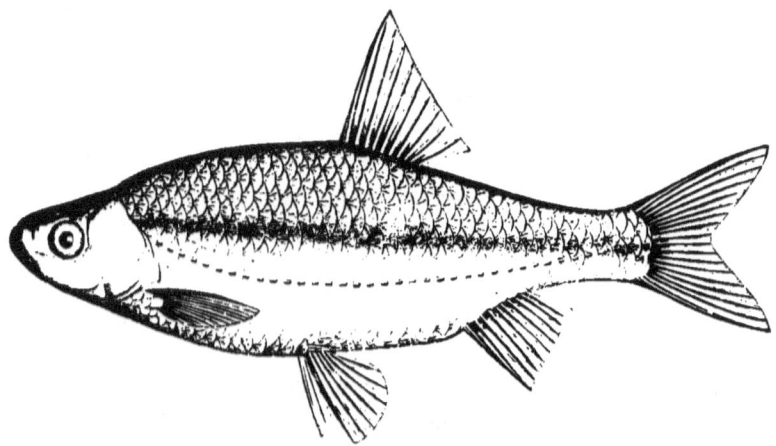

Fig. 70.—Leuciscus aula (Bonaparte).

thick, with a terminal mouth and obtuse snout, but the upper jaw projects slightly beyond the lower. In the young fish the body is four times as long as high, and usually three and a half times as long as high when mature, though the proportions are variable. In old age the height is twice as great as the thickness. The eye is relatively larger in the young, but in maturity its diameter is about one-fourth of the length of the head. The breadth of the frontal bones between the eyes increases with age. The profile of the frontal region is moderately arched. The dorsal fin commences in the middle of the length; it is a little higher than long, truncated behind, so that the last ray is half as long as the first. The anal fin is much less developed in height, and its rays are shorter than those of any other fin. The last ray is two-thirds of the length of the longest one. The pectorals are but moderately developed, and become relatively small in old age. They are too short to reach to the ventrals. The ventral fins extend to the vent. The terminal rays of the caudal fin are longer than the head. The lateral line descends from the upper angle of the operculum towards the ventral fin in a concave curve, rising again to the middle of the tail. Under the dorsal fin the lateral line is in the lower third of the body.

The largest scales on the lateral line are longer than the diameter of the eye; they are similar to those of the Roach, and are strong. Consecutive striæ surround a central thickening, and there is a variable number of radiating rays, usually five to six, but sometimes as many as eleven, between which the rounded border of the scale is festooned. The smallest scales are towards the two extremities of the body, and along the middle of the back.

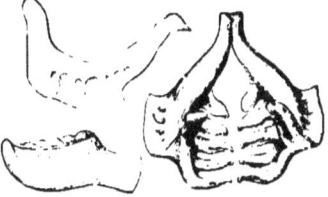

Fig. 71.—PHARYNGEAL TEETH OF LEUCISCUS AULA.

The colour of the back is a pale olive-green, but the sides are sometimes golden-green and sometimes yellow. A broad black band extends about the lateral line from the caudal fin to the operculum. A golden band often exists above it. The abdomen is silvery-white, with more or less of a golden tinge. The fins are colourless, but exhibit some amount of variation with age, season, and locality. They are sometimes grey, and the pectoral, ventral, and anal may have a yellowish or reddish tinge. The colours become a little dim with age.

This fish does not often exceed six inches in length.

According to Fatio, the males are relatively few. At the breeding season they have the first ray of the pectoral fin thickened.

The air-bladder is large and constricted in the middle. There are thirty-six vertebræ. The pharyngeal teeth are in a single row of five on each side, and resemble those of the Roach (Fig 71).

The species is common in the north of Italy in both lakes and streams, and is provincially known as *Triotto*. It is found in Dalmatia, and in Spain and Portugal. It is nowhere valued as food.

There are two varieties of this species recognised by Heckel. The variety which, following Bonaparte, he names *L. rubella* (See FRONTISPIECE) has the form more elongated, and the head higher than in *L. aula*, with the dorsal and anal fins higher than long. The body is said to be always four times as long as high, and five times as long as the head. The dorsal profile forms a flat arch, in front of the dorsal fin; as compared with *Leuciscus aula*, the dorsal fin is relatively higher in this variety, and has the base shorter. The back has a dark brownish-green colour; the pectoral, anal, and ventral fins are reddish;

and the lateral band of a lead-blue tint, so characteristic of *Leuciscus aula*, is either wanting or only partially developed. Its length is from two to six inches. It is not known north of the Tyrol, but is found in Lake Garda and other Italian localities, in Istria and Dalmatia, and at Brusa, in Asia Minor.

Many authors, like Canestrini, Günther, and Fatio, refuse to recognise it even as a variety, because intermediate forms can be found, and individuals of the species *Leuciscus aula* present characters found in this fish; but such a proceeding prejudges the question of specific variation, and it seems to us more convenient to recognise the variety as exhibiting a direction in which the species *Leuciscus aula* tends to become modified. Von Siebold would go so far as to regard the fishes which have been referred to *Leuciscus rubella*, *L. aula*, *L. adspersus*, and *L. basak*, as varieties of *Leuciscus rutilus*.

Heckel's second variety, *L. basak*, is found only in Dalmatia, where it is locally known under the name *Basak*, and occurs in the Lago di Drusino, near Imosky, and in other localities (See FRONTISPIECE).

It is remarkable for the small size of the forehead, and has small scales on the back. Moreover, the body is so attenuated that its height is only equal to the length of the head, and the head is nearly one-quarter of the length of the fish. Hence the head is relatively larger than in *L. rubella*. The eye is large, being one-quarter of the length of the head, and is removed from the rather pointed snout by its own diameter. The dorsal fin is higher than in the previously described variety, and has a shorter base. There is no trace of the longitudinal lateral band.

## Leuciscus adspersus (Heckel).

D. 10, A. 10, V. 9, P. 14, C. 19.  Scales: lat. line $\frac{\text{above } 15-16}{58-60}$
$\text{below } 6-7$

This is another of the fishes of Dalmatia described from a small lake near Imosky, which is deeply sunk, like a crater, among the surrounding rocks. The lake is known as the *Jessero rosso*, and has a subterranean outlet, in which the fish also occurs. The species is locally known as the *Gaorize* (Fig. 72).

In aspect it somewhat resembles the Minnow.

The length of the head is equal to the height of the body; the scales are small; and the entire body is spotted with blackish-brown pigment. There is on each side a single row of pharyngeal teeth. As in the variety *basak*, of the preceding species, the head is one-quarter of the length of the body, and the eye one-quarter of the length of the head, similarly distant from the snout, and also one and a half times its own diameter from the other eye.

The ventral fins in the male extend over the vent to the first ray of the anal fin, but in females they are shorter, and do not reach to the vent. The pectoral fins also are more developed in the males. At spawning-time the females have a fleshy thickening at the base of the dorsal fin, and the anterior half of the base of the anal fin becomes similarly thickened with fatty pads. The terminal rays of the caudal fin are not longer than the rays of the dorsal.

The scales are soft, easily detached, marked with close-set concentric rings, and free from radiating rays. They have a firmer consistence in the males. At spawning-time the scales in the female are, to a large extent, wanting, especially on the back; but even when all the scales are present they never

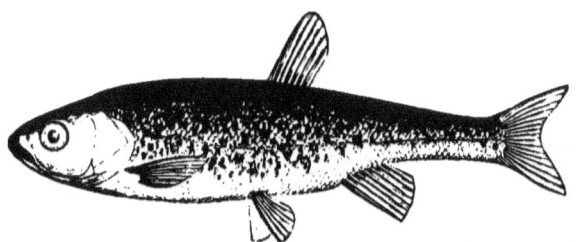

Fig 72.— LEUCISCUS ADSPERSUS (HECKEL).

overlap each other at the sides, and are only just in contact. In character the scales closely resemble those of *Paraphoxinus alepidotus*.

The abdomen is the only part of the body which is not spotted with dark brown pigment. The spots often run into each other, forming a network, and this may be seen on the caudal and dorsal fins, and sometimes on the anal. The pectoral and ventral fins are always colourless.

At spawning-time the region round the vent is much swollen in the females, and behind the vent there is a reddish prominence.

The length varies from two to four inches; and the small size is one of the most constant distinguishing characters of this species.

## Leuciscus rutilus (Lin.). The Roach.

above 7—8
D. 13—14, A. 12—14, V. 10, P. 16, C. 19.  Scales: lat. line 42—44
below 5½.

The Roach (Fig. 73) has the body somewhat elevated, being about as high as the head is long, while the head is between one-sixth and one-fifth of the length. This fish resembles the Rudd in the deep, compressed form of body; and in some parts of Germany the same name is applied indifferently to both.

The mouth is terminal, and the upper jaw scarcely projects beyond the mandible. The diameter of the eye varies, and may be contained four and a half times, or only three and a half times in the length of the head. This difference is not dependent on age, but is peculiar to certain localities. An eye of the same size is sometimes one diameter from the snout, and sometimes distant half as far again. The angle of the mouth, which is sharply inclined, may reach to the front of the orbit, though it sometimes goes no farther back than the nares. The inter-orbital space is equal to from one to two diameters of the orbit, and is rather wider in the male than in the

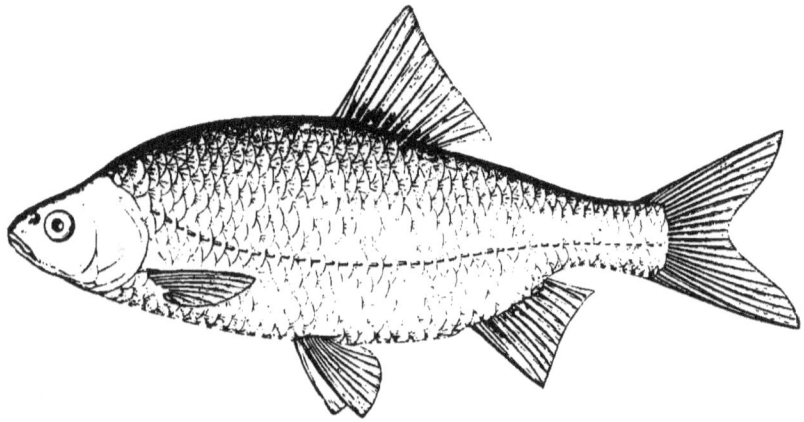

Fig. 73.—LEUCISCUS RUTILUS (LINNÆUS).

female. The sub-orbital arcade is formed of four to six bones, of which two are large and two small. The pharyngeal teeth are in a single row, usually five on each side, though sometimes there may be six on one side, and five on the other. As in all other Cyprinoids, when a tooth is broken, the base is removed, and a new tooth grows in the depression.

The dorsal fin is situate on the convex back, nearly half-way down the length of the body, and behind the origin of the ventrals; but farther forward than in the Rudd. This fin is higher than long, with the last ray half the length of the longest ray, so that the slightly concave posterior border has a truncate appearance. The pectoral fins are usually feebly developed, and scarcely longer than the ventrals. The anal fin commences behind the last ray of the dorsal; it is less elevated, but varies, both in the length of its base and of its rays. The ventrals also vary in form and length. The caudal fin is deeply notched, and the lobes are pointed. The terminal rays are longer than the head.

The greatest height of the body is three times the depth of the tail.

The lateral line runs parallel to the ventral border as far as the anal fin, and is nearer to the ventral border than to the dorsal border; its canal opens by a line of simple pores.

The scales are rounded at the free edges, with the sides horizontal. The attached end, which is firmly adherent, has a median portion and lateral angles. The scales vary in size with the individual, but the largest are at least equal to the diameter of the eye. They are marked with concentric lines of growth and a few rays, which form slight festoons on the free border. The thickened centre of the scale, and the whole of the exposed part is spotted with blackish pigment granules. The smallest scales are found on the anterior part of the back and pectoral region.

As in allied fishes, the colour varies with conditions of existence, and becomes most brilliant at the spawning season, especially in the males. The colour of the upper part of the body is some tint of blue or green, sometimes becoming blackish. The sides are brighter, sometimes yellowish, but more commonly silvery, with a bluish or greenish tinge. The belly is silvery-white.

The ventral and anal fins are red; the dorsal and caudal are grey, with red spots, and they often have a blackish border. The pectoral is greyish-white, though in old individuals this fin, in common with the others, becomes smeared with red. The iris is silvery, but gets dotted with red with age.

The Roach grows to a rather larger size than the Rudd, but commonly weighs less than two pounds. Its length is usually from twenty to twenty-five centimètres, though in the Kurische Haff it reaches a length of thirty centimètres.

Frank Buckland, who was an excellent Roach angler, tells us that the fish is afraid of every float that is the least bit too big, and to catch it the finest tackle must be used. A single hair is much better than the finest drawn gut. The best hair known for Roach fishing is the long peculiarly coloured hair from the tails of Her Majesty's cream-coloured state coach-horses. The best bait is gentles, which must be fattened on sheep's liver.

The Roach is one of the most widely distributed of fishes, being found in small and large streams, ponds, rivers, lakes, and even in the salt waters of the Baltic. On British coasts it thrives best near to the sea. Its habit is gregarious. According to Yarrell, it frequents deeper waters by day, and feeds in the shallows at night. Its food comprises worms, insects, fresh-water mollusca, spawn, and water-plants.

According to Dr. O. Grimm, the Roach in a dried state is quite a national dish in Russia, and the demand appears to be increasing. From

300,000,000 to 400,000,000 of the Caspian variety of the Roach are caught every year. When fresh they weigh about 7,000,000 poods (a pood being thirty-six pounds English), but only weigh half as much when dried. In 1808 one thousand salted Caspian Roach could be bought in Astrakhan for from sixty to eighty copecks, that is, from one shilling and tenpence to two shillings and threepence; but in 1872 the price had so increased that one thousand unsalted Roach cost one pound sterling (seven roubles) at the fisheries. The roe of the Caspian Roach, to the amount of several tons of thousands of poods, is made into caviare, and exported; but it is not commonly separated from the roe of the Bream; the caviare from both together, to the amount of 150,000 poods, is exported from Astrakhan.

The Roach spawns in April and May in Prussia, May in Austria, and June in England, when the scales of the male become rough. The fishes then assemble in weedy places in shoals, and exhibit those lively movements which have given rise to the adage, "As sound as a Roach." It is not often safe to depend on mediæval etymology, but it had been supposed that the Roach was incapable of becoming diseased, and was hence named after St. Roch, the legendary Æsculapius.

Lund says that females ascend the stream preceded and followed by males, and that when the females, tired with these attentions, deposit their eggs, the males rub themselves against the precious deposit, and fecundate the ova. The young are said to breed in the second year.

The eggs number, according to Benecke, 80,000 to 100,000. Day states that they are greenish, but become red when boiled. They are deposited with considerable noise and movement. At this time the fishes lose their shyness, and are easily captured. The eggs are hatched in from ten to fourteen days. The fish are, however, not much valued for food, being full of bones, but are in the best condition in October. They grow quickly, and form excellent food for Trout and Pike. The fish is sociable, and not only herds with troops of Bream, Rudd, &c., but breeds with them, forming several bastard varieties.

In winter the Roach retires to the bottom of the water, and remains there, according to Fatio, in a sleepy or hybernating condition till the spring.

The fish is liable to attacks from a Leech, and a multitude of intestinal worms, which are sometimes so numerous as to give the body a swollen aspect.

The Roach is found throughout Europe north of the Alps, in Britain, Belgium, Holland, France, Germany, Austria, Scandinavia, and Russia, where it occurs in all river basins, and is found in great quantities in the Sea of Azov and the Caspian. In France it is known as *le Gardon*; in North Germany it is *die Plötze*; in Austria, *das Rothauge*, or *Rothflosser*; in Sweden it is *Mort*; and in Holland, *de Blank-voorn*.

Heckel and Kner record that varieties, with the eyes large and forehead small, are obtained from the Oder, from ponds in Lemberg and Lake Constance, and spread over the north of Europe. Another variety, with large eyes and deep body, occurs in the Neusiedler See and Platten See. A variety, with small eyes and a broad forehead, is found in the Danube and its tributaries, and in the Traun and Atter See.

A more marked variety was distinguished as *Leuciscus pausingeri* of Heckel (Fig. 74). It is characterised by having the dorsal fin higher, with a shorter base, and lower anal fin than in the typical forms of the Roach.

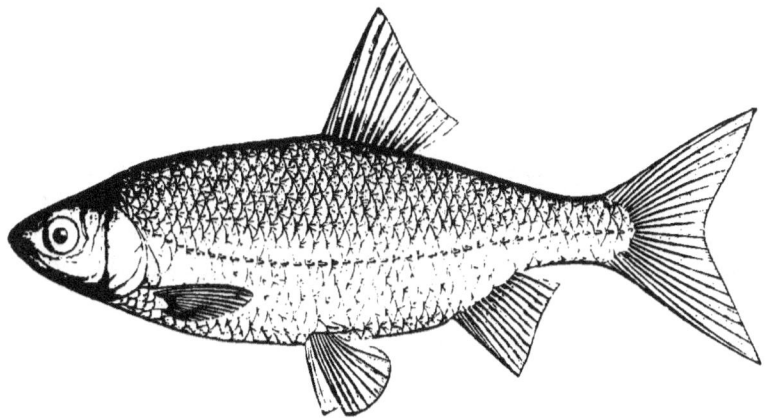

Fig. 74.—LEUCISCUS RUTILUS, VAR. PAUSINGERI.

It is four times as long as high, and five and a half times as long as the head. The frontal profile rises steeply to the back without an intervening depression. The eye is one-fourth of the length of the head.

The beginning of the dorsal fin is exactly opposite the ventral, and is one-third higher than long; its upper border is more concave than in the type. The anal fin is as long as high. The pectoral and ventral fins are of the same length, and the pectoral reaches nearer to the ventral, than does the ventral to the vent.

The eye has a deep-red iris. The largest example is nine inches long. This variety is found at Egel See, in Upper Austria.

## Leuciscus pigus (Lacép.).

D. 12—14, A. 14, V. 10, P. 18—19, C. 19. Scales: lat. line 46—49
above 7½—8
below 4.

In the Italian lakes, especially Como and Lugano, there is a beautiful Leuciscus, highly valued for food, which goes by the name of *Pigo*, a word synonymous with *lover*, so that the Swiss writers have called it "le gardon galant." The fishermen of Lake Como term it in the autumn *Eucobia*.

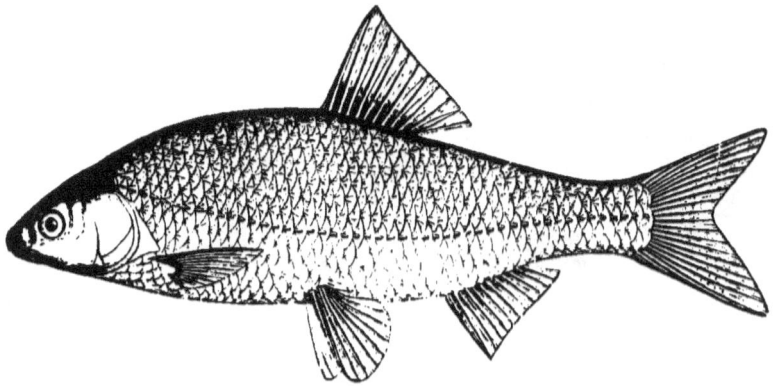

Fig. 75.—LEUCISCUS PIGUS (LACÉPÈDE).

This fish came to be described as a distinct species, and distinguished by some writers from the Leuciscus of the Danube, which was termed *L. virgo*, but

Fig. 76.—PHARYNGEAL TEETH OF LEUCISCUS PIGUS.

they are, beyond doubt, varieties of the same species, and as the name *L. pigus* is the older, we first give some account of the lake variety (Fig. 75).

The form of the body is elongated, with the head shorter and the eye smaller than in *L. rutilus*; the anal fin is longer than high. The head is about one-sixth of the length, and the fish is about four and a half times as long as high. The diameter of the eye is about one-fifth of the length of the head, or a little more. The profile rises quickly over the blunt snout to the hinder part of the head. Fig. 76 shows the pharyngeal teeth.

The dorsal fin is scarcely higher than long, is opposite to the ventral, and begins in the middle of the length. The length of the anal fin relatively to its height is a distinction from *L. rutilus*. The terminal rays of the caudal fin exceed the length of the head. The largest scales on the lateral line have a greater diameter than the eye; they have numerous diverging rays, as in *L. rutilus*.

At spawning-time the otherwise naked head, and the scales are covered with prickly growths, so characteristic of Cyprinoids (Fig. 77). The back is green becoming bronzy towards the sides; the belly is silvery; the anal and ventral fins are nearly black, and the other fins greyish. It reaches a length of fifteen inches, and often weighs three pounds.

Fig. 77.—HEAD OF LEUCISCUS PIGUS, SEEN FROM BELOW.

The variety called by Heckel *L. virgo* is known to the fishermen of Vienna as *Frauenfisch*. It is also known in Austria as *Nerfling*. It belongs to

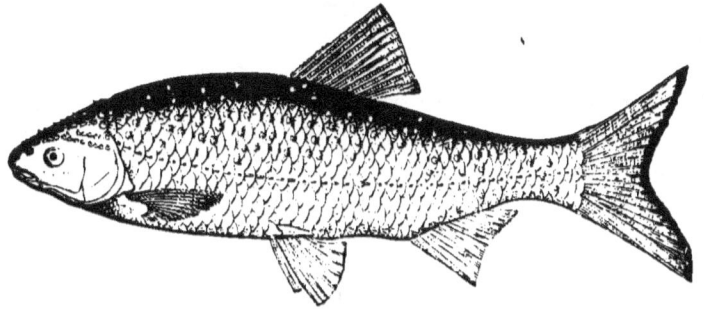

Fig. 78.— LEUCISCUS PIGUS (LACÉPÈDE), VARIETY VIRGO (HECKEL).

the group of the so-called white fishes. It is less common than the Ide and some other species. Its flesh is well-flavoured and wholesome, but is not prized as food. In the Danube and its tributaries its length varies from six to fifteen inches, and the weight may reach about two pounds (Fig. 78).

The head is small and sub-conical. The breadth between the eyes is equal to half the length of the head. The hinder border of the orbit is in the middle of the length of the head. The aperture of the mouth is small, reaching under the nares, but the large nares are rather nearer the orbit than the snout. The mouth is slightly protractile, and the nose is inflated a little over it. The sub-orbital arcade consists of five or six bones. The operculum is more rounded at the base than in the typical *L. pigus*. The pharyngeal bones are larger than in the Roach, and the pharyngeal teeth form a single row, six on the left bone, and five on the right.

The dorsal fin is one-fourth higher than long. Its soft rays divide into four. The anal fin is entirely behind the dorsal, has a shorter base, is slightly higher than long; its anterior soft rays divide four times, its posterior rays divide three times dichotomously.

The ventral fins are rather in advance of the dorsal, are rounded, and rather shorter than the pectorals, which are not so long as the head. The pectoral fins are more rounded than in the typical *L. pigus*. The caudal fin is well developed; its lower lobe is rather the longer; its dichotomous rays divide three or four times.

The scales are large, and firmly attached, largest in the anterior half of the body, and over and under the lateral line, which descends ventrally less suddenly than in the Roach. These scales have a diameter of one and a half times that of the eye. They are four times as large as the smallest scales, which are found on the breast. The scales have the usual wavy concentric striping, and on their free border a fan of five to seven rays.

At the beginning of the spawning-time in spring, the males develop lenticular spots which rise into white points. Their width and height are often more than the diameter of the eye; sometimes the blunt cone becomes a sharp, curved spine. On the sides of the body they are larger and thicker set than elsewhere, and cover the whole of the free part of the scale. They are rarely seen below the lateral line.

The head is covered with similar growths, which often form two longitudinal rows, extending, from the upper angle of the gill-opening, forward over the eyes. The hinder head and frontal region is crowned with them, and they extend round the nares. They are frequently so thick-set on the principal rays of the dorsal and caudal fins as to give a notched aspect to the rays; and the middle rays of the caudal fin are covered with tubercles as large as poppy seeds. After spawning, these tubercles are shed; and there remains for some time a depression like a scar in the spot where each stood. In life their bases are full of salts, but when kept in spirit the growths acquire a horny consistence.

In winter the colour of the back is pale greenish-brown, the sides are bluish, the belly is silvery, and the throat milk-white. The upper part of the head is dark; and the iris pale-yellow, with dark spots.

The dorsal fin is paler than the back; the pectoral fin is transparent. The ventral and anal fins are half-white, half-red, while the caudal is pale-red in the middle, and deeper red at the extremities of the lobes, which are bordered with grey.

At the end of March, when the first trace of the wart-like growths appears, the back acquires a darker green, and the sides put on a violet tint, which becomes bluish-red as it extends forward. The opercular region and all the scales shine with an opalescent splendour, which is not equalled by any other European Cyprinoid. The upper part of the dorsal fin becomes reddish, the pectorals yellowish, and the pale red of the ventral and anal fins changes into deep orange, while the caudal becomes a yellowish-red. These colours increase in brilliance to the end of April or beginning of May, the back becoming darker, the snout bluish-red, while the cheeks and operculum have a silver ground with a play of colour which changes to rose-red, violet, azure-blue, green, or yellow; and all the fins become bright red. After spawning, the colours fade, and the cicatrices left where the pearly growths fell off gradually disappear.

There are twenty-seven vertebrae in the dorsal region, and nineteen in the tail.

The *Leuciscus roseus* of Bonaparte is probably a variety of *L. pigus*. It is known only from the Lakes of Piedmont, and is so rare that Canestrini has never seen a specimen. The differences of proportions are all small and unimportant. The rose colour is its most distinctive attribute; and Fatio suggests that it is a local pale-coloured variety of *L. pigus*, comparable to the pale-coloured varieties of the Roach.

The *Leuciscus pigus* is captured by the Italian fishermen chiefly in the early morning. During the breeding season it is taken in great numbers with the net. As soon as the fishes are seen on the surface the net is extended round them, and they are sometimes driven into shallow water. It is a voracious fish, to which nothing comes amiss which can be swallowed, though its foul feeding does not affect its quality as food. When taken in great quantities the Italians salt and dry it in the sun for winter use.

## Leuciscus friesii (NORDMANN).

D. 11—12, A. 12—14, V. 10—11, P. 20, C. 19.
Scales: lat. line 61—67, transverse 11 3.

This species is found in the mountain lakes of Bavaria, like the Chiem See; in the Atter See and Mond See in Upper Austria, and is known from the Lake of Derkos, near Constantinople, and is met with in the Sea of Azov and the less salt parts of the Black Sea, and in the central and southern parts of the Caspian, where it ascends the rivers Koora and Terek.

The species is well defined by its elongated, nearly cylindrical form, and small scales, though the proportions vary with the locality (Fig 79). The height of the body at the beginning of the dorsal fin is equal to the length of the head, and is one-sixth of the entire length of the fish.

The head has a bluntly-conical form, but is not quite so broad at the operculum as high. The eye is rather small (one-sixth of the length of the head), twice its diameter from the snout, and half the length of the head from the other eye, so that the frontal region, which is convex, is remarkably broad.

The dorsal contour is arched as far as the dorsal fin, but the ventral contour is nearly straight from the pectoral to the anal fin.

The dorsal fin commences well in advance of the middle of the body; it is higher than long, the base being two-thirds of the length of the longest ray; it is truncate behind, the terminal rays being reduced to about one-

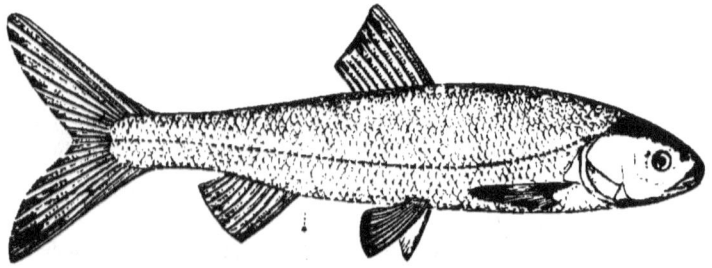

Fig. 79.—LEUCISCUS FRIESII (NORDMANN).

third of the height in front. The anal fin is far behind the termination of the dorsal; it is as high as long, and its base is as long as the base of the dorsal. The ventral fin is broad, of moderate length, a little shorter than the pointed pectoral, and opposite the commencement of the dorsal. The caudal fin is well developed, evenly lobed, and deeply forked.

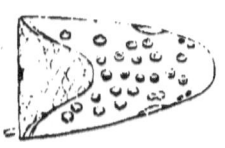

Fig. 80.—HEAD OF LEUCISCUS FRIESII, SEEN FROM ABOVE.

The scales are nearly circular, but longer than broad, and marked with many radiating rays. They form a fan of fifteen to twenty lines on the free edge of the scale. The largest scales are smaller than the diameter of the eye.

The lateral line descends a little below the middle of the fish in a concave curve. Its mucus-canal opens by simple pores.

During life the colour of the upper part of the head and back is blackish-green, the sides are blue-grey, and the throat and abdomen milk-white. Every scale of the abdomen shows behind the margin where it is over-lapped, a black spot of half-moon shape. The dorsal, pectoral, and caudal fins are blackish, though the two latter become grey towards their extremities. The ventral and anal fins are bluish-white at their bases, sometimes becoming reddish. At spawning-time the males develop remarkably large conical pointed tubercles, which grow from rounded bases, like the tubercles of *Leuciscus pigus*. The white colour of these thorn-like processes has suggested the name of Perlfisch for this species. The largest tubercles are found on the upper part of the head (Fig. 80) and fore part of the back; the smallest on the fins. When this growth takes place the entire abdomen acquires a delicate red colour.

The Perlfisch grows to a larger size than the Nerfling, measuring over twenty inches when adult, and Von Siebold mentions some examples which measured twenty-six inches; the weight may be about ten pounds.

It commonly lives in deep water during the whole year, and in the lakes comes to the surface only at spawning-time, in the first half of May, when it may ascend the tributary brooks. It then feeds on earthworms, cockchafers, and small fishes. During the spawning-time—which, according to Von Siebold, lasts a fortnight, and, according to Heckel, three weeks—the Perlfisch are chiefly caught with the line, and then are to be seen in the Munich fish market, though not much valued for food. In lakes the young fishes commonly live at a depth of fifteen fathoms, but the oldest may be taken in six fathoms.

The skeleton, according to Günther, consists of twenty-six vertebræ in the dorsal region and seventeen in the tail, which is one fewer in the back and two more in the tail than in *L. pigus*. The pharyngeal teeth are powerfully developed; they are in a single row, six on one side and five on the other; they are club-shaped, owing to a constriction towards the base; and the teeth are described by Von Siebold as more compressed than in the Nerfling.

## Leuciscus cephalus (Linnæus).—The Chub.

D. 11, A. 11—12, V. 10, P. 14—15, C. 19.

Scales: lat. line 42—46, trans. 7½ 6½.

The Chub (Fig. 82), which is widely spread over Europe and Asia Minor, belongs to the division of the genus Leuciscus in which the pharyngeal teeth are in two rows, a type which most Continental writers regard as of generic importance, and as forming their genus Squalius.

The body is thick, but slightly compressed, with a very broad head. The body is more than five times the length of the head; the greatest thickness is less than half the height. The head is longer than high, and the breadth of the frontal region between the eyes is nearly half the length of the head. The eye is between one-fifth and one-sixth of the length of the head; its hinder border is in front of the middle of the head. The upper jaw is scarcely longer than the lower, which is thickened at the symphysis in a fold. The nares are rather large, nearer to the snout than the eye, and less widely separated than the orbits. The dorsal profile is similar to the ventral profile, as far as the dorsal and ventral fins.

The dorsal fin, which begins nearly in the middle of the body, is much higher than long; its base measures less than half the length of the head. It is somewhat truncate behind, but the last ray is more than half the length of the longest ray. The anal fin is well behind the dorsal; its base is as long as the base of that fin, but it is not so high, and its margin is slightly convex. The ventral fins are in front of the dorsal, and do not reach back to the vent. The caudal fin is broad and strong, not deeply forked, with the lower lobe slightly the longer, and its longest ray is nearly as long as the head. The jointed rays of the dorsal, anal, and caudal fins sub-divide three or four times.

The scales are large and strong; the largest, which are over the lateral line, measure one and a half times the diameter of the eye. On the back, tail, and abdomen, the scales are smaller; the smallest, which are on the throat, are one-quarter of the size of the largest. The scales are finely marked with concentric lines, and have a fan of eight or ten rays, which scarcely indent the free margins. The lateral line is sub-parallel to the ventral margin, and its mucus-canal opens by short, simple pores. A branch of the cephalic mucus-canal is indicated on the pre-operculum by large pores, and another branch extends over the eye to the nares.

The number of vertebræ in the skeleton varies; according to Dr. Günther, there may be twenty-five or twenty-six thoracic, and eighteen or nineteen caudal vertebræ, so that some varieties have two more than others. The facial bones of the skull are weak, in comparison with its other parts. The ethmoid bone is as broad again as long. The supra-orbital and hindermost sub-orbital bones, which in other species of the genus are small, are well developed in this species. The third sub-orbital bone is small. The bones associated with the gill apparatus are strong. The pharyngeal bones are small, elongated, and in the form of a pot-hook. The pharyngeal teeth (Fig. 81) are long, compressed, and hooked, with the formula 5·2—2·5.

The colour of the back is blackish-green, changing on the sides to yellow or shining-silver; on the throat it becomes reddish-white. The cheeks and operculum have a brilliant rose-red and golden glimmer, the lips are red; the iris is golden yellow, with dark green spots. The dorsal fin is blackish, and, like the caudal, red at the base, though the latter is sometimes olive, and sometimes margined with blue. The anal and ventral fins are deep red, but the rays are more intense in colour than the interspaces. The spine-like scale over the base of the ventral has a similar red colour. All the other scales have a black pigment spot towards the middle of their free margins. The colours of the young fish, especially those of the lower fins, are pale, or yellowish, and sometimes spotted with black, while the caudal fin is quite black.

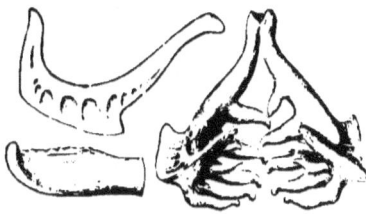

Fig. 81.—PHARYNGEAL TEETH OF LEUCISCUS CEPHALUS.

In England the Chub attains a length of upwards of twenty inches, and may weigh between five and six pounds. In the Danube and Neckar its weight is commonly four to five pounds; but it reaches a weight of eight or

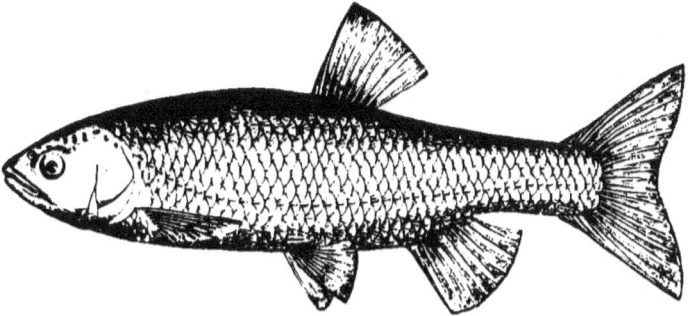

Fig. 82.—LEUCISCUS CEPHALUS (LIN.).

nine pounds, in the Austrian lakes. In France and Germany the size is much the same as in England. In the colder and higher Swiss lakes the fishes are smaller.

The Chub is found indifferently in the lakes of mountainous and low-lying country, and in brooks and rivers. It rather prefers a gravelly or sandy bottom. Hundreds of young fish may often be seen swimming together, till a shadow of a cloud or some noise disturbs them, when they dart away like arrows. This species is more voracious than any other Cyprinoid. It lives on worms, flies, and other insects when young, but as it grows older frequents deeper water and larger streams, when it becomes predatory, and lives on small fishes captures the fresh-water crayfish (*Astacus fluviatilis*) when casting its shell, and if frogs are abundant feeds on them, while, exceptionally, it eats mice. At Strassburg the Chub is known as the Mouse-eater, and is reputed to eat water rats, which is not improbable, since one individual at least has been found in England with a rat partly swallowed in its throat. Chub increase in weight about one pound a year when well fed, and live for eight or nine years. According to Buckland, they are very much like the horse that was hard to catch, and good for nothing when caught. In any case the fish is not much sought after by English fishermen. But national taste varies, and Blanchard tells us that in some localities in France the Chub is a not unimportant source of food supply. It is excellent food for Pike. It is shy but often cunning. It remains motionless in the stream till its prey appears, and then rushes on some Gudgeon or other small fish, which instantly disappears down its huge gullet. When in company it will advance slowly along the banks on an exploring expedition, but under the influence of pleasure or fear, moves with surprising speed.

Its mouth is not sensitive, and its appetite is always ready for what the angler has to offer. Fatio mentions that in Switzerland the hook is baited for it with bread, cheese, cherries, gooseberries, or plums. It is also taken with the net, and is said to be sometimes speared with a trident.

Heckel and Kner state that during the period when the elder is in bloom, these fish suffer from an eruption of the skin and often die. In ponds they are liable to a sickness in which they cease to grow, become thin, and get large heads with sunken eyes; when they must be removed from the ponds, as the sickness is infectious. They are infested with parasites, internal and external.

Chub spawn during the month of April or May in England, May or June in France and North Germany, and June in Austria; the date varying with the temperature. The spawning occupies eight or ten days, or a fortnight. The eggs are small and numerous, deposited near to the banks, on stones and gravel. Large troops of fish assemble and spawn together, pressing against each other with lively movements, often jumping out of the water, till the spawn is deposited; when they leave the locality. The males at spawning-time develop tubercles, which are largest on the head, but extend down the scales of the back, and upper parts of the sides.

The air-bladder is large, somewhat constricted and slightly bilobed in front. The intestine forms two great curves, and its length exceeds that of the fish; the stomach is large.

This species is exceedingly variable, and some of the varieties have a local importance as geographical representatives of the type described. Thus the *Leuciscus caredanus* of Bonaparte (Fig. 83), is essentially a southern variety of *L. cephalus*, though Steindachner attributes it rather to age and individuality. Since, however, this type is met with in Spain and Portugal, especially in the Tagus, Douro, Mondego, and Albufera Lake, throughout Italy and Dalmatia, and is not recognised in the north of Europe, it may be useful to mention some of the characters in which it differs from the typical Chub.

According to Heckel and Kner the southern fish has a longer and flatter head, smaller frontal region and larger mouth; the eyes are separated by two and a half times the orbital diameter, but this amounts only to two-fifths of the length of the head. The nares are farther apart; the dorsal profile is flatter than the ventral contour. The dorsal fin is behind the middle of the length; its longest rays are longer than its base. The anal fin is rather farther back than in the northern form. The caudal is evenly-lobed, and the terminal fork penetrates to half the length of the fin. The jointed rays of the dorsal and anal fins divide three times. The back is green with pale

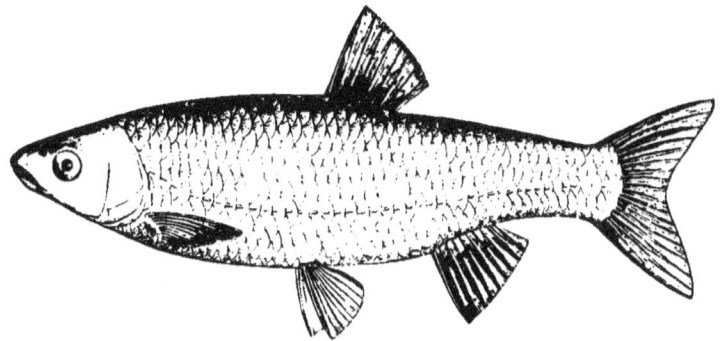

Fig. 83.—LEUCISCUS CEPHALUS, VAR. CAVEDANUS (BONAPARTE).

bronze tinting on the scales, and the belly is silvery. All the fins are transparent at the bases, becoming darker towards the extremities; and the anal, ventral, and caudal fins, are often bordered with black. This variety of the Chub frequents deep water, and hides under stones: it does not readily take bait, but it is said to be partial to the living pupa of the silk-worm. It is not valued for food, but is sometimes taken in nets which are put down for better fishes.

It is found in the Po and smaller streams; in Lake Garda, and other northern lakes.

A more elongated variety of the Chub is found in Dalmatia. It was regarded by Heckel and Kner as the *Leuciscus albus* of Bonaparte, as Günther is disposed to think, though Canestrini differs. But all modern writers agree in regarding it as a slender Chub (Fig. 84).

The most striking characteristics of this variety are the elongated form, large head, and long pectoral fins. The body attains its greatest depth far in in front of the dorsal fin; its length is five and a half times its greatest height, and four and a half times the length of the head. The greatest thickness is in the pectoral region, and is equal to half the length of the head. The eye is fairly large, being one-fifth of the length of the head; it is separated from the other eye by twice the orbital diameter. The angle of the oblique mouth reaches nearly as far back as the anterior border of the orbit. The dorsal fin, which commences in the middle of the length, is much higher than long. The anal fin has a similar length of base, but is shorter and less truncated behind. The ventral fins are rather in front of the dorsal fin, but behind the middle of the body. They do not reach back to the vent. The long pectoral rays are equal to the longest rays of the caudal fin. The lateral line descends rapidly over

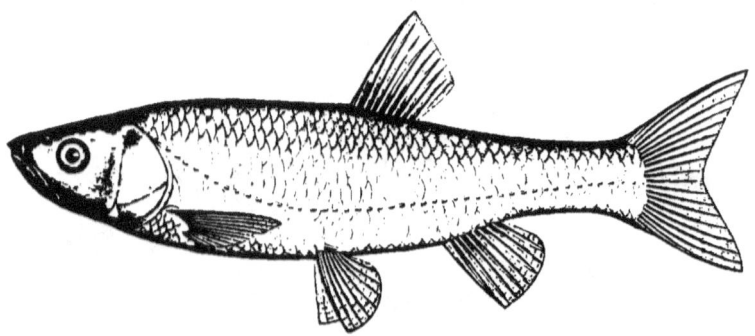

Fig. 84.—LEUCISCUS CEPHALUS, VAR. ALBUS (BONAPARTE).

the pectorals to the lower third of the body, and rises again to the middle of the tail. The length of the fish is from four to nine inches. It is recorded by Heckel and Kner from the Kerka, near Scardona, in Dalmatia. Canestrini regards this fish as essentially an elongated form of the *L. cavedanus*; but, in our view, it is a good geographical variety.

## Leuciscus vulgaris (Fleming).—The Dace.

D. 10, A. 11, V. 9—10, P. 17, C. 19. Sq. lat. 47—52, transverse $8\frac{1}{2}/6$.

The Dace is characteristic of the parts of Europe north of the Alps. It is also known in Britain as the *Graining*. In France it is *la Vaudoise*; in North Germany it is *der Hasling*; in Austria, *der Hasel*; in Sweden, the common name is *Stämm* (Fig. 85).

It is a very variable fish, and distinguished naturalists have separated its varieties as distinct species. The form of the body is elongated, and rather compressed, with a blunt snout, and similar ventral and dorsal contours in the anterior part of the body. The greatest height, in advance of the dorsal fin, exceeds the length of the head, which is nearly one-fifth of the entire length. The body has a compressed aspect, but the greatest thickness is rather more than half the height. The head is rather small, rounded above and flat at the sides. The mouth is small, with the upper jaw wider than the mandible; and the jaws do not perfectly meet, so that, according to Günther, a triangular interspace is included at the angle. The angle of the mouth extends back as far as the nares. The length of the head is four and a half times the diameter of the eye, and the eyes are separated by one and a half times their own diameter. The dorsal fin is placed over the ventrals in the middle of the body, and is conspicuously higher than long;

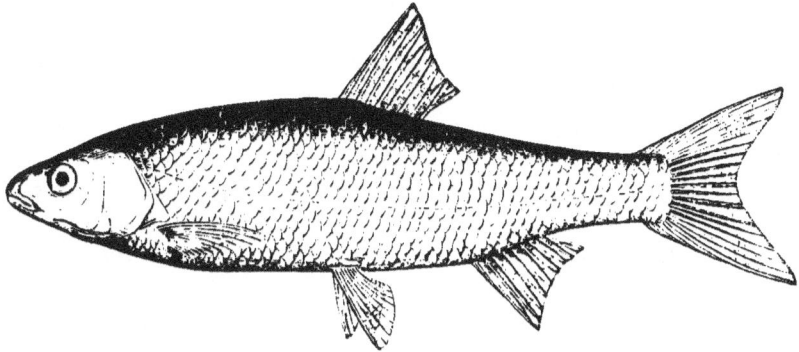

Fig. 85.— LEUCISCUS VULGARIS (FLEMING).

its base is longer than the head, and the free margin is concave. The anal fin is behind the dorsal, scarcely higher than long, with a base shorter than the length of the head. The ventral fins are broad and well-developed; the pectoral fins are rather longer. The caudal fin is strong, evenly-lobed, and as long as the head. The lateral line descends in a regular concave curve.

The scales are very similar to those of the Chub; they are rather higher than long; measure one-third the diameter of the eye in the young fish, and four-fifths the diameter of the eye in the adult. Each has the posterior border somewhat truncated, with the upper and lower angles concavely excavated.

In life, the colour resembles polished steel; the back and upper part of the head, and the dorsal and caudal fins, may be black or steel-blue. The sides are silvery, often with a bluish tinge. The pectoral and anal fins are pale red, and the anterior part of the former may be spotted with orange; the iris is green, yellow, and red.

The Dace frequents tranquil tributary streams and brooks, preferring shallow water, but is found in rivers and lakes. It swims quickly and often skims over the surface. It lives on worms, insects, and vegetable substances; in some waters being almost entirely a vegetable feeder, but Günther remarks that where animal food is abundant the fish are better nourished. Its flesh, however, is not valued for food, though in season from October to January. Heckel and Kner mention that in Upper Austria, according to old custom, any one is at liberty to catch it from four to five o'clock in the evening, between Ascension day and St. John the Baptist's day.

The Dace commences to spawn when the warm weather begins, as early as March in Germany, but usually in May and June in England. The eggs are numerous and deposited on stones, where they are preyed upon by many enemies, so that the species is nowhere very plentiful. Its habit is gregarious, and the fish commonly swim in shoals.

The young remain for a time in the stagnant or tranquil water in which they are hatched, and as they increase in size become more adventurous, and move to rapid parts of the stream. Yarrell mentions that it is often taken by fly-fishers when fishing for Trout. It is tenacious of life, and valued as bait for Pike; in Germany it is used as bait for Salmon. Buckland records that the late Duke of Wellington invented an apparatus of india-rubber bands by which the Dace is held secure to the hook. The length rarely exceeds eight to ten inches; and may reach a weight of one pound. It is found in most of the large rivers of England, but is not recorded from Scotland or Ireland.

The best known variety in England is the *Graining*, for which Yarrell used the name *Leuciscus lancastriensis*; it is distinguished by a smaller head, which is one-sixth of the length, while the depth is one-fifth of the length. The nose is rounder; the scales are rather larger than those of the Dace, with a greater vertical diameter, and fewer radiating rays. The fins are a little longer, and the upper parts of the body are a pale drab, tinged with bluish-red, and sharply defined from the paler-coloured lower part of the body. All the fins are pale yellowish-white.

In France the varieties which have been distinguished are more numerous. Blanchard describes a fish which has the aspect and colour of a Roach, found in the lake Mariscot, near Biarritz, which in the pharyngeal teeth and rays of the dorsal fin more resembles the Dace. The body is more compressed than in typical Dace, and the back higher. This variety is named *Leuciscus bearnensis*, Blanchard. Another variety, found in the Garonne, distinguished by Valenciennes as *L. burdigalensis*, has the head less obtuse and more elongated than in the common Dace. The dorsal fin is black, and the fins of the lower part of the body are orange at the base.

In Switzerland and North Germany the varieties merge in each other, so that most writers have followed the example of Von Siebold in uniting them without an attempt at defining their geographical distribution.

Heckel and Kner endeavoured to distinguish several of these varieties as species. Thus the *Leuciscus chalybæus* has the head directed rather more downward so as to give a relatively convex profile to the back, and was re-

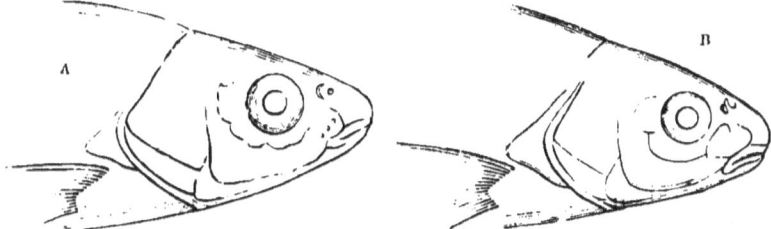

Fig. 86.—HEADS OF LEUCISCUS VULGARIS: A, OF TYPE; B, OF VAR. RODENS.

garded by them as peculiar to the little river Kamp in Lower Austria, where it frequents deep water (*see* FRONTISPIECE).

Another variety, *Leuciscus rodens*, is remarkable for its elongated form and rather short head (Fig. 86, B). The depth of the body exceeds the length of the head, and its thickness is more than half its height. The nose is thick and

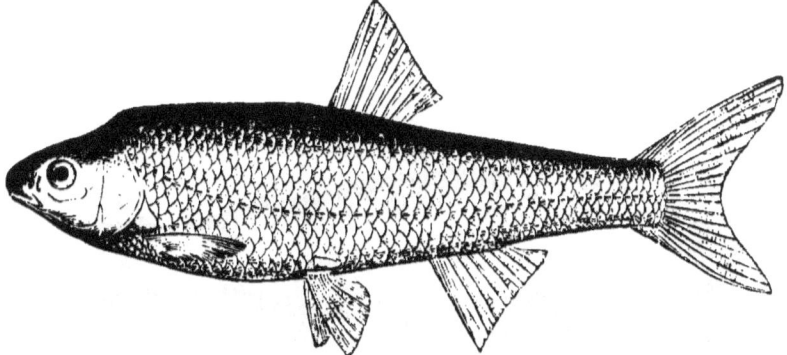

Fig. 87.—LEUCISCUS VULGARIS, VAR. ROSTRATUS (HECKEL).

arched. The scales are rather small, the largest being only half the diameter of the eye. Below the lateral line there is a clear silvery or golden longitudinal band. All the fins are transparent, and the back is greenish-blue. At spawning-time the males develop black pigment spots, which afterwards vanish. This form is found in Lake Constance, and is probably the fish found in the Neckar, in which Günther describes a similar development of pigment at the spawning period. Fatio mentions the association with these spots with white granulations on the top of the head and sides of the body in the Dace of Switzerland.

Attempts have been made to introduce Dace into the Lake of Geneva, but hitherto without success.

The variety which was described by Linnæus as *Cyprinus leuciscus*, and which has, therefore, good claims to be considered the type of the Dace tribe, is most widely distributed, being found throughout European Russia, though absent from the Caucasus. According to Canestrini, it is absent from Italy, and it is exceedingly rare in Austria; but throughout Germany, Switzerland, France, and Belgium, it is generally present.

*Leuciscus rostratus* (Fig. 87), is a variety known in the Tyrol as the *Marzling*. The dorsal profile is nearly horizontal, but the back rises suddenly from the head.

The vertebral column varies a little, there being from twenty-three to twenty-four abdominal, and nineteen to twenty caudal vertebræ.

## Leuciscus illyricus (Heckel and Kner).

D. 11, A. 12. Scales: lat. line 49—54, transverse $\frac{9-10}{4-5}$.

This fish belongs to a small group of species which are found in the South of Europe. It has the high compressed body and large scales characteristic of the preceding species, but is especially distinguished by the

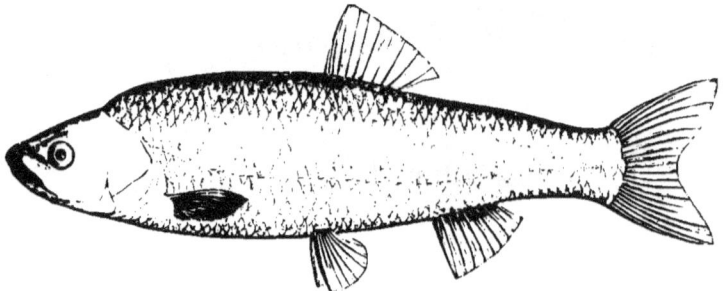

Fig. 88. LEUCISCUS ILLYRICUS (HECKEL AND KNER).

upper jaw being the longer, by shortness of all the fin-rays, by the numerous rays on the scales, and by the small size of the eye (Fig. 88).

The head is one-fifth of the length of the fish, and the height of the body exceeds the length of the head. The eye is one-sixth of the length of the head, is separated from the snout by twice its diameter, and the eyes are separated by about two and a half times the orbital diameter. The mouth extends downward at an angle of about 45°, and reaches under the nares. The nose is blunt, but more pointed than in the Dace or Chub. The fleshy lips of the premaxillary extend forward over the lower jaw; the nares are nearer to the eye than to the snout. The profile over the head is straight, and then rises abruptly in a sharp curve to the back, which is highest in front of the dorsal fin. The ventral profile forms a more regular curve.

The dorsal fin is opposite the ventral, as long as high, rather truncated behind, with the last ray two-thirds of the length of the longest ray. The proportions of the anal fin are similar, but it is shorter and lower. The pectoral and ventral fins are moderately developed, and the caudal fin is smaller than in any of the species described. Its longest rays are only two-thirds of the length of the head. It is evenly-lobed, but the fork is short.

The size of the largest scales exceeds the diameter of the eye; the free edge of a scale shows about twenty or more line rays, while the concentric striping is almost invisible to the naked eye. This character of the scales forms a useful distinguishing difference from the Chub (Fig. 89).

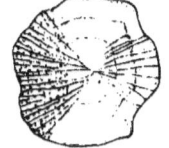

Fig. 89. LARGE SCALE OF LEUCISCUS ILLYRICUS.

It would appear to be a rare species, since the descriptions are drawn from specimens in spirit; and it is impossible, therefore, to speak with certainty of the colours. The back appears to be blue-grey, and only a little darker than the side. The operculum is coppery. Black pigment margins the free edges of the scales, so that the upper part of the body has the aspect of being covered with a network.

This fish reaches a length of thirteen inches in the River Isonzo, which flows into the Adriatic Sea. It is also known from Knin and Sign, in Dalmatia, where it is distinguished by the provincial name "Klien."

## Leuciscus pictus (Heckel and Kner).

D. 11, A. 11—12, V. 8. Scales: lat. line 43—44, transverse 8½/6.

This species is known only from the River Rieka, in Montenegro, where it reaches a length of five to six inches. Dr. Günther describes it as having remarkably thick lips, with the lower lip extending across the median symphysis as a continuous fold.

The height of the body is equal to the length of the head. The diameter of the eye is two-ninths of the length of the head, and is rather more than half the width of the inter-orbital space. The cleft of the oblique mouth does not extend back to the front of the orbit; the mandible is overlapped by the upper jaw. The dorsal fin is higher than the anal. The origin of the former is opposite to the ventrals, and much nearer the base of the caudal fin than the end of the snout. The bases of the dorsal and anal fins are equal in length. The pectoral fin terminates at some distance in advance of the ventral. The scales are marked with numerous striæ. There are three longitudinal rows of scales between the lateral line and ventral fin. The body is irregularly mottled with brown. There are no spots on the head or the fins. The habits of the species have not been observed.

## Leuciscus svallize (Heckel and Kner).

D. 12, A. 12—13, V. 9.  Scales: lat. line 46, transverse 8,8.

*Leuciscus svallize* (Fig. 90), known only from Dalmatia, is a southern representative of the Dace, from which it differs in having more rays in the anal fin. It is closely allied to the Chub, and has an elongated form, with wide

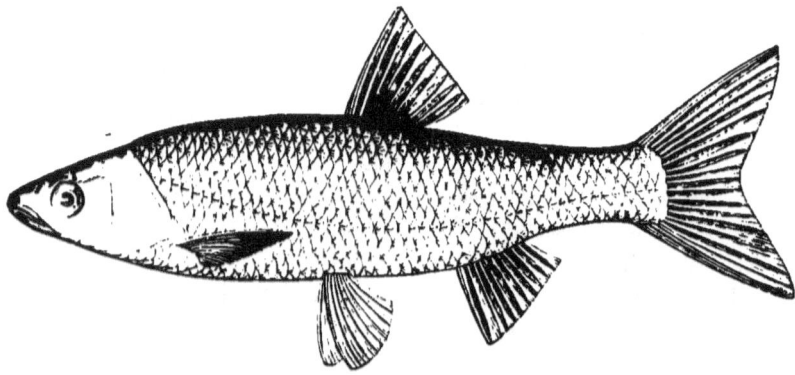

Fig. 90.—LEUCISCUS SVALLIZE (HECKEL AND KNER).

head, large eye, mandible almost as long as the upper jaw, a high dorsal fin, and scales with few rays. The head is one-fifth of the length of the fish, and its length is nearly equal to the height of the body. The greatest thickness is half the length of the head. The head is about four and a half times as long as the eye; the eye is separated from the snout by its own diameter, and the width of the frontal region between the eyes is fully one-third of the length of the head. The mouth is rather narrow, less oblique than in many allied species, and reaches back to the front margin of the orbit. The hindermost sub-orbital bone is not larger than the first, and the intermediate bones are very narrow.

The ventral fins begin half-way down the body. The dorsal commences opposite to the middle of the base of the ventral; it is truncate behind. Its longest rays are almost as long as those of the caudal fin, which exceed the length of the head. The anal fin is lower than the dorsal, and nearly as long as high.

The scales are rather large, being from three-quarters to four-fifths of the diameter of the eye; the free margin has a fan of from six to ten rays; and they are marked with fine concentric striping. Their borders show an edging of black pigment, as in *L. illyricus*; but, owing to the larger size of the scales, it forms a network with wider meshes. There are more longitudinal rows of scales between the lateral line and ventral fin. The colour is known only from specimens in spirit; the back is dark steel-blue, and the sides, according to Dr. Günther, are of shining silver. The fins are transparent. The species is found in the River Narenta and a lake near Vergoraz. The pharyngeal teeth are hooked and slightly denticulated, and are arranged in two rows, with the formula 5—2, 2—5.

## Leuciscus ukliva (HECKEL).

D. 10—11, A. 11—12. Scales: lat. line 62—65, transverse 11—9.

This small fish (Fig. 91) is about six inches long, and has been caught hitherto only in the River Cettina at Sign, in Dalmatia. Its head is short and

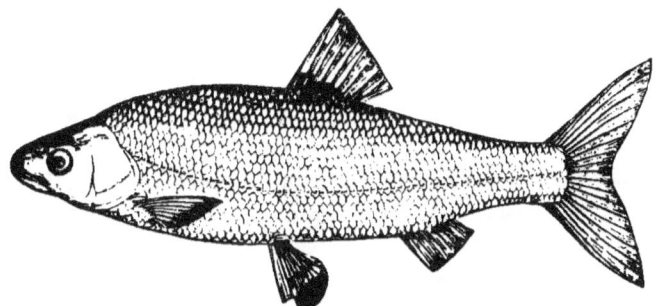

Fig. 91.—LEUCISCUS UKLIVA (HECKEL).

thick, with the nose arched, and projecting slightly over the small mouth. A narrow black band runs along each side, above the lateral line.

The fish is five and a half times as long as the head; but the height of the body is only five-sixths of the length of the head, and in the tail the height is diminished to one-third of this amount: the width at the operculum is half the length of the head. The diameter of the eye is one-quarter, and the inter-orbital space is about one-third, of the length of the head. The snout is convex, with the upper jaw overlapping the mandible. The angle of the small mouth does not reach back to the anterior margin of the eye. The dorsal fin commences opposite to the middle of the base of the ventral fin. Its longest ray is equal to the length of the pectoral fin. The anal fin is far behind the end of the dorsal, is less elevated, is truncated horizontally, and its rays are as long as those of the ventral fin. The deeply-forked caudal fin has its longest rays nearly equal to the length of the head. The largest scales have only one half the diameter of the eye; a fan of ten to twelve fine rays is seen on the free border of each scale. There are five longitudinal series of scales between the lateral line and the ventral fin. The swim-bladder of the females is remarkably small and short, and the well-developed ovary extends in front of it. The pharyngeal teeth are in two rows with the formula 5·2—2·4 or 5. The colour of the back is blackish-green; the sides are yellow, and the abdomen silvery. The black longitudinal band over the lateral line is sometimes broken into fine points. All the fins are orange-coloured at their bases, and paler at the extremities.

## Leuciscus turskyi (Heckel).

D. 10, A. 11, V. 8. Scales: lat. line 70—72, transverse 15/5—6.

This is a beautiful little species (Fig. 92), about six inches long, which Heckel found in the back water of the river Cicola, at Dernis, in Dalmatia, and which was named in honour of General von Tursky, to whom Heckel and Kner were indebted in making their studies of Dalmatian fishes. In several respects this species resembles the last described, the scales and mouth being small, and the body being marked at the side with a broad longitudinal band, which extends over the caudal fin; and Professor Canestrini, who prefers to unite species which are not markedly distinct, has proposed not only to regard *L. turskyi* as a variety of *L. ukliva*, but associates with them *L. microlepis* and *L. tenellus*, although the number of scales is stated by him at 62—64 in *L. ukliva*, 70—72 in *L. turskyi*, 73—75 in *L. microlepis*, and 78—80 in *L. tenellus*. But while we fully concur with Canestrini in recognising the near affinity between these forms, they may be accepted as varieties which it is convenient to separate.

The body is four and a half times the length of the head, and the height of the body is three-fourths of the length of the head. The diameter of the eye is one-fifth of the length of the head. The eyes are distant one and a half times their own diameter from each other, and from the snout. The angle of the small mouth reaches back to below the front of the orbit, but is directed obliquely downward, as in the allied species. The first bone of the sub-orbital ring is remarkably broad and large. The ventral fins are in the middle of the body. The dorsal fin is slightly farther back, but its longest rays have only the length of the head; it is truncate behind. The anal fin is less obliquely truncate, and is lower. The ventral and pectoral fins are both well-developed. The latter is as long as the terminal rays of the caudal fin, being one-sixth of the length of the body. The ventrals reach back to the vent. The largest scales are two-fifths of the diameter of the eye; they have numerous rays

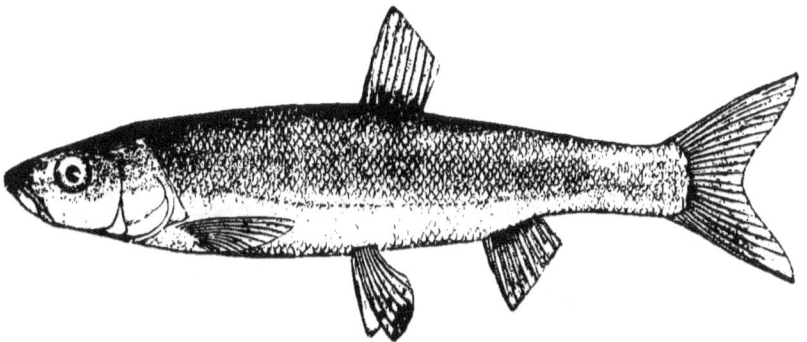

Fig. 92.—LEUCISCUS TURSKYI (HECKEL).

on the free border, but the concentric markings are feeble. There are five or six longitudinal series of scales between the lateral line and ventral fin. The colour of the back is blue-green, the sides are shining golden yellow, and the abdomen is silvery. The black longitudinal band extends down to the lateral line throughout its length. All the fins, except the dorsal, are deep orange at the base.

## Leuciscus microlepis (Heckel).

D. 11, A. 11, V. 9.  Scales: lat. line 73—75, transverse 15—7.

This fish has very much the form of the Dace, with the head long and pointed, the mouth large and oblique, scales very small, and all the fins pale yellow (Fig. 93). It is probably only a variety of *L. turskyi*. It is from four to four and a half times as long as the head, and, as in the allied forms, the length of the head exceeds the depth of the body in front of the dorsal fin. The eye is one-sixth of the length of the head; in old age it becomes relatively a little smaller. The frontal interspace between the eyes is twice the orbital diameter in width, and in the adult fish the eye is the same distance from the snout, but in the young fish the snout is a little less elongated. The nares are near to the orbits; the jaws are long, reaching under the eye. The cephalic canals are large and open, with numerous pores, and are particularly conspicuous in the sub-orbital and frontal areas. The dorsal profile is a flat arch.

The dorsal fin commences behind the middle of the body over the thirty-fifth to thirty-seventh scale of the longest longitudinal row. It is higher than long. The ventral fins are a little in front of the dorsal; their longest rays are shorter than those of the dorsal fin. The anal fin commences behind the termination of the dorsal; its rays are shorter than those of the ventral. The longest rays of the pectoral fin are equal to those

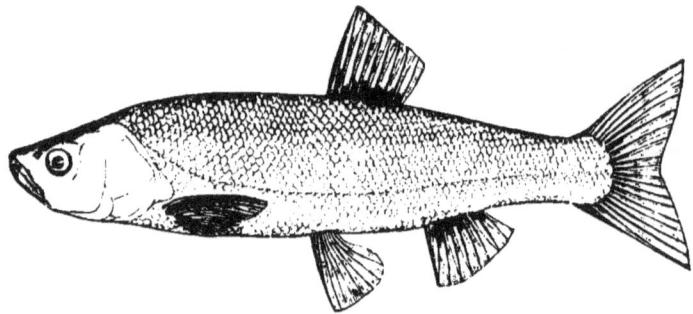

Fig. 93.— LEUCISCUS MICROLEPIS (HECKEL).

of the caudal fin. As in the allied species, the scales have rays diverging in a fan shape at the free border; they usually number ten to twelve, but as the largest scale is scarcely half the diameter of the eye, the rays appear to be numerous. The colour of the back is brownish-green, but the sides towards the abdomen are of burnished silver. All the fins are yellowish, without a black border, and there is no longitudinal lateral band as in *L. turskyi;* but, as Canestrini remarks, this lateral band frequently disappears in *L. muticellus, L. aula,* and other forms.

This species is only found in Dalmatia, in the River Narenta, near Vergoraz, and in the Lake of Dusino, near Imosky, where it usually reaches a length of six or seven inches, and is occasionally a foot long.

At Vergoraz it is known as the *Makli.*

## Leuciscus tenellus (Heckel).

D. 10—11, A. 11.   Scales: lat. line 78—80, transverse 18 8.

This species (Fig. 94) has been recorded only from the neighbourhood of Livno, in Bosnia, where it reaches a length of six inches. It has a broad head, which is long and bluut. The lower jaw rises abruptly, the mouth is large, and the scales are smaller than in any of the species hitherto described. The fish is about four and a half times as long as the head, and the greatest height of the body is two-thirds of the length of the head. The diameter of the eye is between one-fifth and one-sixth of the length of the

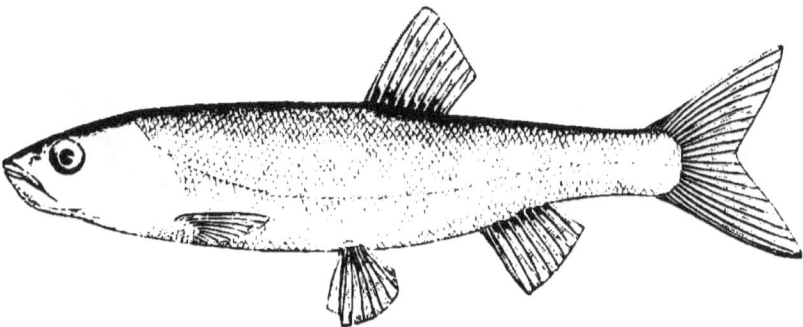

Fig. 94.   LEUCISCUS TENELLUS (HECKEL).

head, and the frontal interspace between the eyes is one and a half times the orbital diameter. The jaws are of equal length; the angle of the mouth reaches back to below the anterior margin of the orbit. The dorsal fin is behind the middle of the fish, and its longest rays are exceeded only by those of the caudal, which are two-thirds of the length of the head. The pectoral fin is as long as the dorsal, and the ventral as long as the anal.

The largest scales, which are on the lateral line, are scarcely one-quarter of the diameter of the eye. They are delicate, have a remarkable concentric striping, and each shows on the free edge a fan of six to eight diverging rays.

The lateral line descends somewhat rapidly, and runs in the lower third of the height.

The colour of the back and sides shades from iron-grey into steel-blue. The belly is silver-white. All the fins are yellow except the dorsal, which is grey.

## Leuciscus muticellus (Bonaparte).

D. 10—11, A. 11—12, P. 9—10.
Scales: lat. line 46—60, transverse 9—10/5—6.

Many Continental writers have referred this species to a section of the genus Leuciscus, which Heckel named Telestes, and distinguished by the number of its pharyngeal teeth, which follow the formula 5·2—2·4. But Von Siebold found that in seventy-two individuals thirty-three had the teeth 5·2 on the left and 4·2 on the right, two individuals reversed this arrangement, and thirty-seven had 5·2 on each side. Since this character is not associated with

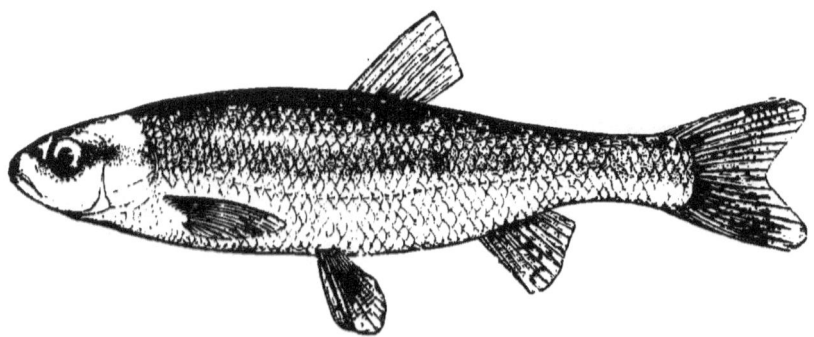

Fig. 95.—LEUCISCUS MUTICELLUS (BONAPARTE).

any other of importance, and is so inconstant, we follow Dr. Günther in grouping Telestes in the genus Leuciscus. The most obvious character of the species is the broad black band which extends from the arched snout over the eye down the body to the tail (Fig. 95).

The fish is from five to five and a half times as long as the head. The height of the body exceeds the length of the head. The diameter of the eye is one-fourth or one-fifth of the length of the head, but the length of the nose and size of the eye vary with age and sex. In the young fish they may be more than one-third the length of the head. Hence the distance of the eye from the snout varies, but is commonly about equal to the orbital diameter. The width of the interorbital space exceeds the diameter of the eye, and may be about one-third of the length of the head. The snout is rather arched, and projects somewhat over the lower jaw. The cleft of the mouth is more

Fig. 96.—PHARYNGEAL TEETH OF LEUCISCUS MUTICELLUS.

horizontal than in other species; it does not quite reach back to a vertical line drawn from the front of the orbit.

The dorsal fin begins half way down the length of the fish, and in specimens from the Neckar is over the insertion of the ventrals. Its base is only half the length of the head; and it is higher than long. Its first jointed ray is the largest, and it is twice as long as the last ray. The anal fin has the shortest rays, but its first jointed ray is similarly the longest, and twice as long as the shortest. The pectoral fin is small, but its length exceeds that of the ventral and the dorsal fins. The ventral fins do not extend back to the vent. The caudal fin is deeply forked; its longest rays are as long as the head. The number of scales is remarkably variable in different localities. There are from forty-eight to forty-nine scales in the longitudinal rows in the variety from the Danube which Heckel distinguished as *L. agassizii*, and fifty-six scales in the lateral line in the variety which he named *L. rysela*. Günther found in specimens from the Neckar that the number varied from fifty-four to sixty. The middle scales are about the size of a pupil of the eye, though some scales are large enough to cover the yellow ring round the pupil. The free margin of the scale shows a fan of eight to ten rays. The scale is as high as long. The lateral line is yellow.

The colour of the back is dark grey, varying into steel-blue; the abdomen is of shining silver. A black, longitudinal band, which often commences at the tip of the nose, goes through the eye and over the operculum to the tail. It is formed of densely-grouped black pigment spots. At the fore part of the body it has the width of three scales; behind the anal fin the width is reduced to one scale and a half; but the band widens again, so that at the base of the caudal it is nearly five scales wide.

All the fins are transparent and unspotted in Austrian specimens, but in examples from the Neckar the fins of the lower part of the body are yellow at the base, and this colour is occasionally seen in the dorsal and caudal. Bavarian fish have much black pigment in spots on the dorsal and caudal fins. In the tributaries of the Danube the largest specimens have a length of about five inches; but the fish is larger in the Neckar, Dr. Günther obtaining specimens nine inches long at Heilbronn—and it is equally large in Bavaria.

The pharyngeal teeth have a compressed form, are hooked, and slightly denticulated (Fig. 96).

This fish feeds on animal and vegetable substances, and Günther found small molluscs and beetles in the stomach. The lining membrane of the visceral cavity is black. The number of thoracic vertebræ is twenty to twenty-one, and of caudal vertebræ nineteen to twenty-one. There are sixteen to eighteen pairs of ribs.

The tongue is adherent to the surface on which it rests. The stomach is not very definitely divided from the intestine, which makes two folds, and is as long as the fish. The liver has three lobes, and its lower lobe lies between the intestine and ovary. It has a large green gall-bladder. The eggs are rather large, and number 6,000 in the two ovaries. The kidney has a transverse thickening which fits into the constriction which divides the air-bladder. The air-bladder is small, with its two portions similar.

The fish spawns in March and April, and its colouring is then most brilliant. The eggs are deposited in three or four days, and placed upon stones. After spawning, the black band at the side becomes paler.

The flesh is nowhere valued for food, but is esteemed as fodder for other fishes, and is sold in Munich as food for Salmon.

In Austria the fish is known as *Laugen*, in Bavaria it is the *Stromer*, in

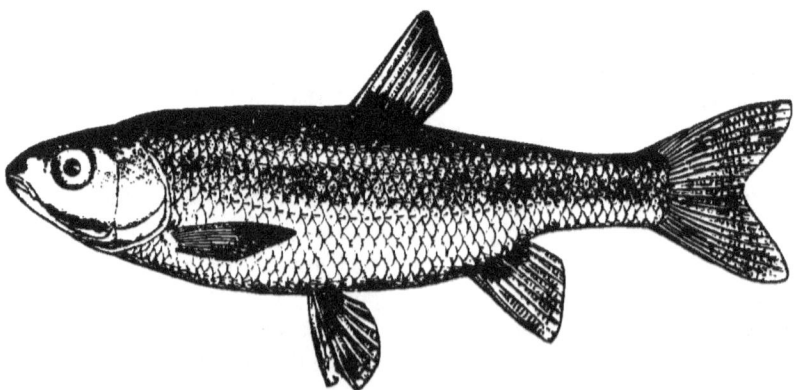

Fig. 97.—LEUCISCUS MUTICELLUS, VAR. SAVIGNYI (BONAPARTE).

the Neckar it is known as *Gangfisch*, in Switzerland as *Riesling*, in France as *Blageon*, and in Italy as *Vairone*.

A variety of this species was distinguished by Bonaparte as *L. savignyi*. It is well known in the north of Italy, from Milan, from Lake Lugano, the Ticino, Lambro, and the Olona, where it is common (Fig. 97). Its body is remarkably deep, and the depth greatly exceeds the length of the head. The anal fin has only eight jointed rays. The thickness of the body is only half the height. The nose is more rounded than in the type. The diameter of the eye is one-quarter of the length of the head. The angle of the mouth reaches as far back as the hinder nares. The anal fin is more developed than in the typical forms of *L. muticellus*, and the ventral fins have the shortest rays. The pectoral fins are well formed, and extend back almost to the ventral. The terminal rays of the caudal fin are as long as the head.

The largest scales are more than one-half the diameter of the eye. The smallest scales are on the fore part of the back and pectoral region. The scales have a fan of more than ten rays. The colour and longitudinal band are similar to those of the type *L. muticellus*, but the lateral band is less distinct, and often scarcely visible. All fins are of a pale yellow colour. The length is about four and a half inches. This variety lives in shallow water, under stones. It spawns in May, June, and July, and is then described by Filippi as covered with small tubercular growths on the head and scales. The pharyngeal teeth are stronger than in *L. muticellus*, and often less denticulated, usually there are four in the outer row instead of five. The vertebræ are rather less numerous. This variety is found south of the Alps. *L. muticellus* is distributed in an oblique band of country in Central Europe. It is found in France, Switzerland, Würtemberg, Bavaria, some parts of Germany and Austria, and especially in the tributaries of the Rhine, the Rhône, and the Danube.

## Leuciscus phoxinus (LINNÆUS).—The Minnow.

D. 10, A. 9—10, V. 9—10, P. 16—17, C. 19.

The Minnow, which is widely distributed throughout Europe, is one of the best known inhabitants of the fresh waters of England and Scotland, and in Ireland is met with in Dublin and Wicklow (Fig. 98).

It has many local names, which vary with almost every country In France it is called *le Vairon*, in Sweden, *Elritzor*, in North Germany, *Elritze*, and in South Germany and Austria, *Pfrille*, in Italy it is known as *Fregarolo* or *Sanguinerola*—the last name, like the English provincial name *Pink*, obviously referring to its colour.

The body is five to six times as long as high, and the height is equal to the length of the head, but the proportions vary with age, since the head becomes longer in old specimens. The thickness of the body is from two-thirds to three-fourths of the height, giving it a sub-cylindrical appearance. The eye is one-fourth of the length of the head, and the breadth of the frontal region, between the eyes, is one and a half times the orbital diameter. The average distance of the eye from the snout is equal to the diameter of the eye. The mouth is small and terminal; its cleft reaches under the nares. The nose is arched and thickened; in fact, the whole profile from the nose to the back of the head is conspicuously arched, while the back is comparatively flat. The dorsal fin is opposite to the space between the ventrals and anal, and in about the middle of the body. Its base is short and its greatest height is equal to the length of the pectoral and anal fins. Its shortest ray is half the length of the longest. The dorsal fin terminates opposite to the vent. The anal fin may be truncated posteriorly to half its length. The length of tail between the anal and caudal fins is as long as the head. The ventral fins reach to the vent, and, like the slightly longer pectorals, are rounded. The caudal fin is evenly-lobed, but not conspicuously forked. The lateral line does not descend so low as in other species of Leuciscus. It is most clearly defined over the ventrals and is rarely to be traced as far as the caudal. It opens with simple pores, but the last fifteen or twenty scales are not perforated with mucus canals. The perforated scales are relatively higher than the scales adjacent to them; the perforation is large and occupies about two-thirds the length of the scale. The scales are remarkably delicate and small; they are longer than wide, and

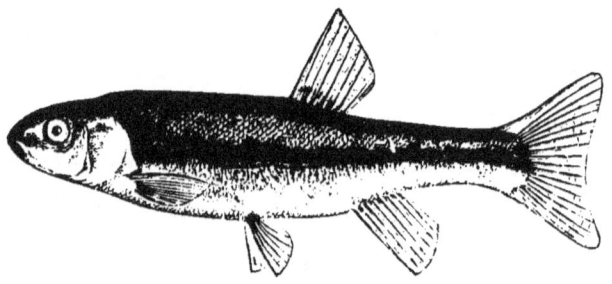

Fig. 98. — LEUCISCUS PHOXINUS (LINNÆUS).

marked with a radiating fan of fifteen to twenty rays. They do not overlap conspicuously, but the degree of imbrication varies with the individual and locality. Scales are frequently absent from the under surface of the body as far as the anal fin. The largest scales are in the tail. The number along the lateral line varies from eighty to ninety.

Fig. 99. — PHARYNGEAL TEETH OF LEUCISCUS PHOXINUS.

The Minnow varies its colour with the locality. It is more brilliant when taking food; and duller by night than by day. The variability of colour is due to the existence in the skin of two layers of pigment cells, which are superimposed. One layer, most abundant on the back, has the cells star-shaped, black, and contractile; the other deeper-seated layer has the cells smaller, somewhat rounded, of varied colour, with the majority yellow. By the expansion and contraction of the first layer the second is more or less covered or exposed, and this change is accompanied by a variation in the quantity of blood injected into the tissues.

The colours are most brilliant at spawning-time, particularly in the males. The back is dark brown-green, often decorated with a black line which may extend to the caudal fin. The sides are greenish-yellow with a metallic lustre. The angles of the mouth are carmine; the throat is black. The breast and pectoral fin are red, and there is usually a blackish spot at the base of the caudal. Sometimes the black line down the side is broken into a line of spots. The dorsal fin is banded with black or dark-grey; the pectoral, ventral, and anal often have a rose tint at the base. The iris is silvery but clouded with gold. At spawning-time many males become sooty-black, with the exception of the fins, the operculum, and abdomen, which are spotted with black; the females are always paler. Both sexes develop sharp or pointed warty tubercles, which are characteristic of the genus. After spawning the colours become pale, and after death they vanish. The males are rather shorter than the females, and have the frontal profile more arched, while the females have the dorsal fin placed farther back than in the males. In the same brood the males would measure from two to three inches, while the females would be commonly three and a half to four inches. But in some localities circumstances seem to favour their growth, so that giant fish are found. One such from Poprad in Hungary measured five inches, and specimens have been recorded, in the Lake district of England, fully seven inches long. The eggs even in large females are exceedingly small, some measuring a millimètre and a quarter in diameter; they number 700 to 1000. Spawning usually occurs in May and June, but they have been taken in Central Europe in April in full bridal colouring when snow was on the ground. The young usually appear in twelve to fifteen days, but the period varies with temperature. They increase very rapidly in number, but the individual growth is slow; reproduction does not take place till the third or fourth year. They are sometimes bred artificially as food for Trout and Salmon. Being caught in a net, the female is taken in one hand, and the male in the other; with slight pressure the eggs and milt are deposited, and the water is then stirred with a feather. The fecundated eggs are placed in perforated boxes in a running stream. When deposited naturally the eggs are found adhering to the interstices of stones, and sometimes in masses measuring as much as eight inches long by two inches broad.

The accessory gills are comb-shaped. The supra-orbital bone is very large, forming the upper border of the orbit. There are twenty-one vertebræ in the thorax and nineteen in the tail, and fifteen to sixteen pairs of ribs. The hinder part of the swim-bladder is rounded, and twice as long as the fore part, but in females it is thin and pointed. The lining membrane of the visceral cavity is somewhat clouded with black pigment. Fig. 99 shows the pharyngeal teeth.

The Minnow is a voracious fish, feeding on vegetable substances, worms, insects, small species of *Helix*, and freshwater mollusca; but it pays the penalty for this gratification of a varied appetite by becoming infested with many kinds of entozoa. They eat their own dead, and old fish eat young ones. Minnows prefer clear flowing water with a gravelly or sandy bottom, but are found also by grassy banks and places which it would be impossible for other fishes to reach. They are very shy, and disappear at the least noise, but have some enterprise and are very curious. They skim about in large swarms at the top of the water, and are often found in long processions travelling in search of food. In hot weather they ascend the tributaries of rivers, leaping over obstacles wherever the leader takes them. They thus extend from the Danube into the Traun Sea. In autumn they retire into deeper water, hiding under roots of trees or under stones, where they live with Gudgeon. Probably, every person who has ever looked into a small stream, has been surprised by the singular way in which minnows constantly arrange themselves in circles like the petals of a flower, with their heads nearly meeting in the centre, and tails diverging at equal distances. The Minnow is of sociable habit, and is found swimming with the Trout and many other species, and when swimming in isolation it is usually in positions which are not easily accessible to larger fishes. Fatio says : " I have often taken a large number of Minnows in a few minutes, with a single sweep of a butterfly net held in the stream. Attracted by the novelty, the curious little things come to look at the strange object, and, seeing a large opening, approach nearer and nearer to explore it. The most adventurous enter, and all the band follow to study the secrets and discover the riches of the treacherous cavity. By gently lifting the net it may sometimes be taken up a quarter full of Minnows."

Formerly Minnows were valued as food. Dr. Günther mentions that they are well-flavoured, but nowhere are they recorded as important in diet, probably on account of the shyness of the fish, and small size, as well as the bitter flavour imparted by the gall-bladder, which needs to be carefully removed. When plentiful, they are eaten in North Germany. In earlier ages, however, the Minnow was a royal dish; and when William of Wykeham entertained Richard II. and his Queen, in 1394, among other fishes served at table were seven gallons of Minnows. We have no particulars as to the method of cooking, but Frank Buckland states that they are very good simply fried as whitebait.

At Danzig a variety occurs in the stagnant waters of low-lying districts, which has the sides brassy-yellow, and the back darker, with fine black spots. Sometimes the entire fish is covered with elevated black spots, which, according to Fatio, are due to a skin parasite.

In Russia it is found in all swift mountain rivers and brooks, and often occurs in great quantities; it is recorded in all European countries, from Norway to Italy, with the exception of the Spanish peninsula. It is found in the Alps almost up to the limit of the snow line.

## Leuciscus hispanicus (STEINDACHNER).

D. 10, A. 11, V. 9, P. 12.   Scales: lat. line 62—65.

Steindachner caught this fish near Merida, in Spain, in rivulets which flow into the Guadiana, and at present it is known only from this locality. It has an elongated body and pointed head. The height of the body is one-sixth of the total length. The head is about one-fifth of the length of the fish. The diameter of the eye is one-third of the length of the head, and is equal to the width of the frontal interspace. The distance from the eye to the snout is less than the diameter of the eye. The mouth descends obliquely. The jaws are equal in length. The profile rises rapidly to the dorsal fin, and then descends rapidly. The dorsal fin is twice as high as long, and commences in the middle of the body. The ventral fins are in advance of the dorsal and reach back to the vent. The pectoral fins are longer than the ventrals, and, like those fins, are only moderately rounded. The anal fin is higher than long, and has eight jointed rays. Its lower margin is concave. The caudal fin is longer than in the Minnow and more deeply notched. Its rays are longer than the head.

There are four pharyngeal teeth in a single row on each side, but Steindachner suggests that since his specimens were small a second row may have been developed later, as happens in the Chub. The scales are very delicate, cover each other very partially, and are larger in the lower part of the body than above. The largest scales are below and behind the shoulder-girdle. In one specimen the lateral line is absent; in another it is absent on one side, and in a third it is present on both sides of the body, but terminates in front of the ventral fins. The dorsal region is coloured golden-brown or grey; the lower larger portion of the body is silvery. These two parts are divided by a strongly-marked black line. There are five black spots over the upper part of the body. This fish is three inches long.

## Leuciscus arca (STEINDACHNER).

D. 10—11, A. 10—11, V. 9—10.   Scales: lat. line 40—46, trans. $\frac{7-8}{3-4}$.

This species is limited to the Spanish peninsula. It is known in Portugal as *Ruivaca*, and as *Escalo* on the Minho in the north of Spain. Like the Chub it varies in form of body with the locality. It is extremely elongated in the mountain streams of Central Spain and Portugal, and even in the Ebro. But the better-fed individuals in the deeper waters of the Minho, where plants abound, are rounder in body; the height of the body in these well-nourished specimens is two-ninths of the length of the fish. The body is five to five and a half times the length of the head. The back is moderately arched. The dorsal fin is usually one and a third, and the anal fin one and a half, times as high as long; there are seven to eight jointed rays in each of these fins. The caudal fin is fully as long as the head. In the short form there are forty to forty-four scales in the lateral line; and below the lateral line there is often one row of scales more on one side of the fish than on the other. The mouth is terminal; the jaws are nearly equal; the angle of the mouth is under the nares. The pharyngeal teeth are 5—5. There is a greyish band along the side.

This species is common in the small streams about Madrid; it is rare in the Tagus and Douro, and is absent from the south of the peninsula.

It is very closely allied to the *Leuciscus aula*, of which it may be regarded as a geographical representative.

## Leuciscus alburnoides (STEINDACHNER).

D. 10, A. 10—12, V. 8, P. 13. Scales: lat. line 39—40, transverse $\frac{8\frac{1}{2}-9\frac{1}{2}}{2-2\frac{1}{2}}$.

The body is elongated and greatly compressed, with the depth equal to the length of the head. Its terminal obliquely-descending mouth recalls the genus Alburnus, but it differs from the latter in having but one row of pharyngeal teeth, and in having a short base to the anal fin. The head is a little more or a little less than one-fifth of the total length. The eye is a fourth of the length of the head. The inter-orbital space is wider than the orbital diameter, and the eye is its own diameter behind the snout. There are five pharyngeal teeth in a single row on each side. The dorsal profile is less arched than the ventral contour.

The dorsal fin commences a little behind the middle of the body. Its free margin is slightly convex. It is about one and two-third times as high as long, and is higher than the anal fin, though their bases are equal. The ventral fin is conspicuously in front of the middle of the body. The pectoral fin is as long as the dorsal is high. The longest rays of the deeply-forked caudal fin exceed the length of the head. The lateral line descends rapidly for the first eight scales, and then runs nearly parallel with the ventral outline. The scales are relatively large; they are largest on each side of the lateral line in the middle of the body. Each has its free border festooned by rays which vary in number.

A blackish longitudinal line extends from the snout down the middle of the body to the base of the caudal fin. The colour of the back is bluish-green, with a metallic lustre, and the abdomen is silvery.

The fish reaches a length of four and a quarter inches. It is found in the southern parts of Spain and Portugal, in the tributaries and main waters of the Guadiana, Guadalquivir, Genil, and the Guadiaro, near Gibraltar.

## Leuciscus macrolepidotus (STEINDACHNER).

D. 10, A. 10—11, V. 9—10. Scales: lat. line 33—36, trans. 7 4½.

This fish is limited to Portugal, where the name *Ruivaca* is applied indifferently to *L. arcasii*, and to it. The fin rays are alike, and the form of the body is similar. The head is relatively larger in *L. macrolepidotus*. The mouth is terminal and the upper jaw slightly overlaps the mandible. There is an indistinct greyish band along the side. According to Dr. Günther, there are two and a half longitudinal series of scales between the lateral line and ventral fin. It reaches a length of three to four inches.

Professor Steindachner is disposed to regard this fish as a large-scaled variety of *L. arcasii*, a conclusion which seems to be justified by his specimens.

## Leuciscus arrigonis (STEINDACHNER).

D. 10—11, A. 11—13, V. 10, P. 14—15.

Scales: lat. line 46—52, transverse $\frac{8}{4\frac{1}{2}-5}$.

This Leuciscus has an elongated body, with the greatest height equal to about the length of the head. The body is about five and a half times as long as the head. In the young fish the eye is relatively large, but in the mature form it is one-fourth of the length of the head. The snout is rounded, and projects a little beyond the mouth. The inter-orbital space is always wider than the length of the snout, and increases in width with age, becoming equal to one and two-thirds the orbital diameter. The cleft of the mouth is curved, as in *L. arcasii*, and the angle of the mouth reaches back to the orbital margin.

There are six pharyngeal teeth in one row on the left side, and five on the right. The dorsal fin is in the middle of the length, with its base equal to two-thirds of the height of the fin. The ventral fin is opposite to the dorsal, or slightly in advance of it; its terminal edge is rounded. The base of the anal fin is longer than that of the dorsal, but the height is less. The caudal fin is shorter than the head; its lobes are pointed, and in young specimens the lower lobe is sometimes slightly the longer. The pectoral fin is considerably longer than the ventral.

The scales are rather long and strong. The hinder margin of each is delicately festooned by six to fourteen rays. There is an elongated scale at the base of the ventral fin. The lateral line is parallel to the ventral margin. There are four and a half to five series of scales between the lateral line and ventral fin. The back is grey, shading into steel-blue. The abdomen is golden yellow. The fins are reddish-yellow, but the caudal is spotted with black, especially at its hinder margin. The blackish longitudinal lateral band commences at the extremity of the snout, in the fore-part of the body; it extends over three scales, and is prolonged to the tail. In spawning females both dorsal and ventral profiles are remarkably arched. This species spawns in April. In individuals four and three-quarter inches long the eggs are as large as millet seed. Some specimens exceed a length of six inches. It is found in the Xucar, and in Lake Uña, in Spain.

## Leuciscus lemmingii (Steindachner).

D. 10, A. 10, V. 9, P. 14—15.  Scales: lat. line 59—63, trans. $\frac{12-13}{6}$.

The body is elongated, and compressed, with a moderately arched back. When young the height of the body is less than the length of the head, but in mature individuals the height may exceed the head-length. The head is between one-fifth and one-sixth of the length of the fish. The mouth is small, semicircular, and inferior, so that the strongly-arched snout projects more or less visibly beyond the lower jaw. The angle of the mouth is below the anterior nares. The lips are moderately fleshy. The diameter of the eye is one-quarter of the length of the head, or a little less. The frontal interspace between the orbits in the old fish never equals twice the orbital diameter, and in young individuals exceeds the orbital diameter only by some small fraction. The ventral profile is more markedly curved than the back. The dorsal fin begins in advance of the middle of the entire length; it is usually opposite to the ventral fin. Its height is at least half as much again as the length of its base. The anal fin has nearly as long a base, but its rays are not so high. Its free margin is more convex than in the dorsal fin. The deeply-forked caudal fin is evenly lobed, and longer than the head. The pectoral fin is shorter than the caudal. The ventral is rounded.

The scales are small, round, marked with many concentric rings, and a fan of eight to thirteen rays, which slightly festoon its free border. The lateral line is more or less parallel to the ventral contour. The pharyngeal teeth show in most cases the formula 6—5, though the formula 5—5 is occasionally met with.

The body is dark blue-grey, with metallic lustre, and the abdomen is silvery. The upper sides of the body and head have innumerable black flecks. A dark blue-grey band extends from the eye to the base of the tail, but is distinct only in specimens in which the black spots are few in number. It recalls the form of *L. adspersus* of Dalmatia.

It is abundant in the Guadiana, and less frequent in the Guadalquivir at Seville. Its length is six inches.

## Leuciscus heckelii (NORDMANN).

D. 13, P. 17, V. 9, A. 12—13, C. 19. Scales: lat. line 43—46, trans. $\frac{8}{6}$.

This large species of Leuciscus has the height and form of a Bream, but is more like the Rudd. Its length is usually from eleven to fourteen inches. In a full-grown individual the height is one-third of the total length. The head is scarcely one-sixth of the total length; it is thick, and rather flattened at the top. The snout is rounded and thick, and extends beyond the small mouth, which in consequence has a somewhat inferior position. The eyes are of medium size, and placed rather higher than in the Rudd.

The scales are large, adherent, and harder than in the Rudd; they usually have from three to five rays.

The dorsal fin is above the middle of the base of the ventral fin, and consequently placed farther forward than in the Rudd. Its first branched ray is somewhat longer than the head; it is twice as high in front as behind, and its terminal border is concave. The proportions of the anal fin are similar. There are four and a half longitudinal series of scales between the lateral line and ventral fin.

The fish is entirely silver-white, with some bluish-grey on the head and back, and pale yellow on the sides. The eyes are yellow, with silver and red tints. The lower fins are pale yellow, the dorsal fin is darker; all are bordered with black.

This fish is known only from the Crimea.

## Leuciscus pyrenaicus (Günther).

D. 11, A. 11—12, V. 10.   Scales: lat. line 14—16, trans. 8 6.

This is a very common species in the middle and north of Spain, but also occurs in the south in the Guadiana and Guadalquivir. In the north it reaches a length of nine to eleven inches, but in the south varies from nine to five inches in different streams. The proportions vary a little with almost every river, and when the northern and southern forms are contrasted the differences almost amount to varieties. In mountainous districts specimens are longer, thinner, and darker. The height of the body is equal to the length of the head, and is about one-fourth of the total length, exclusive of the caudal fin. The head is twice as long as broad. Its upper surface is slightly convex. The inter-orbital space is contained twice and two-thirds in the length of the head, but may be one-third of the length of the head in some specimens; it is less than twice the diameter of the eye. The diameter of the eye varies from less than one-fifth to one-sixth of the length of the head. Less than its own diameter separates the eye from the snout. The jaws are nearly even in front. The cleft of the moderate mouth scarcely reaches below the front of the orbit. The hindermost sub-orbital bone is rather larger than the first. The intermediate part of the sub-orbital ring is broad; the third bone is scarcely narrower than the fourth.

The origin of the dorsal fin is opposite to the fifteenth or sixteenth scale of the lateral line, and a little behind the root of the ventral fin. The anal fin is but little higher than long, and has its free margin straight. The caudal fin is forked. The pectoral fin terminates far in advance of the ventral. The ventral fin is more than half the length of the head. There are three longitudinal series of scales between the lateral line and ventral fin. Some small specimens have only thirty-nine to forty scales in the lateral line, and many of these are only half the usual size. Other examples have as many as forty-six scales. The scales show numerous radiating striæ. Dr. Günther states the colour to be more or less uniform, sometimes with a well-defined brown spot at the base of each scale, so placed that the spots form longitudinal series.

In the middle of Spain it spawns in April, but earlier in the south. The young when only two and two-third inches long already breed.

In Portugal it is known as *Bordalo* or *Roballinho*, and in Spain as *Cacho*. It is common in all the rivers of the south of Spain and Portugal.

Steindachner who first described these fishes, regarded them as Chub.

And he mentions also some very round specimens, with short noses, protruding eyes, thick leathery skin, with rough scales, slender intestine, and sexual organs feebly developed, which he regards as sterile variations comparable to those of the Carp.

## Leuciscus polylepis (STEINDACHNER).

D. 10, A. 11, V. 9, P. 15.  Scales : lat. line, 68—71, transverse $\frac{12-13}{5-6}$.

This little fish, four and a half inches long, is known only from Croatia. The body is long and compressed. The fish is rather less or rather more than five times as long as the head. The greatest height is equal to the length of the head. The eye is one-quarter of the length of the head, or a little less, and is separated from the snout by its own diameter. The arching of the snout increases with age. It projects a little beyond the rounded mouth opening. The cleft of the mouth does not reach back to below the anterior border of the orbit. The inter-orbital space is rather convex, and its width slightly exceeds the diameter of the eye.

The origin of the dorsal fin is in advance of the middle of the fish. The fin is considerably higher than long. Its base is less than half the length of the head. The height of the anal fin exceeds the length of its base by one-fifth. The ventral fins are shorter than the pectoral, are placed in front of the dorsal, and do not reach the vent. The caudal fin is deeply forked; its longest rays are shorter than the head.

The scales are small, remarkably delicate, and thin; they are rather higher than long. The largest are near the lateral line, in advance of the ventral fin; the smallest are on the neck. The free margin shows a fan of eight to fourteen rays on each scale. The scales in the anterior ventral region do not overlap, but lie next each other ; above the lateral line they overlap one-half. The lateral line runs parallel to the contour of the abdomen, except for a few scales in front, where it descends rapidly. There are five to six rows of scales between the lateral line and ventral fin.

The colour of the back is dark grey, shading into steel-blue. There is a dark band, three to five scales wide, along the side of the head and body, above the lateral line, formed of black spots. The abdomen is silvery. The fins are orange-coloured at the base, and yellow at the extremities, but the rays of the caudal and dorsal fins show black points. The pharyngeal teeth are in two rows, 5. 2—2. 4.

Spawning takes place in June and July.

### Genus: **Paraphoxinus** (Bleeker).

Paraphoxinus is a small genus comprising two species of fishes which are confined to the Adriatic region of the Balkan peninsula. It is distinguished by having the body free from scales, or covered with scales which are rudimentary and do not overlap. Scales, however, extend along the lateral line, but this line is more or less incomplete. The mouth is terminal. The pharyngeal teeth are in a single row on each side, with the formula 5—4. The other characters are, for the most part, similar to those of the Phoxinus type of the Genus Leuciscus.

### Paraphoxinus alepidotus (Heckel).

This remarkable species (Fig. 100) has much the same relation to Leuciscus that Aulopyge has to Barbus, in being free from scales. In the adult the height of the body is equal to the length of the head, but in old females it is

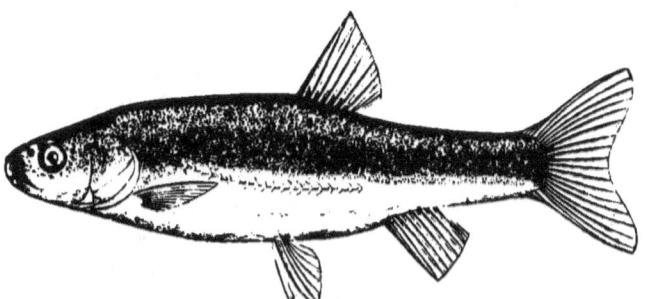

Fig. 100.—Paraphoxinus alepidotus (Heckel).

greater. The thickness of body is not less than half the height, nor more than two-thirds of the height. The back is rounded. The length of the head is less than two-ninths of the length of the fish. In males it is a little shorter. The diameter of the eye is from one-fourth to one-fifth of the length of the head. Its own diameter separates it from the snout, and the frontal interspace between the eyes is one and a half times the diameter of the eye. The angle of the oblique mouth scarcely reaches back to the anterior nares, which are near to the eye, and rather elevated in position. The jaws are equal, and the lips round. The profile is very like that of the Minnow, but with age the body becomes higher and more compressed. The dorsal fin begins behind the

the middle of the body, and somewhat behind the origin of the ventrals; it is higher than long, and more truncated behind than in the Minnow. The anal fin is lower, and nearly as high as long. The ventral fins are more feebly developed, and do not reach back to the vent. The pectoral fins are also small, and but little longer than the ventral. The longest rays of the caudal fin are not more than four-fifths of the length of the head. The vent is behind the termination of the dorsal fin.

The most remarkable characteristic of the fish is the scaleless body, though the lateral line, as far as it extends, is formed by a single row of easily-detached scales. The lateral line descends over the ventral fins to the lower third or fourth of the height of the body, and rises towards the tail. It resembles the lateral line of the Minnow in its variations. In some individuals it reaches to the end of the tail, while in others it may terminate in the middle of the body, or even earlier. The cephalic mucus-canals are well marked by an unusual number of pores thickly-placed in rows. Fig. 101 shows the pharyngeal teeth.

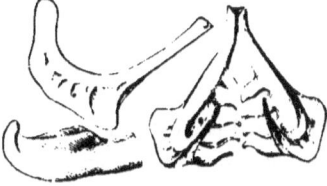

Fig. 101.—PHARYNGEAL TEETH OF PARAPHOXINUS ALEPIDOTUS.

The colour of the back is blackish-brown, the sides are paler, with steel-blue longitudinal bands, and are marked by fine points of black pigment. The belly is silvery. All the fins are yellow and unspotted. The length varies from three to nearly four inches. This species is found in Dalmatia at Sign, and in the Narenta, and also occurs at Livno, in Bosnia.

## Paraphoxinus croaticus (Steindachner).

D. 3/7—8, A. 3/7—8, V. 2/6—7.

While the *Phoxinellus alepidotus* (Heckel) has scales only along the lateral line, this species is covered with remarkably small delicate ovate scales with concentric ornament, so arranged in shallow pits as to leave slight interspaces between the adjacent scales.

In the lateral line the scales number sixty-three to seventy, are closer together, and pointed behind; but the lateral line, which is nearly parallel to the abdominal contour, does not always reach to the caudal fin, and may terminate in the region of the anal fin.

The proportions of the body vary with age. The height in old individuals, six or seven inches long, exceeds the length of the head; while in younger specimens, five inches long, it is only two-thirds or three-fourths the length of the head, and the profile of the back is then flatter. When fully grown the length of the head is to the length of the body as one to four and five-sixths; and in the young the relative length is as one to four and a half. The fish increases in thickness with age, and the thickness is at first equal to half the length of the head, while in the adult it becomes two-thirds of the length of the head, and the body is then nearly round. The eye is placed in the anterior third of the head. The snout is blunt. The mouth is oblique, and its angle extends back to below the line of the orbital border. There is a wide interspace between the anterior and posterior nasal apertures. The dorsal fin has short rays; it commences behind the middle of the body, and is higher than long. The anal fin is some distance behind the vent, and is behind the end of the dorsal. Its base is as long as that of the dorsal, but the fin is not so deep. The genital papilla is placed in front of the anal aperture, and is wider and shorter in the female than in the male. The ventral fin is small and fan-like. The pectoral fin is larger and pointed. The caudal fin is deeply notched and evenly lobed.

The course of the cephalic canal is defined by numerous pores, which form white spots on the thick skin of the head, which is greenish-black or brown. The back is dark green, with a metallic lustre. The scales have the aspect of golden spots on a dark ground. The side of the body is paler; and the abdomen is silvery with a yellowish tinge. The broad blue longitudinal band, which extends along the middle of the side, coincides with the lateral line only at its commencement and at its termination.

This species is found only in the subterranean brooks and rivers of Croatia which occasionally emerge from the rock. Its Croatian name is *Piuri;* in German it is called *Grundel.*

## Genus: **Tinca** (Cuvier).

The Tench type, which had been recognised under the generic name *Tinca* by many of the earlier naturalists, was first accurately defined in Cuvier's "Règne Animale." Its small scales are embedded deeply in the thick skin; the lateral line is incomplete. The fins are short; the dorsal is opposite to the ventral, and has no spine. The mouth has a barbel at its angle; the lips are moderately developed. There are short lanceolate gill-rakers, and rudimentary pseudo-branchiae.

The pharyngeal teeth are wedge-shaped, slightly hooked at the ends, in a simple row, with five on the left side, and four or five on the right.

## Tinca vulgaris (CUVIER).—THE TENCH.

D. 11—12, A. 9—10, V. 10—11, P. 15—17, C. 19.

The Tench (Fig. 102) has the body about four times as long as high. Its head varies in length, and may be a little longer than the height of the body, or a little shorter. The thickness of the fish is about one-half its height, but all measurements vary with age and sex. The diameter of the eye is about one-sixth of the length of the head. The eye is one and a half times its

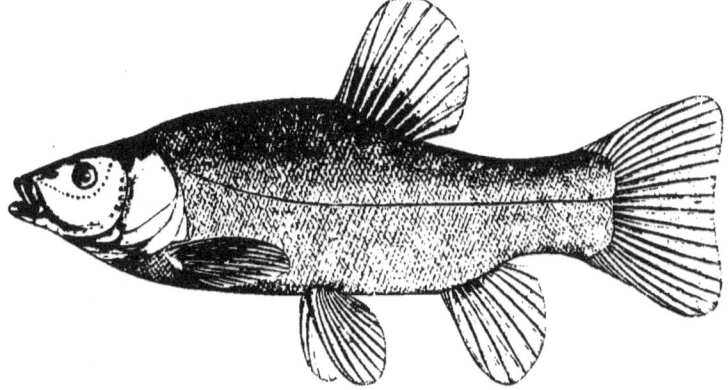

Fig. 102.—TINCA VULGARIS (CUVIER).

own diameter behind the extremity of the snout, and distant two and a half diameters from the eye on the other side of the head. The nasal apertures are midway between the orbital ring and the extremity of the snout. The mouth is rather small; it opens at the extremity of the snout, and its hinder angles incline upwards under the nares. The jaws are equal, the lips fleshy, and the small barbels have a length equal to one-half the diameter of the eye. Fig. 103 shows the pharyngeal teeth.

The contour of the body ascends in an even, moderate curve from the snout to the dorsal fin, with a slight indentation in the occipital region, which is, however, more marked in males. The height of the tail is greater than in the previously described genera, and may be equal to half the greatest height of the fish.

All the fins are rounded. The dorsal fin is rather nearer to the end of the tail, than to the snout; its first ray is rudimentary, and is counted as an osseous ray by the Austrian writers and by Von Siebold, but is ignored by Dr. Günther; the second ray is imperfectly developed. The longest rays of this fin are equal to two-thirds of the height of the body. The anal fin is not close to the vent; it is as deep as the tail, and as long as the pectoral fin. The fifth and sixth rays are the longest. The ventral fins are a little in advance of the beginning of the dorsal; the first two rays are often rudimentary; sometimes the fin is as deep as the dorsal is high; but the depth varies with sex. At first, according to Von Siebold, the fins are alike in male and female, but after development of the reproductive organs has taken place, the first ray becomes thick in the male and very thin in the female. These fins reach back to the vent. The pectoral fins have thinner rays, the first scarcely stronger than the others; the longest are the fourth to sixth, and they do not reach to the ventrals. The longest rays of the tail fin are almost equal to those of the dorsal fin; the tail is evenly rounded, with a very slight emargination.

Fig. 103.—PHARYNGEAL TEETH OF TINCA VULGARIS.

The remarkably small scales shine through the thick greenish-brown epithelium of the skin as a number of glistening golden points. There are over a hundred on the lateral line. The scales are much longer than they appear to be, for only their extremities are seen while they are still attached to the fish. They are narrow, rounded at both extremities, marked with numerous striæ, which radiate longitudinally from a point near the base. The concentric striæ are very wavy, and arranged round the basal point. There are from thirty-two to thirty-three scales above the lateral line, and eighteen to twenty-five below, to the base of the ventral fin. The lateral line at first descends in a curve from the operculum as far as the middle of the body, and is then prolonged in a straight median line to the middle of the tail. The mucus-canals are short and cylindrical. The mucus-canal system of the head is well developed, with numerous pores. There are twenty-three or twenty-four of these apertures on the pre-operculum and lower jaw. They are even more abundant on the sub-orbital ring, and not less numerous on the anastomosing branches of the frontal and occipital region. The secretion of mucus from these canals and the lateral line is so great that it forms a thick stratum over the body, and renders the Tench as slippery as an Eel.

The colour of the Tench varies with the water. When it lives in muddy water its colours are dull; but when found in clear water the fish is often brilliant. Its common tint is a dark olive-green, shading into black, with a remarkable transparent brassy glitter. The colour is paler on the sides, and on the belly becomes greyish-white. The fins are colourless, reddish-brown, or violet, their colours, like those of the fish, varying with the conditions of existence. The commonest type is the brightly-coloured one with the dull golden glimmer; but the so-called Golden Tench, *Tinca aurata*, of Cuvier, has a pale golden hue, with red lips, flesh-coloured fins, and dark spots on the body. This variety is one of the most beautiful of European fishes. The eye has a golden circle to a red or brown iris.

The males are more brightly-coloured than the females. Like the young fish they have the head larger, the fore part of the back more arched, and all the fins except the caudal more strongly developed. This character is most marked in the ventral fin.

The Tench reaches a length of one foot and a half, and, in Carp ponds in Germany, often grows to a weight of six, and, occasionally, eight pounds, but south of the Alps a weight of twelve pounds is recorded, as from Lago Maggiore; but in rivers it is rarely heavier than three pounds.

It grows quickly, attaining a quarter of a pound weight in the first year, three quarters of a pound in the second, and three pounds in the third. It lives for six or seven years. Yarrell, quoting from Daniel's "Rural Sports," tells of a Tench which measured thirty-three inches in length from the eye to the caudal fork, weighed eleven pounds nine ounces, and had grown into a hole among the roots of trees so that escape was impossible.

The Tench is singularly tenacious of life, and in the eastern parts of England, as at Peterborough, it is often brought to market and exhibited for sale, and if not sold put back again into water. It is said to live an entire day out of its element, and can exist in water which contains the minutest quantity of oxygen gas, and may exist during summer in the dried-up mud of ponds.

Professor Canestrini has described a singular arrangement of the bones and muscles attached to the base of the ventral fins in the male Tench, by which the abdomen may be drawn inward and upward. He regards this arrangement as aiding in the discharge of the ripe milt, a view which Fatio supports by remarking that the stones and hard substances against which many fishes rub themselves in the breeding season are absent from the muddy habitat which the Tench delights in. In ponds and marshes the fish spawns earlier in the year than when inhabiting lakes and rivers. And it is asserted that the old females spawn earlier than the younger ones. The spawning season lasts from May to August, and the eggs are deposited near the banks. The fish assemble in great numbers, and each individual of three pounds' weight deposits about 300,000 small yellow or greenish eggs. The eggs may fall to the bottom, but are usually attached to water plants, especially *Potamogeton*, sometimes known as Tench weed. There are at least two males attendant on one female. The young are hatched in a week.

The Tench is an idle or meditative fish which passes much of its life on the bottom, resting on its fins as upon four feet. But towards the breeding season it takes more exercise, and occasionally comes to the surface, when it emits a sound as though smacking the lips. In spite of its habitual apathy, on an emergency, it moves with the swiftness of an arrow, and may bury itself in the mud with the help of its fins. In the winter Tench commonly burrow into the mud as a protection against cold, and pass the season in a motionless condition which closely approaches hybernation. It enters the mud, fat, in the late autumn, and re-appears thin, with the abdomen flat, in the spring.

In the summer, according to Von Siebold, the Tench enjoys a siesta or day-sleep, during which it may be seen on clear days motionless on the bottom, and may be brought to the surface with a stick without being disturbed, when it lies on its side as though dead; but if sufficiently roused by blows, it awakes and swims away to bury itself in the mud. The fish has numerous parasites and enemies, but its mucus is repugnant to many carnivorous fish. It feeds on vegetable substances, insects, mollusca, and mud which contains much organic matter. It is captured with the rod or net, or weir basket, without difficulty, in summer.

In the fen country of England Tench are often highly esteemed. Some farmers are content to wash off the mucus with warm water before the fish is cooked; others prefer to remove the skin, but this depends upon the mode of cooking. Its flesh is white and firm, and prized from Sweden to Italy. Dr. Badham states that at Florence it is rightly held superior to any fish food which enters the market, but mentions a prejudice of the old women of Italy, that it is so impregnated with marsh malaria as to infect any one with ague who eats it. Nevertheless, to tempt us to eat it, Badham mentions the following as a good way of serving Tench. The fish is cooked in a rich gravy sauce, containing raisins, currants, pine cones, kernels, with the other ingredients of an "Agrodolce" stew. It is then brought hot to table, and eaten with the juice of lemon. Or, take enough water to just cover the fish, add a quarter of a pint of vinegar, a bunch of thyme, an onion, some lemon peel, a little scraped horse-radish, and salt. Put in the Tench before the water boils. When cooked serve with a sauce made by dissolving two anchovies in water over the fire, and add a half a pint of stewed oysters, a quarter of a pint of shrimps and melted butter. Garnish with pickled mushrooms and lemon.

The Tench was formerly used for external application in medicine, being held to absorb poisons from the body, or act as a poultice. The older writers regarded the Tench as the fishes' physician; and believed that it was therefore spared by the Pike and other voracious fishes, but anglers find it to be an excellent bait; and more than twenty young ones have been taken from the stomach of a Trout.

It is distributed in every European country, not only in the small sluggish streams or muddy districts in which it especially abounds in England, but in all rivers large or small, and in most of the lakes, if not in all. It is also recorded from Brusa, in Asia Minor.

### Genus: Chondrostoma (Agassiz).

This genus is almost limited to Europe. Distinct species occur north and south of the Alps and in Spain, Italy, and the Balkan peninsula; while another species extends its range farther east into Asia Minor, and the valley of the Tigris. The characters which unite these species together into a genus are, first, the presence of a hard, brown, horny covering to the lower jaw, which forms a kind of beak; and secondly, the pharyngeal teeth are knife-shaped, not denticulated, and arranged in a single row, which may include from five to seven teeth on each side, though occasionally the number of teeth on the two sides is unequal. The scales are small. The lateral line terminates in the tail. The dorsal fin, which is inserted above the root of the ventral, has never more than nine branched rays. The anal fin has a rather elongated base with not fewer than ten rays. The gill-rakers are short and fine. The lining membrane of the abdomen is black.

### Chondrostoma nasus (Linnæus).

D. 12, A. 13—14, V. 11, P. 15—16, C. 19. Scales: lat. line 57—62, trans. $\frac{9}{9}$.

This species, which is absent from Britain, is widely distributed on the continent of Europe, north of the Alps. It is also found in the Turkish rivers, and in Russia throughout the river systems of the Black Sea, the Sea of Azov, and the Caspian, and does not appear to extend beyond the Caucasus. It is met with in the rivers which flow into the southern part of the Baltic. It occurs throughout Germany, and in most parts of Austria, where it is generally known as the *Näsling*. In France it is popularly recognised as *le Nez.* It occurs in Belgium, and, according to Graells, in Spain. It is found in rivers and lakes, in plains and mountain districts, and is present in the principal lakes of Switzerland as well as those of Hungary, and is as common in the elevated streams of Transylvania as in the low lands by Danzig (Fig. 104).

It prefers deep water, and feeds on all kinds of aquatic plants, but especially on confervae, and vegetation growing on stones, which is neatly removed by the scissor-like action of its transverse horny lip. But it readily takes animal substances, preferring the spawn of other fishes, worms, and many kinds of small animals, though the diet varies with the district. In some places it is reputed to eat mud and decomposing substances, and it certainly burrows in

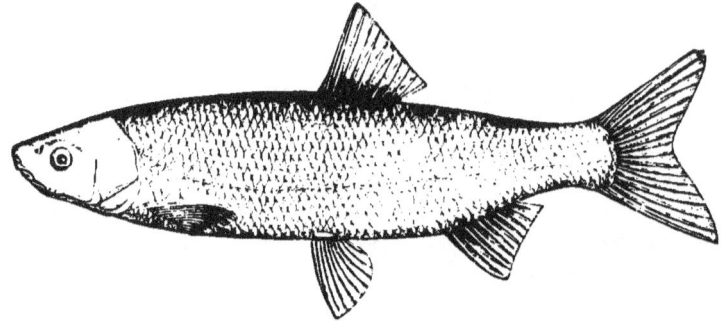

Fig. 104.—CHONDROSTOMA NASUS (LINNÆUS).

mud in search of food. At Würzburg, in Bavaria, it is popularly known as *the Spitter*, because at the moment of being caught it vomits the triturated food which was between the pharyngeal teeth.

The length of the head is one-sixth of the length of the entire fish. The height of the head is two-thirds its length; the depth of the body in front of the dorsal fin exceeds the length of the head. The length of the head is four and a half to five times the diameter of the eye. The eye is one and a half times its diameter from the snout, and the width of the frontal region between the eyes is twice the orbital diameter. The mouth is remarkably horizontal; its cleft being nearly straight in the transverse direction, while the lateral cleft reaches back to the anterior nares (Fig. 105). Both jaws have horny margins, and that of the upper jaw forms a thin edge. The nose, which is arched, and conspicuous enough to have been noticed by all peoples, extends markedly over the mouth. The dorsal profile is much less convex than the ventral contour. The dorsal fin

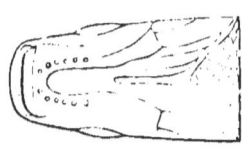

Fig. 105.—HEAD OF CHONDROSTOMA NASUS, SEEN FROM BELOW.

commences slightly behind the ventral, half way down the body, and is one-fourth higher than the length of its base. The anal fin, which commences far behind the termination of the dorsal, has as long a base as the latter, and its height is equal to its length. The ventral fins do not reach the vent. The pectorals are feebly developed, though their longest ray exceeds those of the ventrals. The caudal fin is evenly-lobed, deeply-forked, and about as long as the head.

The scales are firmly attached. They vary in form and size in different parts of the body, being sometimes higher than long, sometimes longer than high. The common form of the long scales is sub-hexagonal or sub-ovate, but there are always two remarkable festoons on the fixed border. On the free border there is a fan of six to ten diverging rays. The usual size of the scale in young fish varies from one-fifth to one-fourth the diameter of the eye; but in old fish it amounts to three-fourths or nine-tenths of the orbital diameter. In fish of medium size the scale is half the diameter of the eye. There are five longitudinal series of scales between the lateral line and ventral fins. The lateral line descends so as to be slightly concave.

The colour of the back is a metallic blackish-green; the sides are paler, becoming silvery towards the abdomen.

The dorsal fin and upper lobe of the caudal are of the same colour as the back, but the lower lobe of the caudal is red, like the anal, ventral, and pectoral fins; in some localities these fins are orange or purple. The iris is brassy, often with a red speck. The skin, but especially in the abdominal region, is covered, like that of many other allied fishes, with points of black pigment.

Dr. Günther described from Tübingen a beautiful coffee-coloured variety, in which the colour was most intense at the tail. The whole fish had a golden lustre, but the scales—which were larger than usual—were bordered at the free edge with silver. Similar fishes were previously described by Schæfer from the Moselle as *Ch. auratus*.

This species is usually about eighteen inches long, and its weight rarely exceeds three pounds; but in Central Europe it is brought to market from nine inches to a foot long, weighing one pound and a half. It is not greatly esteemed for food, being one of the most bony fishes known, but reaches the fish markets of France and South Germany in great quantities, where it is bought by the poor at a low price.

About fifteen tons are taken every year in two or three weeks near Augsburg.

It spawns in April and May in Germany, Switzerland, the Rhine, and France, when it collects in shoals, and seeks the bottom in gravelly places where the stream flows swiftly. Bloch states that the whitish eggs number nearly 8,000. Fatio found them to be as large as millet-seed, and of a greenish tinge. Heckel and Kner mention that this species spawns in August in the Danube, but the statement is open to some doubt in face of the spring spawning in other parts of Europe, though it is well known that spawning may be delayed by unfavourable weather. At the breeding season the male acquires the conical warts which are usual in Cyprinoids; but in the female they are limited to the skull and the sides of the snout. The colours then become more intense, and, according to Benecke, in North German specimens the abdomen becomes black. When the spawn is being deposited, the fish grow extremely active, pressing against each other, and often make a rustling noise with the fins as they jump along the surface of the water.

The Näsling in Switzerland is infested with many entozoa, but in the Neckar Dr. Günther found it to be free from parasites.

The bones of the face are wider than in other Cyprinoids. The infra-orbital bones are small, owing to the size of the second, third, and fourth elements in the orbital ring, and a fifth bone unites these to the skull. The pharyngeal bones, in harmony with the teeth they carry (Fig. 106), are thick and strong. There are twenty-seven thoracic vertebræ, and twenty-one in the tail. There are twenty-one pairs of ribs.

The intestine is nearly three times as long as the entire fish. After making a fold in the lower part of the abdomen, it forms a spiral of several coils, and then runs back to the vent. The liver has two lobes, and surrounds the intestine. The gall-bladder is long and yellow, and situate in the upper part of the right lobe. The milt is blood-red. The posterior portion of the air-bladder is twice as large as the anterior part.

This species is very variable, but the varieties are limited in their geographical range. Thus the coffee-coloured variation already referred to as found in the Moselle and Neckar, originally distinguished as *Chondrostoma auratus*, and popularly known in Germany as Gold Nose and Golden Mackerel, varies still further at Trier. The back becomes golden olive-brown, and the abdomen is dull brown. All the scales are broadly bordered with pale golden-green. White scales in patches are scattered over the body. The dorsal fin and upper lobe of the caudal are golden green, and the other fins are red, as in the ordinary Näsling. Both the dorsal and anal fins have fifteen rays.

A bluish variety, *Chondrostoma cærulescens*, has been recorded in Western France by Blanchard, from the Rivers Doubs and Ognon. The whole body has a blackish-grey tint, with a bright steel-blue iridescence. Below the lateral line the scales are spotted with large black points, and similar spots occur round the eye and in the opercular region, though in some individuals the black punctation is faint. The dorsal and caudal fins have a grey tint, and the other fins are slightly yellow. The mouth is rather narrower than in the Näsling. The scales are more pointed. The dorsal fin is less elevated. There are six pharyngeal teeth on each side. This fish reaches a length of nine and a half inches.

Blanchard distinguishes another variety as *Chondrostoma dremæi*, which is found in Southern France, in some tributaries of the Garonne, as well as in that river at Toulouse. Its colour is very similar to that of the last variety. The upper part of the body is a bluish-grey, with a slaty longitudinal band above the lateral line, though it is faint in some individuals. The black spots on the scales are dense about the lateral line. The pectoral and anal fins are dotted with black on the rays. The head is shorter. The operculum is nearly square, and larger than in other types. The scales are short. The body is much more compressed. The form of the pharyngeal bones is distinctive, but the number of teeth is six on each side.

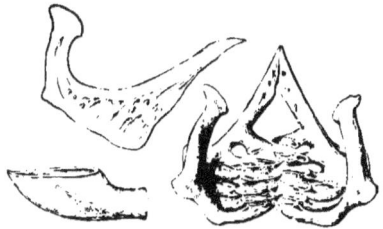

Fig. 106.—PHARYNGEAL TEETH OF CHONDROSTOMA NASUS.

In the Rhône there is another *Chondrostoma* with six pharyngeal teeth, closely resembling the foregoing varieties, but distinguished by Blanchard as *Ch. rhodanensis*. The body is yellowish, with fine black spots scattered over the head and upper parts of the body. The mouth is much narrower. The concentric striæ on the scales are less numerous. The pharyngeal teeth are more slender. It is taken at Lyons, Avignon, and many other places.

## Chondrostoma phoxinus (HECKEL).

D. 11, A. 10—11. Scales: lat. line 88—90, trans. $\frac{17}{10}$.

This delicate little fish at first sight closely resembles the Minnow. It has an elongated body, with extremely small scales, and six pharyngeal teeth on each side (Fig. 107).

The greatest depth of body in front of the dorsal fin is equal to the length of the head, but the fish is five and a half times as long as it is high. The breadth of the head is not quite half its length. The nose is moderately arched, and projects a little over the upper jaw. The small mouth is horse-shoe shaped, rather than semi-circular; and its cleft extends under the nares (Fig. 108). On each side of the lower jaw are three small pores.

The eye is one-quarter of the length of the head, and is one orbital diameter from the snout. The profile of the back is a greater curve than the outline of the abdomen. It rises rapidly to the dorsal fin, and then sinks to the tail. The dorsal fin is in front of the middle of the body. The anal fin is far behind the termination of the dorsal. The bases of these fins are equal in length, and about half as long as the head; but the rays of the anal fin are only two-thirds as long as those of the dorsal. The ventral fins are under the dorsal, are rather wide, and do not reach to the vent. The longest rays of the pointed pectoral fins are four-fifths of the length of the head. The evenly-lobed caudal fin is as long as the head. The lateral line, which com-

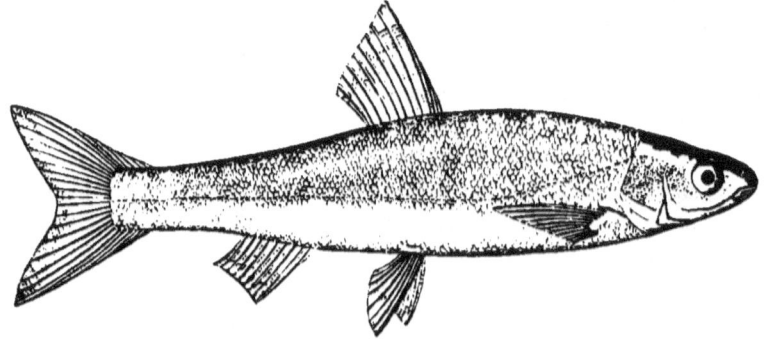

Fig. 107.—CHONDROSTOMA PHOXINUS (HECKEL).

mences somewhat high up, descends to the termination of the pectoral, and then continues in a straight course to the middle of the tail. The delicate small scales are broader than long, but their broadest diameter is only one-third of that of the eye. They generally show six diverging rays, with relatively coarse concentric striping, and each has a large indefinite area at the base. The colour of the back is blackish, the sides shining silver, the belly white. The pectoral, ventral, and anal fins are a beautiful pale yellow. Black points occur upon the part of the head behind the eye, on the shoulder-girdle, and on all the scales above and below the lateral line. At both ends of the body these black pigment spots become diffused, so as to form below the lateral line an obscure longitudinal grey band.

Fig. 108. — HEAD OF CHONDROSTOMA PHOXINUS, FROM BELOW.

This species is found at Sign, in Dalmatia, and at Livno, in Bosnia. The largest specimens do not exceed a length of five and a half inches.

## Chondrostoma genei (Bonaparte).

D. 11, A. 11, V. 9, P. 16, C. 19.   Scales: lateral line 52—56, trans. 9/7.

In Italy this fish (Fig. 109) is commonly called *Lasca*, but in Lombardy it has the provincial name *Strie*, and in the Tyrol is known to fishermen as *Lau*. It is almost limited to Italy, though specimens are obtained from the River Inn and Bavarian tributaries of the Danube. Dr. Günther also records it from the Rhône; and it has been supposed to occur in the upper Rhine, though its presence there is very doubtful. The length varies from five and a half to about eight inches. The species is distinguished by its semi-

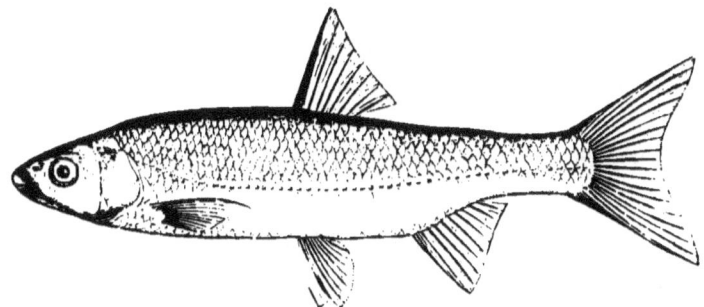

Fig. 109.—CHONDROSTOMA GENEI (BONAPARTE).

circular mouth, by a dark band which extends over and along the lateral line, and by the pharyngeal teeth, which are 5—5 and sometimes 6—5. The height of the body is equal to the length of the head; the fish is five and two-thirds times as long as it is high. The diameter of the eye is one-quarter of the length of the head, but its relative size diminishes with age. The eyes are separated by a frontal interspace of one and a half orbital diameter, and they are the same distance from the snout. The nose is narrower and blunter than in the Näsling. As in

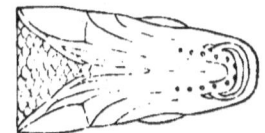

Fig. 110.—HEAD OF CHONDROSTOMA GENEI SEEN FROM BELOW.

other species, the semi-circular mouth (Fig. 110) descends sharply to the angles, which do not reach back to the nares. There is a row of six mucus pores along the lower jaw. Both dorsal and ventral profiles are flat arches. The dorsal fin, which begins half way down the body, is truncate behind; its base is as long as the head. The anal fin has the base as long as the dorsal fin; it commences in the hinder third of the body length, it is not higher than long, and its rays are shorter than in other fins. The ventral fins are placed under the beginning of the dorsal, and are only a little shorter than the pectoral. The terminal rays of the evenly-lobed caudal are as long as the head. The lateral line descends in a moderate curve, sinking below the middle of the side. The largest scales are equal in size to half the diameter of the eye; the free edge of each shows a fan of ten to twelve rays. The concentric striping is rather coarse, and forms a network in the centre.

There are four longitudinal series of scales between the lateral line and ventral fin.

The colour of the back is a clear green-grey, with golden iridescence. The sides become silvery towards the belly, with black pigment spots. A broad grey band extends over and along the lateral line. All fins are yellowish-white, with beautiful orange borders, and the same colour is seen in the region of the mouth and on the skin of the opercular plates. The iris is yellow, with brown and silvery specks.

There are forty-two vertebræ, and sixteen pairs of ribs. The intestinal canal is very long, and makes three curves.

## Chondrostoma soëtta (Bonaparte).

D. 12, A. 14—15. Scales: lat. line 57—60, transverse $\frac{9}{8}$.

This *Chondrostoma* is known only in Italy and the Balkan peninsula, and is regarded as a representative of *Chondrostoma nasus* in rivers south of the Alps (Fig. 111). It has seven pharyngeal teeth on each side, though occasionally specimens are found with the formula 7—6. The teeth are long, straight, and knife-like.

The body is high; the height is less in youth than in age; it may be more than one-fifth of the length. The greatest height, as in all other species of *Chondrostoma*, is in front of the dorsal fin. The fish is from five and a half to five and three-quarter times as long as the head. The greatest thickness is scarcely half the height. The head is about four and a half times as long as the eye. The orbit of the eye is separated by its own diameter from the snout, and the frontal interspace is one and a half times the orbital diameter. The arch of the mouth is intermediate between the condition in *Chondrostoma nasus* and *Chondrostoma genei*. It is a broad, rather flat arch, the cleft of which reaches under the nares. The snout is thick and arched, and projects a little over the mouth.

The dorsal fin is in the highest part of the dorsal profile, in the middle of the body; the length of its base is about four-fifths of its height. It is truncated, so that its longest ray is more than twice as long as the shortest. The anal fin begins behind the end of the dorsal; it is as long as high, and in it, too, the last ray is half the length of the longest ray. The ventral fins commence slightly in front of the dorsal, and reach to the vent. They are but little exceeded in length by the pectoral fins. The terminal rays of the caudal fin are at least as long as the head.

The scales are similar to those of the Näsling, but larger. The largest scales are over the lateral line, and have two-thirds of the diameter of the eye. They have very dense concentric striping, and few rays. In young specimens the diameter of a scale is only one-quarter of that of the eye, while

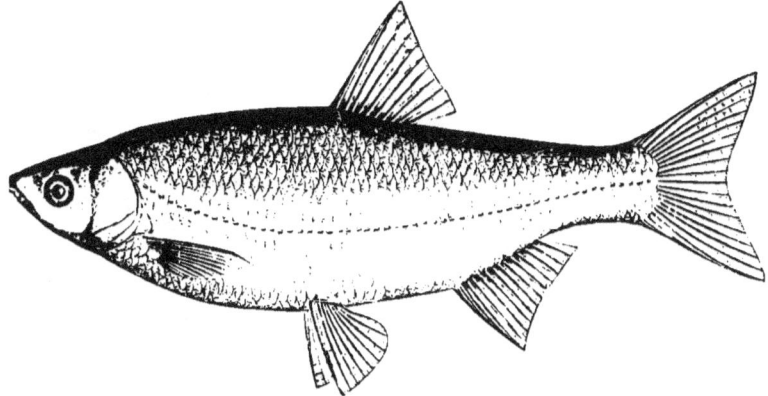

Fig. 111.—CHONDROSTOMA SOËTTA (BONAPARTE).

in old specimens the lateral scales may be nearly as large as the eye. The number of scales below the lateral line varies, being frequently five or six, but sometimes seven or eight.

The comb-shaped accessory gills are large. The gill-arches have very numerous thick short rake-teeth, which have a knife-like form, and end in fine curved points.

The colour resembles that of the Näsling except that there is more of a green tint on the back and sides, and the fins of the lower part of the body, instead of being intensely red, are orange. In young specimens the anal fin is grey. During spawning the males become covered with tubercles on the head and on the scales of the anterior part of the back.

The length is usually about one foot, but sometimes fifteen inches. The weight is one pound and a half to two pounds. There are forty-four vertebræ.

In Italy it is known as the *Saretta*. Its habits are like those of the Näsling. It is taken indifferently in large lakes, like Garda, and in rivers. It is captured with the line and net, and is sufficiently abundant to be a cheap food for the poorer classes, both in Italy and Montenegro.

## Chondrostoma knerii (Heckel).

D. 11, A. 12. Scales: lat. line 52—54, transverse 9/6.

This little fish, which does not exceed six and a half inches in length, is limited, so far as is known, to the river Narenta, in Dalmatia (Fig. 112).

It is characterised by a rather blunt head, with four pores on each side of

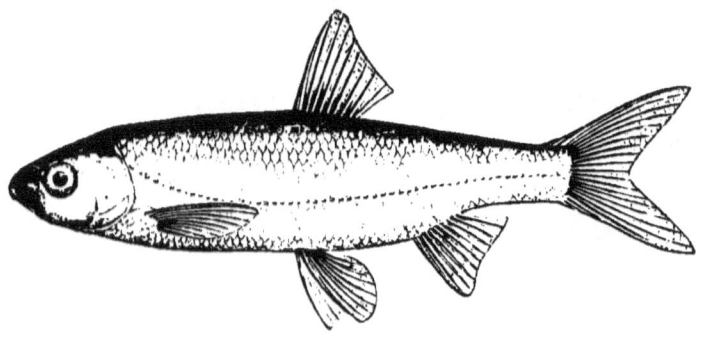

Fig. 112.— CHONDROSTOMA KNERII (HECKEL).

the lower jaw, long pectoral fins, semi-circular mouth (Fig. 113), and six pharyngeal teeth on each side. The strongly-curved mouth and fewer rays in the dorsal and anal fins distinguish it from *Ch. nasus* and *Ch. soëtta*, while the blunter head, longer pectorals, and number of pharyngeal teeth, separate it from *Ch. genei*.

The fish is five and two-third times as long as the length of the head, and five and a quarter times as long as high. The height of the head is four-fifths of its length.

The diameter of the eye is not quite one-quarter of the length of the head; the breadth of the frontal interspace between the eyes exceeds the diameter of the eye, and the eye is a similar distance from the snout. The angle of the mouth does not reach to the nares. The dorsal profile is a more regular and somewhat higher arch than the ventral outline.

The fins resemble those of *Ch. genei*, both in position and form; but the rays of the anal are longer, so that the length of the base is two-thirds of the height. The base is shorter than that of the dorsal. The pectoral fins are pointed, and quite as long as the head. The shorter ventral fins are under the dorsal, and behind the middle of the body. The longest rays of the caudal exceed the length of the head. The lateral line descends somewhat concavely over the ventral fin.

The scales are marked with wavy concentric striping, with an indefinite spot in the middle. The free margin of a scale shows a fan of twelve to fourteen rays. The largest scales are two-thirds of the diameter of the eye. There are five rows of scales between the lateral line and ventral fin.

The pectinate accessory gills are rather large.

The cephalic canals are wide, and well developed. The anterior sub-orbital branch can be traced by the sub-orbital bone to the end of the nose. It is very wide, and the canal opens by two or three pores. The inner mucus skin of the mouth forms numerous compact folds on the palate as in all other species of the genus, but the margins of the mouth in this species look as though finely toothed, for the parallel free edges of the furrows and folds of the skin of the palate terminate laterally in a serration.

Fig. 113. -- HEAD OF CHONDROSTOMA KNERII (HECKEL), SEEN FROM BELOW.

The colour of the back is a brown or greenish-black. It becomes silvery towards the abdomen, and in the anterior half of the body the scales are covered with black pigment spots both on their free edge and at the base behind the ventral fins. These spots are found only on the scales which lie nearest to the lateral line. Before the tail terminates they cover the entire height of the scales.

*Chondrostoma meigii* (Steindachner), from Spain, is closely allied to this fish, if it is not a variety of it. The pectoral fins are shorter. Its mouth is described as semi-circular, and figured as transversely straight; and we agree with Dr. Günther in hesitating to regard it as a distinct species until it has been more fully described, though the more pointed form of head, more arched back, and transverse mouth, render its specific difference probable.

## Chondrostoma rysela (Agassiz).

D. 11—12, A. 12—13. Scales: lat. line 50—60, transverse $\frac{8-9}{3-6}$.

The fish which is thus named is generally regarded as a hybrid between *Chondrostoma nasus* and *Leuciscus muticellus*. It is very variable, and sometimes resembles one of these species and sometimes the other.

The cleft of the mouth is always more arched than in *Ch. nasus*. The pharyngeal bones usually have a shallow notch on the front part of their external margin; and the pharyngeal teeth are generally 5—6, or 5—5; but there is sometimes an additional tooth in a separate series. The nose projects but slightly over the mouth, is blunt and rounded. The fish closely resembles *Ch. genei* in the fins, snout, and mouth; but the body is less elongated. There are five scales between the extremity of the pectoral fin and the ventral. This type most resembles *Leuciscus muticellus* in colour, being slaty-grey on the back, silvery-white on the sides, and white below. A band formed of spots of black pigment, giving a grey effect, runs from the head to the tail, but is sometimes wanting. The fins are orange, but the dorsal and caudal are margined with black, and there is a similar but fainter border to the pectoral. Sometimes the angle of the mouth and operculum are orange. The length of this fish varies from eight to fourteen inches. It is very rare, and has been found in Bavaria, in the rivers Inn and Isar, and is stated to occur in the Danube, and in the Rhine at Basel. It has no distinctive provincial name.

## Chondrostoma polylepis (STEINDACHNER).

D. 11, A. 12. Scales: lat. line 68—75, transverse $\frac{12}{9}$.

This fish, when taken from the clear mountain waters of Spain, is darkish-green on the back, with a wonderful metallic glimmer, and is golden on the belly. The fins, except for their black margins, are reddish. The head is covered with blackish-brown flecks, and there is a blackish fleck on every scale. Specimens from the Tagus and the rivers flowing in the low-lying country are much paler, olive-green, or brownish in colour, and there are only fine black points on the body scales. At the upper part of the body the scales are darker at their edges than in its middle. The fish reaches a length of sixteen inches, but is rarely more than ten inches long in the fish markets of Toledo and Madrid. It is popularly known in the Spanish peninsula as *Boga*, and also as *Madrilla*.

The form of body is elongated. The greatest height in front of the dorsal fin, in young individuals, is equal to the length of the head, while in old individuals the head is relatively shorter. The fish is five and a half times as long as the head in the young, while in the adult the head is one-sixth of the length. It is more than twice as long as wide. The eye varies, becoming relatively smaller with age, and is between one-fifth and one-sixth of the length of the head in the adult. The snout is arched and broad, and is prolonged in advance of the transversely horizontal mouth, the angle of which reaches back to the anterior nares. In old specimens the length of the snout is twice the diameter of the eye. The frontal interspace is convex, widens with age from one and three-quarters to two and three-quarter times the diameter of the eye. The dorsal fin is opposite the ventral, in the middle of the body; but is farther back in the young. It is higher than long; the free margin is concave. The base of the anal fin is longer, but its height is less. In proportion and character it resembles the dorsal. The ventrals do not reach back to the vent, but end four scales in front of the genital papilla. The pectoral fin is a little shorter than the head. The caudal is deeply forked, with the lower lobe rather longer than the upper, and is longer than the head. The scales are firmly adherent, and have a fan of twenty rays at the free margin. The larger scales are longer than high, and about two-thirds of the diameter of the eye. There is an elongated scale at the base of the ventral fin.

## Chondrostoma willkommii (Steindachner).

D. 12, A. 12, V. 10.   Scales: lat. 63—67, transverse $\frac{10-11}{4\frac{1}{2}-5}$.

There is but little difference in form of body and mouth between this species and *Chondrostoma polylepis*; but there is a constant difference in the greater number of pharyngeal teeth, which are on each side 6—6 (Guadalquivir), or 7—6 (Guadiana), though in the latter river the formula 6—6 is sometimes found. The body is compressed, with the ventral contour rather more arched than the dorsal profile. The snout is usually conically pointed, but in old fishes it becomes very blunt. The transverse cleft of the mouth is slightly curved, its width is about the diameter of the eye, but it widens with age to one and one-third times the orbital diameter. The eye is usually one-quarter of the length of the head, but in old individuals the head length is four and a half times the diameter of the eye. The breadth of the frontal interspace between the orbits is one and three-fifths to twice the diameter of the eye. The length of the snout only slightly exceeds the diameter of the eye. The relative length of the head is much the same as in *Ch. polylepis*; and the fins are very similar in proportion and character. Along the lateral line, which is parallel to the abdomen, there are usually 64 to 66 scales. The free posterior border of a scale may be a little pointed or rounded; it has numerous diverging rays.

The fish is found in great numbers in the Guadiana and its tributaries, and is plentiful in the Guadalquivir at Cordova, though rare in that river at Seville. The back is dark bluish-grey, or brown in muddy streams. Sometimes there is a trace of a blue-grey longitudinal band in the middle of the body.

## Hybrid between Chondrostoma polylepis and Leuciscus arcasii.

Steindachner found the fish which he thus accounts for as a hybrid in the River Tara, a tributary of the Douro, in the north of Spain. In form, pharyngeal teeth, and bony covering of the lower jaw, it agrees with *Ch. polylepis*, but the mouth is semicircularly curved as in *Leuciscus arcasii*. The snout is not so prolonged as in the former. The dorsal fin is variable; in some specimens it is like *Leuciscus arcasii* in having seven jointed rays, and a convex upper margin; others are like *Ch. polylepis* in having eight jointed rays and a concave upper margin to the fin. The jointed rays of the anal fin vary from seven to nine. The number of scales in the lateral line is not constant; one specimen agrees with *Ch. polylepis* in having sixty-two scales, while the others are intermediate between fifty-two and fifty-nine, and resemble *L. arcasii* in the character of the scaling. The scales are larger than in *Ch. polylepis* with strong concentric striping and a fan of ten to sixteen rays. There are eight and a half to nine and a half rows of scales above the lateral line, and four to five rows below. In the example with sixty-two scales in the lateral line, there are eleven rows above it. The colour is the same as in *Leuciscus arcasii*; a dark blue-grey back, longitudinal bands above and along the lateral line, and orange-yellow bases to the fins.

## CHAPTER VI.

### FRESH-WATER FISHES OF THE ORDER PHYSOSTOMI (*continued*).

GROUP RHODEINA—GENUS RHODEUS: The Bitterling—GROUP ABRAMIDINA—GENUS ABRAMIS: The Bream—A. vimba—A. elongatus—A. ballerus—A. sapa—A. leuckartii—A. bjorkna—Hybrids—A. bipunctatus—A. fasciatus—GENUS ASPIUS: A. rapax—GENUS ALBURNUS: A. lucidus—A. alburnellus—A. mento—A. chalcoides—GENUS LEUCASPIUS: L. delineatus—GENUS PELECUS: P. cultratus—GROUP COBITIDINA—GENUS MISGURNUS: M. fossilis—GENUS NEMACHILUS: The Loach—GENUS COBITIS: The Spinous Loach.

## GROUP : RHODEINA.

### GENUS : **Rhodeus** (AGASSIZ).

RHODEUS is the type of the Rhodeina, Dr. Günther's ninth group of the Cyprinidae. This is a small assemblage of three genera, limited to Japan and China, with one species, the *Rhodeus amarus*—the *Bitterling* of Germany, and *la Bouvière* of France—characteristic of Central Europe. The lateral line is developed only in the anterior part of the trunk. The dorsal fin has nine to twelve branched rays, and extends between the ventral and anal fins. The anal fin has about twelve rays. The mouth is small without barbels; the pharyngeal teeth, in a single row of five on each side, are compressed, with a simple furrow on the bevelled surface. During the spawning season the male develops tubercles on the snout, and the female grows a long external urogenital tube, or oviduct. The genus contains three species, which are all of small size ; two of them are limited to China.

### Rhodeus amarus (BLOCH).

D. 3/9—10, A. 3/9, V. 2/6, P. 1/10, C. 19.
Scales : lat. line 34—38, transverse 10—12.

The Bitter Carp is a very small fish, intermediate in aspect between the Common Carp and the Crucian Carp (Fig. 114). Its body is about three times as long as high, nearly five times as long as the head, and about three times as deep as thick. The head is nearly as high as long, but is short, with an obtuse snout. The eye is about one-third of the length of the head, and is placed towards the middle of its side. The angles of the small mouth, which is rather oblique, reach back under the nares. The nares are double, with the anterior opening round, and the hinder opening oval.

The contour rises from the back of the head to the dorsal fin in a sharp curve, and again descends even more suddenly towards the tail, where the height is scarcely one-third of the height of the middle of the body.

The dorsal fin begins in the middle of the length of the body; the anal commences before the termination of the dorsal, and both have the longer rays equal to half the height of the fish. The vent is placed midway between the anal and ventral fins, and is covered with scales which extend round it like a sheath. The ventral fin reaches back to the anal. The ventral and pectoral fins are equal; the caudal is evenly-lobed and moderately notched.

The scales are thin and elongated, approaching a sub-oval form; their longer diameter is equal to that of the eye in the adult fish. The scales in the young, according to Fatio, are relatively only one-fourth or one-third as large as those in the adult. The scales partially overlap each other, and are marked

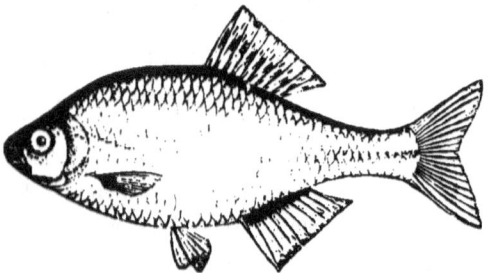

Fig. 114.— RHODEUS AMARUS.

with radiating lines, ten to thirty-five in number. The free edge is rounded; and the attached margin is sub-convex. The scales on the back, breast, abdomen, and root of the tail, are smaller and more circular than those on the sides. In the lateral line the scales are transversely wide, and only from one or two to seven are perforated.

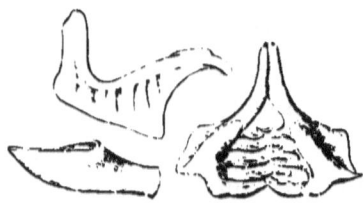

Fig. 115.— PHARYNGEAL TEETH OF RHODEUS AMARUS.

The accessory gills are free and pectinate. The entire upper division of the gill-arches is covered with membrane, and it has been supposed that this separation of the mouth aperture and the peculiar respiration it implies may not be unconnected with the tenacity of life for which this fish is remarkable. Fig. 115 shows the pharyngeal teeth.

The colour varies with sex and season of the year. The males at spawning-time in summer develop every year two circular patches of warty tubercles above the mouth, which periodically disappear, a character in which they are paralleled by many other Cyprinoids, like *Chondrostoma*. The operculum, back, and sides, above the anal fin, have a beautiful violet colour. On each side of the tail is an emerald-green stripe, which extends half way along the length of the body to the caudal fin, and ends in a fine streak. The pectoral region and abdomen are silvery, with rose-coloured or orange-red patches. Behind the operculum is a broad, vertical silver band, covered with dark violet spots. The dorsal fin is brown, spotted, almost banded, with black. The anal fin is bright red, with black edges; the pectoral and caudal fins are pale in colour and transparent. The iris is very red above.

The females are smaller, their height being about four-fifths of that of the males. The head is rather longer, but all the fins are shorter than in the male. At ordinary times there is but little difference between the sexes in colour, but when the male puts on his metallic beauty of varied colour, the female has the back greenish-brown, with the sides, abdomen, and pectoral region, silvery. The line at the middle of the side of the tail varies from steel-blue to black, is rarely half as wide as in the male, and often disappears. All the fins are pale in colour except the dorsal, which is bordered with a broad black margin. The iris is pale yellow, with orange spots. There are no bony tubercles on the snout.

There is some difference in the air-bladder, for although it is generally divided in both sexes into unequal parts, the anterior third taking a pear shape, this part is usually much more slender in males than in females.

The female fish has none of the brilliant colouring of the male, but possesses a long orange-red oviduct, which often extends some two-thirds of the length of the fish, hanging free like a worm in front of the anal fin, and sometimes extending back beyond the caudal fin. It was discovered by Herr Krauss, of Stuttgart. The eggs are yellow, and, when they distend the external oviduct, give it a beaded aspect, as they move down in succession one behind another. The diameter of the oviduct is less than that of the eggs, which are elongated to a cylindrical form within it, regaining their oval shape when laid. The oviduct, after performing its function, is gradually shortened until it disappears in a papilla.

The long intestine exhibits two principal convolutions, which lie below the liver. Von Siebold found the entire intestine filled with diatomaceæ and fragments of algæ. The heart-shaped bladder opens into the base of the oviduct, so that were it a permanent structure it might be termed the urogenital canal.

No one has observed the mode of deposition of the eggs, or ascertained whether this remarkable tube, which is furnished with its own blood-vessels and nerves, is made use of to place the eggs in a secure position.

This is one of the smallest of European fishes. The females are usually from one and a half to two inches long, and the males do not exceed three inches.

The Bitterling prefers clear flowing water with a stony bottom, and lives in rivers and brooks. In the Seine it spawns from May to August, in Austria it spawns in April. The vitality of the male is greater than that of the female, but both survive conditions which destroy other fishes. It is less useful as a food fish than any other fresh-water fish, on account of its small size and bitter flavour. In Austria the Perch will eat it, and it is used in the Danube by fishermen as bait for eels; but it is rejected by most fishes, showing that bitter flavours are not grateful to them.

It is absent from Great Britain and Scandinavia, but is widely distributed in Central Europe, especially through France, Holland, and Germany; and is found at Brusa, in Asia Minor. It lives in several warm springs, such as those of Töplitz, in Croatia.

## Group: ABRAMIDINA.
### Genus: **Abramis** (Cuvier).

The Bream family comprises about sixteen genera of fishes, characterised by having the anal fin elongated; and the body, especially the abdomen, com-

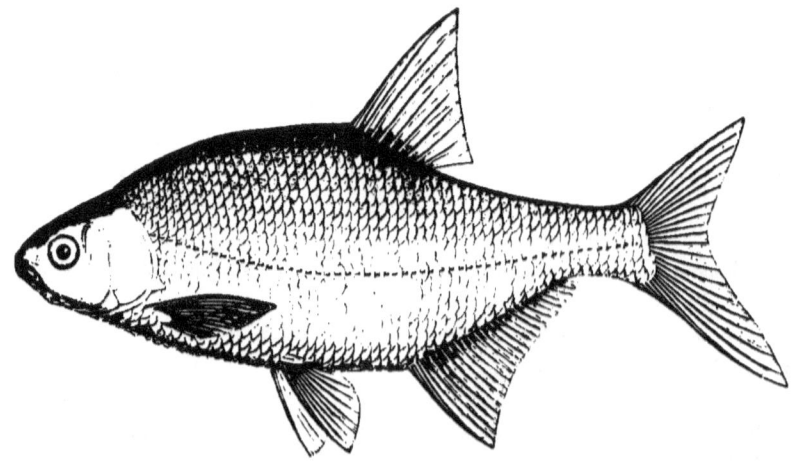

Fig. 116.— ABRAMIS BRAMA (LINNÆUS).

pressed into a ridge or edge. Several of the genera are limited to China, India, Western Asia, and Eastern Africa, though the group Abramidina is represented in Europe by Breams, Bleaks, a species of Aspius, and a species of Pelecus.

Abramis has the belly compressed into a sharp edge behind the ventral fins, and the scales do not cross this edge. The pharyngeal teeth may be in one or two series. Both jaws have simple lips; the lower lip is interrupted in the middle line in front, and the lower jaw is generally the shorter. The dorsal fin has no spine. The lateral line runs below the middle of the body and tail. The body is characterised by its compressed and elevated form. The genus belongs to the northern parts of Europe and adjacent parts of Asia. It is represented in North America by several species, which have a smaller number of rays in the anal fin than the corresponding European types.

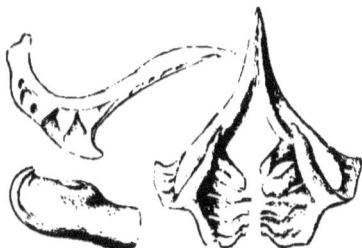

Fig. 117.— PHARYNGEAL TEETH OF ABRAMIS BRAMA.

## Abramis brama (Cuvier).—The Bream.

D. 3/9, A. 26—31, V. 2/8, P. 1/15. Scales: lat. line 51—57, transverse $\frac{12-13}{6-7\frac{1}{2}}$.

This species is remarkable for its deep body, compressed from side to side. It is about three times as long as high, and more than three times as high as thick (Fig. 116). The head is about one-sixth of the total length; and the eye is one-fourth or one-fifth of the length of the head. The mouth is curved in a horse-shoe form. The snout is blunt. The pharyngeal teeth, as in all species of the genus, are notched at the extremity; and, as in all typical Breams, are in a single row of five on each side (Fig. 117). The dorsal fin commences behind the middle of the body; its base is much shorter than its height. The fin is abruptly truncated behind, with the first ray four times as long as the last. The anal fin is much longer than high, and is between one-third and one-fourth of the length of the body; it is situated behind the dorsal. The ventral fin is placed in front of the dorsal, and scarcely reaches back to the vent. The pectorals reach back to the ventral fins. The caudal has the lower lobe longer than the upper, and equal to the length of the base of the anal fin, while the shorter upper lobe is as long as the longest ray of the dorsal fin.

The scales are firmly adherent, soft, higher than long, and rounded at the free end, where they are marked with a fan-like pattern of ten or twelve diverging rays. The basal border is irregularly festooned. The scales on the throat and fore part of the abdomen are small, but still smaller scales occur on the sides of the back of the skull, which is free from scales. The lateral line extends in a curve parallel to the ventral outline, and the mucus-canals of the head are obvious in their extension to the eyes and nares. The comb-shaped accessory gills are free, and well developed.

The colour of the back and upper part of the head is black or olive-green; the sides are yellowish-white, with a silvery lustre, and dark scattered pigment spots; the throat is reddish, the abdomen silvery. The iris is golden yellow, with black spots, and all the fins are bluish-black; but sometimes the early rays have a scarlet tinge.

The Bream varies in size in different parts of Europe. In Russia, where it occurs as far north as latitude 63°, large specimens weigh eight pounds, though the fishermen speak of Bream weighing fifteen or twenty pounds. The average weight at Novgorod is one pound and a half, and at Astrakhan a little over three pounds. But the weight of the fish, as well as the number caught, has diminished everywhere, owing to the increase of fishing and the long time which the Bream needs to attain to a large size. The Bream is one of the larger fishes in the Danube and Lake Constance, where it may be a foot and a half long, six inches high, and weigh five or six pounds. In the Atter See the weight is ten pounds; and various German writers refer to heavier fishes, in which the weight has reached as much as twenty pounds. In North Germany, the length reaches sixty to seventy centimètres, and the weight ten to twelve pounds, though Günther mentions fifteen pounds and a length of two or three feet. In France the size and weight are less, not exceeding eight pounds. In Norfolk a Bream known to be fifty years old weighed eleven pounds and three-quarters, and measured two feet two inches. We saw a Bream taken once from the Serpentine, in Hyde Park, which weighed between seven and eight pounds.

The Bream lives in rivers, lakes (like Thun and Wallenstadt), shallow ponds, and marshes; and, according to Eckström, is found in the sea off Norway and Sweden. It prefers waters with a strong stream and muddy or clay bottom. In summer it remains at the bottom, between water plants, and burrows in the mud, often betraying its existence to the angler by the disturbed state of the water. Its food consists of water plants, worms, insects, and mud.

It is a fish of social habit, often assembling in large numbers, and apparently following the leadership of some able fish, known in France as the King Bream. It is a cautious, timid fish, with the faculty of hearing so acute, that it flies away at any noise, and is disturbed by noise even during spawning. It is preyed upon by most carnivorous fishes and water birds. It suffers much from intestinal worms, the sufferers becoming thin, and pale in colour. It is easily transported, and grows well in ponds, but is good for nothing when put there. Chaucer, however, seems to have thought otherwise; for we learn from the prologue to the "Canterbury Tales":

> "Full many a fair partritch had he on mewe,
> And many a Breme and many a Luce in stewe."

Blanchard mentions that before the palate had become educated by the introduction of Carp, the Bream was held in great honour, as we might have inferred from the French proverb, "*Qui a brême peut bramer ses amis.*" The fish is sold at a low price. As its flavour is in no way disagreeable, it is not to be disdained; but Dr. Badham, who investigated the virtues of fishes with many kinds of cookery, rejects its

> "Flabby solids filled with treacherous bones."

In Norfolk Bream sell at half-a-crown a bushel, and are chiefly eaten in the middle of England, during the Hebrew Passover, and in Lent.

Benecke says that in winter and spring the larger Prussian specimens are very tasty, and preferred by some people to Carp.

Bream spawn generally in May or June, but possibly earlier in favourable localities. The males travel first on a voyage of discovery to find out the soft grassy places on the banks, and the females arrive later. The spawning is usually carried on at night, and is accompanied by considerable noise, for the fish move rapidly, and smack their lips. The old fish lay their eggs first. In favourable weather the spawning is accomplished in three to four days, but if the weather is unfavourable the fish descend again to deeper water without depositing the spawn. The eggs are one and a half millimètre in diameter, yellow, are deposited on water-plants to the number of 200,000 to 300,000, and hatched in a few days.

The scales of the male at the breeding season become rough, with little asperities, which may be light or dark. At the same time, tubercles, like those seen in many other Cyprinoids, are developed on the head; in old fish they remain for a long time, and have been observed as late as October. Bream readily takes bait at spawning-time, and is then caught with a net, but not easily, since the fish disappears at the least noise. In Prussia it is captured in winter by a net inserted through holes in the ice. Benecke states that as many as one hundred tons of Bream are sometimes obtained in this way.

In the Caspian the Bream is **remarkably** abundant, the annual take amounting to 28,500,000, according to Dr. Grimm; but this probably includes the other species of the genus. The value of the fish varies from sixty copecks to one rouble per pood, and the large fishes sell at twenty-four to twenty-seven roubles a thousand. About four hundred thousand poods are exported annually from Astrakhan, though some writers have stated the quantity at three times that amount.

The species is generally distributed throughout the north and middle of Europe, but it is not found south of the Pyrenees and Alps.

Heckel, as long ago as 1835, described a Bream from the shallow Austrian lake called the Neusiedler See, which has been found nowhere else, and is rare even there. It has no distinctive name among fishermen, and, in

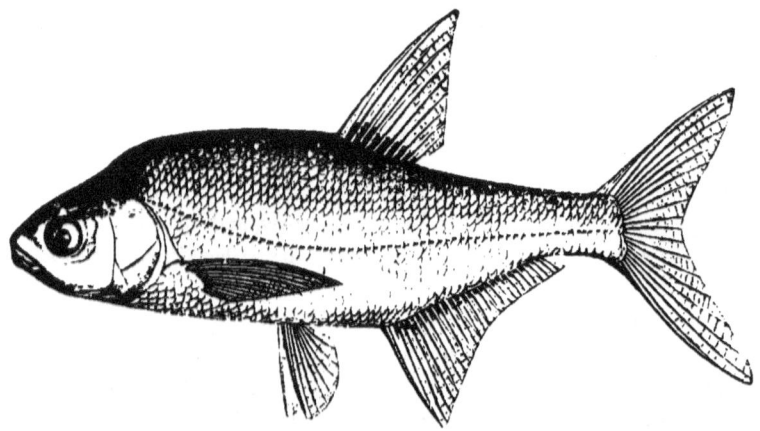

Fig. 118.—ABRAMIS BRAMA, VAR. VETULA (HECKEL).

common with the other forms, is termed Bream. This variety was named the *Abramis vetula*, and appears to be a local race, distinguished by having the body more elongated, the head large and long, with larger eyes, and it has the fins longer and more pointed (Fig. 118). The greatest height of the body is one-quarter of the entire length. The eye is its own diameter from the extremity of the nose. The pectoral fins reach half way down the ventrals; and from the sharp outline of the fins Heckel and Kner term this fish the Pointed Bream. All the fins are spotted with black. The general colour of the body is greenish-grey, reddish-brown on the upper part of the head, and a shining lead colour on the sides and abdomen. It is usually about eight or nine inches long.

Other local races are found in various parts of Europe. The Moselle yields a form common in some seasons about Metz, which Blanchard has named *Abramis gehini*. It closely resembles the Austrian variety, *Abramis vetula*, in size, elongation of body, and development of fins. The scales are said to be shorter than in the Common Bream. The pharyngeal teeth are more strikingly re-curved. In colour the upper part of the body is grey slaty-blue, with the rest of the body silver-white, with scales finely spotted with black. Mr. Gehin says it is a very good fish.

## Abramis vimba (LINNÆUS).

D. 11, A. 21—23, V. 10—11. Scales: lat. line 55—60, transverse 10½ 9.

This species, characteristic of Eastern Europe, is easily recognised by the keeled back of the tail behind the dorsal fin, and by the projecting snout (Fig. 119). The greatest height of the body in front of the dorsal fin is about

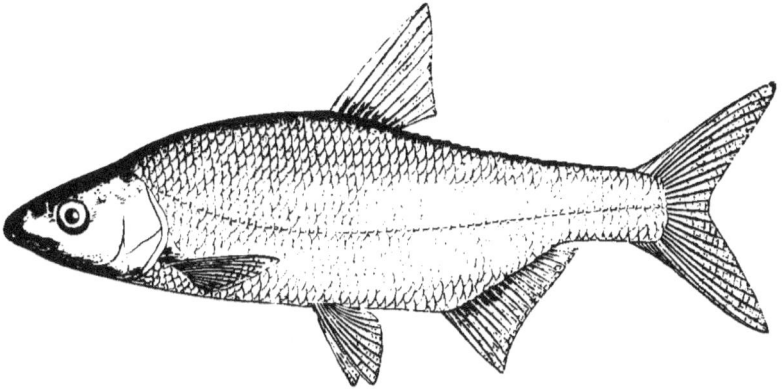

Fig. 119.—ABRAMIS VIMBA (LINNÆUS)

one-quarter of the length of the fish, but the thickness is two-fifths of the height. The head is one-fifth of the length of the fish.

The eye is one and a half times its own diameter from the extremity of the snout, and the same distance from the other eye. It is moderately large, and about one-fifth of the length of the head. The nasal apertures are nearer to the eyes than to the extremity of the snout. The mouth is oblique, and its angle reaches back below the anterior limit of the orbit for the eye. The upper lip is fleshy, and thickens with age; it is arched, and projects beyond the mouth. The elongated profile of the head and snout is the most striking character of the species. From the back of the head the contour ascends in an arch to the dorsal fin, where it begins to sink towards the tail. Where narrowest the tail is two-fifths of the greatest height. The abdominal contour is less convex than the dorsal profile. The dorsal fin, which begins half way down the length, commences rather behind the ventral, and is twice as high as long. Its longest ray is about one-fifth of the length of the body, and about as long as the base of the anal fin. The pectoral fins scarcely reach to the ventral fins, and the ventral fins extend to the vent.

The lower lobe of the caudal fin is scarcely longer than the upper lobe.

The scales are ornamented with fan-shaped rays, which are fairly regular, and number eight to twelve.

The largest scales are along the lateral line, and the scales on the back of the skull are scarcely smaller than on the rest of the body.

The colour is usually greyish-blue on the snout, head, and back, and in the dorsal and caudal fins. The pectoral, ventral, and anal fins are pale yellow, with a deeper tinge at the base. The anal fin is margined with black. The sides and abdomen are silvery. In Upper Austria this fish is popularly known, from its colour, as the Blue-Nose, and in Bavaria, according to Von Siebold, as the Smut-Nose.

At the breeding-season, towards the end of May or beginning of June, the colours change in both sexes. The entire upper surface, down to well below the lateral line, is covered with a deep black pigment, which has a silky lustre. The under surface from the lips to the tail becomes a rich orange-red colour. As in *Abramis brama*, the male at this time develops tubercles, or warts, on the skull, scales, and some of the fins.

The iris is yellow, with dark spots.

This species is always smaller than *Abramis brama*. In Prussia it measures twenty to thirty centimètres. In the Danube it rarely weighs more than half a pound, and in the Atter See may weigh three-quarters of a pound.

In the Atter See it remains at a depth of ten fathoms, and in winter retires to twenty fathoms. If many individuals live together, they burrow in the muddy bottom, stirring it up, and clouding the water. In the spawning-time they are crowded together, like Salmon Trout, in large shoals, for about a fortnight, when they frequent shallow waters. In the Haffs, on the coast of Prussia, they have spawned as late as October. The eggs are deposited upon the bottom, or upon plants, to the number of 200,000 to 300,000. In the Haffs they are captured with nets all the year round.

Like the Common Bream, this species is full of bones, but when fat, is commonly roasted on a spit in Prussia, and esteemed well flavoured. In Russia it is found in the Don, Dnieper, and Bug, and adjacent parts of the Black Sea and Sea of Azov, and is well known in Sweden and the Baltic. Indeed, according to Fries and Eckström, it remains in the depths of the sea in winter, and goes up rivers in the spring to spawn. In Sweden it is a crafty fish, difficult to capture.

## Abramis elongatus (Agassiz).

The rays of the fins and arrangement of the scales are the same in this species as in the Snout-Nose, and Günther remarks that the skeletons are identical in both fishes, there being twenty-three thoracic vertebræ, and the same number in the caudal region; so that *Abramis elongatus* has been regarded as a variety of *A. rimba*, which, instead of descending annually to the sea, remains stationary in rivers. It is so like that species as to be easily confounded, but the nose is blunter and less projecting, and the head thicker (Fig. 120). The greatest height is one-fifth of the total length, so that it is rather more

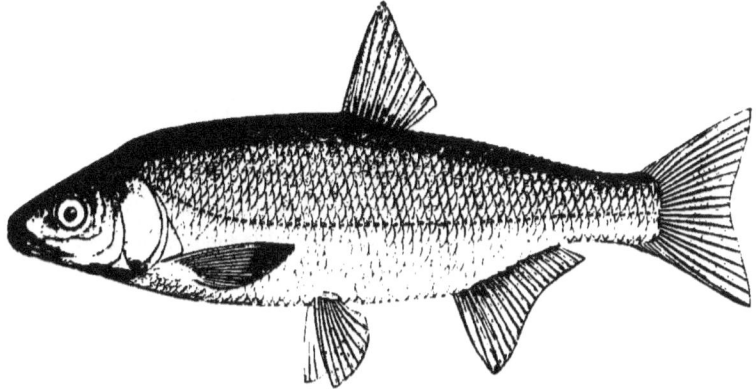

Fig. 120.— ABRAMIS ELONGATUS (AGASSIZ).

elongated, though in old age the body becomes somewhat higher, and more approaches the proportions of *Abramis rimba*. The eye is large, about a fourth the length of the head. It is at first separated from the snout by its own diameter, but with age the distance augments a little; and it is one and a half times its diameter from the other eye. The angle of the mouth reaches back below the nares. The dorsal fin commences rather farther back than the ventrals. Its height is twice the length of its base. The anal fin begins at the end of the dorsal, and, as in all the species, is longer than high. The ventral fins do not quite reach the vent, and are themselves rarely reached by the longest ray of the pectoral fin. The resemblance to *Abramis rimba*, which is seen in the details of the pharyngeal teeth, scales, and mucus-canals, extends to the colour; though broad, bright longitudinal lines, formed of small spots, often extend along the side of the trunk to the caudal fin. The anal and pectoral fins are white, and the other fins are spotted with black dots. Von

Siebold records that the colours change in the breeding season, as in *A. rimba*, and on the whole considers that so many intermediate forms bridge over the interval between these species that *A. elongatus* cannot claim specific rank. It is, however, a smaller fish, commonly about ten inches long, and not exceeding thirteen inches. It is well known as an article of food in most of the fish markets of Northern Germany, and is met with throughout Russia, in the lakes of the Austrian Alps, and along the Danube, both in Hungary and Roumania.

*Abramis tenellus* (Nordmann) is from the rivers of the Crimea, and Kessler thinks that it is possibly the *Abramis persa* of Pallas. Günther remarks that it is closely allied to *A. rimba*. Its pharyngeal teeth are 5—5. There are nineteen to twenty-one rays in the anal fin, fifty to fifty-seven scales in the lateral line, with nine rows above the lateral line, and seven rows below it.

## Abramis ballerus (Linnæus).

D. 11, A. 39—42, V. 10. Scales: lat. line 69—73, transverse $\frac{15}{12}$.

This species is distinguished by the remarkable length of the base of the anal fin, which commences opposite the dorsal, and extends almost to the tail, and by the position of the mouth, which is oblique (Fig. 121). The proportions

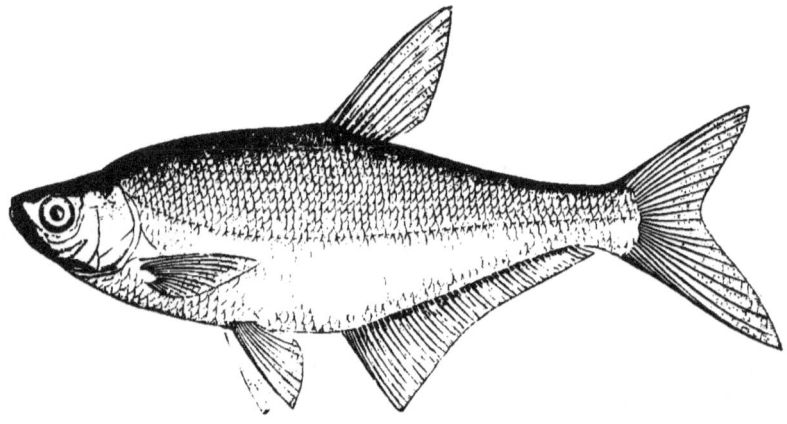

Fig. 121—Abramis ballerus (Linnæus).

of the body are not very unlike those of preceding species, the greatest height in front of the dorsal fin being one-quarter of the length, and the thickness is one-third of the height. The head is only about one-sixth of the entire length; the eye has the usual proportion of a quarter the length of the head. The angle of the mouth does not reach so far back as the anterior nares. The upper jaw and nose scarcely project beyond the mandible. The contour of the top of the head is flat, while behind the head the back rises in a curve. The abdominal outline is more convex than the dorsal profile. The dorsal margin of the tail is not keeled.

The dorsal fin begins half-way down the body, and is usually more than twice as high as long, and is about equal in length to the upper lobe of the caudal fin. Its hinder outline is truncated. The base of the anal fin commences a little behind the beginning of the dorsal. It is two-fifths of the length of the body, is two and a half times as long as deep, and is truncated. The pectoral fins extend over the base of the ventrals; the ventrals reach to the vent. The lower lobe of the caudal fin is somewhat elongated.

The scales are ornamented with a few rays, commonly six or seven, but sometimes only two to four. The scales become largest on the sides, towards the back of the head, but even these are scarcely half the diameter of the eye, and, on the whole, the scales are smaller than in other species. The lateral canal is nearly horizontal.

The colour is like that in the other species: the upper part of the head is brown, the back is dark blue or green, the sides yellow, shading into shining silver; the abdomen is reddish; the pectoral and ventral fins are yellow, while the fins generally are dotted with black.

The largest specimens are only one foot long, and the weight is seldom more than a pound and a half.

According to the Danube fishermen, this fish lives four or five years. It spawns in April and May, and in ways of life and food resembles the other species. It is found in Sweden, in Germany as far as the Rhine (but not in Holland), in the Danube, and in Russia. It frequents the entire coast of the Baltic Sea, and is found in the Haffs. It ascends the Oder and Vistula. It is found in the fresher parts of the Black Sea, the Sea of Azov, and Caspian.

In this species the pharyngeal bones are much more slender than in *Abramis brama*, and have the anterior process very long.

There are twenty-two thoracic vertebræ, and twenty-six in the tail.

## Abramis sapa (Pallas).

D. 11, A. 11—48, V. 10. Scales: lat. line 50—52, transverse $\frac{9-10}{8}$

This species resembles *Abramis ballerus*, but the anal fin contains a few more rays; the scales are a little larger, and, therefore, fewer, and the snout is remarkably blunt and thick (Fig. 122). The form of the head is like that of

*A. vimba.* The height of the body is more than one-quarter of the length of the fish. The length of the head is one-sixth of the length of the fish. The diameter of the eye is one-third of the length of the head, and it is separated from the other eye by one and a half times its own diameter. The eye is near to the extremity of the snout; the nose scarcely projects over the mouth, which

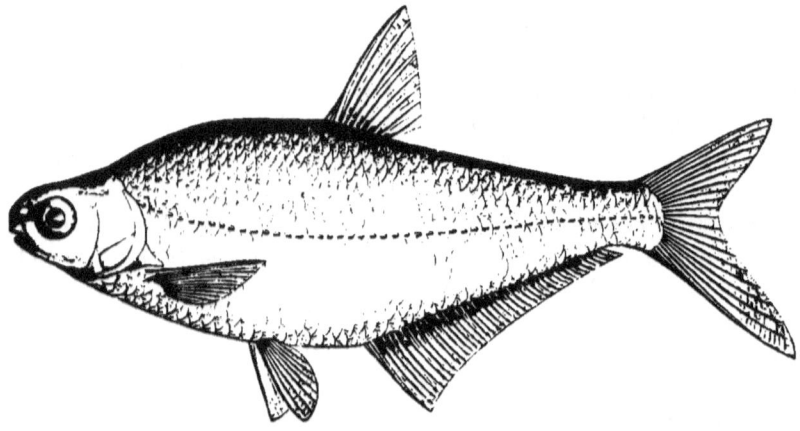

Fig. 122.—ABRAMIS SAPA (PALLAS).

is very small and oblique, so that its angle does not extend back to the anterior nares. The dorsal fin begins in the middle of the length; it is fully twice as high as long, and pointed, owing to the posterior truncation. The last ray is one-quarter of the length of the longest ray. The pectoral fins extend back over the base of the ventrals, which reach to the vent, or beginning of the anal. The elongated lower lobe of the caudal fin is nearly one-quarter of the entire length. In the hinder part of the body the lateral line runs in the middle of the side. It descends so as to make a slightly concave curve in its middle portion.

The scales are ornamented with ten or fifteen remarkably fine rays, forming a fan. The largest scales, which are in the anal region, are about the diameter of the eye.

The accessory gills are well developed. The pharyngeal bones are intermediate in form between *Abramis brama* and *A. vimba*, the anterior process is less elongated than in the Bream.

The back of this fish is scarcely darker than the rest of the body, the entire surface being a bright silver, with mother-of-pearl lustre. The fins are whitish, but the dorsal, anal, and caudal are spotted with black; the iris is pale yellow.

It is seldom more than one foot long, and a pound in weight. It frequents the rapid parts of large rivers like the Dnieper, Dniester, Volga, and Danube. It is found in the fresher parts of the Black Sea, Caspian, and Sea of Aral. Below Vienna it is common, and is sometimes found in the south of Bavaria, but is not known in the Tyrol. It is found in the River Volchoff, of the Baltic system. It lives for about four or five years. On account of the number of bones it is not esteemed for food; but the shining silver scales are used, like those of some other fishes, in the manufacture of artificial pearls.

Spawning begins early in April, when the males develop the characteristic white tubercles on the skin. They are arranged in a single row on the free posterior edges of the scales of the upper part of the body. There are dense longitudinal rows of tubercles on the pectoral and ventral fins. The head is ornamented with discoid tubercles.

## Abramis leuckartii (HECKEL).

The Pomeranian Bream is probably a cross between the Bream and the Roach, and is described as a Roach-like Bream or a Bream-like Roach. This fish closely resembles *Abramis brama* in general aspect, but in the situation of the mouth is like *A. ballerus* (Fig. 123). It differs from all the Breams in the small number of rays, eighteen or nineteen, in the anal fin, and has the caudal fin evenly-lobed. The head is rather less than one-fifth of the entire length, and the body is three and a half times as long as high. The eye does not differ in size or position from the usual condition in Breams. Both jaws are of the same length; the terminal oblique mouth is rather small, and its angles do not reach back to the nares; the nose is rather blunt. In the tail the height diminishes to one-third of the height of the body. The dorsal fin is behind the middle of the body, and is one-third higher than long. The shortest ray measures one-third of the longest. The anal fin begins nearly opposite the end of the dorsal, and is only a little longer than high. Its base is about one-sixth of the entire length of the fish. The ventral fins are somewhat in front of the dorsal fin, and in advance of the middle of the body; they do not reach back to the vent, nor do the pectoral fins reach back to the ventrals.

The scales show three or four rays at most. They are largest along the lateral line, but scarcely attain two-thirds of the diameter of the eye. As in species of Abramis, the males have warts on the scales under the dorsal fin. The lateral line sinks to the lower third of the side.

The pharyngeal teeth are commonly in a single row; exceptionally, Von

Siebold found an additional tooth in a second series. The teeth most resemble those of *A. vimba*.

The colour is silvery, though the upper part of the head and back is greenish. All the fins are white, and only the dorsal and caudal are spotted with black; the iris is yellow. This fish is found in Transylvania and in Austria, especially in the Danube. In Bavaria the length varies from seven to twelve inches. It occurs in tributaries of the middle Rhine, and M. Selys-Longchamps mentions it in the Somme and Moselle. Yarrell quotes it as a British fish from Belfast, and from Essex and Middlesex. It is well known in the Haffs on the Prussian coast, and is found singly in all Russian rivers.

Although Von Siebold regarded this fish as a bastard, he, nevertheless, formed a new sub-genus to receive it, which was named *Abramidopsis*. He

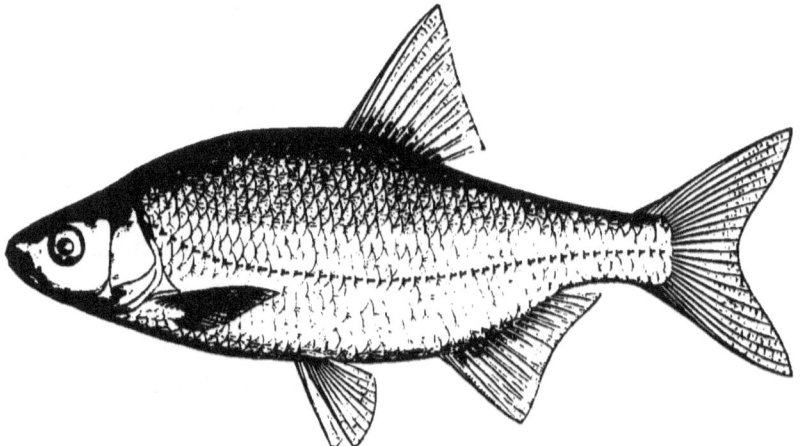

Fig. 123.— ABRAMIS LEUCKARTII, HYBRID BETWEEN LEUCISCUS RUTILUS AND ABRAMIS BRAMA.

was influenced in this eccentricity (1), by the form and arrangement of the pharyngeal bones and teeth, about half the individuals examined having six teeth on one side and five on the other; (2), by the short anal fin; (3), by the low dorsal fin; (4), by the less compressed body; and (5) by the scales being angularly bent, so as to cover the compressed abdominal edge behind the ventrals. In the lateral line there are forty-eight to fifty scales, with five longitudinal series of scales between the lateral line and ventral fin.

## BLICCA.

The White Breams were separated by Heckel as a distinct sub-genus on account of the presence of two rows of prehensile teeth upon the pharyngeal bones, which are arranged, five in the outer row, and two in the inner row on each side. But since the other characters of the genus are in all respects those of Abramis, we prefer to follow Dr. Günther in regarding Blicca only as a convenient division of the Breams, which has no special importance in classification. Like the typical Breams, the Bliccas form hybrids with Leuciscus. Their geographical distribution is similar, being found in Northern and Central Europe, in the countries about the Black Sea, and in the United States.

### Abramis (blicca) bjorkna (LINNÆUS).

D. 11, A. 22—27, V. 9—10.  Scales: lat. line 43—48, transverse $\dfrac{9-10}{7}$

The White Bream or Bream-flat, of the English, *la Bordelière* of the French, *Blicke*, or *Güster*, or *Zobelpleinze* of the Germans, *Bjelk* or *Bjorkna* of the Swedes, is widely distributed in Europe north of the Alps. In

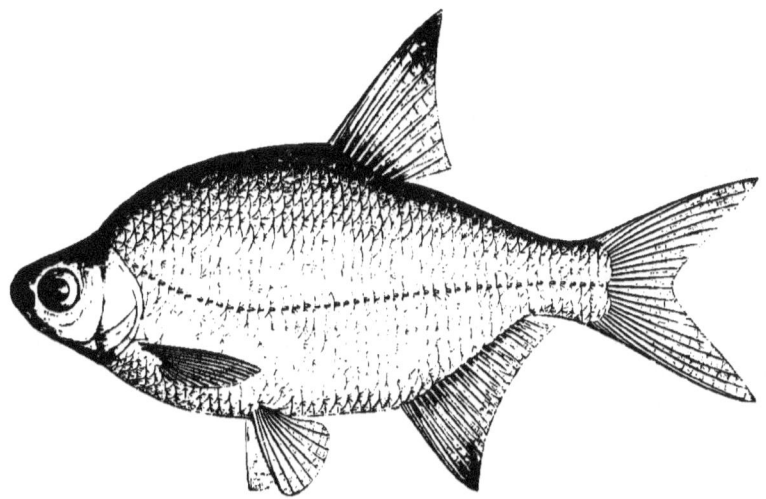

Fig. 124. ABRAMIS BJORKNA (BLICCA).

England it is found in the Trent, Cam, and other rivers of the eastern counties, but is less commonly met with than on the Continent, where it is not limited to fresh water, being plentiful in the Haffs of Prussia. The Rev. Richard Lubbock records that on taking the hook it rises in the water, so that the float, instead of being drawn down, is laid on the surface (Fig. 124).

In general shape this fish resembles *Abramis brama*; the body is rather high, the greatest height being one-third of the length, though the height varies. The thickness is one-third of the height. The head measures about one-sixth of the length of the fish; the eye is between one-third and one-fourth of the length of the head, and is larger than in the Bream. The nasal apertures are near to the eye. The mouth is small, oblique, and directed upward; its angle scarcely reaches back under the nares, and the lower jaw is rather the shorter of the two. The dorsal fin commences behind the middle of the body, and is less than twice as high as long; the third or fourth ray of the fin is four times as long as the last ray. The anal fin commences opposite to the termination of the dorsal. It is shorter than in the Bream. Its depth, which is three-fifths of its length, is equal to the height of the dorsal; according to Von Siebold, it may contain as few as nineteen rays. The ventral fins are in front of the dorsal, and do not reach back to the vent. The pectoral fins do not reach the ventrals; the terminal lobes of the caudal fin are equal.

The scales are rather longer than in the Common Bream; their exterior border is rounder, and the basal border more festooned. Scales are absent from the anterior part of the back, and from the ventral ridge.

The colour of the back is brownish-blue; the sides are blue, with a silvery lustre; the belly, which is white in life, becomes reddish after death; the dorsal, anal, and caudal fins are bluish-grey, but the pectoral and ventral fins are always red at the base, and become redder as the fish advances in age. The iris is silvery, with green flecks.

This species never measures more than one foot in length, and seldom weighs more than one pound. It is one of the commonest fishes in Central Europe, and frequents lakes and ponds, but prefers gently flowing water with a sandy bottom. It lives in somewhat deep water in the autumn and winter, but in the spring haunts shallow places near the banks. It lives on water plants, insects, and worms, is less timid than the Common Bream, and often remains in one place for a long time. The fully matured ovary may contain 100,000 eggs, which are each about two millimètres in diameter.

The fish spawns in May, preferring shallow places overgrown with bulrushes. The older fish begin to lay their eggs first, and the spawning commonly takes from three to four days, but may be performed more rapidly if the weather becomes colder. It is always a noisy proceeding, and the disturbance of the water may be seen from some distance.

The medium-sized individuals spawn a week later, and so on at intervals down to the youngest, which are ten to twelve centimètres long.

When spawning the fishes are very active, and so heedless that in Austria they are easily caught with the hand. They are nowhere valued as food, and kept only for the support of predaceous fishes. They are frequently afflicted with tape-worm.

At the breeding season the colours become darker in both sexes, and the fins acquire an orange tint, but Fatio thinks the colour may to some extent depend upon the food. The male frequently has small tubercles developed on the scales.

The pharyngeal bones are characteristic, being much more rounded than in other Abramidina. Opposite to the anterior tooth, the outer side of the anterior process is inflated. The formula, according to Von Siebold, is 3·5—5·3 in Bavaria, and, according to Heckel, 2·5—5·2 in Austria, and in Switzerland Fatio finds both types as well as 1·5—5·1, 2·5—5·3, and 2·6—5·2; so that we can only infer some variation with geographical distribution (Fig. 125).

The fish which Nordmann described as *Abramis laskyr*, and records as common in the Black Sea and rivers flowing into it, is a variety in which the fins are more developed.

There are twenty vertebræ in the thorax, and the same number in the tail.

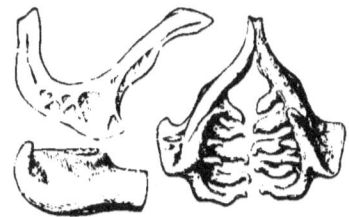

Fig. 125.—PHARYNGEAL TEETH OF ABRAMIS (BLICCA) BJORKNA.

The *Abramis laskyr* of Pallas has been thought worthy of recognition by many writers (Fig. 126). Its greatest height is more than one-third of the length, and the thickness is less than one-quarter of

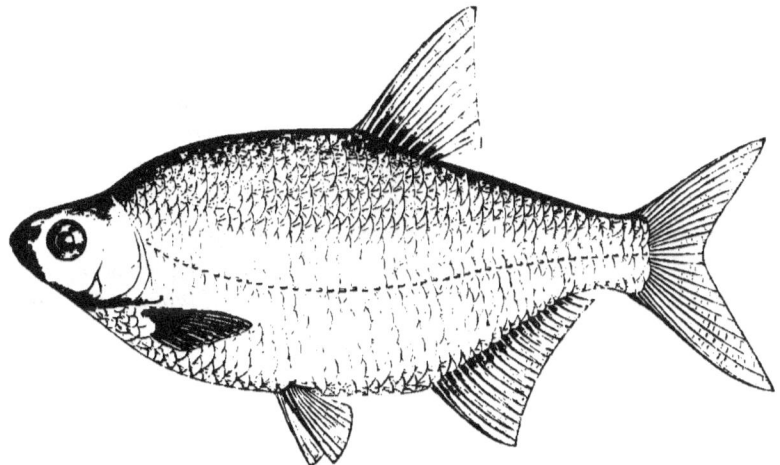

Fig. 126.—ABRAMIS BJORKNA, VAR. LASKYR.

the length. The forehead is smaller than in the preceding type. All the fins are brown, and the red base is wanting in the pectoral and ventral, but Nordmann states that the colour varies locally, and with the season. It is common in the fish market of Odessa, where it is brought from the Bug and Dnieper. In common with the type, it is distributed in the fresh waters of Russia, but in Finland does not occur north of lat. 62°. Several varieties have been described. *Abramis bjorkna* breeds with other fishes, and forms bastard races, to one of which Von Siebold has applied the name *Bliccopsis abramorutilus*.

### BLICCOPSIS ABRAMORUTILUS (HOLANDRE).

The fish thus designated by Von Siebold is a hybrid between *Abramis blicca* and *Leuciscus rutilus*, and is very similar to the hybrid between the Bream and the Roach. The pharyngeal teeth, according to Günther, are sometimes in one row, sometimes in two; but Von Siebold records that among the specimens examined he found the formula 3·5—5·3, 2·5—5·3, 3·5—5·2, 2·5—5·2. The pharyngeal bones are more delicate than those of *Abramis bjorkna*, and differ in form. The snout is tumid, and rather projects over the mouth. This hybrid has a larger eye than the Roach.

There are forty-three to forty-six scales in the lateral line, with eight rows of scales between it and the dorsal fin, and four rows between it and the ventral fin. The fish is seven to ten inches long. The colour is olive-green on the back and coppery-yellow on the sides. The anal, ventral, and pectoral fins are dark grey, but the pectoral is sometimes red, and all the fins have always a reddish colour at the base. The anal fin contains fourteen to eighteen rays. In its high back this fish resembles the Roach.

The spawning-time is at the end of April, and the males then have a crescentic warty development on the skull, and on the inner side of the rays of the pectoral fin. This bastard is found over Belgium, Holland, Austria, and Germany, is not known west of the Rhine, but occurs in the Vistula, and is found singly in all Russian rivers, according to Dr. O. Grimm.

The variations which it puts on are very numerous, and presumably result from the inter-breeding of the bastard with the parent types, as well as to a varying predominance of characters of the male or female fish.

## Abramis bipunctatus (HECKEL AND KNER).

The generic position of this fish is somewhat controverted. It is characterised by having every scale upon the lateral line marked with spots of black pigment, so as to form a shining band, which separates the silvery side below it from the darker back (Fig. 127).

The body is five times as long as the head, and four and a half times as long as high, though in young fishes the relative length is a little more. The angle of the mouth, which is moderately inclined, reaches back under the nares. The ventral margin between the ventral fin and vent, forms a sharp edge, but the scales do not extend across it. The dorsal fin begins behind the ventrals, but in front of the middle of the length of the fish; it is higher than long, and as high as the head is long. The anal fin is longer than high; the ventral fins scarcely reach to the vent. The largest rays of the pectoral fin are equal to those of the dorsal fin.

The colour of the back is dark green, the sides are silvery-green, and the belly bright silver. By the side of the lateral line, which is conspicuous, with its margin of black dots, there is a horizontal blue-black band extending

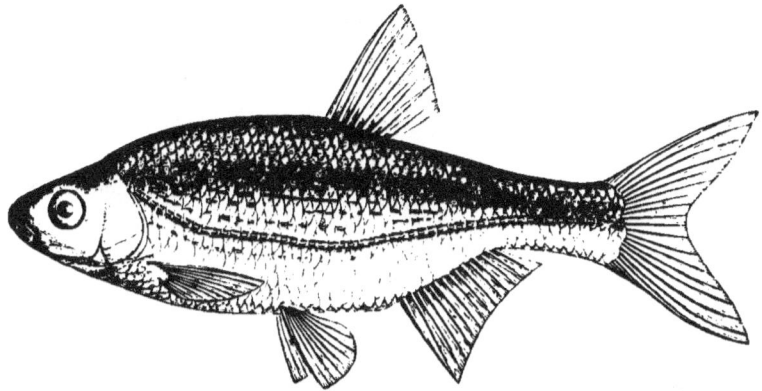

Fig. 127. ABRAMIS BIPUNCTATUS (HECKEL AND KNER).

from the gill-aperture to the caudal fin. Its width is equal to about three scales, and it extends above the lateral line.

According to Blanchard, the colours become vivid at spawning-time, when the green tint descends lower down the sides. The dorsal, caudal, and pectoral fins are normally spotted with black, and the ventral and anal fins are yellow and reddish. At ordinary times the black colour may nearly vanish. There is no difference of characters between males and females, except that the lateral line is often a little lower in the male.

The male usually has the swim-bladder larger, especially in the hinder portion, while the anterior part is smaller and shorter than in the females. Its posterior part is shaped as in the Bream, and the anterior part, according to Fatio, is slightly bi-lobed.

This fish spawns in May and June on gravelly bottoms in streams. The largest number of eggs found by Fatio was 1,915; the eggs are relatively large. They are hatched in about eight days.

The species frequents clear water. Its length varies from four inches in the Danube to six inches in the Neckar. Its scales are sometimes collected for the manufacture of artificial pearls.

The skeleton includes thirty-eight to forty vertebræ, of which eighteen are in the thorax and twenty-two in the tail.

The intestine is about as long as the fish, and forms two folds. The food is partly vegetable, but consists mainly of insects, worms, and small molluscs. The pharyngeal teeth are two in the outer row, and five in the inner; the formula, according to Dr. Günther, is 5·2—2·4[5]. The teeth are smooth, and have their extremities hooked.

This species is not known in Britain, though Blanchard speaks of it as a common English fish; but is found in France, Belgium, Holland, North and South Germany, Switzerland, and Austria, and extends into the south-west of Russia, being found by Kessler in the Dnieper and its tributaries.

The flesh is not delicate, though the flavour is not bad. It is eaten fried.

In Paris it is commonly known as the Smelt of the Seine, though belonging to a different family. In every country it has many local names. In some parts of France it is termed the *Sperlin;* and Fatio has referred it to a new sub-genus, which he terms *Sperlinus*, regarding it as a connecting link between Abramis and Alburnus. The number of pharyngeal teeth on the right side, which is almost constant, separates it from Alburnus.

The fish is liable to a disease in which blisters or gassy vesicles appear on the opercular plates, at the base of the fins, at the back of the mouth, and over the eyes. Another disease is due to a parasite allied to *Diplostomum cuticola*, which produces black spots over the body and in the mouth. It is also liable to tape-worm and other parasites.

## Abramis fasciatus (Nordmann).

D. 10, A. 14. Scales: lat. line 40—45, transverse 9/4.

This fish is found in the mountain brooks of the Crimea, and in the countries east of the Black Sea, especially the Caucasus and Turkestan. It is about four and a half inches long.

The body becomes more compressed with age; it is moderately elevated, its height being one-fourth of the total length, or less than one-third, exclusive of the caudal fin. The head is relatively small, and is one-sixth of the total length. The snout is obtuse, rounded, and a little shorter than the lower jaw. The eyes are large, and the orbital diameter is one-quarter of the length of the head. The mouth cleft is oblique. There is a scarcely apparent median prominence at the symphysis where the two sides of the lower jaw unite.

The origin of the anal fin is vertically below the end of the dorsal; it is as long as the pectoral fin; its base is equal to half the length of the head; it is truncated behind. The lower lobe of the forked caudal fin is a little longer than the upper.

The scales are relatively large, and the free margin shows a fan of twelve to fourteen very fine striæ. Each scale of the lateral line has two blackish dots.

The pharyngeal teeth are conical and curved, arranged in two rows, with five in the outer row and three or four in the inner. Besides these teeth there are several others much smaller, which are not attached to the pharyngeal bones, but placed on the mucous membrane of the pharynx, and are consequently absent in dry specimens.

The colour on the upper part of the head is greyish-green. The rest of the body is silvery, becoming yellowish on the sides in old individuals. Above the lateral line, and parallel with it, are two blackish bands, which become more distinct towards the tail. The bands are still visible in individuals deprived of their scales. The iris is silvery. The fins are transparent, with a pale red spot at the base of the pectoral, ventral, and anal.

## Genus: **Aspius** (Agassiz).

The genus Aspius is intermediate in character between Abramis and Alburnus, and the form of body is similar to the fishes of these genera. The lower jaw projects more or less beyond the upper jaw, which is slightly protractile; the lips are thin, but divided in the median line in front. The pharyngeal teeth are in two rows, and hooked, with the number 5·3 constant on the one side, and variable on the other, in which the number of teeth may be the same, or reduced by one in each row to 2·4. The dorsal fin has no spine, and is between the positions of the ventral and anal fins. The anal has never fewer than thirteen rays. Behind the ventrals the abdomen is compressed, and the edge is covered by scales.

Only one species is found in Europe; the others are recorded from the Tigris and the mountain region of Hong Kong.

## Aspius rapax (Agassiz).

This fish is found only in the eastern part of Europe; it is known in South Russia as *Belezna*, in Germany as the *Schied*, in Bohemia as the *Bolen*, in Austria as the *Rapfen*, while in Sweden it is the *Asp*. It is characterised by the pointed snout, long head, small eyes, and small scales (Fig. 128).

The body is deepest over the ventral fin, and broadest at the operculum, where the width is one-half the height. The eye is nearly twice its diameter from the extremity of the snout, and is one-sixth or one-seventh of the length of the head. The nares are near to the eye. The angle of the mouth reaches

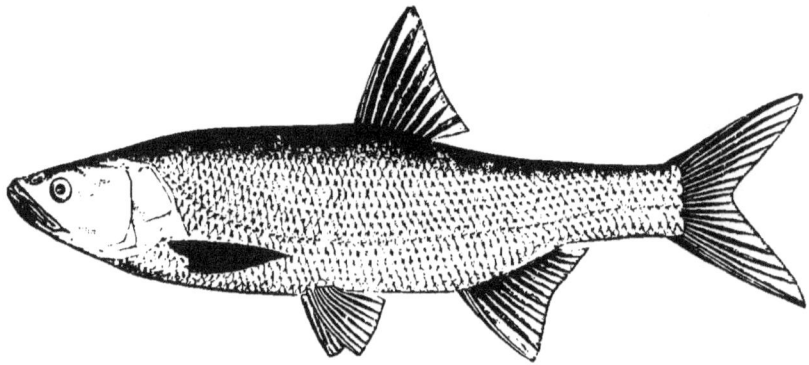

Fig. 128.—Aspius rapax (Agassiz).

under the anterior part of the orbit. The extremity of the lower jaw is thickened in the middle, and this knob is received in a corresponding excavation in the upper jaw.

The contour of the back is a flattened bow.

The dorsal fin commences behind the middle of the body; it is much higher than long; its base is shorter than the head. The fin is truncate behind; its longest ray is two and a half times the length of the shortest ray. The anal fin is as high as long, and the proportions of its rays are like those of the dorsal. Its terminal edge is somewhat concave. The ventral fins are in front of the dorsal, and extend back to the vent, though they are not long. The pectorals are longer, but they do not reach back to the ventrals. The caudal fin is deeply forked, has equal lobes, with the terminal rays as long as the head.

The scales are soft and very delicate, with scarcely visible concentric ribs.

Those on the back and abdomen show numerous regular diverging rays, but the scales on the sides have only two to four rays. The largest scales are found over the lateral line, but they do not attain the diameter of the eye. Between the anal and ventral fins the abdomen forms a blunt keel, which is not covered with overlapping scales, as in Alburnus, but the scales meet each other irregularly in the median line.

The mucus-canal of the head is well marked, especially in the frontal region, along the pre-operculum and the lower jaw. The sub-orbital branch extends in front of the nares. The accessory gills are large and fringe-like.

The rake teeth of the gill-arch are rather short and far apart. Fig. 129 shows the pharyngeal teeth.

The back of the fish is of a blue-black colour, the sides are bluish-white, and the abdomen quite white. The dorsal and anal fins are blue; the other fins have a reddish tinge. The iris is yellow, with some green stripes. The fish does not exceed a length of two to three feet, or weigh more than twelve pounds.

It is commonly found in lakes and rivers flowing through level country, for although it requires pure water, it objects to water flowing rapidly. It lives upon vegetable matter, worms, and small fishes, especially of the genus Alburnus. It spawns in April and May, when it enters shallow streams. From eighty to one hundred thousand eggs are deposited on stones.

It is easily taken in the autumn and at spawning-time, either with the net, or rod and line, baited with a small fish. The males at spawning-time have

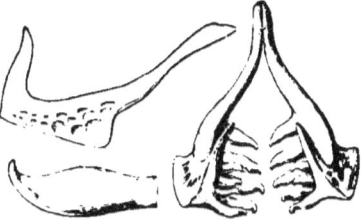

Fig. 129.— PHARYNGEAL TEETH OF ASPIUS RAPAX.

the scales largely covered with white conical spots, which, as in other species, characterise the hinder border of the scale.

The fish grows quickly, and lives for seven or eight years.

It is common in Norway and in all streams flowing into the North or Baltic Sea, extends through Russia, and is found in Pomerania, and the Haffs, Central Europe, and Austria. A nearly allied species is found in the Tigris.

## Genus: **Alburnus** (Heckel).

This genus has a form of body which resembles the genus Pelecus, and has the lateral line running below the middle of the side. The dorsal fin is short, and opposite to the space between the ventral and anal fins, in which there is a general resemblance to Abramis; only the fins are placed farther back, so that the dorsal is well behind the middle of the body, but not so far back as in Pelecus. The anal fin is extended, and never contains fewer than thirteen rays. The lower jaw projects beyond the upper, which is protractile; the lips are thin. The gill-rakers are slender and dense; pseudo-branchiæ are present. The pharyngeal teeth are in two series, and hooked (Fig. 130). Behind the ventral fins the abdomen is compressed into a sharp keel, which the scales do not cross. This genus extends through Europe, but is better represented in the south-east than in the west, while some Russian species range into Persia, and other species are peculiar to Asia Minor, Syria, the Tigris, Kurdistan, and some Persian streams.

## Alburnus lucidus (Heckel and Kner).—The Bleak.

D. 10–11, A. 19—23, V. 9—10, P. 15, C. 19.
Scales: lat. line 47—53, transverse 8—9, 4—3.

The Bleak is eminently a fish of the region north of the Alps. It is *l'Ablette* in France, *Laube* in South Germany, *Uckelei* in Prussia, and *Löja* in Sweden (Fig. 131).

It is a small fish, generally four or five inches long, and not exceeding seven inches in any part of Europe.

It is four and a half to five times as long as high, and five and a half times as long as the head. The eye is separated from the snout, as well as from the opposite eye, by its own diameter. The oblique mouth reaches back under the nares. The lower jaw, partly owing to the oblique position of the mouth, stands in front of the upper jaw when the mouth is closed; and the projecting point of the lower jaw is received into a depression in the pre-maxillary bone, as in the genus Pelecus.

Fig. 130. Pharyngeal teeth of Alburnus lucidus.

The profile of the back is very slightly convex, but the back of the head is defined by a ridge, in which the curve of the back begins.

The dorsal fin is shorter than high, but about as high as the pectoral is long. The ventral fins do not reach to the vent; the base of the anal is more than equal to the length of the head. The caudal is deeply forked, and its slightly longer lower lobe is one-fifth of the length of the fish.

The scales are remarkably soft and delicate, and finely rayed. The outline is vertically sub-ovate; the basal border is slightly bi-lobed, according to Blanchard, but this condition is only to be expected in the lateral line. The free border is slightly festooned between the rays. The diameter of the scale is less than that of the eye, and the scales overlap, so that one-half is exposed.

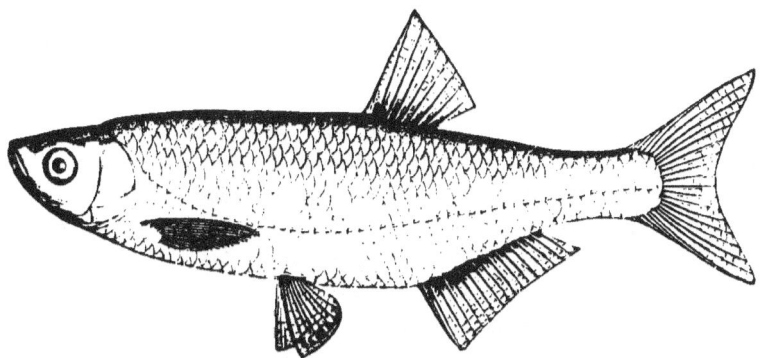

Fig. 131.   ALBURNUS LUCIDUS (HECKEL AND KNER).

The lateral line descends, so as to be roughly parallel to the ventral contour. It is distinctly defined.

The colour is steel-blue, with belly and sides of bright silver. The dorsal and caudal fins are grey, but the other fins are colourless.

The Bleak lives in rivers, lakes, and ponds. It is remarkably plentiful in Lake Constance. Frank Buckland noticed its abundance at Oxford, where the city sewer entered the Thames; but though common in the Thames, Lea, and New River, it prefers clear water. It is a lively fish, and moves about near the surface, in calm, warm weather, catching insects, and hence falls an easy prey to the young fly-fisher. It rarely descends into deep water, is rather inquisitive than shy, and sufficiently greedy to pounce upon anything thrown into the water, but discriminating enough to reject unsatisfactory food.

The Bleak spawn in May and June, depositing the spawn either on reeds or stony bottoms. The fish assemble in large shoals, move rapidly, as though to escape enemies, but often skim the surface of the water, when some fall a prey to gulls and sea-swallows, while others are seized by Perches. The spawning is intermittent, and takes place at two or three periods; the older fish spawn first, and the youngest last. They increase rapidly; but their lives are short, since they are a principal prey of carnivorous fishes and water birds, which follow their shoals. In some places they are prized as bait, but are nowhere valued as food, though captured in great numbers for the bright silvery pigment upon the scales. The species is found in Europe only, north of the Alps.

The fishes are frequently infested with tape-worm, which Yarrell records to be often longer than the fish; and it has been supposed that the irregular swimming which has gained for some fishes the name of Mad Bleak, is due to the annoyance of internal parasites. Sometimes, however, the bird which feeds on the fish inherits the parasite, which makes itself at home in the new abode when its former host is dissolved.

The Chinese are said to have been the first to have discovered the art of using the silvery pigment from fish scales for ornamental purposes. In the sixteenth century the Venetians, who introduced the art to Europe, imitated pearls so successfully with this preparation that their industry was suppressed by the Government as fraudulent.

More than two hundred years ago the art of making artificial pearls found a home in France, and has reached such perfection that its results, exhibited in the International Exhibitions, almost defied detection of the manufactured pearl. Yarrell mentions that formerly a considerable trade was carried on by Thames fishermen in supplying the French market with scales of the Bleak for the manufacture of the "Essence d'Orient," but even in his time the French had learned to obtain the scales from fishes in their own rivers. In England there was formerly a considerable demand for imitation pearls, which about the time of the 1851 Exhibition still made a conspicuous show in shops devoted to articles of personal adornment. Small glass beads were lined with pigment, and then filled with wax. The cheapest sorts were prepared from the scales of the Roach and Dace, were dull in lustre, and rather yellow. A better kind of pearl was made from the Bleak, but the scales of the Whitebait were most valued, and sold at a price varying from one guinea to five guineas a pound. At the present day, according to Blanchard, the scales of the Bleak sell at from twenty to twenty-four francs a kilogramme, and 8,000 fishes are necessary to provide this quantity. From this one-quarter of the weight of the "Essence d'Orient" is obtained.

The method of preparation of this substance is to scrape off the abdominal scales with a knife. The scales are washed and triturated, so as to separate the pigment, which has a metallic aspect, and falls to the bottom of the vessel in microscopic particles. This sediment is then washed with ammonia, to remove the organic matter. Seen under the microscope, the mass which then remains consists of oblong particles of rectangular form four times as long as wide. Some authors determine this substance as phosphate of lime, others as phosphate of magnesia. The iridescence has been attributed to the presence of guanin.

Mons. Selys-Longchamps described a variety of the Bleak in Belgium in which the rays of the anal fin are shorter than in the German types; but intermediate forms are found. Heckel and Kner mention a variety known as *Alburnus lacustris*, in which the height of the body is conspicuously greater, especially in the young fish. It acquires its varietal name from inhabiting the Niensiedler See and Platten See.

This variety closely resembles a fish from the Danube, which the same authors name *Alburnus breviceps*, which shows no important differences (Fig. 132).

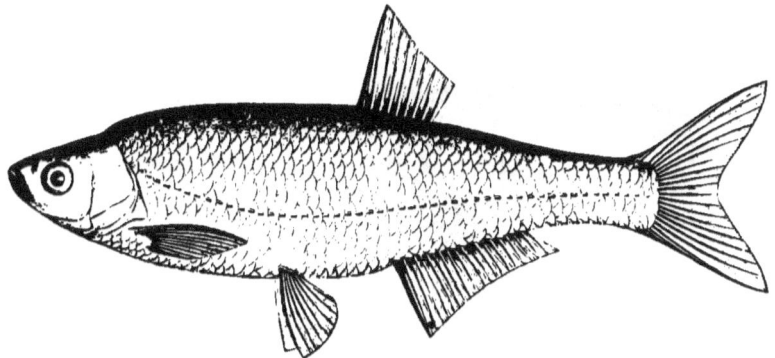

Fig. 132.—ALBURNUS LUCIDUS, VAR. BREVICEPS (HECKEL AND KNER).

In it the silver tint is less bright. The forehead is broader, so that the eyes are more distant from each other; and the head is a little shorter. The common length is about five inches. The *Alburnus fabræi* of Blanchard, found up the Rhône, is a similar fish, with a well-rounded back.

In the Lakes of Geneva and Bourget a Bleak occurs, which Blanchard names *Alburnus mirandella*. Its body is more elongated than in the Common Bleak. The scales are stronger, and there are fifty-seven to fifty-eight in the lateral line. The back is nearly horizontal. The lower jaw is a little longer than the upper. The colour recalls that of the Sardine, by which name it is known in Savoy.

These varieties are instructive, as showing the capacity of the species to vary in different details of organisation.

## Alburnus alburnellus (Martens).

D. 10—11, A. 16—19, V. 9.  Scales: lateral line 44—48, transverse $\frac{7-8}{4}$.

This species differs from *Alburnus lucidus* in having fewer scales, fewer rays in the anal fin, and fewer vertebræ; and as it is characteristic of Southern Europe, we may, with Heckel and Kner, regard it as a southern representative of that species (Fig. 133).

The head is proportionately longer and larger than in *Alburnus lucidus*, and is one-fifth of the length of the fish. The eye is separated from the end of the snout, and from the opposite eye, by its own diameter. The

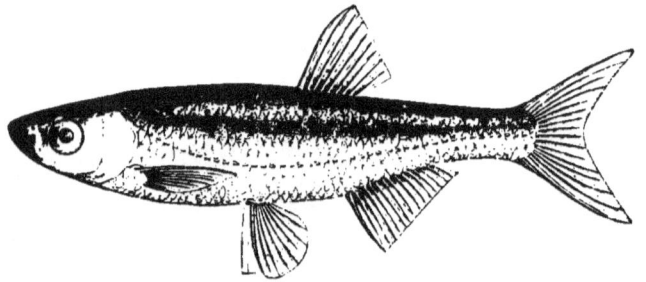

Fig. 133.—ALBURNUS ALBURNELLUS (MARTENS).

position of the dorsal fin, which begins half-way down the length, is a striking difference. The ventral fins commence well in front of the middle of the body. The longest rays of the dorsal and pectoral fins are equal, as well as the longest rays of the ventral and anal.

It is a rather smaller fish than *A. lucidus*, the largest specimens measuring but little more than four inches. It is found in shoals in lakes and rivers, spawns in June and July, and furnishes food for Pike, Perch, and Burbot. The colour, scales, and lateral line are similar to those of its northern representative.

The most northern point at which it is found is Botzen, in the Southern Tyrol, but it is well known from the lakes of Northern Italy—Garda, Maggiore, &c.—and from many Italian rivers.

*Alburnus fracchia*, of Heckel and Kner (Fig. 134), is regarded by Dr. Günther as a variety of this species, a view in which he follows Canestrini, though the latter is disposed to regard the fish as a hybrid between the type just described and *Leuciscus aula*. It is regarded by Heckel and Kner as a southern representative of *Abramis bipunctatus*, differing only in the profile of the snout and back. Its dorsal profile is nearly horizontal. A broad grey-blue band runs down the body from the snout to the tail. The body is five and a half times as long as high. The eye is about one-third of the length of the head. It is more than its own diameter from the opposite eye, and less than its own diameter from the snout.

The dorsal fin commences far behind the ventral. It is higher than long. The anal fin commences in advance of the end of the dorsal, and is longer than high. The ventrals do not reach the vent, and the pectorals scarcely reach

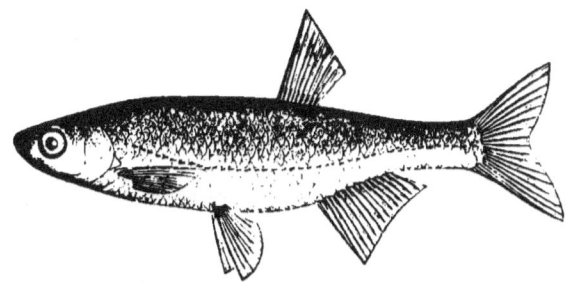

Fig. 134.—ALBURNUS ALBURNELLUS, VAR. FRACCHIA (HECKEL AND KNER).

the ventrals. The caudal fin is deeply forked, with its lobes pointed and equal. It is about three inches long, and is characteristic of the north of Italy.

A hybrid between *Alburnus alburnellus* and *Leuciscus uklira* is described by Dr. Günther as having sixteen rays in the anal fin, forty-four striated scales in the lateral line, with the base of the fins orange-coloured, and a dark band along the sides of the body. The fish is four inches long. It is found in the River Narenta.

## Alburnus mento (AGASSIZ).

D. 11, A. 17—19, V. 9—10, P. 15, C. 19.
Scales: lat. line 65—68, transverse 11—7.

The body of this fish (Fig. 135) is rather elongated, from five and a half to six times as long as high. The greatest breadth is about half the height. The snout elongates a little with age, but is never longer than the diameter of the eye. The eye is about one-quarter of the length of the head, and separated from the other eye by one and a half times its diameter. The nares are nearer to the eye than the snout. The mouth is oblique. Its angle scarcely reaches back to the anterior nares. The lower jaw is more thickened than in the other species, and so has the aspect of projecting in front of the upper jaw.

The dorsal contour is a very flat arch, which in old age becomes nearly horizontal. The dorsal fin commences in front of the middle of the length. It is placed between the ventral and anal fins, is much higher than long, and truncated posteriorly. Its longest ray is twice the length of the shortest.

The base of the anal fin is as long as the head; its border is slightly concave; the last ray is half the length of the longest. Though the ventral has its longest rays equal to those of the dorsal, it does not reach the vent. The pectoral fins do not reach the ventrals; the caudal fin is evenly lobed. The anterior anal rays commence behind the last dorsal rays.

The largest scales are on the lateral line, but measure less than half the diameter of the eye. The smallest scales are on the breast and anterior part

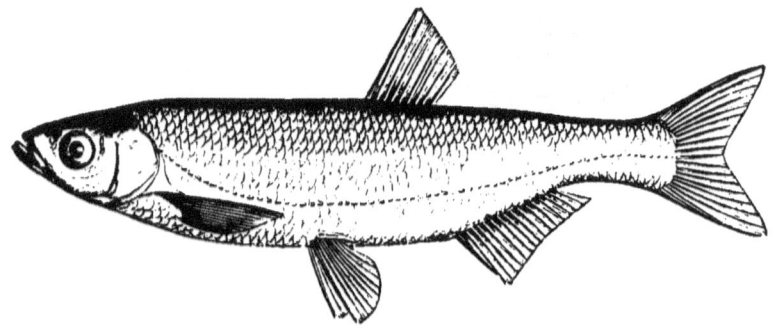

Fig. 135.—ALBURNUS MENTO (AGASSIZ).

of the abdomen. They are marked with many radiating rays and thick concentric lines, have a pale silver lustre, are soft, and easily fall off.

There is a spur-like scale extending on the base of the ventral fin, which in old age becomes as long as the fin.

The cephalic canals extend on the sub-orbital ring, and to the nasal apertures, and are especially developed on the pre-operculum.

The colour of the head and back varies from dark green to steel-blue; the sides are pure silver.

All the fins are grey and transparent, but the dorsal and caudal fins are marked with black spots. The iris is silvery, sometimes spotted with black.

This is the largest species of the genus, and reaches a length of ten inches, though Bavarian specimens are rather larger than those from Austrian waters. It frequents mountain streams, and streams which flow into mountain lakes, preferring clear cold water with a stony bottom. It often lies for a long time perfectly still against the stream, like a Trout, and then moves with astonishing speed. It is rare in the Danube, but is found in the Atter See, Traun See, Gmundner See, in Austria; the Ammer See, Chiem See, and other lakes of Bavaria, and occurs, according to Kessler, in the rivers of the Crimea.

It spawns in shallow water during the month of May, preferring places with a stony bottom. According to Heckel and Kner, the fishes pack themselves close together, with their heads downward in a nearly perpendicular position, and disencumber themselves of the spawn by slashing movements of their tails. The spawn deposited, they swim away; when a second swarm takes their place, and spawns, and they are usually followed by a third swarm. Small white conical warts appear on the male at spawning-time, both on the head and scales.

It is associated with fishes which are much more valued, and hence is not much sought after by fishermen; but it is good eating, and appears in the Munich and Vienna fish markets, where it is known as the *Mai-Renke*.

## Alburnus chalcoides (GÜLDENSTÄDT).

D. 11, A. 18 [? 24], P. 15, V. 9, C. 19.
Scales: lat. line 50, transverse 11—5.

This Bleak is found in Southern Russia, in the Black Sea, the Caspian, the Sea of Azov, and Sea of Aral. It ascends the rivers flowing into those waters, and is found in Northern Persia. The pharyngeal teeth are compressed laterally, and curved at the points. The teeth of the inner row are larger than those of the outer row, and their upper part is deeply grooved. Similar, but smaller, teeth are attached to the membrane of the palate. The body, excluding the caudal fin, is four times as long as deep, and the length of the head is equal to the depth of the body. The lower jaw projects beyond the upper. The pectoral fins are well developed, but do not extend to the ventrals. The anal fin begins immediately behind the end of the dorsal. The colour of the body is uniform silver.

The fish is fat, and taken in great numbers in the Bug, Dnieper, Don, Terek, and Koora. About 2,500,000 are taken annually in the Koora and the Sea of Azov. It is smoked and used for food throughout Russia, under the name of Chamaïka. On the shores of the Black Sea it is known to the fishermen as Salaiva.

Kessler has described *Alburnus tauricus* from the River Salghir, in the Crimea; but Günther regards this form as a hybrid between *Alburnus lucidus* and *Leuciscus dobula*.

## Genus: **Leucaspius** (Heckel).

Leucaspius is a genus which was regarded by Heckel and Kner as intermediate between Leuciscus and Alburnus. Its chief claim to generic distinctness from Alburnus rests upon the incomplete development of the mucus-canal forming the lateral line, which in the single species known, hardly extends beyond the pectoral fin. The scales are distinctive in their ovate form, and their freedom from diverging fan-like rays, though, as Benecke points out, some faint indication of two or three rays may be detected.

Behind the ventral fins the abdomen is compressed into an edge. The dorsal fin is short, and has no spine, and is opposite to the space between the ventrals and anal. The anal fin includes more than thirteen rays. The lower jaw projects beyond the upper, and its thickened median termination is received in a corresponding depression. The upper jaw is protractile. The gill-rakers are short and slender. The pharyngeal teeth are compressed and hooked, sometimes in one row, sometimes in two. This type is known only from Central and South-eastern Europe.

## Leucaspius delineatus (Heckel).

D. 11, A. 14–17, V. 9–10, P. 11, C. 19.
Scales: lat. line 48, transverse 15.

Heckel and Kner referred this species (Figs. 136, 137) to two different genera—Leucaspius and Squalius, owing to variations in the pharyngeal teeth. When these teeth are arranged in a simple row of five on each side, it found a place in the genus Leucaspius, but when one of these teeth was displaced, so as to form a second row, it was referred to the so-called genus Squalius. This circumstance may be a caution against attaching too much importance to the characters of the pharyngeal teeth.

The head is one-fifth of the length of the fish; its length is equal to the greatest height of the body; and the thickness of the fish is equal to half the height. The eye is large, its diameter being about one-third of the length of the head; and it is separated from the snout and from the other eye by its own diameter. The mouth is very oblique, and its angle reaches below the anterior nares. The lower jaw resembles that of Alburnus and Aspius in the thickening of its median portion in front; it slightly projects when the mouth is closed. The sub-orbital ring is broad. The profile is nearly straight on the head, and is then continued as a flattened arch on the back. The abdominal curve is rather greater. Behind the anal fin the tail tapers somewhat, and is two-fifths of the greatest depth of the body. In this region there is a space as long as the pectoral fin, which is free from scales. The ventral fins are in the middle of the length; they have shorter rays than the other fins, but nearly reach to the vent.

The dorsal fin commences behind the ventral, and the anal fin begins under

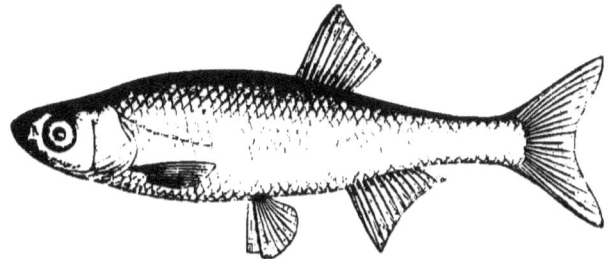

Fig. 136. LEUCASPIUS DELINEATUS (HECKEL).

the end of the dorsal. All the fins are well developed, considering the size of the fish; but the longest rays of the dorsal are scarcely longer than those of the anal. The lobes of the caudal fin are as long as the head.

The typical distinguishing character of the species, as of the genus, is the

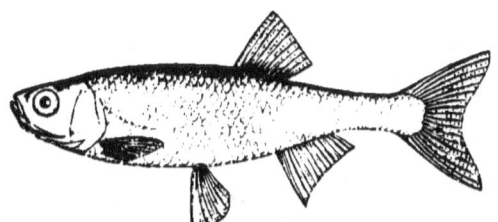

Fig. 137.—LEUCASPIUS DELINEATUS (HECKEL).

absence of the lateral line (Figs. 136, 137). Sometimes it extends only to two or three scales, and rarely to as many as six or seven, which open by simple pores.

The largest scales are scarcely equal to half the diameter of the eye. They are soft, and easily fall off. On the anterior part of the body the scales are marked with fine concentric stripes, but are free, or almost free, from rays; though in the hind part of the body the rays are distinctly visible in scales from the side of the tail. The mucus-canals are strongly developed in the head.

The colour varies with the locality; the back and upper part of the head may be green, brown, yellow, or like polished steel. On each side of the tail there is often a clear silver or steel-blue band, which is sharply defined by a dark edging. The sides of the head and body and the abdomen are silvery. The fins are all transparent, and free from spots; the iris is silvery, with yellow-green flecks. This species is extremely variable, and almost every locality has a local variation. Von Siebold records that in thirty-six individuals eight had the pharyngeal teeth in two rows; in ten fishes there was a double row of teeth on the left side only, and six fishes had the double row of teeth on the right side only, while the remaining twelve had only one row of teeth on each side. Spawning takes place in April and May. According to some Russian observers, three papillæ appear behind the vent at spawning-time.

A common size is two to three inches, and the largest specimen does not exceed three and a half inches in length.

The species is found in all the rivers of Southern Russia, in Greece, Dalmatia, the Danube, and its tributaries, in Moravia and other parts of Austria, and throughout Germany, especially in the Kurische Haff. It extends as far west as Brunswick.

## Genus: **Pelecus** (Agassiz).

The genus Pelecus was formed by Agassiz for a fish of the Bream type, having the body oblong and greatly compressed, with the entire abdomen forming a sharp edge. Its scales are small. The lateral line descends in an unusual way behind the pectoral fin towards the abdominal border. The mouth is nearly perpendicular. The dorsal fin is placed far back, and is opposite to the commencement of the anal, which resembles that of the Bream in its numerous rays. The pectoral fins are very long. The pharyngeal teeth are in a double series, and strongly hooked (Fig. 139). The intestine is short. It was united with the genus Leuciscus by Valenciennes, on account of similarity in the pharyngeal teeth, and placed in Abramis by Nilsson; but most modern writers recognise it as a distinct genus. It is found only in the waters of Eastern Europe, but extends into the Baltic; it occurs in Sweden, but not in Russia; and is characteristic of Germany and Austria.

## Pelecus cultratus (LINNÆUS).

D. 9—10, A. 28—31, V. 9, P. 16, C. 19.
Scales: lat. line 92—108, transverse 14—6.

This species (Fig. 138), known in Sweden as the *Skarbraxen*, in Germany as the *Sichel* or *Ziege*, and in Austria as the *Sichling*, is so unlike other fishes of the Carp family that some of the older writers, like Klein, included it with the Pikes, while Wülf placed it with the Herrings; and it certainly approximates the Carps towards the Herring group. The dorsal profile is almost horizontal from the forehead to the dorsal fin, while the abdominal outline is convex, like a bent bow. The greatest height is midway between the pectoral and ventral

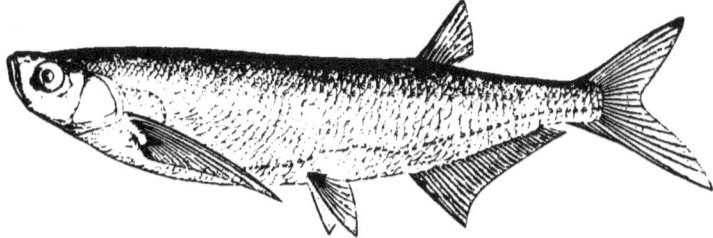

Fig. 138.—PELECUS CULTRATUS (LINNÆUS).

fins, and the fish is about four and a half times as long as high. The greatest thickness at the operculum does not exceed one-third of the height. The head is one-fifth of the length. The eye is large, and approximates towards the corners of the mouth. Owing to the direction of the mouth, the lower jaw lies in front of the upper. There is a median protuberance in the lower jaw which fits in to a corresponding depression in the pre-maxillary bone. The lips are fleshy only at the corners of the mouth. The double nares are near the eye. The scales on the head commence just over the eyes, corresponding with the extension of the dorsal muscles, which cover the hinder part of the skull. The dorsal fin is on the hinder third of the body; it is rather truncated behind, somewhat small, and only one-third higher than long. The anal fin is only one-third as high as long; but its length exceeds that of the head. The ventral fin commences behind the middle of the body, and nearly reaches to the vent. The prolonged point of the sabre-shaped pectoral fin, which is more than one-quarter of the length of the fish, nearly reaches to the ventral fin. The lower lobe of the deeply forked caudal fin is the longer of the two, and nearly as long as the base of the anal.

The lateral line is at first nearly horizontal, then curves suddenly downward, and at every pore gives off an oblique transverse branch, and, curving down to near the abdomen, it rises again midway between the ventral and anal fins; but descends again, and then continues parallel to the anal fin to its termination in the caudal. Every pore in a scale gives off one, or occasionally two, short lateral branches. The cephalic canals are strongly developed, and the sub-orbital branch occupies the entire width between the eye and the nares. The inferior maxillary branch extends along the free operculum.

The pharyngeal bones and teeth are slender. The tooth formula is 5·2—2·5 (Fig. 139).

Fig. 139. — PHARYNGEAL TEETH OF PELECUS CULTRATUS.

The scales are not firmly adherent, very delicate and flexible; their margins are entire. They are marked with numerous rays, which are not regular, but very distinct, especially on the ventral surface. The smallest scales are behind the skull; the largest are on the sides, but they never exceed half the diameter of the eye. The scales forming the abdominal edge overlap one another, so as to form a small, soft, flexible keel. There are no scales on the base of the pectoral fin. Elongated, blunt-pointed scales occur on the bases of the ventral fins. The rake teeth are long and pointed in the anterior row of the first gill-arch, but are short in the succeeding rows.

This species is commonly from six to twelve inches long, and does not exceed a foot and a half in length, or a pound and a half in weight. It lives from four to five years, spawns in May, multiplies slowly, and is not often met with in rivers. It was known to Linnæus as inhabiting the Baltic, but it prefers pure running water, where it usually keeps near the banks and lives like other Cyprinoids. Bloch states that the ovary contains one hundred thousand eggs.

The colour of the neck is steel-blue or bluish-green; the back is a brown-grey; the cheeks have the lustre of mother-of-pearl; and the sides are silvery. The dorsal and caudal fins are grey; the other fins reddish; the iris is silvery. The shining silvery coat attracts its enemies, and makes it an easy prey to water birds and predaceous fishes. Being scantily clothed with flesh, which is soft and stored with bones, it is very little valued as food, and hence is seldom fished for.

The Austrian authors record that in early times it was a fish of bad omen, presaging war, famine, and pestilence, a reputation which was maintained by its irregular appearance at intervals of several years.

It is widely distributed in Europe; is found in the Crimea, and in the Danube and rivers of European Russia which empty into the Black and Caspian Seas, as well as in those seas. In Prussia it is common in the Haffs and estuaries, where it is taken in great numbers with nets.

## GROUP : COBITIDINA.
### GENUS: **Misgurnus** (LACÉPÈDE).

The genus Misgurnus belongs to the fourteenth, or last, great group of Dr. Günther's arrangement of Cyprinoid fishes. This group is named COBITIDINA. Three of the genera, Misgurnus, Nemachilus, and Cobitis, are represented in Europe, while the remaining genera are from the East Indies, China, Japan, and the Malayan Archipelago. All these types agree in having the mouth surrounded by six or more barbels; the air-bladder is partly or entirely contained in a bony capsule; the pharyngeal teeth are in a single series; the dorsal and anal fins are short; and the scales may be small, rudimentary, or entirely absent.

In the group thus defined, Misgurnus is distinguished by an elongated compressed form of body. It has ten or twelve barbels, of which four belong to the mandible. The caudal fin is rounded; the dorsal is opposite the ventral. The air-bladder is enclosed in a bony capsule. The species of the genus are curiously distributed, in being absent from Western Asia. The species *Misgurnus fossilis* is found only in Europe. One species occurs in Bengal, while the others are limited to China and Japan.

## Misgurnus fossilis (LINNÆUS).

D. 9, A. 8, V. 7, P. 9—11, C. 16.

This fish, often known as the Pond Loach in France, and *Schlammpitzger* or *Bissgurre* in Germany, has a very elongated body, with small head, and a mouth surrounded by ten barbels, of which four belong to the mandible. The caudal fin is rounded, and the body ornamented with blackish-brown longitudinal bands, which may be replaced by a yellow band (Fig. 140).

The body deepens with age, so that while in the adult the height is equal to the length of the head, it is less than the head length in the young. The total length of the fish is between seven and a half and eight and a half times the length of the head. The thickness of the head is equal to half its length, and its height is the same as that of the tail behind the anal fin. The body has a sub-cylindrical aspect, its thickness being two-thirds of the height. The mouth is small, and rather inferior, the small blunt lower jaw being beneath and overlapped by the projecting upper jaw, and covered by its lip and barbels, so that the mouth aperture is not seen. There are four barbels at nearly equal distances on the upper jaw, and a longer barbel at each angle of the mouth. On the under jaw there are two short barbels in front on each side of the median symphysis. The barbels at the corners of the mouth are little less than half the length of the head; so that the barbels on the lower jaw decrease in length anteriorly. The eye is very small; its diameter is less than one-seventh of the length of the head. It is placed high up, is separated from the other eye by twice the orbital diameter, and is more than this distance from the snout. The suborbital spine is completely covered by the skin; and, though its position may be detected, it is not erectile.

The nares are divided, are placed near to the eye, and the anterior narine

Fig. 140 —MISGURNUS FOSSILIS (LINNÆUS).

forms a short projecting tube. The gill-aperture is completely closed on the throat, and only open laterally from the level of the eye to the base of the pectoral fin.

All the bones of the face and operculum are covered with skin. The suborbital ring is seen to be well developed in the skeleton, and has a strong spine behind.

All the fins are more or less rounded at their free borders. The dorsal, which is opposite to the ventral fin, begins behind the middle of the body. Its height is less than the height of the body. The anal fin is similar to the dorsal in length, height, and form, and commences some little distance behind the vent. The rays of the pectoral fin are as long as the anal fin rays. The ventral fins are shortest, and do not reach to the vent. The middle rays of the caudal are longest, and exceed the length of the head. The caudal fin consists of fourteen jointed rays, with an unjointed terminal ray on each side; in front of which there are about twenty truncate rays, which reach far up the tail, forming an edge or membranous ridge in the vertical median line both above and below. A somewhat white line extends on the abdominal surface from the throat to the prominent vent. The lateral line extends nearly in the middle of the side. The cephalic mucus-canals are traced with difficulty, as they have only a few pores. The small size of the fins gives the fish an eel-like aspect. Von Siebold states that the only parts of the body which are free from scales are small areas in the median line behind the dorsal and anal fins.

The scales on the back, throat, and abdomen are the smallest; they are larger and more numerous on the sides. Their form is vertically ovate, or nearly circular. They but slightly overlap each other, and under the microscope show numerous concentric lines of growth and radiating rays in every direction. The pharyngeal teeth are twelve to fourteen in number. The rake teeth of the gill-arches are compressed, pointed, and short. There are thirty-one vertebræ in the thoracic region, and nineteen in the tail.

The intestine is short, without pyloric appendages, and appears to have a respiratory function, absorbing oxygen from the air swallowed, and liberating carbonic acid.

The air-bladder is divided into right and left portions, which are similar and adjoin the first three or four vertebræ, processes from which enclose them in lateral bony capsules. The air is said to be sometimes expelled from the air-bladder with a piping noise, though this is improbable.

The head is marbled with brownish-black flecks, which often run together in bands.

The colour of the back is dark brown, with blackish flecks. The abdomen is orange, dotted with black. A broad blackish-brown band extends along the lateral line, which is conspicuous with its thick compressed pores. Above and below this is a broad yellow longitudinal stripe, and this is divided from the orange abdomen by a blackish-brown band, which extends as far as the vent, and then becomes broken up on the tail into scattered points and flecks. The yellow band over the lateral line extends forward to the orbit, and may be traced in front of the eye. The ground colour of the dorsal and caudal fins is a darker brown than that of the other fins; all the fins are dotted with black. The iris is golden.

The length often exceeds a foot, but is commonly less. As a rule, the fish lives in flat regions, and prefers muddy brooks, marshes, and ditches, though also found in large rivers and lakes. It is known on the Continent as the Mud-fish, from its habit of burrowing and burying itself in the mud, particularly in cold weather. Dr. Badham remarks, "The favourite pastime of *M. fossilis* is to roll and wallow in the mire of his pond, whither he retreats for warmth and cover when the air is chilly; and so fondly is he attached to this soft duvet that on leaving it, as he always does on the approach of bad weather, it is only to grub up and disperse the ooze till the water has been rendered congenially dirty to his taste." In Germany and Austria it is regarded as a weather prophet, and sometimes is called the Weather-fish, because it usually comes to the surface about twenty-four hours before bad weather, and moves about with unusual energy; this habit has led to its being sometimes confined in a glass globe as an animated barometer.

It swallows air, and is said to expel it again with noise. It seldom takes a bait, and is commonly caught with nets or baskets. Its flesh is white and flabby, having an odour and taste which suggest the stagnant pool; but Benecke asserts that when kept for some days in running water it becomes well-flavoured. It lives on insects, worms, fish-spawn, and organic matter contained in the mud which it swallows.

It leaves its hiding-place in the mud in spring to spawn. It lays 140,000 brown eggs on water-plants, in April and May in Western Europe, and from April to June in Eastern Europe.

Its growth is rapid, and it is tenacious of life; and, probably, on account of the narrow gill-aperture, can remain a long time in dry places, being sometimes found in the mud of dried-up ditches. It inhabits Holland, Belgium, the eastern border of France, North and South Germany, and is found in all fresh waters in Russia, except in the basin of the White Sea, and the region beyond the Caucasus. In Switzerland it is met with in Lake Constance. Dr. Günther regards *Misgurnus anguillicaudatus* (Castor) as a representative of this species in China, Japan, and Formosa, differing only in colour, shorter pectoral fin, and other minor characteristics.

GENUS: **Nemachilus** (VAN HASSELT).

This genus, which comprises about forty species, is characterised by having only six barbels, none of which are on the mandible. It is closely allied to Misgurnus; has no erectile sub-orbital spine; has the air-bladder enclosed in a bony capsule, and the dorsal fin opposite to the ventral. Some of the species have more than twelve rays in the dorsal fin; others fewer, and the latter are defined according as the body is striped with brown bands or narrow yellow bars, or is without any transverse bars at all. Some of these latter species have the caudal fin rounded, while in others it is distinctly emarginate. The genus ranges over the Malay Islands and Asia, especially Tibet, India, Syria, one species being found in the Lake of Galilee. The only European species is *Nemachilus barbatulus*.

## Nemachilus barbatulus (Linnæus).—The Loach.

D. 10, A. 7, V. 7.

The Loach is supposed to have acquired its name from the French verb *locher*, to fidget. It is known in France as *la Loche franche*, a name which varies in the provinces. About Paris its habits of dabbling in the mud have gained it the name of *Barbotte*. In Germany it is *Schmerle*, or *Grundel*, or *Bartgrundel*. In Italy it is *Cobite barbatello* (Fig. 141).

As compared with *Misgurnus fossilis*, the Loach has a less elongated body, relatively thicker, with a longer and broader head; an evenly truncate caudal fin, with rounded corners. The length is seven to eight times the height, and

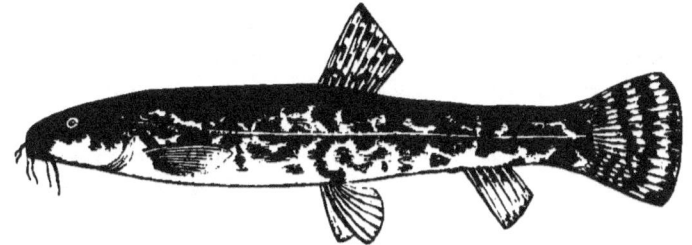

Fig. 141. NEMACHILUS BARBATULUS (LINNÆUS).

the body is five or five and a half times the length of the head. The height of the body only slightly exceeds its thickness, but the head, on the other hand, may be a little wider than high. The mouth is rather larger than in *Misgurnus fossilis*, but, in position and shape, it is similar. The upper jaw projects, and of the six barbels the four middle ones are more distant from the angles of the mouth, and, therefore, nearer together; the front pair is smallest. The under lip carries no barbels; it is notched in the middle, and, according to Günther, each side is again divided by a notch. The cleft between the lips is transverse. The barbels at the angle of the mouth are about one-third of the length of the head, and reach back nearly, if not quite, to the pre-operculum.

The eye is midway between the end of the snout and the gill-aperture, and placed high, near to the frontal profile. Its diameter is one-sixth or one-seventh of the length of head, and the breadth of the frontal interspace is between one and a third and twice the diameter of the eye. The skin does not go over the eye as in *Misgurnus fossilis*, but forms a flat movable band towards the opercular border. There is a row of mucus pores over and under the eye. The gill-aperture is quite as narrow as in the Misgurnus.

The dorsal fin begins in the middle of the length. Its fourth and fifth rays are the longest and as long as the pectoral fin; the last dorsal ray is only half as long. The anal fin is lower than the dorsal, but as long as the ventral. The ventral fins are opposite the middle of the dorsal; their free border is rounded; their longest rays do not nearly reach the vent, which is nearer to the anal fin, and, consequently, far behind the end of the dorsal. The pectoral fin is rounded, but is less deep than in *Misgurnus fossilis*. The caudal is as long as the head, and carries only six or seven truncated rays on the upper and lower membranous border of the tail, which, in consequence, is less distinct than in the Misgurnus.

The scales are remarkably small and delicate, and are to be seen only with the aid of a magnifying glass, when they show radiating rays. On the sides they are separated from each other and never overlap. They are altogether wanting on the head, breast, and abdomen. They become thicker behind the dorsal fin upon the tail, and though these are still smaller than those on the anterior part of the body, they overlap each other. The lateral line is nearly horizontal in the middle of the side, and opens by a number of membranous tubes. The mucus-canals of the head are more distinct than in *Misgurnus fossilis*.

The colour of the back is dark green, the sides are yellow, the abdomen is grey; while brownish-black irregular flecks, points, and stripes cover the head, back, and sides, and give a marbled aspect to the fish.

At the base of the caudal fin there is commonly a black spot like an eye. The dorsal, caudal, and pectoral fins are flecked like the body, and the anal and ventral fins are yellowish-white.

The size is always much less than that of the Misgurnus, rarely exceeding four inches; the largest specimens measure five inches and a half.

The Loach is an inhabitant of mountain regions, as well as plains, and is found in rivers, but especially in clear brooks with a gravelly bottom. It is found in lakes and in the Haffs on the Prussian coast. It does not thrive in stagnant water. It is as shy as the Minnow, and loves to hide under stones. The flesh is delicate, well-flavoured, and easily digested, but the fish requires to be eaten as soon as captured. In former times it was much prized. At the present day the Loach is preserved in fish ponds in Bohemia, where it is fed on sheep-dung, linseed-cake, and poppy seeds. It is voracious. Its ordinary food consists of insects, worms, spawn, and when the supply of these is insufficient, it feeds on plants. It is caught with difficulty, on account of its activity and the slimy covering of the skin. It dies at once when removed from the water.

Its capture furnishes a favourite sport for children, who pursue it barelegged in shallow streams, and frequently spear it with a fork.

According to Dr. Badham, Loach are not infrequently found frozen alive, and then the warmth of the hand is said to be sufficient to thaw them; but he adds the effect is still more expeditiously produced by putting them into the frying-pan, when care must be taken lest they fulfil the proverb by leaping into the fire.

The habit of the fish in remaining motionless and in hiding during the greater part of the day is, apparently, connected with the small size of its airbladder, for although it comes to the surface with great effort in the evening and in showery weather, it is unable to remain there.

The bony capsules which contain the air-bladder are placed on each side of the first and second vertebræ; they are smooth on the inside, spherical in form, connected together beneath the bodies of the vertebræ, and give attachment to the pectoral fins. The air-bladder communicates by a slender canal with the œsophagus.

The fish spawns in England in March or the beginning of April, in North Germany in April and May, and, according to Günther, the spawning may be prolonged in warm seasons as late as August in the Neckar. The eggs are numerous and small; they are deposited between stones, or in holes which the fish has excavated. After deposition they are watched over by the males.

The skull differs from that of other Cyprinoid fishes in wanting the jugal arch. There is one infra-orbital bone in front of the eye. Yarrell describes an interspace between the parietal bones, which in the fresh fish is occupied by cartilage. There are thirty-nine thoracic vertebræ, sixteen caudal vertebræ, and fifteen pairs of ribs. The sac-like stomach is well defined from the intestine. The right lobe of the liver is larger than the left. The kidney is large and extends forward to the connexion between the parts of the air-bladder.

The Loach is found in all fresh waters in Russia, except beyond the Caucasus, where it is replaced by an allied fish, *Nemachilus brandti*. It is rarely found on the north frontier of Italy; has not yet been recorded from Spain; and is absent from Denmark and Scandinavia; though it was introduced into Sweden by Frederick I. Otherwise it is spread over Central and Western Europe, and is common in the British Islands.

### Genus: **Cobitis** (Artedi).

The genus Cobitis as restricted comprises a few species which are found in Bengal and Assam, Japan and Europe. It has the elongated form of body seen in Misgurnus and Nemachilus, to which it is closely allied; but the body is more or less compressed, and the back is not arched. Below the eye there is a small bifid sub-orbital spine, which is capable of being erected. There are six barbels, which are limited to the upper jaw. As in the other genera of the group, the air-bladder is contained in a bony capsule. The insertion of the dorsal fin is opposite to the ventral, and the caudal is truncate, or convexly rounded.

### Cobitis tænia (Linnæus).—Spinous Loach.
D. 10, A. 7, V. 6, P. 6—7, C. 13.

The distinctive pre-orbital spine, which constitutes one of the marked generic distinctions of this fish, has secured for it the popular name of the Spinous Loach; and its habit of frequenting the bottom, and hiding under

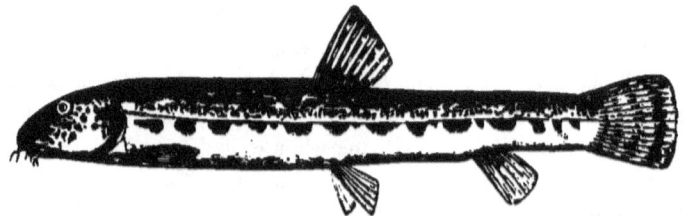

Fig. 142.—Cobitis tænia (Linnæus).

stones, has caused it to be known in many parts of the country as the Groundling. In France it is *la Loche de Rivière*. In Germany it is *Dorngrundel, Steinpitzger,* or *Steinbeisser*. In Holland it is the *de Bleine Modder Kruider*. In Italy it is the *Cobite fluviale* (Fig. 142).

Its distribution is remarkable, being found throughout Europe from Scandinavia to Spain. It, however, appears to be absent from Ireland, and it is probably absent from the White Sea basin in Russia, while beyond the Caucasus it is replaced by the *C. aurata* of Filippi, and *C. nohenackeri* of Brandt. Though absent from continental Asia, it is recorded by Dr. Günther from Japan.

In this fish the head and body are compressed. The body is between seven and eight times as long as high, and the thickness is only half the height. The length of the head exceeds the height of the body. The mouth is inferior; the six barbels are short; two of them are close together in front in the middle of the upper jaw. The two longer ones are at the angles of the mouth. The mouth is small; its opening is transverse, and the cleft reaches under the nares. The small eye is in the middle of the length of the head; its diameter is between one-sixth and one-seventh of the length of the head. It is near to the frontal profile, and separated from the other eye by its own diameter. The thin skin of the head goes over it, as in *Misgurnus fossilis*, without forming a fold.

Both points of the bifid sub-orbital spine are curved backward; the anterior one is short, but the hinder and inner spine is as long as the diameter of the eye. The spine is not seen when the fish is at rest, but is capable of being erected as a defence. The gill-opening is narrow, only reaching from the base of the pectoral fin to the eye. The profile rises in a strong curve over the head, and is nearly horizontal on the back. The dorsal fin commences somewhat in advance of the ventral, and is in the middle of the length; its third and fourth rays are as long as the body is high; but the last ray of the fin is only two-thirds of this height. The anal fin is as long as the rounded ventral, but its free border is less sharply truncate. The pectoral fins are as deep as the ventrals, and a little longer. In the male the upper ray is enlarged and flattened. The caudal fin is truncate, often with the corners rounded. The lateral line is visible only for a short distance over the pectoral fin; farther back it is continued as a dark-coloured, longitudinal furrow.

The scales are very small, scarcely to be counted with the naked eye. There are forty to forty-two vertebræ, and fifteen pairs of ribs. The intestine is nearly straight; the lining membrane of the abdomen is pearly white, with black spots.

The Spinous Loach has a striking colouring. The general tone is orange-yellow, diversified with rows of somewhat round black spots, of which two rows are fairly regular. One of the larger bands of flecks runs in the middle of the height of the body. A smaller row is in the upper third of the height; between these two are irregular points, flecks, and lines. The throat and abdomen are free from spots. On the head a brownish-black line extends from the eye to the upper lip; a similar line runs as a wavy curve from the back of the eye to the opercular point, and a third is seen under the eye; running forward across the cheeks. When these bands are wanting the body is more irregularly spotted. The eye-like spot at the base of the upper lobe of the caudal fin is sharply defined and intensely black. There are also large black flecks in the median line of the back. The caudal fin is vertically striped; the pectoral, ventral, and anal fins are uniformly pale. The iris is pale yellow. This is the smallest species of the genus, reaching a length of two inches and a half in Austria, four inches in Germany and France, and three inches in Great Britain.

Like its allies it is very prolific. It spawns in March and April in France, and in April or May in Germany and Austria, though Marsigli mentions June. Its growth is slow. It is not sought after for food, on account of the toughness of the flesh and the number of bones in it. When captured it emits a peculiar sound.

As in the preceding allied forms, the intestine has a respiratory function. Coming to the surface the fish swallows atmospheric air, which is passed

Fig. 143.—COBITIS TÆNIA, VAR. ELONGATA (HECKEL AND KNER).

through the intestine with some noise. Siebold suggests that this remarkable habit has resulted from the way in which the fish stirs up the mud, and frequents places in which the supply of oxygen is small. It burrows like a rabbit. The bony capsule for the air-bladder extends along two vertebræ next the head.

There is an elongated variety found in Croatia, which Heckel and Kner describe as *Cobitis elongata* (Fig. 143). The height of the body may be only one-tenth of the entire length. The double sub-orbital spine is small, and lies under the skin.

It further differs from *C. tænia* in attaining a larger size, being never smaller than five inches, and usually six; yet in all general characteristics and colour it agrees with the species under which we group it. The body is five and a half to six times as long as the head. The snout is more pointed than in *C. tænia*, so that the eye is farther from it than from the angle of the operculum, while the distance between the eyes is scarcely half the orbital diameter. Hence the forehead is smaller, and the head more compressed. The eye is similarly covered with smooth skin. The barbels are longer, and those at the angle of the mouth reach back to the front of the orbit, and are nearly midway between the eye and the extremity of the snout. The sub-orbital spine appears to be relatively short, on account of the large size of the fish.

Another variety, named by Filippi *Cobitis larvata*, is found in Piedmont. The head is one-fifth of the length of the body, and uniformly brown. There are six pharyngeal teeth in each bone. A brown band extends along the trunk, and becomes broken into spots on the tail. There are two brown spots at the base of the caudal fin. The bony capsule for the air-bladder is larger, and the fish swims better than *C. tænia*. The sub-orbital spine is capable of being raised from a cutaneous fissure.

## CHAPTER VII.

### FRESH-WATER FISHES OF THE ORDER PHYSOSTOMI (*continued*).

FAMILY CLUPEIDÆ.—GENUS CLUPEA: Shad—Twaite Shad—Whitebait—Black Sea Herring—Caspian Herring.

## FAMILY: CLUPEIDÆ.

### GENUS: **Clupea** (CUVIER).

THE allies of the Herring comprise about eleven genera, which are found in all seas, sometimes with species entering rivers. This, however, is never the case with the Anchovies, though some of their Eastern allies, like the genus Coilia, are found in the Ganges and Irawaddy. Another genus, Chatoërsus, enters rivers on the coasts of North and Central America, Australia, and the East Indies, and has become naturalised in some lakes.

The only genus with fresh-water representatives in Europe is the numerous and widely distributed genus Clupea. The fishes of this type have the head naked. The body is covered with scales; it is compressed, and the abdomen forms a serrated edge, which extends forward into the region of the throat. The margin of the upper jaw is formed in front by the pre-maxillary bone, and at the side by the maxillary bone, which consists of three elements. The upper jaw does not project beyond the lower. The cleft of the mouth is of moderate width; the teeth are rudimentary, but their variable condition aids in distinguishing the multitudinous species into groups. Thus the typical genus Clupea, which comprises only the Common Herring, the Californian Herring, and Herrings of the Black and Caspian Seas, has a conspicuous oval patch of minute teeth on the vomer; while a large number of species from the Atlantic, Indian Ocean and Archipelago, China, and Australia, have minute teeth on the palate, and none on the vomer. The European fresh-water representatives belong to a third group, in which there are no teeth at all, though some of their allies have teeth on the tongue. These fishes have the dorsal fin opposite to the ventral. The caudal is forked; the anal fin has fewer than thirty rays. The stomach has a blind sac, with many pyloric appendages.

## Clupea alosa (Linnæus).—The Shad.

D. 19—21, P. 15—16, V. 19, A. 21—24, C. 19.   Scales: lat. line 70.

The Shad, or Allis-Shad, of England, is a visitor to the rivers rather than a true fresh-water fish. Yarrell mentions that it has occasionally been taken above Putney Bridge, and opposite Hampton Court Palace; it does not, however, commonly come so high up the Thames. It ascends the Severn when the water is clear, but when it is flooded keeps below Gloucester; it goes up the river in May, returning to the sea in July. The chief English fishery is at Newnham, near Gloucester, whence Shad are sent to the Forest of Dean, and many local markets, as well as to London. The English rivers are by no

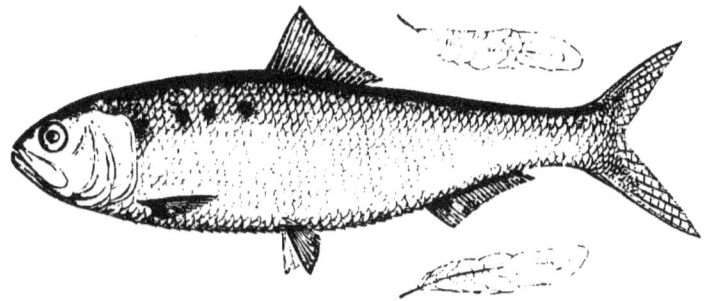

Fig. 111.—CLUPEA ALOSA (LINNÆUS), WITH CAUDAL SCALES.

means exceptional in receiving visits from this fish. It is common in France, where it is known as *l'Alose*. It enters the Rhône, and ascends the Saône as far as Graz, and is found in the Isère. It is also abundant both in the main stream and affluents of the Garonne and Loire; and, according to Blanchard, occurs in most of the rivers of France (Fig. 141).

In Germany it is known as *Maifisch*, because it ascends the rivers in May. It goes up the Rhine as far as Basel, and is especially common in the Main and Neckar. It ascends the Danube, and is occasionally taken at Pesth. It is the *Sabalo* and *Saboga* in Spain, and *Savel* in Portugal. It occurs in the Tagus and Minho. In Upper Italy it is found in the Lakes Como and Garda, where it arrives in May and June. It has been seen at Milan returning to the sea in September; and is taken in the Tiber. It is apparently absent from Russia, where the Black Sea Herring similarly ascends the Dnieper, the Bug, and the Dniester. The *Thrissa* of the ancient Greeks appears to have been the modern Shad.

Dr. Badham states that when taken at sea in winter, it is a dry poor fish, but after a month's sojourn in fresh water in spring, becomes very rapidly plump and delicate. This estimable author remarks in his work " Prose Halieutics," "the *Allis* forms one of an elaborately-finished group of mosaic fish in a house at Pompeii. It was therefore known to the Balbi and their contemporaries, but whether they appreciated it as much as the *bon vivants* of ancient Greece, or disparaged it like Ausonius as a mere *solatium pauperis*, 'the pauper's Alose from the sputtering stall,' is more than we can undertake to settle; but as it occupied the skilful fingers of the ancient mosaicists, and is still considered a fine fish at Naples, we are inclined to think it was held in like estimation by the *connoisseurs* of the same *regno* under the ancient *régime*. A love for music and dancing has been imputed to this fresh-water herring. Aristotle affirms that he no sooner catches the sound of music or sees dancing, than, like Crabbe's sailor 'who hears a fiddle and who sees a lass,' he is irresistibly led to join the sport and to cut capers and throw somersaults out of the water."

The only justification for this statement is the circumstance that the Shad assemble in great numbers near the surface of the water and are noisy in their movements, like so many other fishes at spawning-time. Buckland remarks that they are said to thrash the water with their tails, and that on a calm evening the noise they make may be heard at some distance. Ælian improves on Aristotle's statement by assuring us that the fishermen being well aware of the habit, fasten little bells to their nets so that by the tinkling above the surface, the Shad are attracted to the spot, and netted.

The body is elevated and compressed laterally, and is so like the smaller fish known as the Twaite Shad, that they are not easily distinguished by external characters. The most important distinction, according to Günther, is found in the gill-rakers, which are very fine and long in the Allis Shad, and number from sixty to eighty on the horizontal part of the outer branchial arch, while the Twaite Shad has thick gill-rakers, but they are numerous on the other arches, the numbers being ninety-nine to one hundred and eighteen on the first, ninety-six to one hundred and twelve on the second, seventy-four to eighty-eight on the third, and fifty-six to sixty-five on the fourth. Other distinguishing characters are the compressed ventral edge in front of the anal fin, armed with a row of thirty-seven to forty-two spines, and the low anal fin. There is also a conspicuous dark spot behind the operculum, and the fish has a curious eyelid.

The proportions of the Shad vary somewhat with age. In old females at spawning-time the height of the body exceeds the length of the head, and the body is only three and a half times as long as high. The height is much less in males and in half-grown females. In the young state the body is four and a half times as long as high, and the height is less than the length of the head, which is one-fourth of the length. The head is conspicuously longer than high. The breadth between the opercula is less than half a head length, but in the young fish the thickness is one-quarter of the length of the head. In old females the eye is one-sixth as long as the head, and the inter-orbital space has widened from two-thirds of an orbital diameter to twice the diameter of the eye.

The eye is covered both anteriorly and posteriorly with a fold of skin, which forms a transparent eyelid over the middle of the eye, opening in the centre with a vertical slit. This skin includes cartilaginous plates, which have a half-moon shape, so that the pupil is not completely covered, and the anterior plate can pass over the posterior plate. One eyelid extends under the nares, and the other to the opercular bone. These eyelids have been compared to the fold seen at the hinder margin of the orbit in Salmon.

The nares are divided by a small membranous band, so that the posterior narine is the longer and larger. The mouth is terminal, rather wide, and very oblique. A notch divides the small pre-maxillary bones from each other. The jaws are equal. The cleft of the mouth extends back under the orbits. The maxillary bone resembles that of the Grayling, except in being composed of three pieces, since it reaches as far back as the hinder border of the orbit. Both the maxillary and pre-maxillary are furnished at their margin with extremely fine-pointed short teeth which are easily lost, and disappear early. The sub-orbital ring consists of six bones, is broad, and the maxillary bone extends under its wide anterior portion. The mandible is overlapped by the wide upper jaw; it is toothless, but teeth exist in the young, and are soon shed. The gill-aperture reaches higher than the upper border of the orbit, and the opening extends far into the throat. Of the eight branchiostegal rays, the last two end in a bony plate which unites with the inter-operculum and sub-operculum to form a peculiar cavity in front of the base of the pectoral fin. The pre-operculum is the largest opercular element, and extends higher than the sub-operculum. There are thirty-three thoracic vertebræ and twenty-five in the caudal region. The number of bones of the nature of ribs attached to the vertebral column is unusually large.

All the fins are relatively small. The dorsal fin is longer than high. It begins in front of the middle of the body; its last ray is more prolonged than those which immediately precede it. When the fin is laid back it is partly hidden in a furrow formed by the scales, which are raised on each side of its base. The vent is situate, three times the length of the head, behind the snout. The anal fin is immediately behind it; its base is as long as the height of the head; its rays are similar to those of the dorsal fin, but lower, and, like that fin, partially enclosed between lateral borders of projecting scales; the last rays are a little elongated. The ventral fin is somewhat in front of the middle of the body; its rays are not so short as those of the anal, but are shorter than in any other fin. In the angle over the base of the anal is a scale prolonged like a spine; it has a soft point, and extends for half or two-thirds of the length of the fin. The pectorals are pointed, measure more than half the length of the head, and are near to the ventral edge; the scales overlap them at the base. The terminal rays of the deeply-forked caudal fin are shorter than the head; the lower lobe is slightly the longer. The middle rays of this fin have very numerous and finely-jointed branches.

There are eighty scales in the lateral line, and there are twenty or twenty-two scales in vertical series over the ventral fins. The scales vary much in size. At the sides they are largest, square, and nearly equal to the diameter of the eye. On the dorsal part of the tail the scales are elongated, their free edges are marked by parallel delicate rays which do not reach the middle of the scale. The scales on the caudal fin have a peculiar character. The middle rays are covered with larger scales, while on the upper and lower lobes the scales become smaller. The upper surface of the large scales is again covered by three or four smaller scales; each of the long scales recalls the form and structure of the wing of an insect; or, as it has a stalk firmly fixed in the skin, it might be compared to a dicotelydonous leaf. As the mid-rib runs through the leaf, so the stalk extends through the scale and sends out on each side numerous accessory branches which spread over the surface of the scale (Fig. 114). In the middle of the tail are two shorter scales, perforated by fine canals. The stalk of each scale and its branches are regarded as offshoots of the mucus-canal of the lateral line, which in the region of the shoulder sinks deeply under the scales and appears to be wanting in the body, but reappears at the base of the caudal fin, branching in its upper and lower lobes with the finest ramifications. The cephalic canals are well developed and extend on to both sides of the fore part of the trunk. The shoulder region behind the angle of the gill-opening is ornamented with a black spot on which skin extends, which, like that of the cheeks and operculum, shows innumerable dendritic branches, none of which penetrate the substance of the scales.

Another peculiarity of this fish is seen in the structure of the saw-like ventral edge, for though the serrations are formed by the projecting scales, the sternal ribs extend down to them, and Dr. Günther describes this arrangement as recalling the structure of the sternal ribs in the crocodile.

The Shad feeds on small fishes and on vegetable substances. The intestinal canal makes two convolutions. Behind the stomach there is an immense number of pyloric appendages of unequal length, some being four times as long as others. They are longer and more numerous in the male than in the female. The liver has two lobes, but the right lobe subdivides into two; the gall-bladder is large, dark green, and on the right side. The air-bladder is elongated to a point at both ends; a short thick pneumatic canal connects it with the stomach.

The organs of reproduction are enormously developed. The two ovaries are symmetrical, and at spawning-time extend over the pyloric appendages, liver, stomach, and intestine, between which they squeeze their way while distending the ventral cavity. The eggs are of uniform size, and are estimated to number hundreds of thousands.

The young cock is smaller than the hen, and arrives in the rivers earlier, but the big cock is late in arriving. The small fish sell in the spring at six shillings a dozen, but later they are four shillings a dozen, and one hen counts for two cocks.

The length is about two feet. The fish seldom weighs less than four pounds, and in France the weight varies from two to three kilogrammes, but in Germany and Austria it is smaller.

The back is a pale olive-green with golden iridescence. The colour is paler on the sides; the throat and abdomen incline to sea-green; but the scales have a gold and silver and frequently a pinkish lustre. The upper part of the head inclines to brown, and the operculum is golden. At the sides the scales are dotted with black and the large spot on the shoulder is dark olive-green.

The dorsal, caudal, and pectoral fins are blackish-grey. The anal is grey and spotted with black; the ventral fins are white.

## Clupea finta (Cuvier).—Twaite Shad.

D. 18—20, A. 20—24, V. 9, P. 15—16, C. 19. Scales: lat. line 60—75.

The general form and proportions of the Twaite Shad are similar to those of the Allis Shad, so that there is no difference in the head except that the gill-rakers are stout and bony, and number only from twenty-one to twenty-seven on the inner side of the horizontal part of the outer branchial arch. The body is a little more elongated.

The fins are alike, but the anal fin is smaller. The basal half of the caudal is similarly covered with small scales, and has the two elongated scales parallel to each other towards the middle of the fin, which we have already seen to characterise the Allis Shad. The resemblance extends to there being fifteen of the remarkable three-pronged abdominal scales, or spine-scutes, behind the base of the ventral fin. The colour is the same. The large blackish blotch behind the upper part of the gill-opening is usually followed by a series of from four to eight similar spots, but in the young fish these are more numerous, and may extend to the tail; but the character is not an invariable distinction from the Allis Shad, because there may be, according to Dr. Günther, similar spots in that species.

In view of this close resemblance, the different geographical distribution is remarkable.

The habits of the Twaite Shad are similar to those of the allied species. It enters the Severn and the Thames to spawn during the month of May; and Yarrell states that great numbers are taken every season below Greenwich opposite the Isle of Dogs, and it was formerly abundant at Millbank and above Putney Bridge. Young specimens taken in October measure two and a half inches; in the following spring they are four inches long.

Blanchard states that it arrives in the rivers in France some weeks later than the *C. alosa*.

Steindachner found it in the Minho, in Spain, in October, and noticed it in Lisbon fish market in November.

In the east of Europe it replaces *Clupea alosa*. In North Germany it is known as the *Perpel*, but appears to be taken only occasionally, though it comes into the Haffs from the Baltic in great numbers. It is in no demand for food, though Benecke differs from other authors in regarding its flesh as well flavoured; but this may be due to crustaceans being its common food on the Prussian coast. It is a characteristic fish of Scandinavia and Denmark, but does not appear to occur in Russia. It is found in the Nile.

It is a smaller fish than the Allis Shad, not often exceeding a size of thirteen to fifteen inches, and the weight is about two pounds.

The skeleton includes fifty-six vertebræ.

## Clupea harengus (LINNÆUS).—Whitebait.

Pennant, who wrote his "British Zoology" in 1776, states that during the month of July there appear in the Thames, near Blackwall and Greenwich, innumerable multitudes of small fishes, which are known to Londoners by the name of Whitebait. They are esteemed very delicious when fried with fine flour, and occasion, during the season, a vast resort of the lower order of epicures to the taverns contiguous to the places they are taken at. Since Pennant's time, other orders of epicures have learned to appreciate Whitebait, so that during the season it is in universal demand. But, though naturalists have not wanted opportunities for studying the fish, it furnishes a curious note on the views of older writers that many of the best observers regarded it as a distinct species. Dr. Günther, however, discovered that the Whitebait is the young of the common Herring, mixed it may be with some other young fishes. Indeed, Eels, Pipe-fish, Sticklebacks, Gobies, and many other types which have nothing in common with the Whitebait, are captured with it; and Mr. Day found that while Whitebait consisted entirely of Herrings in the autumn, there were about seven per cent. of Sprats mixed with them in the summer; but even then the Sprat is easily recognised, for it has no teeth in the vomerine bone of the palate, which are present in the Herring, only seven or eight cæcal appendages to the intestine, instead of about twenty in the Herring, while the abdomen has a sharper keel in the Sprat, and the form of the body is different.

Usually, Whitebait are about two inches long, but they necessarily grow much larger, though they cease to be sold as Whitebait when the length exceeds three or four inches.

The spawn of the Herring appears to be shed at sea in March, and hatched in from one to six weeks, according to temperature. Then the young frequent the estuary of the Thames, Southampton Water, mouths of rivers on the opposite coast of France, and other localities, such as the Frith of Forth.

Whitebait are taken from a boat moored in the river with the net kept at the surface; so that as the fish come up with the tide they enter it, when the small end is from time to time drawn into the boat, and the fish shaken out. They do not reach Woolwich till the tide has been running three or four hours, and the water has become a little brackish, but lower down, where the water becomes salter, they may always be taken during the season. The Whitebait feed on minute crustacea.

## Clupea pontica (EICHWALD).

D. 2—3/13—14, P. 16, V. 1/8 A. 1—3/17—18, C. 5—6/19,5—1.

The Black Sea Herring ascends the Dniester, Bug, and Dnieper in shoals every spring. In the Dnieper it is found as far up as Kiev, but the only tributary in which it is recorded by Kessler is the Psiol. It is taken in bag nets in enormous quantities. It is a thin fish, and does not salt well; but is dried and smoked as food for the agricultural population of the surrounding steppes. Its usual weight is from a quarter to half a pound.

The body is as high as the head is long, and measures one-quarter of the length of the fish without the caudal fin. The cleft of the mouth extends to the orbit. The lower jaw is rather prominent. Teeth of small size are found on the tongue, vomer, and palatine bones. The operculum is striated. The ventral fins are behind the commencement of the dorsal. The dark spot on the shoulder is not very distinct. The caudal fin is yellowish-grey, with a broad black border. There is a dark spot at the anterior extremity of the dorsal fin. There are about forty pyloric appendages to the intestine, of which those in front are short. The fish is known in Russia as *poozanok*.

This species, as Günther remarks, is nearly related to the *C. harengus*.

## Clupea caspia (EICHWALD).

The larger Caspian Herring, *C. caspia* (Eichwald), is somewhat intermediate between the Herrings and Shads. It similarly ascends the Volga. Thirty years ago, says Dr. Grimm, the fish were caught only for oil, but now 250,000,000 a year are salted in Astrakhan. Each fish weighs more than a pound. They are salted in boarded pits, each pit holds one hundred thousand fishes. Packed in barrels, each holding from one thousand to five thousand, they sell at twelve roubles the thousand, a rouble being two shillings and tenpence.

This species has thirteen rays in the dorsal, and eighteen rays in the anal fin. The body, exclusive of the caudal fin, is three times as long as high. There are small teeth on the vomer and palatine bones, but no teeth on the tongue. The maxillary bone extends below or behind the orbit.

The pectoral fins are of a deep black at the base. Four or more black spots extend in a line on the upper part of the side of the body.

## CHAPTER VIII.

### FRESH-WATER FISHES OF THE ORDER OF PHYSOSTOMI (*continued*).

FAMILY SALMONIDÆ — GENUS SALMO: Salmon — Salmon Fisheries — Salmon Disease. Trout; Gillaroo — Salmo hardinii — Sewin — S. argenteus — S. venernensis — S. mistops — S. nigripinnis — S. obtusirostris — S. microlepis — River Trout — S. ausonii - Galway Sea Trout — S. autumnalis — S. lemanus — S. carpio — S. rappii — S. dentex — S. spectabilis — S. marsiglii — S. lacustris — S. schiffermülleri — Loch Leven Trout — Orkney Trout — S. polyosteus — Grey Trout — Salmon Trout — Great Lake Trout — S. bailloni — S. genivittatus — S. labrax.

## FAMILY : SALMONIDÆ.

### GENUS : **Salmo** (ARTEDI).

ALL the Salmon tribe have the head naked, and the body covered with scales; though in the genus Salanx, of China and Japan, which is the Whitebait of Eastern seas, the scales are small, thin, and deciduous.

The upper jaw is formed by the pre-maxillary bones in front, and maxillary bones at the sides, and carries no barbels. There is always a small fatty fin behind the dorsal fin. The belly is relatively wide, and more rounded than in most fresh-water fishes. The intestine is generally remarkable for the number of its pyloric appendages. There is a large simple air-bladder. The ova fall into the abdominal cavity before they are deposited. Typical Salmon, and many Trout, live part of the year in fresh water, and part in the sea; but there are many species of the genus, known as Trout and Charr, which are confined to lakes and rivers. Two genera of the family, Argentina and Microstoma, however, never enter rivers; they both range from the Mediterranean to the Arctic seas, and the former at least is an inhabitant of deep water.

The genera of the Salmon tribe which are found in the fresh waters of Europe are Salmo, Osmerus, Coregonus, and Thymallus, all of which belong to the division of the family termed SALMONINA, in which the dorsal fin is opposite, or nearly opposite, to the ventrals.

In all species of Salmo the scales are small, and cover the body. The cleft of the mouth is wide, and the long maxillary bone often extends beyond the eye. The teeth are strong, and well developed, conical in form, and are not limited to the jaws, but are found on the tongue, the median bone of the palate named the vomer, and on the palatine bones. The short anal fin never contains more than fourteen rays. All species have a multitude of pyloric appendages. The eggs are large. The young are generally distinguished by having dark cross bands; and in old fish the caudal fin becomes less deeply notched. The lower jaw is longer in males than in females.

Dr. Günther remarks on the almost infinite variations of these fishes with age, sex, development of the reproductive organs, food, and the properties of the water they inhabit; and later observations have thoroughly confirmed these views. Some of the species interbreed, and it is probable that these hybrids mix again with the parent species, increasing the variability. The colour is particularly changeable. The young fish lose their transverse bars with maturity, and the males during and after spawning-time become more brilliantly and variously coloured than the females.

The water influences the colour in developing tints which harmonise with its character. In clear, rapid rivers Trout acquire dark rounded spots; in large lakes with pebbly bottoms the round spots are replaced by crosses, and the fish are brighter and more silvery. In muddy or peaty pools or lakes, the colour becomes darker; and, where there is little light, may be almost black; while in the sea the spots are rarely developed, and the coat acquires a silvery brightness.

Food rather influences the colour of the flesh than of the skin, though when the fish is in good condition, the skin colour is more uniform and bright than at other times. The pink tint of the flesh has been attributed to the pigments of fresh and salt water crustacea on which these fishes feed.

The size depends upon food, and species which are extremely small in small mountain pools, become heavy with abundant food in rivers or lakes.

The proportions of the different parts of the body vary so much as often to make the identification of species difficult. The snout becomes more pointed and produced at maturity in the male, so that the head increases in length with age, but when the fish is well fed, the relative size of the head is much smaller than when it is badly fed. The fins are also variable, and while the caudal is commonly deeply notched in the young, it becomes truncate in the adult; and individuals which live in rapid streams have all the fins more rounded than those which live in lakes. The skin, too, in old males, thickens remarkably about the spawning season, so that the scales become embedded in it, and invisible.

But while these external characteristics afford no aid in separating species, Dr. Günther relies on the following characters:—1, the form of the pre-operculum in the adult fish; 2, the width and strength of the maxillary bone in the adult; 3, the size of the teeth, though the pre-maxillary teeth have little value; 4, the arrangement and permanence of teeth on the vomer; 5, shape of the caudal fin in relation to size, age, and reproductive condition; 6, size of the pectoral fins; 7, size of the scales, as shown in the number of rows above the lateral line; 8, number of the vertebræ; and 9, number of pyloric appendages to the intestine. All these characters, however, are more or less inconstant.

The genus Salmo is divided into two sub-generic groups—*Salmones* and *Salvelini*. The former have the teeth not only on the front of the vomer but along its length, though the hinder teeth in some species are lost with age. These fishes are usually known as Salmon and Trout. The *Salvelini* have teeth on the front of the vomer only, and are known as Charr. Both groups have representatives in many northern countries, peculiar species being found in the British Islands, France, the Alpine region of Europe, Hungary, Algeria, Italy, Russia, and Tartary, the Scandinavian peninsula, and Finland; while many species are common to Northern Asia and North America, and others are limited to the Rocky Mountains, to Greenland, and Labrador.

## Salmo salar (Linnæus).—The Salmon.

D. 14, A. 11, P. 14, V. 9. Scales: lat. line 120—130, trans. $\frac{22}{19}$ to $\frac{26}{22}$

The form of the Salmon is rather elongated. The length of the head is nearly equal to the greatest height of body; the snout is pointed. The ventral contour is very much arched, while the dorsal profile is flatter. The

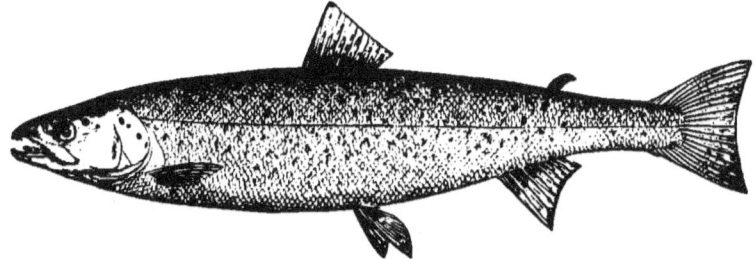

Fig. 145.—SALMO SALAR (LINNÆUS).

body is deepest under the dorsal fin, and five and a half to five and three-quarter times as long as high. The greatest thickness of the body is about one half of the height; the diameter of the eye does not exceed the length of the head. The length of the snout is twice the orbital diameter, and the nares are midway in the length of the snout. The breadth of the inter-orbital frontal space is two and a half times the width of the eye. There is a half-moon-shaped furrow in front of the eye (Fig. 145).

The lower jaw is elevated in the middle, so as to form a blunt projection in front. The teeth are directed backward (Fig. 146). The head of the vomer is pentagonal and toothless. The stalk of the bone usually has three or four teeth in a single row. In old fish they are gradually lost from behind forward. There are three or four teeth on each side of the tongue; they are as strong as those in the maxillary bone; the sixteen or seventeen teeth in each palatine bone are weaker, and the row is but little curved. The opercular plates are rounded at their edges. The pre-operculum has a distinct lower limb. There are eleven branchiostegal rays on each side; the large pseudobranchiæ are comb-shaped.

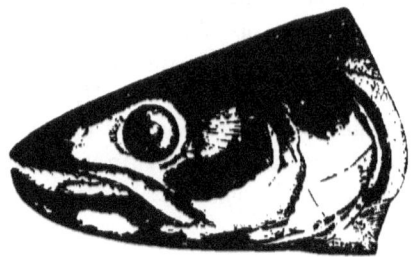

Fig. 146.—HEAD OF SALMO SALAR.

The dorsal fin begins in front of the middle of the body; it is at least twice the length of the head behind the snout, is longer than high, and truncated behind so that the last ray is two-fifths of the length of the third and fourth rays. The adipose fin is opposite to the end of the anal, and twice as high as long. The last ray of the anal fin is only one-third of the length of the longest ray. The pectoral fin is comparatively small and scarcely one-tenth of the length of the fish. The caudal fin is deeply cleft in young individuals, which have the longest rays of the fin more than twice the length of the middle rays. The forked or emarginate condition may remain till the fish is more than two feet long, when the fin becomes more or less truncated.

The hinder part of the body is elongated, and covered with relatively large scales, there being constantly eleven or sometimes twelve in a transverse series running from behind the adipose fin obliquely forward to the lateral line. The scales are rather small; they are ovate and marked with wavy fine concentric lines. Günther counts 120 in the lateral line; Day 120 to 125, and Heckel and Kner 130; and the number of scales seems to vary between these limits in the Salmon of North Germany. The course of the lateral line is horizontal and rather above the middle of the height of the body. The cephalic canals open on the head, with distinct rows of pores. There are eight or nine pores on the sub-orbital branch, and five or six pores on the mandibular branch. The number of vertebræ is fifty-nine or sixty. The pyloric appendages, which are at first very slender, vary in number from fifty-three to seventy-seven, and as the fish comes up from the sea these organs are invested with fat, so that fully five ounces of fat were removed from them by Buckland in large Rhine Salmon.

The colour of the back varies from a dark steel-blue to bluish-black, becoming paler on the sides in the adult, while the under side of the body is silvery, with a mother-of-pearl lustre. The upper part of the head is bluer than the back. The cheeks are silvery; large black spots are scattered over the frontal region above the eye and the operculum, and there are smaller black spots which may be round, X-shaped, or star-shaped, on the back and sides. Yarrell states there are usually more spots on the female than on the male. About four or five may be counted about the lateral line in the fore part of the body. Benecke mentions that in North Germany the head of the male Salmon is sometimes spotted with red. At spawning-time, and in old males, these red spots become confluent in wavy lines, and the entire abdomen is red.

The dorsal fin is grey, with a dark border, and often has a row of small black points at its base. The pectoral and caudal fins are similar in colour, but in British fishes the fins are usually darker than on the Continent. The ventral fins are white, but dusky on the inner side, while in Germany they are commonly pale red at the base. The anal fin is white.

In the young stage, termed Parr, the sides of the body are marked with about ten or eleven dusky cross bars which are vertically ovate in form. There are some orange spots. The black spots on the body are then few, and almost entirely above the lateral line.

A little later the fish is termed a Smolt, when the dusky patches become fainter and the black spots more numerous; and when the dark patches are vanishing and become more elongated vertically, the Grilse stage is reached, and then the pectoral fin loses its pointed character, and the scales of the lateral line become paler. Under normal circumstances the Smolt is on its way to the sea, and the Grilse on its first journey back from the sea to the spawning-ground. In the sea the colours are bright, and the silver scales exceptionally shining, but on entering fresh water all the tints become duller; and fish have occasionally been captured which have become brown or coppery.

Not much is known of the food of the Salmon while they are in the sea, but they feed greedily on sand-eels, Smelts, sea-urchins, starfishes, shrimps and other crustacea, and the fry of all kinds of sea fishes. Very little food, if any, is found in the intestines of Salmon caught in rivers, yet from their taking bait, such as lug-worms, they certainly feed, and it is quite possible that the increase of weight of the fish bears no proportion to the weight of solid food taken. After spawning they develop cannibal propensities, and the old feeble Kelts or Kippers, which remain in the rivers grow voracious, and consume immense quantities of Salmon fry, before they recover strength enough to go down to the sea. The old feeble fishes are not palatable or commonly used for food, though occasionally salted or smoked and known as kipper salmon.

The size varies with the food and river. Buckland attributed the large size of Salmon in the Tay and Rhine to the abundance of Smelts which the fish finds at the entrance to those streams, but the principal growth takes place in the sea. This has been proved by gradually changing the water in which Salmon have been hatched and kept in an aquarium. As the water became more salt the fishes grew more active and voracious, taking many times the quantity of food which satisfied them when the water was fresh, and growing with marvellous rapidity. A Grilse-kelt of two pounds, marked on going down to the sea, returned, and was captured four months later, weighing eight pounds; while marked Salmon of ten pounds have similarly increased in six months to seventeen pounds. Pennant records a marked fish weighing seven and three-quarter pounds on the 7th of February, which weighed seventeen and a half pounds when taken on the 17th of March following.

The age to which Salmon live is not known, but was estimated by Buckland at not less than fifteen years, and the weight steadily augments with age. The heaviest Tay Salmon recorded by Buckland weighed seventy pounds. Mr. Lloyd noticed one which passed through the hands of a London fishmonger and weighed eighty-three pounds; and Von Siebold saw Salmon at Grodno, in Russia, which weighed ninety-three pounds. A large proportion of the fish seen in London weigh from twenty to forty pounds, but fish of all sizes are sent to market.

Salmon fishing is prohibited between the 1st of November and the 1st of February, and for a month earlier net fishing is unlawful, though anglers are permitted to catch Salmon during September, and in some localities they are allowed to fish during November.

There is also during the fishing season a close time every week, which varies in duration, but is usually from noon on Saturday till six o'clock on Monday morning, during which time free passage is allowed to the fish to ascend the rivers to spawn.

The Salmon grows to a length of from four to five feet, though the female becomes mature at a length of about fifteen inches, and the male at a length of seven or eight inches. Salmon are entirely absent from the Mediterranean basin, but widely distributed in the rivers of northern and central Europe, as far south as the north of Spain; and they are found in North America. The fish has popular names in all countries. In Sweden it is known as *Lax*, in Finland as *Lohi*, in Germany as *Lachs*, and sometimes as *Salm*, in Holland it is *Zalm*, in France it is *Saumon*, though the hook-nosed males are designated *Bécard*. In Great Britain the name differs with the age, although there are numerous local names in the different Salmon districts. The young fish during its first and second year is known as Parr, Pink, and Smolt. On its first return from the sea it is a Grilse, or Salmon-peal. After spawning it is a Kelt, though the male fish is then generally termed a Kipper, and the female a Shedder, or Baggit.

Salmon are very common in all Scandinavian rivers, from Lapland to Scania, though fishes found in the streams which flow into the North Sea and Cattegat are reputed to be fatter than in rivers which enter the Baltic.

Carlyle assures us a sheep would be unable to distinguish between the brightness of the sun and the brightness of a scoured pewter tankard, only recognising that both were of incomparable splendour. And, according to Lloyd, the Swedish fishermen, like English Salmon poachers, having discovered the Salmon's love for bright objects, make use of the torch at night to convey the impression of sunrise, in order to lure the fishes to their nets. But the Norwegian fishermen, improving on this method, whitewash the rocks in the vicinity of their nets, or, where there are no rocks, erect white boards, or suspend sheets, which are termed "Salmon attractors," designed to represent the foam of the cataract, which the Salmon is seeking to ascend. But while the white colour is found attractive, the fishermen believe that the fish avoids red colours, so that red clothing is carefully discarded, and, according to Bishop Pontoppidan, even red tiles have been removed for this reason from a fisherman's house. This objection, however, does not extend to food, for Buckland mentions that boiled prawns are an excellent bait in Scotland.

Salmon are said to fear shadows, being driven to the bottom by the shadow of a bird crossing the water. Their habits, indeed, depend a good deal on the weather. When the wind blows from the west or north-west, Salmon enter the Randers Fiord, in the east of Denmark, and ascend a distance of twenty-eight English miles in less than four hours; but if the wind changes to the east or south, little or no progress is made, so that the coming of Salmon is looked upon as a kind of storm warning. They always go up against the wind, so that in Jutland, an off-shore wind is sometimes termed a "Salmon-wind." During a thunder-storm the fishes lie still, or, at least, are not seen; and even at ordinary times, according to Danish fishermen, the Salmon is visible for only about six hours in the twenty-four, namely, from five to six in the morning, from eight till nine, and from eleven till twelve; it appears again in the afternoon from five till six, and from eight till nine, and in the night from eleven till twelve, and one till two.

The spawning-time varies with the country. In the south of Sweden and North Germany the period is commonly at the end of October, or the beginning of November. But in Denmark spawning is later, and may not take place till February or the beginning of March. In Scotland the usual months are November and December; but Day mentions having seen a male Salmon in the Severn in full breeding livery in July.

The female selects comparatively shallow water, often with a depth of only six to eighteen inches, and avoids currents. It was formerly believed that the male and female excavate a furrow or hole, in which the eggs may be deposited; but, according to the observations of Mr. Alexander Keiller, the male has nothing to do with this part of the work. According to his observations in the River Save, in Sweden, the eggs are not placed in a hollow at all, but dropped on a comparatively smooth surface. This, however, may not always be the case. According to Mr. Keiller, the female fish is near to the bottom, and may actually touch it, and after a few eggs are deposited she lies on her side, and with a blow of the tail throws up a mass of dirt and stones, which may conceal the eggs. When spawning has commenced, the female cannot retain the roe, or the male the milt, though the female is said to leave the spawning ground at intervals during the day.

The fish never spawn on the bare rock, or among very large stones. The nest which a single Salmon may accumulate has often about the bulk of a cart-load of gravel, yet by the succeeding summer this heap is usually carried away by the stream. The male in these observations was never seen in actual company with the female, but about six feet behind her, just beyond the heap of stones which she accumulates, so that the only part which the male takes in the breeding is the deposition of milt, where the stream will bring it in contact with the ova. Behind the male, at a distance of twelve or fifteen feet, there are always Trout and other fishes, ready to pounce on the eggs when the female sets them moving with her tail.

At some little distance to the right and left of the male are other males, and the one in attendance on the female is occupied nearly all his time in incessantly charging these younger interlopers; and, owing to these movements, the milt of the male becomes widely distributed, so as to fertilise the eggs which the female scatters.

It is well known that the jaws of the male become elongated during the breeding season by a cartilaginous projection, and the lower jaw turns up in a hook, which fits between the pre-maxillary bones. Whether this growth is stimulated by incessantly ramming his nose against his rivals is a question that may be worth investigation. When the spawning-season is over the growths from both jaws are gradually absorbed. The contests are desperate; the force of the charge often throws the Salmon which is struck out of the water; the fishes are gashed in every direction, and when the fighting-season is ended the males are covered with scars, and have a very battered appearance, much as though they had been mauled by otters. In striking an enemy or rival, the jaws are always clenched like a man's fists.

Pennant gave, on the authority of Mr. Potts, of Berwick, a somewhat different account of Salmon spawning in the Tweed. At the latter end of the year, but chiefly in the month of November, the Salmon ascend the rivers to spawn, as far as they can travel, often for hundreds of miles, making their way against the most rapid currents, and rising over waterfalls with amazing agility; but when the spawn is developed they search for a suitable place for spawning, and the male and female unite in excavating a receptacle, about eighteen inches deep, in the sand or gravel. After the fishes have spawned, it is said that they cover the fertilised eggs carefully with their tails, for after spawning they are observed to have no skin on that part. Shaw states that the female digs the gravel with her tail, and covers the eggs, as asserted by Keiller, without help from the male.

Yarrell quotes Ellis on the "Natural History of the Salmon" as stating that a pair of fishes form a furrow by working up the gravel with their noses against stream. The furrow made, the male and female throw themselves together on their sides, and rubbing against each other, shed their spawn into the furrow, during a period of eight to twelve days.

Mr. Charles St. John, in his "Tour in Sutherlandshire," says that the process of preparing the spawning-beds is curious. The two fishes come up together to a convenient place, shallow and gravelly. Here they commence digging a trench across the stream, sometimes making it several inches deep. In this the female deposits her eggs, and, having left the bed, her place is taken by the male. When the male has performed his share of the work, they both make a fresh trench immediately above the former, thus covering up the spawn in the first trench with the gravel taken out of the second. The same process is repeated till the whole of their spawn is deposited.

From these different accounts we must infer either that the method of spawning varies in different localities with the nature of the rocks in the stream, or else that the difficulties attendant on observation have not allowed the facts to be recorded in so uniform a way as to ensure perfect confidence. That considerable nests are built up is certain, and we may accept Mr. Shaw's and Mr. Keiller's account of the work being chiefly performed by the tail of the female fish, and yet desire fuller records of the spawning process.

The number of eggs commonly produced is estimated at from eight to nine hundred for every pound weight of the fish, though the eggs are sometimes much more numerous. Buckland states that the total weight of the eggs varies from one-fifth to one-quarter of the weight of the fish. Stoddart estimated that for every 30,000 eggs deposited by a large Salmon under natural conditions, only four or five develop into fishes fit for table.

The eggs of Salmon vary in size, according to Mr. Day. Those of small fish are two-tenths of an inch in diameter, and in larger fishes the diameter is three-tenths of an inch; but even in the same fish the eggs vary in size, and it has been shown by Sir Maitland Gibson's experiments that the young fishes raised from large eggs are stronger, and grow faster, than those raised from small eggs. The small eggs from puny fishes, yield fry which are liable to a greater percentage of deaths than those taken from older fishes.

The time required for hatching depends somewhat on the season of the year and the river, varying from ninety to about one hundred and forty days, but the period is influenced a good deal by temperature.

Signs of life appear in about forty days, and in a few days before the eggs are hatched the eyes become visible as dark spots. The eggs require cold, and do not thrive well if the temperature is above 45°. They are commonly hatched in February or March, when the young are five-eighths of an inch long. When the fry are about two months old the umbilical sac, which has the aspect of a pale red currant, is entirely absorbed; and at first their growth is extremely slow. Shaw recorded that by the middle of May they were an inch long, with large heads and wedge-shaped bodies, on which the transverse bands of the Parr were already marked. In a year they increased to a length of three and a half inches, and in two years had a length of six and a half.

In the early months of their existence they attract but little attention, darting under stones and seeking gentle eddies; but in June they begin to scatter themselves over the shallow parts of the river, rarely leaving their birthplace while they remain Parrs, until, at two years of age, they become silvery Smolts, and flit from their native river. The young Smolt descends the river slowly and cautiously; when it comes to a rapid or waterfall, its head is turned up stream, until, carried to the brink, a determination is taken to descend. All through May the Smolts are travelling. They revel in the sea, and when they come back in a few months it is as Grilse, with boisterous energy. They find out the parent river in some instinctive way, and possess a homing faculty like that of pigeons, or some dogs, which enables the fishes to find their way home rapidly, even when taken far away by boat. Buckland suggests that they smell out their home. The Grilse certainly travel back to the land they left as Smolts; and, like the great romping Salmon they travel with, learn to leap up weirs, and waterfalls, and other obstacles in the way. They ascend the Rhine to the falls of Schaffhausen, and go up by the Aar into the Lake of Zurich, 1,230 feet above the sea, and pass on into the Wallenstadt Lake, seventy-eight feet higher. They reach the Lake of Lucerne by the Reuss. They were formerly taken in the Main, at Bamberg, and in the Neckar, at Heilbronn. Ascending the Elbe and its tributaries, Salmon reach the Fichtelgebirge. They reach Moravia by the Oder, and the Carpathians, in Galicia, by the Vistula. Their abundance is sometimes incredible. Von Siebold states that one thousand fishes, averaging thirty pounds each, have been captured in a day at one fishery on the Memel, by nets spread across the river, and taken up every two hours. They do not, however, go up the river in a mass, but in small troops. Fishermen state that females lead the way, followed by the old males, while the young come up last. This is confirmed by the Scottish fisheries taking most Salmon in July, and more Grilse in August. Benecke tells us that sterile fishes do not ascend rivers, but are taken in great numbers on the Baltic coast. In Britain, Salmon enter the northern rivers earlier than rivers farther south. Buckland states that a female full of eggs, a fresh run fish, and a Kelt, have sometimes been taken in one haul of the net, and other evidence points to the conclusion that there is sometimes a spring, as well as an autumn, migration; but fishes are always later in going up rivers which become muddy and swollen in spring by the melting of mountain snows; and streams which intercept the sediment by lakes receive the earliest supply of Salmon.

The fishes ascend when the river is full. In Sutherlandshire, the two rivers Oykill and Shin enter the sea by a common mouth five miles long. In early spring all the fishes which enter the stream diverge into the Shin, which is clearer and warmer, while later in the season they ascend the Oykill. Similarly in Cumberland, Salmon prefer the Eden to the Esk.

The perseverance of the fishes in surmounting obstacles is a remarkable sight. Their efforts are renewed again and again until the bound made by straightening the bent tail enables them to leap over the obstruction.

The limit of their perpendicular spring is about twelve or fourteen feet, and when they attempt greater leaps they are frequently killed by the violence of their exertions. Hence, as Salmon are a valuable property, and a single large fish may sometimes be worth as much as three sheep, ladders and staircases have been invented and erected in Salmon rivers to enable the fishes to surmount difficult waterfalls.

The value of Salmon fisheries is not inconsiderable to the landed proprietors, though much of the Irish fishing ground is of the nature of commons. Alexander Russel stated the rental value of the Salmon fisheries in the Spey, Tay, and Aberdeen rivers at £40,000 per annum, of which the Duke of Richmond received £13,000 a year for his fisheries in the Spey. Expenses, however, are not inconsiderable, and when Russel wrote, in 1864, seven hundred men employed on the Tay fisheries received about £9,000 a year; but the Tay furnishes eight hundred thousand pounds' weight of Salmon, equal to the weight of eighteen thousand sheep, and of three times the value.

With this enormous amount of excellent food supplied by Nature, it is marvellous that the fisheries should have been neglected and ruined in so many English rivers. The number of fishes has been diminished by nets at the mouths of streams, which prevent the fishes from ascending to spawn; and thus proprietors, if we may use the figure, have killed the goose which laid the golden eggs. Moreover, such fishes as escaped the nets and reached the higher parts of the river too often fell victims to the poacher. Besides, the increase of population and of the manufacturing industries have poisoned the rivers in England, so that Salmon avoid them. The last Salmon was killed in the Thames, according to Yarrell, in June, 1833.

Buckland notices that eighteen out of twenty-five cathedral towns were built upon Salmon rivers, from six of which the fish has been exterminated. These towns, which Salmon know no more, are Canterbury, on the Stour; London and Oxford, on the Thames; Rochester, on the Medway; Winchester, on the Itchen; and Bath, on the Avon.

Attempts have been made of late years to establish a race of land-locked Salmon by artificially fecundating the eggs, and breeding Salmon from parents which have never descended to the sea. Earlier experimenters had shown that Salmon can be kept in fresh water for some time: certainly till they are five years old. But the most important experiments have been carried on by Sir J. R. Gibson Maitland, and their history is recorded by Mr. Francis Day. Salmon were hatched in March, 1881; at the age of two years and four months most of them were of a golden tint, spotted, and in the banded Parr livery; while others were beautiful silvery Smolts, such as are seen going down to the sea, and still showing the Parr bands in certain lights. A few months later, the fishes began to leap from their ponds, and were found dead, but with the breeding organs well developed; and in November, 1884, one of these fishes, which weighed one pound and a quarter, was found dead, when a hundred eggs were taken from it and milted from a Loch Leven Trout. Eighteen of these hatched, thus demonstrating that a visit to the sea is not necessary to develop the reproductive function. But at the beginning of December the experiment was tried of fertilising the eggs of females bred in the pond with the milt of males raised with them, and considerable numbers of young fishes were hatched. Thus far, Grilse, or fishes with the Grilse livery, have been bred from the eggs of Salmon without making the usual journey to the sea which precedes the development of the Grilse characteristics, and these fishes have proved fertile, so that a second generation is being reared from them under similar land-locked conditions.

Mr. St. John tells us that Mr. Young, who manages the fisheries of the Duke of Sutherland, thinks that Sea Trout and Salmon interbreed, and records the following evidence :—" A pair of Salmon, male and female, being seen forming their spawning-bed together, the male Salmon was killed with a spear and taken out of the water. The female immediately dropped down the stream to the next pool, and, after a certain interval, returned with another male. He having shared the same fate as his predecessor, the female again went down to the pool, and brought up another male. The same process was gone on with of spearing the male, till the widowed fish, finding no more of her own kind remaining in the pool, returned at last accompanied by a large River Trout, who assisted her in forming the spawning bed, etc., with the same assiduity that he would have used had she been a Trout instead of a Salmon."

Sir J. Gibson Maitland has endeavoured to produce hybrids between Salmon and Loch Leven Trout, his experiments being fully recorded by Mr. Day, in a series of communications to the Zoological Society; but, although multitudes of fish are reared successfully, hitherto the hybrids have proved sterile. At an age of eight months the young Parr have from ten to thirteen finger marks along the sides. At twenty months old some have twelve marks on both sides, and occasionally ten on one side, and eleven on the other.

Professor Malmgren believes that certain Salmon in a lake in Finland are descendants of the common Salmon, which have ceased to have communication with the sea owing to elevation of the land, and the result has been a dwarfed race, with smaller eggs than those of *Salmo salar*.

For some years Salmon have been liable to an epidemic disease which has caused a large number of deaths in the rivers of Scotland, the north of England, and North Wales. According to Huxley, the first symptom of the malady is the appearance of greyish patches upon the skin of the top and sides of the head, the adipose fin, or the bases of the other fins; in fact, upon those parts of the body which are not defended with scales. When first observed the patch may be a circular spot no bigger than a sixpence, but it soon increases in size, becomes confluent with other patches which may have appeared near it, and extends over the healthy skin. The central part, which had at first a raised softer centre, acquires the consistency of wet paper, and the true skin beneath it ulcerates, and an open bleeding sore is formed, which may extend down to the bone. The disease gradually spreads over the whole of the back and sides, and extends into the mouth. This is a contagious and infectious disorder, comparable to ring-worm, or the potato disease, or muscardine among silk-worms. It is the work of a minute fungus named *Saprolegnia*, which is closely allied to the *Peronospora*, which causes potato disease—the one type always being limited to plants, and the other to animals.

Professor Huxley proved the nature of the disease by cultivating the fungus on the bodies of dead flies, and therefore demonstrated that neither pollution, nor drought, nor over-stocking, will produce Salmon disease if the *Saprolegnia* is absent. By a simple arithmetical calculation of the number of spores, it is shown that a single diseased fly might render a shallow stream dangerous to Salmon for several days, while forty large fully diseased Salmon would furnish one spore to every gallon of the 380,000,000 gallons which every day flow over Teddington Weir in the Thames. There is no known remedy for the disorder.

Multitudes of Kelts die annually from the exhaustion consequent on spawning.

## TROUT.

We have arranged the Trout according to the number of fin rays, and give in the following table a summary of characters which exhibit the resemblances of the chief types with each other.

TABLE OF THE PRINCIPAL EUROPEAN SPECIES OF TROUT.
*Arranged according to the number of rays in the fins.*

| Species. | Number of Fin Rays. | | | | Scales. | | | Pyloric appendages. | Number of Vertebræ. |
|---|---|---|---|---|---|---|---|---|---|
| | Dorsal. | Anal. | Pectoral. | Ventral. | In the lateral Line. | Above. | Below. | | |
| **SALMO.** | | | | | | | | | |
| stomachicus | 15 | 12–13 | 13 | 9 | 125 | 28·33 | | 44 | 59–60 |
| hardinii | 15 | 12 | 14 | 9–10 | 122 | 22·30 | | | |
| cambricus | 14 | 11–12 | 16 | 9 | 120 125 | 27·40 | | 49–61 | |
| argenteus | 14 | 11 | 15 | 10 | 123 | 26·30 | | 61 70 | |
| venernensis | 14 | 11 | 14–15 | 9 | 120 | 28·34 | | 59 62 | 58 60 |
| mistops | 14 | 12 | 14 | 9 | 118 | 26·34 | | 43 52 | 59–6) |
| nigripinnis | 14 | 12 | 13 | 9 | 120 125 | 28·30 | | 33–42 | 57 59 |
| obtusirostris | 14 | 11–12 | 13 | 9 | 101–103 | 20·21 | | | |
| microlepis | 13–14 | 11 | 14 | 9 | 115–140 | 29·30 | | 72 | 60–61 |
| fario | 13 14 | 11 12 | 14 | 9 | 120 | 27·30 | | 33–46 | 59–60 |
| ausonii | 13 14 | 10 11 | 13 | 9 | 120 | 26·30 | | 38 51 | 57–58 |
| gallivensis | 13 | 11–12 | 15 | 9 | 125 | 25·38 | | 44 | 59 |
| autumnalis | 13 | 12 | 15 | | | | | | |
| lemanus | 13 | 12 | 14 | 9 | 115–118 | 26·36 | | 45–52 | 57–59 |
| carpio | 13 | 12 | | | 123 | | | 40–50 | |
| rappii | 13 | 12 | 13 | 10 | 120 | 27·35 | | 48–54 | 53–60 |
| dentex | 12–13 | 12 | 13 | 9 | 118 126 | 24·35 | | | |
| spectabilis | 12–13 | 11 | 13 | 9 | 120 | | | | |
| marsiglii | 13 | 11 | | | 120 | | | 90–103 | |
| lacustris | 13 | 11 | 13 | 9 | 120 | 24·30 | | 60–74 | 60–61 |
| levenensis | 12–14 | 10–12 | 12 14 | 9 | 120–130 | 28·26 | | 60 80 | 59 |
| orcadensis | 13 | 11 | 14 | 9 | 115 | 25·30 | | 50 | 56–57 |
| polyosteus | 13 | 11 | 14 | 9 | 128 | 24·30 | | 40 | 61 62 |
| brachypoma | 13 | 10–11 | 14 | 9 | 118 123 | 27·30 | | 45–47 | 59 |
| trutta | 13 | 11 | 15 | 9 | 120 | 24·34 | | 49 61 | 59–60 |
| ferox | 13 | 10 11 | 16 | 9 | 125 | 26·30 | | 44 49 | 58–59 |
| bailloni | 13 | 10 | 12 | 9 | 120 | | | | |
| genivittatus | 11 | | | | 120 | | | | |
| labrax | 10 | 9 | | | | | | | |

All Trout are very variable. The characters which Dr. Günther considered of most importance in defining species of Trout were the number of vertebræ, and the number of pyloric appendages to the intestine. On evidence of this kind he separated northern and southern forms of *Salmo fario*, because the southern fish had fewer vertebræ; but Mr. Day has met with individuals having the larger number of vertebræ as far south as Cardiganshire, while Dr. Cobbold found only fifty-six vertebræ in a *Salmo fario* in Scotland. *Salmo fario* has vertebræ numbering from fifty-six to sixty, and all the other British fresh-water Trout which do not migrate have a number of vertebræ which is between these limits. The variation of the pyloric cæcal appendages in this species in Britain ranges between thirty-three and forty-seven.

The common Brook Trout was taken from the Itchen, near Winchester, and from the Thames, to Tasmania, and thence to New Zealand. Under the different conditions of food, these Trout in Otago became fat and plump almost to deformity. Instead of increasing, as in Scotland, about one-third of a pound a year, they increase from one to two and three-quarter pounds a year, and reach a weight of twenty-one pounds or more. They vary, too, in proportions and colour, while the cæcal appendages vary from thirty-three to sixty-one in the female, and from thirty-seven to fifty-five in the males examined. In all British Trout, excepting the Loch Leven Trout, the variation in number of these appendages ranges between thirty-three and fifty, so that we are compelled to conclude that the number of these appendages is dependent on food, and valuable as a specific character only in so far as species depend upon the conditions of existence. Even the colour of the flesh is variable.

Mr. Day mentions that in the lower part of the Itchen beyond Alresford crustaceans abound, while they do not occur in the upper part of the stream; and that the flesh of the cooked Trout from the lower part of the river cuts pink, while in the upper part of the Itchen it is nearly white. This author, in evidence that the colour of the flesh varies with the food, mentions that where the American Charr, *Salmo fontinalis*, has been turned out into the rivers of Cardiganshire they are good and rich, with a peculiar gamboge colour. The same species in Perthshire is fat, with the firm flesh of a beautiful pearly white, while in other localities it is said to be pink. Hence we can only conclude that the colour of the flesh may vary; the same species of Trout is known similarly to vary in contiguous streams in Wales.

The external colours change with the nature of the soil or bottom of the water, the character of the current, the extent and depth of the water, as well as with temperature, light, and food. Mr. Day tells us that clear water in rapid rivers or lakes with a pebbly bottom frequently produces silvery fish with X-shaped black marks. The colour is imitative, and the fish rapidly adapts itself to its surroundings. A living Black Trout, according to Mr. C. St. John, placed in a white basin becomes pale in half an hour, and in some days becomes absolutely white; and, conversely, a white fish put into a black vessel, becomes in a quarter of an hour as dark-coloured as the bottom of the jar, and is, therefore, almost invisible. Hence, almost every river possesses variations of colour peculiar to itself, and we are led to conclude that colour cannot be of much value as a specific characteristic.

Mr. Day further shows that a small race of Trout transferred to a lake where food is plentiful attains a large size, thus confirming the experience of the Austrian naturalists.

## Salmo stomachicus (GÜNTHER).

D. 15, A. 12—13, P. 13, V. 9. Scales: lat. line 125, trans. $\frac{28}{33}$.

This Trout is limited to the loughs of Ireland, where it is known as the Gillaroo. It was first observed, as Mr. Day remarks, by the Hon. D. Barrington, in 1773, who found no exterior marks to distinguish it from the Common Trout. There is a red Gillaroo, with black spots on it, and a white Gillaroo, with black spots, which is smaller, and said to be better eating.

Dr. Günther has termed this species *stomachicus*, in allusion to the remarkable thickening of the middle muscular layer of the stomach, which is compared by Day to the gizzard of a bird. The stomach increases in thickness as the fish reaches the adult condition; and since this Trout feeds on *Limnæa*, *Ancylus*, and other fresh-water mollusca, the stomach appears to have developed—as the character of a local race—the power of crushing these shells. It is interesting to note Mr. Day's statement that he has seen examples of the *Salmo ferox* from various localities with a muscular stomach; showing, apparently, that the tendency in this organ to vary is by no means peculiar to the Gillaroo, though Sir Humphry Davy states that he caught fish no longer than his finger, and found the stomach as hard as in the larger fish, and with walls as thick in proportion to its size. He further remarks that the Charr, which feed at the bottom, in the same way as the Gillaroo, have a stomach of the same kind, but not quite so thick. Yet although the food and conditions of nutrition have changed the stomach in this curious manner, there are still forty-four pyloric appendages.

The back and sides are marked with reticulated black spots or, according to Yarrell, with dark reddish-brown spots on a yellow-brown ground. On or below the lateral line there is a row of red spots. The free margins of the dorsal, anal, and caudal fins are white, and the dorsal has black spots.

The fish prefers a gravelly bottom. It reaches a length of twenty-nine inches in Lough Earne, and is said to spawn in November, though Dr. Günther has some doubt whether this large fish belongs to the same species as the smaller fish from Lough Melvin. The teeth on the vomer are in a double series, and are persistent throughout life. The fins are well developed, and the pectoral is pointed.

According to Mr. Day, the Trout of Swaledale has a thickened stomach like the Gillaroo, and thirty-five pyloric appendages.

In all probability many varieties of Trout remain to be described, for when taken by fishermen the internal anatomy is not always observed, nor are the technical variations of fin, scale, colour, and proportions always recorded. And a better knowledge of these intermediate forms is likely to show that several types, which resemble each other in fin rays and scaling, to which names are here given, should be grouped together, as we have grouped many Carp, as geographical varieties of a few species.

## Salmo hardinii (Günther).

D. 15, A. 12, P. 14, V. 9—10. Scales: lat. line 122, trans. 22—30.

The *Silfrer lax*, or Silver Salmon, of Sweden, is recorded from Lake Wener and the River Gotha. Mr. Lloyd mentions that it is a splendid fish in appearance, and though the flesh is of a paler colour, is held in nearly the same estimation as Salmon. He found it always in the finest possible condition, even in early spring, when it is chiefly taken. The common weight is seven to nine pounds, but it occasionally weighs as much as fourteen pounds. There are specimens in the British Museum twenty-three inches long. It ascends the rivers which flow into Lake Wener, to spawn.

The colour of the back is greenish, the sides and belly are bright silver, with some scattered black spots on the head, body, and dorsal fin. The pre-operculum has a distinct inferior limb. The maxillary bone is strong, and extends behind the orbit. The head of the vomer is toothless, but its body carries a few teeth in a single series. The caudal fin is deeply emarginate. Thirteen scales descend in a transverse series obliquely forward, from behind the adipose fin to the lateral line.

It is said to have a peculiar habit of jagging the line when hooked. Mr. Lloyd suggests that it may originally have been a Sea Trout, which has become cut off from the sea.

## Salmo cambricus (Donovan).

D. 14, A. 11—12, P. 16, V. 9.   Scales : lat. line 120—125, trans. $\frac{27}{38-40}$.

A variety of the Salmon Trout, termed Peal and Salmon Peal, is known in Wales as Sewin, and often as the Bull Trout. It is by no means limited to Britain, though characteristic of the Welsh area, Cornwall and Devonshire, being also found in Ireland, while it is known on the Continent from Denmark and Norway.

It is a fine fish, reaching a length of about three feet, with the head rather long as compared with the depth of the body. The length of the operculum as compared with its depth, according to Dr. Günther, in the Parr stage is one-fifth greater, two-fifths longer in the Grilse stage, three-fifths longer in fishes of twenty-two inches, and four-fifths longer in fishes thirty-two inches long. The radiating striæ on the base of the hinder opercular border are more conspicuous than the marginal striæ. The sub-operculum projects backward. The pre-operculum has a distinct lower limb with the angle rounded.

The snout becomes elongated in males, and in the spawning season the male lower jaw is hooked. The strong maxillary bone is longer than the snout, and even in young fish may extend behind the orbit. The dentition of the jaws and palate is well developed, though the teeth on the maxillary bone are smaller than those on the pre-maxillary, palatine bone, and mandible. The head of the vomer is triangular, broader than long; it becomes toothless in the adult fish, and has a few teeth across its hinder margin in young individuals. The body of the bone carries a longitudinal ridge, on the sides of which teeth are placed in a single series so as to point alternately to the right and left, but by the time the fish is twelve or thirteen inches long all but two or three in front have been shed. These may remain for some time, since Dr. Günther records them in fishes twenty-two and twenty-four inches long. The inter-orbital space is very convex, and the orbit is far below the upper profile of the head.

The fins are not conspicuously developed. The dorsal is a little longer than high, with the last ray cleft to the base, as in *Salmo trutta*. The height of the anal fin is half as much again as the length of its base. The ventral fin is shorter than the pectoral, and the caudal varies with age. It is forked when the fish is six inches long, is slightly emarginate in the Grilse, and truncate in the adult.

The scales in young fishes are easily shed; they are thin, short, and rounded, as large on the body as on the tail, with the usual fourteen scales in a transverse series between the adipose fin and the lateral line, though this number is not quite constant, and may occasionally be sixteen in large fishes. There are fifty-nine to sixty vertebræ, and forty-nine to sixty-one pyloric appendages.

The transverse Parr marks disappear when the fish is five or six inches long; it then has a silvery brightness, and is greenish on the back, with a few small round black spots on the head and sides. As the fish becomes older the back may be greenish-brown, the sides silvery, and the belly dark-brown at spawning-time in the male, while in the female the silvery tone remains. The spots become irregular and X-shaped, and are scattered over the sides, both above and below the lateral line. The fins have a blackish tinge, and small round black spots usually occur on the dorsal. The colour, however, of the other fins in the Grilse is only slightly grey, while in the Smolt the ventral and anal are perfectly white.

When in the sea they are known to feed on marine crustacea and fishes. Sterile individuals are met with in Denmark. This fish breeds readily with the River Trout (*Salmo fario*, var. *ausonii*) in Wales, where the people distinguish it as *Teb-y-dail*, which signifies "fall of the leaf," in allusion, as Dr. Günther says, to its reddish colour and the dark-brown spots of the male. The hybrids are fertile, and migrate to the sea, but when kept in fresh-water ponds they never breed, though growing to a length of eighteen inches. Males are relatively numerous.

When young the hybrid is very like the Trout, but in maturity it takes on the characters of the Sewin. There is a series of round red spots along the lateral line. They do not acquire their reddish tinge till after the second return from the sea. The pectoral and ventral fins are yellow in many specimens, and then the lower front margin of the anal, and the corresponding margin of the dorsal may be yellow, while bright orange-coloured spots extend along the lateral line and below it. In Denmark another series of hybrids is formed with the other variety of the River Trout (*S. fario*, var. *gaimardii*). There are fifty-eight to fifty-nine vertebræ in the Welsh fish.

### Salmo argenteus (CUVIER AND VALENCIENNES).

D. 14, A. 11, P. 15, V. 10. Scales: lat. line, 123, trans. $\frac{26}{30}$.

This is a Sea Trout, which ascends the rivers of France, and, according to Dr. Günther, occasionally reaches the British coast. It attains a length of two feet six inches, and is regarded by some modern authors as identical with the Salmon Trout; but we prefer to follow Günther in making it a distinct variety, because it has an extra ray in the dorsal and in the ventral fin, and four to six fewer rows of scales below the lateral line, and a few more pyloric appendages.

The head, exclusive of the caudal fin, is one-quarter of the total length of the fish. It is depressed, broad, and elongated. Radiating striæ are conspicuous on the lower margin of the operculum. The jaws are strong, but the teeth are hardly so strong in proportion. The head of the vomer is wider than long, with a transverse series of teeth behind. The body of the bone has a single series of teeth on a longitudinal ridge, on which three or four are preserved in a fish twenty-six inches long. The thin scales on the slender tail are angular behind, but not smaller than those on the body. Twelve or thirteen scales descend obliquely forward in a transverse series from the adipose fin to the lateral line.

The colour is silvery, with a dark back. There are round black spots on the operculum, and X-shaped spots on the sides.

## Salmo venernensis (GÜNTHER).

D. 14, A. 11, P. 14—15, V. 9. Scales: lat. line 120, transverse 28—34.

This Trout is common in Lake Wener and other localities in Sweden. It is not migratory, and attains a length of more than three feet. It is valued for food; but the flesh is less firm than that of the Salmon, and is orange-yellow rather than red. Mr. Lloyd states that many remain in the River Gotha all the year round, though the majority leave in the late spring, and pass the summer in Lake Wener, coming back again in the autumn to spawn. At spawning-time the snout of the male becomes prolonged, and the mandible hooked.

Mr. Lloyd, who terms it *Wenerns-lax*, regarded it as the Grey Trout, usually known as *Salmo eriox*, which had become confined permanently to fresh water; but Dr. Günther removed it from that species. He states that the eye is very small, being only one-fourth of the length of the post-orbital part of the head.

The head of the vomer is triangular, with a transverse series of teeth across its posterior margin. There is a single series of teeth on the body of the bone. In the young these teeth form a zigzag line, and some of them are shed before the fish acquires a large size.

There are fourteen scales between the adipose fin and the lateral line, running obliquely forward.

The fish is silvery, inclining to brownish, or greenish on the back. There are more or less numerous small black spots, angular, or X-shaped, on the sides; and rounded spots on the operculum.

It differs from *Salmo trutta*, in having a higher body, more deeply emarginate caudal fin, stronger maxillary bone, and smaller eye. There are fifty-nine to sixty vertebræ, and fifty-nine to sixty-two pyloric appendages. It is probably a variety of *S. mistops*, and in fin formula and scaling resembles *S. nigripinnis*.

## Salmo mistops (Günther).

D. 14, A. 12, P. 14, V. 9.   Scales: lat. line 118, trans. 26—34.

This is a migratory Trout, found in the Eidfjord River, in Norway, which Dr. Günther regards as closely allied to *Salmo gallivensis*. Its eye is extremely small, being from one-eighth to one-seventh of the length of the head, and, in harmony with the small size of the eye, the frontal region is convex above, and the snout pointed. The maxillary bone is long, narrow, and weak. The vomerine teeth are in a single series, and gradually disappear, the vomer being toothless in fishes seventeen inches long. The pre-operculum has the posterior margin, the angle, and well-developed lower limb rounded. The tail is rather broad, and covered with rounded scales, which are a little larger than those on the back. The back is greenish, but the general colour is silvery, with black X-shaped spots on the sides, round black spots on the operculum and head, and a few black spots on the dorsal fin. It has fifty-nine to sixty vertebræ; and forty-three to fifty-two plyloric appendages.

The young loses its Parr marks before it is nine inches long.

The Scandinavian Trout are all closely allied.

## Salmo nigripinnis (Günther).

D. 14, A. 12, P. 13, V. 9.   Scales: lat. line 120—125, transverse 28—30.

A Trout is found in the mountain pools of Merionethshire and other localities in Wales, and Lough Melvin in Ireland, which reaches a length of sixteen inches. The female is mature at a length of seven inches.

The upper part of the body is dark, with a greenish or reddish colour, occasionally blackish. The sides and belly are greyish, though each scale, like those of the back, usually has a silvery centre. The spotting of the sides varies; sometimes there are numerous black ocelli, with a whitish rim, sometimes a few rounded reticulated black spots on the sides, with a series of red ones on the lateral line. The dorsal fin commonly has rounded black spots, but all the fins are dark, and in the young fish dorsal, ventral, and anal have a black and white outer margin. The depth of the body slightly exceeds the length of the head. The head of the vomer has a transverse series of teeth across its base, and there is a single series on the body of the bone, all being persistent throughout life. The long and black pectoral fin is a marked characteristic. The eye varies in relative size with age; in fishes with a length of eight inches, it is fully one-quarter of the length of the head; with a length of eleven inches and a half, it is one-sixth of the length of the head.

The vertebræ vary in number from fifty-seven to fifty-nine; and the pyloric appendages number from thirty-six to forty-two. Sterile specimens have been met with.

## Salmo obtusirostris (Heckel).

D. 14, A. 11—12, P. 13, V. 9, C. 7/17/6. Scales: lat. line, 101—103, trans. $\frac{20}{21}$.

The River Trout of Dalmatia and Italy has no distinctive popular name, and is known to the Italians as *Trota* (Fig. 117).

In Dalmatia, from which it was first described, it reaches a length of

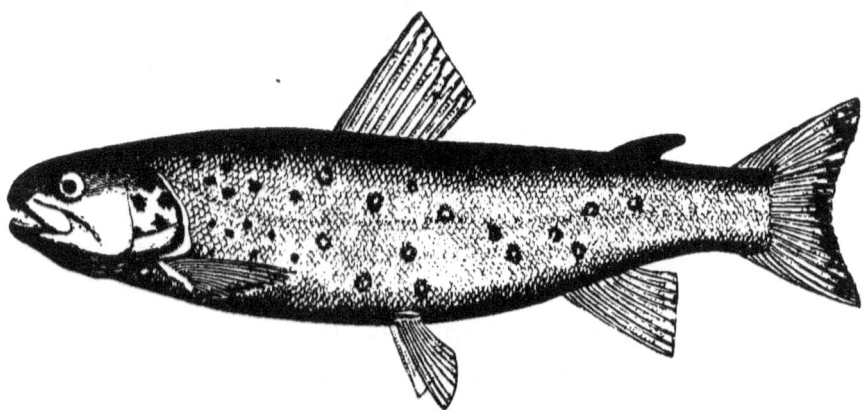

Fig. 117.—SALMO OBTUSIROSTRIS (HECKEL).

fourteen or fifteen inches. Heckel and Kner met with it in the Zermagna, the Salona, and the Verlica, near Imosky, and in these streams it lived on Phryganean larvæ, which occur in countless multitudes. In Italy it is found in the Tiber.

The snout and head in this fish are shorter than in *Salmo ausonii*, and the dorsal and anal fins are higher; but, though it is closely allied to the latter, it shows several differences, especially in the fewer scales, which lead us to regard it as one of those geographical representatives or varieties dependent upon conditions of ancient physical geography which separated the parent type from both before Europe had acquired its present contours.

The height of the body is always more than the length of the head, which is one-fifth of the total length of the fish. The head is shorter, both absolutely and relatively, than in *S. ausonii*. The base of the tail is thicker, and its least height measures as much as half the head length. The forehead descends to the mouth in a sharp curve, so that the snout is thick and rounded. The eye is relatively small —one-fifth of the length of the head—is more than its own diameter behind the snout, and twice its diameter from the other eye. The maxillary bone is broad, and extends below the middle of the eye. The cleft of the mouth is short, and all the teeth in the jaws are smaller than in the preceding type. The teeth on the vomer form two parallel series, with a transverse row of six teeth in front (Fig. 148). The palatine bones are nearly straight, and converge forward, so that the palatine dentary arch is not parallel to the maxillary arch (Fig. 149).

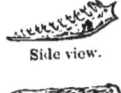
Side view.

Front view.
Fig. 148.—VOMER OF SALMO OBTUSIROSTRIS.

The edge of the pre-operculum is straight, and is more perpendicular than in several forms. The sub-operculum does not extend back behind the operculum, and the lower borders of both these opercular bones are nearly horizontal. There are ten to eleven branchiostegal rays. The rake-teeth on the gill arches are long and pointed, and the last arch carries ten rather long teeth.

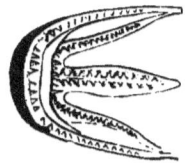
Fig. 149.—VOMER PALATINES AND MAXILLARY ARCH OF SALMO OBTUSIROSTRIS.

The dorsal profile rises less rapidly than in the Common Trout, and sinks less. The dorsal fin begins in the middle of the length, is a third higher than long, and its edge is evenly truncate behind, so that the last ray is about half as long as the longest ray. The adipose fin is somewhat long and high, and reaches farther back than the end of the anal fin. The base of that fin is only two-thirds as long as the base of the dorsal; but the fin-rays are only a little shorter. The ventral and pectoral fins are somewhat longer and more pointed than in the Common Trout, the pectoral reaching back to the beginning of the dorsal. The terminal rays of the emarginate caudal are three-quarters of the length of the head. The scales are rather larger and stronger than in *S. ausonii*, so that along the more strongly-curved lateral line there are only about one hundred and three perforated scales.

According to Heckel and Kner, the pyloric appendages differ from those of the Common Trout. The crown-shaped first row is the longest. The second row runs in a curve on the intestine, recalling the condition in Coregonus.

In colour this species resembles the pale varieties of the Common Trout. The red and black spots are both intense. The red spots cover the operculum, back, and sides of the body down to the caudal fin. After death the red spots become paler, and the black spots of the back and head become a deeper black, and their round form alters to an irregular **X**-shape. All the fins, except the dorsal, show traces of coloured spots, and are edged with blackish-blue. This species does not migrate.

## Salmo microlepis (Günther).

D. 13—14, A. 11, P. 14, V. 9.   Scales: lat. line 135—140, trans. $\frac{29}{30}$.

The Trout thus named is known only from specimens, a little over eight inches long, from Hungary. The body is greenish or brown, with a silvery lustre, and round or reticulate black spots on the side, among which red or orange spots are more or less plentifully scattered.

One of the specimens has twelve transverse Parr marks. The fins are free from black or yellow colour on the margin. There are fifteen or sixteen scales in an oblique series, descending backward from the adipose fin to the lateral line. The head is short, and the body slender. The snout is short, and the broad maxillary bone does not extend to the hinder margin of the eye. The teeth on the vomer are in a zigzag line, and, therefore, presumably in a double row. The pectoral fin is not pointed. The caudal is emarginate. There are sixty to sixty-one vertebræ, and seventy-two pyloric appendages.

This fish differs from the *Salmo lacustris*, in having three extra rows of scales above the lateral line; and in the absence of mature specimens, we must regard the large number of scales in the lateral line as marking an interesting specific variation from that type.

## Salmo fario (Linnæus).—The River Trout.

D. 13—14, A. 11—12, P. 14, V. 9.   Scales: lat. line 120, transverse 27—30.

There are two principal forms of the River Trout. One is characteristic of Scandinavia, Iceland, Scotland, and Ireland, and the other occurs throughout Central Europe, Russia, South Britain, and ranges into Sweden. The former variety is distinguished by Dr. Günther as *S. gaimardii*, this name having been given by Cuvier and Valenciennes to the Trout from Iceland, figured by Monsieur Gaimard. The second variety is named *S. ausonii*, a name given by Cuvier and Valenciennes to a French example of the species. Both varieties occur in Britain, the southern forms ranging as far north as the Rivers Eden and Calder, and the northern form, according to Günther, extending as far south as Shropshire. For this latter we are disposed to reserve the name *Salmo fario*, as it certainly was the fish known to Linnæus.

The Norwegian names for the River Trout are *Fjalloret* and *Fjalloring*. It ranges from Lapland to Scania, being found in the rivers and lakes of the Fjalls. In Germany, where it is known as *Bachforelle*, its distribution has become restricted. The longest Continental specimens, reaching seventeen inches, have been found in Jutland. Specimens fifteen inches long have been taken at Wellington, in Shropshire. The largest Trout recorded in the Tweed and northern rivers weighs from five to seven pounds; but exceptional fishes, taken in Yorkshire, have weighed seventeen pounds, and measured thirty-one inches. The female attains maturity at a length of seven or eight inches.

The body, according to Günther, is rather short and compressed, with the head relatively small and well-shaped. Excluding the caudal fin, the fish is five times as long as deep, and less than four times as long as the head. The pre-operculum is nearly crescent-shaped, and the lower limb when present is indistinct; the snout is blunt, conical, and of moderate length. The maxillary bone is broad and stout, usually longer than the snout, extending as far back as the posterior border of the eye. The teeth are not conspicuously large, and are smallest on the maxillary bone. Those on the vomer are in a double row, sometimes becoming zigzag, but with a single transverse row on the triangular head of the bone, completing the arch of the teeth on the palatine bones.

The inter-orbital space has a conspicuous median ridge, but is rather flattened.

The dorsal fin commences in front of the middle of the body; it is higher than long. The anal fin may be twice as high as long. The ventral fin commences under the hinder part of the dorsal, and does not reach to the vent, nor does the pectoral extend near to the ventral. The caudal fin is truncated, sometimes concave, sometimes with the lower lobe rounded.

There are fifteen rows of scales, extending from the adipose fin forward to the lateral line, but frequently there are more than one hundred and twenty transverse series of scales. The scales are rounded, with concentric striping, and without any median ridge.

The back and sides are greenish-brown, but the belly is a dirty white. Numerous X-shaped or round black spots extend along the top and sides of the head and the middle part of the sides of the fish. The spots are always round on the opercular bones. The dorsal, adipose, and caudal fins are generally filled with black spots, and the dorsal, anal, and ventral each have a white outer edge, margined with black.

The habits of the fish are rather nocturnal, or, at least, it is much more active in the evening and night, when it moves with caution, and yet with boldness, in pursuit of small fishes, insects, and their larvæ. Yarrell mentions battles between Pike and Trout, in which the latter was eventually victorious. The rearing of this fish by artificially impregnating the eggs, now so important in stocking rivers, appears to have been first carried on successfully in Hanover by Mr. S. L. Jacobs.

The flavour of the Trout varies with the stream, being dependent on the food. The flesh is a deeper red than that of the Salmon, and the flavour is almost as rich, being best between June and the beginning of October. Mr. Stoddart records an experiment of feeding Trout in separate tanks on worms, Minnows, and insects. On worms they grew slowly; fed on Minnows they became larger; while those which fed on flies soon weighed twice as much as the others put together.

Trout are long-lived. A specimen kept in a well at Dumbarton Castle for twenty-eight years never increased in weight, probably from deficiency of suitable food. Yarrell mentions another Trout, which is said to have lived for fifty-three years in a well near Broughton-in-Furness.

The number of pyloric appendages in this variety ranges between thirty-three and forty-six. The number of vertebræ is fifty-nine or sixty.

## Salmo fario (LINNÆUS).—Var. ausonii.

D. 13—14, A. 10—11, P. 13, V. 9. Scales: lat. line 120, trans. 26—30.

This fish may be regarded as a southern representative of the River Trout, although in Russia it is found in the smaller tributaries of the White Sea and the Baltic, as well as in the similar streams which flow into the Black Sea and Caspian; but it is characteristic of Germany and Austria, and reaches far south in France, occurs in the rivers of the Maritime Alps, in Italy, and was found by Steindachner in Spain; and Moreau found it in the Pyrenees at a height of seven thousand feet. In Sweden, where it frequents stony brooks, it is popularly known as *Bäckrö* and *Stenoring*; it is said by Nilsson never to occur in the higher mountain regions. In Germany it is distinguished as *Forelle*, in Transylvania as *Förren*, in Hungary as *Pistrang*, in Italy as *Trota*, in France as *la Truite*, and in Britain as the *Trout*, which ranges from the Thames to Penzance, and from Poole and Wareham northward to Troutbeck River, in Westmoreland, and the Eden (Fig. 150).

The head is well-proportioned, and the body is somewhat too deep and full to be elegant. The greatest height in front of the dorsal fin is two-ninths of the length. The thickness of the body is half the height; the length of the head exceeds the depth of body, and, excluding the caudal fin, is one-quarter of the length of the fish. The eyes are separated by one and a half times their diameter, so that the frontal space is rather large. The fold at the outer margin of the eye, or rudimentary eyelid, is broad and thick. The small round nares are nearer to the eye than to the snout. There is an indistinct, very oblique lower limb to the pre-operculum. The maxillary bone is strong and expanded, longer than the snout, and extends below the posterior margin of the orbit. When the mouth is closed the mandible is shorter than the upper jaw, though it is broad and powerful. The snout is produced in the male, and the lower jaw in old males has a terminal hook, which is received into a deep depression of the extremity of the upper jaw, so as to keep the jaws apart.

The teeth are strong and well-developed. They are larger in the mandible and pre-maxillary than in the maxillary and palatine bones. The head

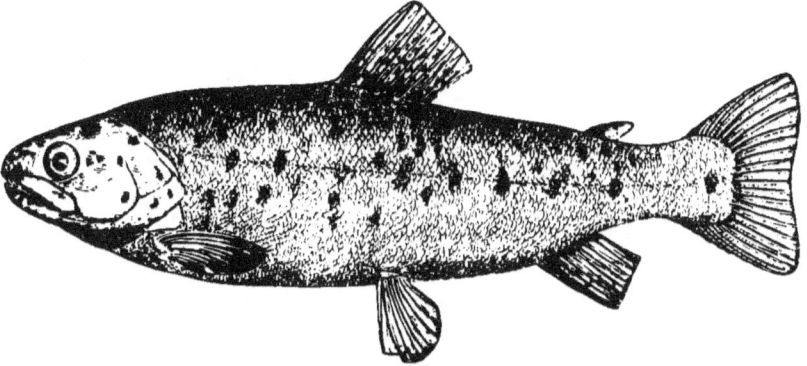

Fig. 150.—SALMO FARIO, VAR. AUSONII (LINNÆUS).

of the vomer is triangular, broader than long, carrying a transverse series of teeth, while on the body of the bone there is a double series of strong teeth, which may be parallel or alternate. All the teeth of the mouth have their points curved inward. There are four or five teeth in a row on each side of the tongue. There is some variation in the dentition with sex.

As in all members of the group, the gill-opening is wide. The number of branchiostegal rays is nine on the right side, and ten on the left. The gill-arches are of moderate length, and their pointed rake-teeth are not numerous. The last arch has usually only seven to eight short, blunt, finely curved teeth.

The ventral and dorsal contours of the fish are similar. The dorsal fin commences a little in front of the middle of the body. It is rather sharply truncate, and nearly as long as high. The four anterior rays of this fin are rudimentary, and covered by the skin. The adipose fin is opposite the termination of the anal fin, which is placed far back, and the vent is only removed from the beginning of the caudal fin by one head-length. The ventral fins are below the hinder half of the dorsal; their rays are shorter than those of the pectoral and anal. The pectoral fins are rounded rather than pointed; the caudal varies with age, and may be forked, truncated, or rounded, and the terminal rays may be of unequal length.

The scales are nearly circular, very small, and thin, without rays, and finely striped concentrically. They are largest on the lateral line, smaller on the back and belly, and smallest at the base of the tail, where they partly cover the middle rays of the caudal fin. There are fully one hundred and twenty scales in the lateral line, which is horizontal for about half its length. The course of the cephalic mucus-canals is traced by round pores, but the branches are not conspicuous.

The number of pyloric appendages to the intestine varies from thirty-eight to fifty-one, acccording to Günther. They are longest in front.

The colour of the Trout varies extremely with food, and condition of the water. Even in the same brook it is rare for many to be absolutely alike, so that no common colouring can be given, beyond saying that the head, body, and dorsal fin generally carry numerous red and black spots, that the black spots often have a pale edging, and that they may be round, irregular, or X-shaped. The anterior part of the dorsal, anal, and ventral fins is generally yellow. As a rule the ground-colour of the body is brownish or brownish-black, though the fishes become darker where the stream is shady; and it is this dark variety, distinguished as *Steinforelle* in Austria, and *Stenoring* in Sweden, that is commonly termed the *Black Trout*. These darker forms often have the bright red spots on the sides surrounded by a light-coloured ring. They reach a weight of four to five pounds.

The Alpine or Mountain Trout has the body covered with much smaller and more numerous brown, black, and red spots, which extend on to the head, but the belly is whitish, and, according to Bloch, the head is green with some golden margins to the mouth, and opercular elements.

The Golden, or Pond Trout, which lives in brooks or ponds with a gravelly bottom, through which spring-water runs, may have the sides glistening with golden-yellow, and covered with bright red round spots, which are often enclosed in a blue ring, and the dorsal fin may be spotted with purple. In lakes the colours are commonly less bright, and grey predominates with large, irregular black spots; but if these varieties, which vary in weight as much as in colour, are put into the same pond, they gradually lose their peculiarities, and become uniform. Exceptionally the red spots may be replaced by white points or black specks; but, as a rule, the purer the water the brighter the colours. The black colour is common in waters that contain iron, as well as in woods, but even then when the water is shallow, and light penetrates through the foliage, the fishes grow pale.

The size of the Trout does not usually exceed one foot, and the weight is half a pound to a pound, though Lloyd believes that he killed it in Sweden of eleven pounds' weight, and Heckel and Kner mention a fish taken in 1851 at Wiener Neustadt, which was thirty-five inches long, nine inches high, and weighed twenty-two pounds. It is not rare for Trout to attain a weight of fifteen to seventeen or twenty pounds in localities in which suitable food is abundant; and Valenciennes refers to Trout of three to four feet in length, though the largest specimen taken recently in France appears to have weighed twelve kilogrammes.

In the young fish the size of the eye, relative length of the head, and other proportions are generally less; but the young are also distinguished by transverse bands, termed Parr markings, of dark-brown, like those of the Perch, which gradually vanish with age. The young fish has also a deeply-forked caudal fin, while in old age the tail becomes rounded. The males are characterised by a larger head, fewer and stronger teeth on all the bones of the mouth, as well as by the hooked mandible.

Trout flourish best in clear cold running water, and when found in lakes are most numerous at the mouths of streams which flow from them. They swim with remarkable speed, but lie still for a long time in deep pools, or in the shadow of overhanging bushes and trees. Like the Salmon, they often make sudden leaps out of the water, and pass over weirs, small waterfalls, and other impediments. They live on insects, molluscs, spawn, worms, and small fishes, and rush impetuously after gnats, which skim over the surface of the water. When their weight is about two pounds they are as greedy as Pike.

They spawn in the autumn, from October to December; and in North Germany, Benecke records spawning as late as January; but this is probably in the mountain districts. The female is mature at a length of eight inches. The urogenital papilla is then enlarged and tumid, and in both sexes there is a thickening of the skin, and, according to Benecke, a swelling of the fins. The eggs in a two-year-old Trout number two hundred to five hundred; in the third year the number increases from five hundred to one thousand; in the fourth to the fifth year the number is two thousand. They are four to five millimetres in diameter, yellowish or reddish, and are laid in small quantities, at intervals of many days, being placed between stones, under trunks of trees, but also in holes, not unlike nests, which the fishes excavate. The spawn when deposited is partially covered with gravel. The period of incubation, according to Blanchard, is forty to seventy days. A temperature of 50° Fahr. is favourable to the growth of the young fish. Many individuals are sterile. The Trout is everywhere one of the most valued fishes, the flesh being easily digested. It is in the best condition from May till the end of September.

In England it is now reared artificially for stocking rivers, and widely distributed; but the quality of the flesh varies with the stream in which it may be placed. Trout-fishing has votaries wherever there are Trout streams; but when taken with the rod the bait varies with the month and time of the day, for the angler needs to be a practical entomologist, and to study the habits of the flies on which the Trout feeds, if they are to be taken with artificial insects. Their feeding-time is usually in the morning and evening, and they are commonly taken where tributaries join a river or lake. The Trout admits of very much more cultivation than it has hitherto received, and may be raised with profit wherever a stream runs through ponds with a sandy or gravelly bottom.

The weight augments rapidly where the food is abundant.

One form of disease to which Trout are liable, the nature of which is unknown, is characterised by an increase in the size of the head and a wasting of the body. Such a fish is known in Austria as an *Adventurer*, or *Quixote*.

Dr. Günther counts fifty-seven or fifty-eight vertebræ, of which twenty-three are in the thoracic region, and the remainder in the tail. The last six, as in all species of the genus, enter into the fan of the tail. There are thirty feeble ribs, and accessory ribs are developed, as in the rest of the Salmon tribe. The stomach is distinct from the intestine; there are thirty-eight to fifty-one pyloric appendages. The intestine makes two curves. The liver is not lobed; there is a large gall-bladder. The air-bladder extends down the entire length of the ventral cavity.

## Salmo gallivensis (GÜNTHER).—The Galway Sea Trout.

D. 13, A. 11—12, P. 15, V. 9. Scales: lat. line 125, transverse 25—38.

Dr. Günther has given names to, and described, a number of British Trout which his predecessors regarded as varieties either of the Sea or the River Trout. Without considering these interesting local forms as species in the ordinary sense of the term, they may be accepted as convenient names for modifications which would have been regarded as species if the chain of intermediate variation had been less closely linked. One of these is the Galway Sea Trout, a migratory species, well characterised by its pointed, though not elongated snout, broad convex forehead, small eye, feeble teeth, slender jaws, and extremely thin and short pyloric appendages, which number about forty-four. It grows to a length of at least eighteen or nineteen inches; and would appear to spawn late, since a specimen of that size, caught in August, was stated to have the eggs still fixed in the ovaries, and of the size of hemp-seed.

The small vomerine teeth are in a single series bent alternately to the left and right. Most of them are persistent to the age of maturity. The teeth on the tongue are very small, not larger than those of the mandible. The pectoral fin is pointed; the caudal is forked in the young, and truncate in mature examples. Its rounded scales on the tail are a little larger than those on the trunk.

The young have nine Parr marks, with ocellated black and red spots, and have white borders to the anal, dorsal, and ventral fins. The mature fish is very dark, with a silvery lustre on the under side. There are black spots on the operculum, and numerous **X**-shaped black spots on the sides; the fins are blackish, with a few indistinct spots on the dorsal, and a black posterior margin to the caudal. There are fifty-nine vertebræ.

## Salmo autumnalis (Pallas).

D. 13, A. 12, P. 15.

Pallas mentioned a fish under this name, which in the month of October ascends into the River Neva in shoals to spawn. It is two feet six inches long, and has a strong scent, which is offensive to Salmon and to the species of Coregonus. The pectoral fins are reddish; the ventrals bright-red; the anal is brownish-red, and both the latter have white margins. The caudal and dorsal fins are brown, in harmony with the colour on the upper part of the body. Below the lateral line the colour is greyish, with small yellowish spots, and the belly is white.

It is remarkable that this fish, which Pallas sufficiently defined as having thirteen dorsal, twelve anal, and fifteen pectoral rays, is entirely unknown.

## Salmo lemanus (Cuvier).

D. 13, A. 12, P. 11, V. 9. Scales: lat. line 115—118, transverse $\frac{26-28}{36}$.

This fish may be regarded as a connecting link between *Salmo fario* and *Salmo rappii*. It is found in the Lake of Geneva and in the Lago Maggiore, but the specimens in the British Museum are sterile, a condition which rather suggests that they may be hybrids. The back is greenish; the belly and sides are silvery. There are numerous very small **X**-shaped black markings scattered over the sides, and there are small black spots on the dorsal fin and operculum, but the other fins are green. The head is well-shaped, and of medium size, while the body is stout. The snout is rather elongated in the male, with a corresponding mandibular hook at spawning-time. The maxillary bone is well-developed. The teeth are in no way remarkable, except that the single series on the vomer alternates to the right and left, and persists through life. The pectoral fin diminishes in relative length with age, and the caudal loses its emargination, and becomes truncated. There are thirteen to fourteen scales descending in a transverse series from the adipose fin to the lateral line. The vertebræ vary from fifty-seven to fifty-nine, and the pyloric appendages from forty-five to fifty-two.

## Salmo carpio (LINNÆUS).—The Trout of Lake Garda.

D. 13, A. 12.  Scales: lat. line 123.

This beautiful Italian Trout was known as early as the sixteenth century, but has been little studied (Fig. 151). It agrees with the larger Lake Trout in form and most characteristics, so that it will be necessary to draw attention only

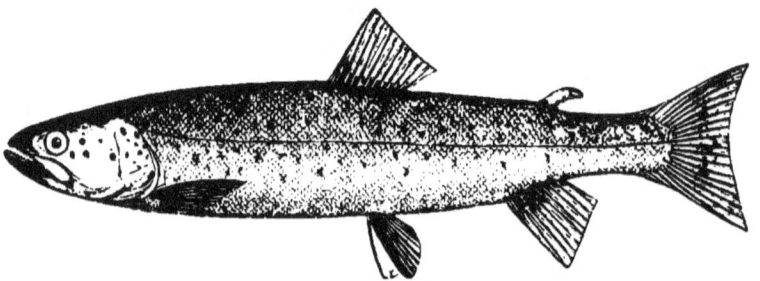

Fig. 151. —SALMO CARPIO (LINNÆUS).

to features in which it differs from them. The maxillary bone in fishes of medium size extends below the hind margin of the orbit. The vomer has three teeth in a triangle on the anterior plate; and on the middle line of the narrow stalk of the bone, there are thirteen teeth in a simple row, with their points turned back, though occasional specimens may be found in which some teeth are bent to the right or the left. The mouth is rather wide.

The snout is longer than the diameter of the eye. The border of the opercular covering forms a regular curve. The number of branchiostegal rays varies by about three on each side, so that the number on the right side is from ten to thirteen, and on the left from eleven to fourteen. The number of short, blunt rake-teeth on the last gill-arch is seven to eight. They often assume a paired arrangement.

The scales in the fore-part of the trunk are relatively large, so that, although the number of scales in the lateral line does not differ materially from that of the Salmon Trout, the number between the shoulder-girdle and the hinder extremity of the pectoral fin is very different, amounting to fully thirty in the Austrian Salmon Trout, while in this fish it is usually fifteen or sixteen, rarely twenty to twenty-two. A part of this difference may be attributed to the longer pectoral fin of the Austrian Salmon Trout, but it is partly due to the larger scales of the Lake Garda Trout. As compared with the Salmon Trout there is one ray more in the anal fin.

The number and length of the pyloric appendages, according to Günther, varies from forty to fifty, and, therefore, more resembles *S. lacustris* than *S. marsiglii*. We are hardly prepared, however, to attach great importance to this character, there being some evidence to show that it varies with food.

The Lake Garda Trout is characterised by the comparative paucity and small size of the black spots which cover the body, though these spots are more numerous and rather larger on the sides of the head. All the fins are unspotted.

The species is not known to attain a greater length than fifteen inches. It is commonly regarded as not migrating, but we learn through Canestrini that while it occurs in the lakes of Lombardy and Venetia, it is reputed to descend the rivers to their mouths, and enter the sea.

It spawns during the month of December. It has long been known under the name of *Carpione*, and highly valued as food. It is also termed by the Italians *Trutta del Lago*; and is similarly known in the Southern Tyrol as *Lachsforelle*.

## Salmo rappii (Günther).—The Bottom Trout of Lake Constance.

D. 13, A. 12, P. 13, V. 10.   Scales : lat. line 120, transverse 27—35.

The *Grund-forelle* is so named from its habit of frequenting the bottom, to distinguish it from the *Schweb-forelle*, which remains visibly suspended in the water. There is very little difference in the fins, and *Salmo rappii* has a ray more in the anal, and often an extra ray in the ventral. There may be one vertebra less, and the number of pyloric appendages is considerably fewer, varying, according to Günther, from forty-eight to fifty-four. It is a fine fish, growing to a length of three feet, and is limited to Lake Constance. The head is rather depressed, and the body has a plump, thick, stouter build than in *Salmo lacustris*. The snout is generally rather obtuse, and becomes elongated only in large males. The male has also a short hook to the mandible. The maxillary bone is strong, expanded in its hinder part, much as in the *Schwebforelle*, and similarly extends back beyond the orbit in old fishes. The preoperculum is crescent-shaped, with the lower limb distinct. The posterior opercular margin is obtusely rounded, the operculum being short. The teeth are strong, and the vomerine teeth are at first in a single series, and then irregularly placed. The fins show no distinctive peculiarities, but the caudal becomes truncate with maturity. There are sixteen rows of scales in a transverse series, descending from behind the adipose fin forward to the lateral line; which is one row more than is found in the *Salmo lacustris*. The sides have a reddish tone with a number of X-shaped brownish spots, and, as usual, the spots on the operculum are round.

## Salmo dentex (Heckel).—The Great Dalmatian Trout.

D. 12—13, A. 12, P. 13, V. 9, C. 7/17/6. Scales: lat. line 118—126, transverse 24—35.

There is probably no other European Salmonoid in which the jaws are so powerfully developed as in this species, but the large size of the pre-maxillary

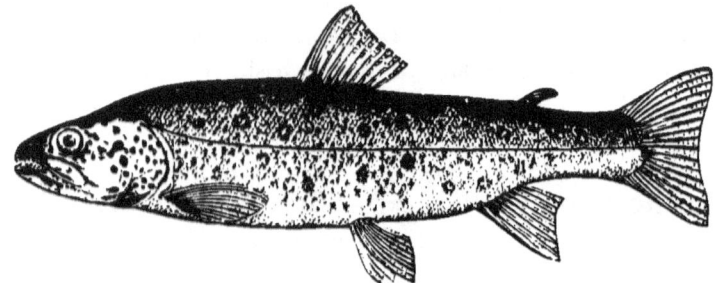

Fig. 152.—SALMO DENTEX (HECKEL).

teeth, which suggested the name *dentex*, is more conspicuous in young than in old specimens (Fig. 152).

The head is small and pointed; the arch of the nose is low, and the whole body is covered with small black marks, which often take on an X-shape. Among these a few larger round red spots are scattered. The length of the head exceeds the height of the body, the fish being five times as long as high, and four and a half times as long as the head. The eye is one-fifth of the length of the head. The breadth of the frontal interspace between the eyes is one and a half times the orbital diameter, and the eye is the same distance from the snout, though the length of the snout varies somewhat with sex and age. The mouth is wide, and the broad maxillary bone extends back below and behind the eyes.

The teeth are remarkably strong. In the pre-maxillary they have a predaceous character, and their length is more than twice the diameter of the scales. The middle teeth in the lower jaw are less developed; but, as usual, the strongest teeth are on the tongue. The teeth number four to six on each side of the pre-maxillary, thirty in the maxillary, twenty on each side of the mandible, twenty to twenty-one on the palatine bone, five on each side of the tongue, and twenty on the vomer. Four of the vomerine teeth form a transverse row in front, and the remainder make an alternating double row behind; but the teeth are easily broken off, and gradually replaced, so that the number may vary in different specimens.

The operculum is large, and extends far back. Its hinder and lower margins form a right angle, but in old males the lower margin becomes curved. The pre-operculum has the lower limb very indistinct. The number of rays on the left varies from twelve to fourteen, and on the right branchiostegal from eleven to thirteen; but there are commonly twelve on the right, and thirteen on the left. The rake-teeth on the gill-arches are of moderate length; on the last arch they number six to seven, and are short and blunt.

The ventral contour forms a flatter arch than the dorsal profile, which is somewhat indented behind the eyes. The dorsal fin begins in front of the middle of the body; it is a little higher than long; its base is equal to half the length of the head, and its last ray is half as long as the longest ray. The rounded adipose fin is opposite the end of the anal fin; that fin is quite as high as the dorsal; its base is one-third of the length of the head. The ventral and pectoral fins resemble those of the Common Trout. The caudal fin is emarginate in young specimens, truncate in maturity, and may be rounded in old age.

The scales resemble those of the Common Trout in size and form. There are thirteen to fourteen scales in a transverse series, descending from behind the adipose fin forward to the lateral line.

The colour is variable; sometimes lighter, sometimes darker than *Salmo ausonii*, but always of a brownish tint on the back, silvery on the sides, and brown on the belly. Old individuals have a dark-brown colour, with coppery metallic lustre, with the belly paler, and throat whitish. There are numerous round spots and irregular markings on the head, and the body is covered with very numerous, small, irregular, black, and often X-shaped dots, which give the young fish a splashed appearance. Everywhere between the black spots are less numerous round spots, of a blood-red colour, most intense on the sides. There are two large pale spots between the eyes and pre-operculum, and white spots mark the under side of the head, especially the lower jaw. The base of the dorsal fin is marked with black spots, but all the other fins are brown; the brown passes into white at the tip of the ventral, and into yellow at the edge of the pectoral. The iris is a coppery-brown. The colour of the flesh is sometimes reddish, sometimes white, from which we can infer only that the colour may vary with food.

This fish does not migrate, and is known only from the rivers of Dalmatia, where it is termed by the inhabitants *Pastrova*. Fine specimens in the River Narenta reach a length of forty-four inches, though the common length is about two feet.

Steindachner has regarded this species as a variety of the Common Trout, *S. ausonii*.

## Salmo spectabilis (Heckel and Kner, Cuvier and Valenciennes).

D. 12—13, A. 11, P. 13, V. 9, C. 17.

This Trout is a somewhat obscure species, Heckel and Kner apparently not intending quite the same fish as Valenciennes. They state that it is spread over Northern Europe, all over Russia, and appears at Teschen, in Silesia, but

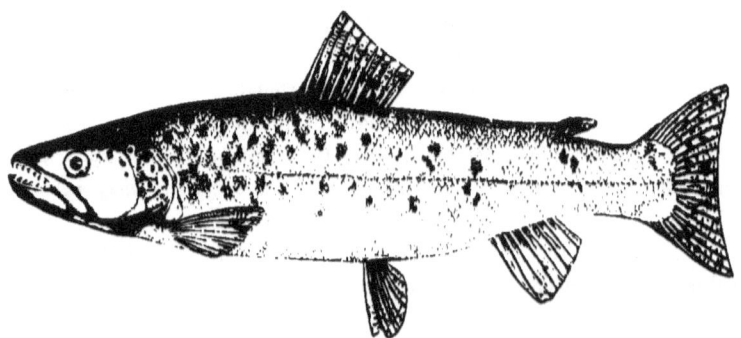

Fig. 153.—SALMO SPECTABILIS (VALENCIENNES).

is not recognised by Russian, German, or Scandinavian ichthyologists; and Dr. Günther remarks that the only point in common between the fishes of the Austrian and French naturalists is that the scales are more conspicuous than usual, a criticism which the published descriptions do not quite justify. Since Valenciennes gave no locality for his specimen, and, so far as is known, no rigid comparison has been made between it and the *S. spectabilis* of Heckel and Kner, we adopt the Austrian nomenclature (Fig. 153).

The fish has a somewhat fusiform shape of body, with the snout pointed, and the head longer than the body is high, though the difference is not great, the fish being over five times as long as high, and four and three-quarter times as long as the head. The eye is small, twice its diameter from the snout, nearly four times its diameter from the hinder edge of the operculum, and more than twice its diameter from the other eye. The mouth is of moderate size, with the jaws equal, and the maxillary bone extending behind the eye. The teeth of the jaws are of uniform size, and similar to those of the palate, though the teeth of the maxillary bone, as usual, become smaller and thicker behind. The body of the vomer has a double row of rather larger teeth, the points of which diverge in a somewhat zig-zag line. The head of the vomer has a simple transverse row of small teeth. The tongue is armed with three or four strong teeth on each side.

The pre-operculum and operculum are rounded. The latter is directed backward. The number of gill-rays is eleven on the left, and ten on the right side. The dorsal and ventral profiles in the anterior part of the body are similar. The rays of the fins are rather short. The scales are somewhat large; they are less embedded in the skin than usual. Valenciennes mentions one hundred and thirty longitudinal series, and Heckel and Kner found one hundred and twenty on the lateral line. The pyloric appendages are long, particularly those of the crown; but the male and female differ in the width of stomach and length of intestine, which are larger and longer in the male.

The colour of the back is steel-blue; the sides towards the belly become silvery; the operculum has large, round, black spots, which are also seen on the dorsal fin and sides, where they frequently take on an X-shape. The other fins are unspotted as a rule, though specimens are sometimes found with spots on the adipose fin and upper caudal lobe, when the black pigment is unusually developed on the sides of the head and body. The size of this fish varies from about one foot nine inches to two feet six.

## Salmo marsiglii (Heckel).—The Salmon Trout of Austria.

D. 13, A. 11. Scales: lat. line 120.

This Salmon Trout (Fig. 154) is found only in the mountain lakes of Austria, such as the Traun See and Atter See, and, according to Von Siebold, in the Chiem See, in Bavaria. Heckel and Kner also say it occurs in Lake Constance.

As compared with the May Trout, the whole form is rather plumper and thicker, but the proportions and measurements of length of head and height of body are much the same, and there is no important difference in size and number of scales, or position and form of the fins. The diameter of the eye is also one-sixth of the length of the head, but its hinder border is farther from the edge of the pre-operculum, and the maxillary bone extends farther back. The teeth on the body of the vomer form a single straight row (Fig. 155); but Heckel and Kner remark that so long as the skin of the palate is not removed, the vomerine teeth appear to form a double row, since their points diverge alternately to the right and to the left. There are three transverse

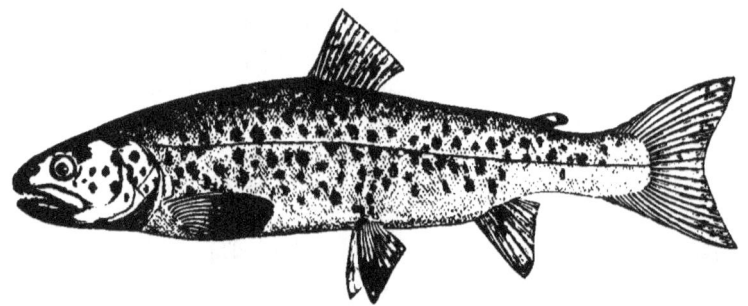

Fig. 154.—SALMO MARSIGLII (HECKEL).

teeth. As in the May Trout, the branchiostegal rays are ten on the right side, eleven on the left. The rake-teeth on all the branchial arches are shorter and less numerous, remarkably short on the last arch, and from seven to nine in number, often near together in pairs, but sometimes at equal distances from each other. The pre-operculum is crescent-shaped, with a very indistinct lower limb; the posterior opercular margin is rather angular. The snout is of moderate length.

Fig. 155.—FRONT VIEW OF VOMER OF SALMO MARSIGLII.

The dorsal profile is more regularly arched than the ventral contour. The scales are always firmly adherent. The number of pyloric appendages is from eighty to a hundred. Those in the first row are not shorter than the others.

The colour changes so quickly after death that the ground-colour is then scarcely to be recognised; but in life the upper part of the head and the back are dark-green, only comparable to the colour of the water of the lakes in which the fish lives. This beautiful tint gets paler towards the lateral line, and is in contrast to the pale blue of the May Trout. After death it becomes opaque brown. The sides are silvery, with a little violet; the throat and belly have the whiteness of pure silver. The upper part of the operculum is green, and the lower part silvery. The skull is marked with a few round black spots, which are larger but more angular on the cheeks and opercular bones. The middle of the back has a few round dark spots, which become larger towards the lateral line. They are both denser and larger than in the May Trout, and towards the tail assume the X-form. Similar spots occur below the lateral line, as far back as the vent, and are largest over the pectoral fins. After death the sides become violet-grey, nearly reddish, and a carmine-red shines through the black spots. The only perfectly black spots then remaining are on the operculum. The greyish-white dorsal fin has a broad dark border, and is marked with many rows of long black spots, between which, after death, a few red ones appear. The caudal fin is violet, and has a broad black band near the margin, succeeded by a narrow band of yellowish-white. The other fins are unspotted, grey or whitish. The iris is silvery, spotted with black.

This fish often attains a length of three feet and more. It is commonly twenty-five to thirty pounds in weight, but has been caught of a weight of fifty to sixty-five pounds. It loves deep water, and generally remains in a depth of twenty to fifty fathoms.

These Salmon Trout seek the haunts of species of Coregonus, which form their chief food, but pursue all kinds of small fishes. When young they remain near the feeding-grounds of species of Leuciscus. If they encounter a shoal the pursuit is eager till the small fishes reach a shallow spot by the shores. The shoal fly through the water swift as an arrow, and often seek to save themselves by springing into the air; but their enemies' movements are no less rapid, and the miserable little Leuciscus, seized first at the tail, is given a dexterous twist, which sends him down head first.

When the Salmon Trout has attained a weight of twenty-five to thirty pounds, it does not bother about the Leuciscus tribe and such small fry any more, but devotes itself to hunting Coregonus, chiefly taking fishes of half a pound to two pounds' weight.

Spawning takes place in November and December, and is the occasion for a grand gathering. Precocious young fishes of a pound weight come up to spawn. They ascend from the lake into tributary streams and brooks, and burrow holes in the ground, which are oblong, and when constructed by fishes of twenty pounds' weight, are sufficiently long and deep for a man to lie in. In these excavations the yellow eggs, which are as large as small peas, are deposited. The same furrows are used by other females spawning later, and are said to be again frequented for spawning purposes in the following year.

This fish is more tenacious of life than the May Trout, does not die so quickly when taken out of water, and can be transported more easily. It will thrive in ponds which are not less than eight to ten feet deep, if the bottom is gravelly, and streams run through them. The flesh is reddish, and highly prized. The method of fishing varies in different lakes. In the Lake of Hallstadt the fish lie in the shadow of a mountain, and if the weather is calm and clear the fishermen follow them to their retreat, and draw their nets in the day-time. The largest and heaviest specimens are caught in the colder months at great depths on lines baited with small Perch, Rudd, or other species of Leuciscus.

## Salmo lacustris (Willoughby).—The Lake Trout.

D. 13, A. 11, P. 13, V. 9, C. 17.   Scales : lat. line 120, trans. 26—30.

There is a Lake Trout found in Lake Constance, which is known as *Schwebforelle* or *Silberlachs*. It has the head long, low, compressed, pointed, and

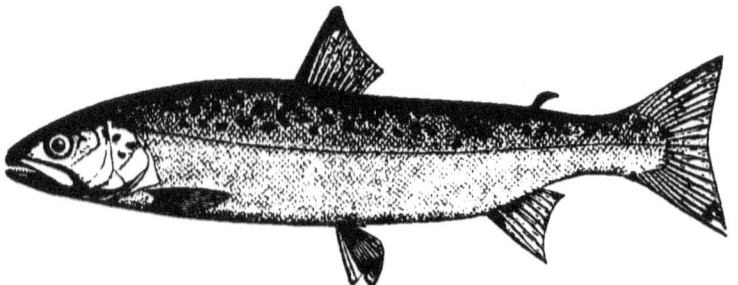

Fig. 156.—SALMO LACUSTRIS (WILLOUGHBY).

rather small, with a slender body. The greatest height of the body is always equal to the length of the head, which is one-fifth of the total length. The eye is one-seventh of the length of the head, and twice its diameter from the snout (Fig. 156).

Fig. 157.—FRONT VIEW OF VOMER OF SALMO LACUSTRIS.

The maxillary bone is much longer than the snout, strong, expanded behind, and in large fishes reaches well beyond the hinder border of the orbit. The teeth are strong; the vomer is comparatively broad, with two somewhat irregular rows of teeth (Fig. 157). The hinder margin of the pre-operculum is nearly parallel to that of the operculum. The lower limb forms a very obtuse angle. The last gill-arch carries seven to eight distant blunt rake-teeth, which are usually shorter than in the Austrian fish known as the May Trout.

The dorsal profile rises in a flattened arch to the dorsal fin. The ventral contour is less convex. The dorsal fin begins half-way down the length, and is rather longer than high. The anal fin is considerably higher than long,

## SALMO LACUSTRIS.

and is truncated behind. The pectoral and ventral fins are pointed, and the adipose fin, which does not reach farther back than the anal, is rather high, and narrower at its base than in the upper part.

The scales are very firm, and are distinguished by a strong silver lustre, which varies in different parts of the body. The colour of the upper part of the head is dark-green, becoming bluish-green on the sides, and bright silver on the belly. Above the lateral line are small black flecks, which may be round, but are usually angular or X-shaped. There are a few round black spots on the operculum and on the dorsal fin. Below the lateral line the spots are few. The pectoral, ventral, and anal fins are bluish-grey. The iris is silvery.

This fish is very voracious. According to Rapp, it enters the rivers Rhine and Ill to spawn. Like other Trout, it is frequently sterile. The British Museum contains sterile specimens, twenty-three inches long. In the backbone there are sixty to sixty-one vertebræ. It attains a weight of twenty-five to thirty pounds.

A variety of this species, known in Austria as the *Maiforelle*, *Mailachs*, or *Maiferche* (Fig. 158), is limited to the larger mountain lakes, such as the Atter

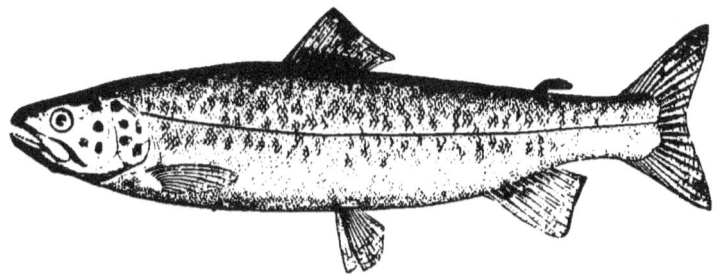

Fig. 158.— SALMO LACUSTRIS VAR. SCHIFFERMÜLLERI (VALENCIENNES).

See, Traun See, and Fuschler See, and may be conveniently distinguished as *Salmo schiffermülleri* (Valenciennes).

In many respects it is similar to the *Salmo lacustris*, the height of the body and length of the head being the same, and both about one-fifth of the length of the fish. Von Siebold, however, regarded the *Schwebforelle* as simply the sterile form of the *Maiforelle*, and while he united both in the *Salmo lacustris*, included other forms under the same name.

Fig. 159. - FRONT VIEW OF VOMER OF SALMO SCHIFFERMÜLLERI.

In the May Trout the diameter of the eye does not exceed one-fifth of the length of the head; it is one and a half times its diameter from the end of the snout. The nares are nearer to the snout than to the eye. The anterior eyelid is broad and large. The maxillary bone has an inclined direction, and extends only slightly behind the eye. The jaws have rather slender short teeth. On the vomer there are three transverse teeth in front, with a simple longitudinal row of four on the body of the bone, and a zigzag row or double row of eight or nine behind these (Fig. 159).

The margin of the pre-operculum expands in the middle of the hinder border. The number of branchiostegal rays of the right side is ten, and on the left side eleven. The rake-teeth are thin, but long and pointed, and the last arch carries nine or ten teeth. The dorsal and ventral profiles are similar.

The dorsal fin begins in front of the middle of the length; it is truncate behind, nearly as high as long; but its height is less than the depth of the body. The anal fin is as high as the dorsal. The adipose fin is opposite the end of the anal. The longest rays of the ventral are equal to those of the dorsal.

The scales are small; smallest in the fore-part of the body and belly, and easily fall off. The stomach is smaller than in the typical forms of *Salmo lacustris*, but the pyloric appendages vary from sixty to seventy-four.

The colour of the upper part of the head and back is black, with steel-blue lustrous scales, the blue becoming paler towards the lateral line, and giving place to silvery-white on the under side of the body. The operculum is also silvery. The sides of the body are marked with irregular deep-black spots, which are generally angular or cross-shaped, and of considerable size. They are rarely round. Red spots are scattered between them; there are similar spots on the operculum, but generally larger and rounder. Below the lateral line the number of spots diminishes, and the belly may have a slight tinge of red. The skin of the dorsal fin is dotted with black, and margined with a broad, dark border. The pectoral fin is yellowish-white at the base, and greyish at the upper side. The anal and ventral fins are whitish.

The usual weight of the fish is from ten to fifteen pounds, though much smaller specimens are commonly caught; and occasionally individuals are captured which weigh from thirty to forty pounds.

During the greater part of the year the May Trout frequents the deep recesses of mountain lakes, only leaving them in May. It then comes to the surface, and pursues small fishes, on which it feeds in the morning and evening. The Trout swim in large circles, and drive their prey into the centre, and then rush at them. When thus engaged they are commonly caught with the net, which is also drawn in a circle. They furnish good sport with the rod, and may be taken in this way during a large part of the year. They die at once when out of water, and live but a short time in ponds, even with a good supply of food. They rarely go out of the lakes into the streams which flow into or drain them.

Heckel and Kner say that their spawning-time is unknown. Though Herr Aigner, of Salzburg, states that they spawn in April and May, yet females caught in May have the eggs scarcely larger than millet seed. Their flesh is white, and more prized in Austria for its flavour than that of the other species of Trout.

## Salmo levenensis (Walker).—Loch Leven Trout.

D. 12—14, A. 10—12, P. 12—14, V. 9, C. 19. Scales: lat. 120—130, trans. $\frac{28}{26}$.

The Loch Leven Trout is exclusively British, being almost limited to Loch Leven, in Fifeshire, and other lochs in the south of Scotland, such as Loch Lomond, and Loch Scone, in Perthshire. It is found in Windermere, and probably occurs in other lakes of the north of England. The largest specimen in the British Museum is twenty-one inches long. Mr. Day has shown that under artificial conditions it may reach a weight of seven pounds in six years. The common size is a little over a foot. This Trout does not migrate.

The upper part of the body is olive-green, becoming paler on the sides, which are marked more or less with X-shaped brown or black spots, as well as some rounded spots. There are round black spots on the sides of the head and operculum. The belly inclines to yellow. The dorsal fin is grey, and, like the adipose fin, has small black or brown spots. The upper half of the caudal may be spotted, and the pectoral is usually blackish at the extremity, and yellow at the base.

This is a well-proportioned fish, more elegant and less deep in body than S. fario. The head, too, is proportionately smaller. The dorsal profile is a little less arched than the abdominal contour.

The fish is rather less than four and a half times as long as high, and five to five and a half times as long as the head. The space between the eyes is convex; the snout is conical, and of moderate length. The mandible is longer than the snout, and occasionally hooked in males. The maxillary bone extends below the hinder margin of the eye. The pre-operculum has generally an indistinct lower limb. The teeth are of moderate strength; those on the body of the vomer usually persist as a single series; but on the head of the vomer, which is broader than long, there is a transverse row of a few teeth. In the specimens reared artificially at Howietown, near Stirling, Mr. Day records a double row of teeth on the body of the vomer. Dr. Parnell mentions eight teeth on the tongue. The fins are well developed, though not large, and all terminate in sharp points. The dorsal and anal are truncated, with concave borders. The caudal is more deeply concave; the pectoral is longer than the ventral, and both are pointed. The pyloric appendages are stated by Dr. Günther at from sixty to eighty, though Mr. Day finds the range of variation to be between forty-seven and ninety. The number of vertebrae does not exceed fifty-nine, and may be one or two fewer.

The scales are thick and have a ridge in the centre. The number of scales in a transverse series, descending from the adipose fin forward to the lateral line, is thirteen or fourteen. Dr. Parnell pointed out that food and season affect the colour and markings of this fish.

In the Parr the number of finger-marks on the sides ranges between eleven and fourteen. The flesh is of a deep red. It feeds largely upon fresh-water mollusca. It spawns in the first three months of the year.

This fish is liable to attacks of fungus, which most frequently affects old males and unspawned females.

Sir J. Gibson Maitland has reared this species successfully in his ponds at Howietown, and demonstrated that hybrids may be produced between it and various Salmonoids. Mr. Day has described some of these experiments. Eggs of a four-year-old Loch Leven Trout were fertilised with the milt of *Salmo salar*. In three years the fishes which resulted from this cross reached a length of eleven inches.

The Loch Leven Trout at Howietown produces eggs of different sizes, according to the size of the parent; and in a seven-year-old fish the bulk of each egg is about one-seventh greater than in a six-year-old, and there can be no doubt that the larger eggs produce the larger fish, since in ten or eleven months the fry of the older parents were three and a half inches long, and those of the younger parents two and a half inches long; but poor food may make the eggs small in an old fish. Mr. Day is inclined to regard Loch Leven Trout as derived from the Brook Trout, but suggests the white variety of Salmon Trout as its immediate ancestor.

## Salmo orcadensis (Günther).—Orkney Trout.

D. 13, A. 11, P. 14, V. 9. Scales: lat. line 115, transverse $\frac{25}{30}$.

There appear to be two races of Trout in the Orkneys, one of which the older writers identified with the Grey Trout (*Salmo eriox*), and the other with the Lake Trout (*Salmo ferox*). One of these, which had previously been regarded apparently as the Grey Trout, Dr. Günther describes from male specimens twelve inches long, taken in Loch Stennis. Its affinity to *S. nigripinnis* is obvious, but it has a broader and stronger maxillary bone, like that of the Brook Trout; larger scales on the tail; and fifty pyloric appendages, instead of about forty, as in that form.

The head is rather short, and the body somewhat slender; the pre-operculum has the lower limb indistinct, as in *S. nigripinnis*. The snout is similarly short and conical. The vomerine teeth are in a single series, and persistent. The pectoral fin is rounded, the caudal truncate. There are about thirteen scales in a transverse series, descending obliquely forward, from behind the adipose fin to the lateral line. The dorsal fin shows some black spots, and the sides have more or less numerous black reticulated spots, with a few red ones interspersed.

It has fifty-six to fifty-seven vertebræ.

## Salmo polyosteus (GÜNTHER).

D. 13, A. 11, P. 14, V. 9. Scales: lat. line 128, transverse 28—30.

This species is founded upon fishes, in indifferent preservation, from Lapland, which reach a length of about seventeen inches.

The teeth on the body of the vomer are in pairs. There are fifteen scales in an oblique series, running from behind the adipose fin to the lateral line. The pre-operculum is somewhat crescent-shaped, with the lower limb indistinct.

With the diminished number of rays in the fins there are ten branchiostegal rays instead of eleven in *S. reneraensis*, and eleven or twelve in *S. mistops*. It is probably a variety of *S. reneraensis*.

## Salmo brachypoma (GÜNTHER).—The Grey Trout.

D. 13, A. 10—11, P. 14, V. 9. Scales: lat. line 118—128, transverse 27/30.

The Grey Trout is a migratory species which frequents the rivers Forth, Trent, and Ouse. It has generally been identified with the *S. erios* of Linnæus; but is distinguished by Günther as having a comparatively small head, though the head increases in length in old males, when it may be a quarter of the length of the body without the caudal fin. The body is rather more slender, and the maxillary bone narrower, than in *S. fario*. The teeth are rather strong in both jaws; but the zigzag double row on the vomer is mostly lost when the fish is little more than half grown. The caudal fin is truncate in individuals only ten inches long, but all the fins are small. The dorsal, ventral, and anal fins have the outer margin black and white in young specimens. The sides of the body show X-shaped or ocellate black spots, with some red spots below the lateral line; and there are round black spots on the dorsal fin. It reaches a length of 30 inches.

## Salmo trutta (LINNÆUS).—Salmon Trout.

D. 13, A. 11, P. 15, V. 9.   Scales: lat. line 120, transverse $\frac{24}{36}$ to $\frac{26}{34}$.

The Salmon Trout, or Sea Trout, is essentially a North-European fish, much more common in Scotland than in England. In Wales and Ireland it is known as White Trout. It is characteristic of Scandinavia, where it is known as *Laxoring*, or *Orlax*. In North Germany it is usually termed *Meerforelle*, but is also called *Lachsforelle*, or *Silberlachs*. It occurs in Russia, chiefly in the White Sea and the Baltic regions, and is common in the rivers and fiords of Jutland.

Its habits are much like those of the Salmon. It commonly lives in the sea, and remains some days in brackish water before ascending rivers. It is voracious, and when fattest the flesh is pale pink; but as the fish gets out of condition the colour becomes nearly white. The flavour, according to the English taste, is inferior to that of the Salmon, but it is quite as much esteemed as Salmon in France.

It is very hardy, and may be kept alive for some time in boxes by fish-dealers. Buckland mentions that it is so abundant in the Tweed as seriously to interfere with the Salmon; and he assures us that the fishwives sometimes clip the tail of the Trout square, and then sell it as Salmon. Jardine states that their abundance is annoying when fishing for Salmon.

They eat small fishes, shrimps, and many kinds of insects. Their stomachs are frequently crammed with the common sandhopper. The body is deeper, and the colour higher, in the Aberdeen rivers and the Tay than in other parts of Scotland. It has been said that if they are prevented from descending to the sea, the head will become large, and the body thin; but McCulloch asserted that the Sea Trout is now a permanent resident in a fresh-water lake in the island of Lismore, off Argyleshire, where it breeds freely. Hence there have not been wanting those who, like Mr. Day, are disposed to regard the migratory Sea Trout as closely related to the non-migratory River Trout, since the most striking difference between them is in the circumstance that the former after a time sheds its vomerine teeth, while the latter retains them through life. In New Zealand the Sea Trout breeds in a fresh-water pond, and, like the Brook Trout, varies the number of its pyloric appendages; for, while in Britain these range in number from forty-nine to sixty-one, this form, when reared in a fresh-water pond, has only thirty-six.

There is no doubt that this species is the White Salmon of Pennant, which he describes from the Esk in Cumberland; and that in the Grilse condition it is known in Scotland as the Phinoc. It is also known in Scotland by various local names, such as, Hireling, Whiting, and Black Tail.

The Salmon Trout grows to a length of three feet, but the female becomes mature when ten or twelve inches long. The head of the fish is rather deep as compared with its length, and slightly exceeds the greatest depth of the body. It is about two-ninths of the total length of the fish, though in old males it may be one-quarter of the length, owing to the increased development of the jaw-bones, and the hook of the lower jaw. The snout is elongated, so that the upper profile of the head is sometimes concave. The maxillary bone is longer than the snout, and does not extend far behind the orbit. The inter-orbital space is convex, and as wide as the snout is long. The hinder opercular margin is obtusely rounded, but with the sub-operculum projecting a little backward. The pre-operculum has a distinct lower limb.

The head of the vomer is triangular, as broad as long, and toothless. The body of the bone has a longitudinal ridge, along which is a single series of teeth, some of the hinder of which diverge outward alternately to each side. Von Siebold and Dr. Günther state that most of these teeth are lost successively at different ages; often only three or four are left in specimens a foot long, while sometimes six or eight teeth may remain in specimens twenty inches long, but in larger specimens still only two or three teeth in front are found.

In all these characters the *Salmo trutta* agrees with the Grey Trout, *Salmo erios*, of the Tweed and the Forth, and we are unable to discover any sufficient specific character to separate the latter, which is the *S. brachypoma* of Günther, from this type. The posterior margin of the caudal fin may be rounded in old specimens, truncate, or slightly emarginate, or deeply forked in the young; but all the fins are short, especially the caudal and pectoral. Both the dorsal and anal fins are as high as long. There are fourteen or fifteen scales in a transverse series, running from behind the adipose fin forward to the lateral line, a character which this species shares with *S. cambricus*, *S. brachypoma*, and other forms. The scales are short, thin, and rounded. The upper parts of the body are blackish, usually with a purplish tinge on the silvery sides, and the under part of the body is silvery. There are generally X-like spots, which may be numerous or few, and there are commonly small round black spots on the sides of the head and dorsal fin. When unspotted it is the *Silberlachs* of the Germans, and *Salmo albus* of some British writers. Each scale is frequently surrounded by minute black dots. In the young state there are nine or ten dusky transverse bars, but as the Parr changes to the Grilse they become lost. When young it often has orange or red spots on the sides, and the Bull Trout Parr of the Tweed is sometimes called "Orange Fins."

The caudal, adipose fin, the inner side of the pectoral, and ventral, are blackish, with the outer side colourless.

There are fifty-nine to sixty vertebræ; and usually from forty-nine to sixty-one pyloric appendages.

Like its allies, this fish is sometimes found sterile, and then remains in the sea, where it is known to the Prussian fishermen as *Strandlachs*. The breeding fishes are believed to remain in fresh water longer than the Salmon, and to spawn from September to November. This Trout makes its way up the small tributaries.

In France it is found only in the Loire, the Seine, and the Meuse. It is recorded by Steindachner from Spain. It goes up the Rhine to the Moselle, and is occasionally taken at Metz; and, like the Salmon, it ascends the Weser, Elbe, and other rivers of North Germany.

We have regarded the Bull Trout, or Sewin of Scotland, which Yarrell identified with the *Salmo eriox* of Linnæus, as only a condition or variety of *Salmo trutta*; but it is at present imperfectly understood, and Dr. Günther records his conviction that it comprises—

(1), Hybrids between the Salmon and Sea Trout;

(2), Salmon which have prematurely returned from the sea after spawning; and

(3), Sterile fish.

Dr. Günther mentions having seen in the month of August, in the Beauly, in Scotland, thirty Salmon to three Bull Trout and one Sea Trout; and adds that the differences between the Bull Trout and Salmon are sometimes so slight as to be scarcely perceptible to an inexperienced eye.

Yarrell regarded the form of the opercular bones as furnishing an infallible distinction, but we have already seen that the opercular bones are liable to unexpected variation. The Welsh fishermen apply the name Sewin to these fish, which are commonly of fifteen to twenty pounds' weight. We may add that if *Salmo eriox* is a hybrid with the Salmon, it is fertile, and is found with the spawn well developed.

Lord Home did not find any difficulty in distinguishing Salmon Trout from Bull Trout, and states that the latter is an inferior fish, which replaces the former in several Scottish rivers.

He mentions that they do not enter the river till the Salmon have commenced spawning, and observes that, though he had killed more Salmon with the rod than any other man, yet he had not killed twenty clean Bull Trout, though Bull Trout Kelts may be killed in thousands.

## Salmo ferox (CUVIER AND VALENCIENNES).

D. 13, A. 10—11, P. 16, V. 9.   Scales: lat. line 125, trans. 26—30.

The Great Lake Trout is entirely British. It does not migrate, and is found in the lochs of the North of Scotland, in Derwentwater, and other lakes of the North of England, in the Lake of Llanberis in Wales, and Lough Melvin in Ireland.

It reaches a length of thirty-one inches, but the female matures eggs when only fourteen inches long. It is a fine Trout, with a deep body, with the tail short, and head rather large in mature males. The pre-operculum is crescent-shaped, so that its posterior and inferior margins do not form an angle. The snout is one-third of the length of the head in the male; but it is elongated with age, and the mandible forms a hook. The maxillary bone is longer than the snout, extends beyond the eye in the adult, though, as usual, the teeth are smaller on the maxillary and palatine bones than on the inter-maxillary and mandible. The head of the vomer is toothless. The body of the bone has a double or zigzag series of teeth, almost becoming a single series behind. They are persistent throughout life.

The fins are rather short, especially the pectoral and ventral. The hinder margin of the caudal is truncate.

The skin is remarkably thick, and completely embeds the scales both on the back and on the abdomen. There are thirteen to sixteen series of scales in a transverse series, descending anteriorly from behind the adipose fin to the lateral line.

The colour of the upper part of the body is a deep purple-brown, sometimes greenish-brown, with the sides lighter, and marked with round deep-black spots, which occasionally have a blood-red margin. During spawning-time the spots have a red centre and black margin. The dorsal fin has small black spots, and the other fins are blackish. Sometimes the spots on the bright silvery sides are reticulated.

There are fifty-eight or fifty-nine vertebræ, and forty-four to forty-nine pyloric appendages.

Mr. C. St. John, in his "Tour in Sutherlandshire," says:—"Loch Awe will always maintain its high repute for its large Lake Trout, which rival the Pike in size and voracity, but are stronger, and far more wary and difficult to catch. A *Salmo ferox* of fifteen pounds' weight is no mean adversary. His first rush, when he finds himself firmly hooked, is nearly strong enough to tow the fishing coble after him. And then comes the tug of war. The monster, held only by a slight line and tapering rod, is one moment deep down, boring his head to the bottom of the lake, with every yard of line run out, and the rod bent into the water; the next he takes a new freak, and goes off near the surface like a steamboat, and before you can wind in, is right under your boat, and close to the bottom of it, your line being you know not where. Again the reel is whirring round so rapidly that you feel your line must break in spite of all your fancied skill. But no! he stops suddenly, and again seems inclined to wind your line round and round the boat; or, by Jove! to upset you if he can, by running against its keel. If there is a projecting nail or notch in the wood he manages to get the line fixed in it. After you have cleared your tackle from this danger, off he darts again. Your Highland boatman swears in Gaelic; you, perhaps, follow his example in English—at least, to a certainty, you blame him for rowing too fast or too slow, and begin to think that you would give a guinea to be honourably rid of the fish without discredit to your skill as an angler. At last your enemy appears exhausted—you have been long exhausted yourself—and floats quietly near the surface. But at the critical moment of placing the gaff in a position to secure him, he flaps his tail, and darts off again as strong as ever, taking good care to go right under the boat again. At last, however, patience and good tackle and skill begin to tell, and, after two or three more feeble efforts to escape, your noble-looking fellow of a Trout is safely lodged in the bottom of the landing-net."

## Salmo bailloni (CUVIER AND VALENCIENNES).

### D. 13, A. 10, P. 12, V. 9, C. 23.

Dr. Günther mentions this fish as found in the Somme. It is believed to be a migratory Trout. He states that it should be compared with *Salmo cambricus*, but the description is insufficient to elucidate its affinities. Moreau says it is distinct from *S. cambricus*, and Blanchard identifies it with the common Salmon Trout. The small number of rays in the anal and pectoral fins appears to define it as a good species.

There are about one hundred and twenty scales in the lateral line. At spawning-time the back is lead-coloured, with violet reflections, and purple spots. The sides are silvery. The caudal fin is grey. The anal and pectoral fins are yellow, and the ventral is white.

## Salmo genivittatus (HECKEL AND KNER).

D. 11. Scales: lat. line 120, trans. $\frac{28}{?}$

Heckel and Kner had but one specimen of this fish, which was eighteen inches long (Fig. 160). It was found in the Sala, a tributary of the Isonzo. Nothing is known of its habits. It is a beautiful species, of elegant form, with a large head, small eye, short dorsal and anal fins, and a long tail. The head, according to Heckel and Kner, is about one-quarter of the length of the body, inclusive of the caudal fin; its height is two-thirds of the length, and

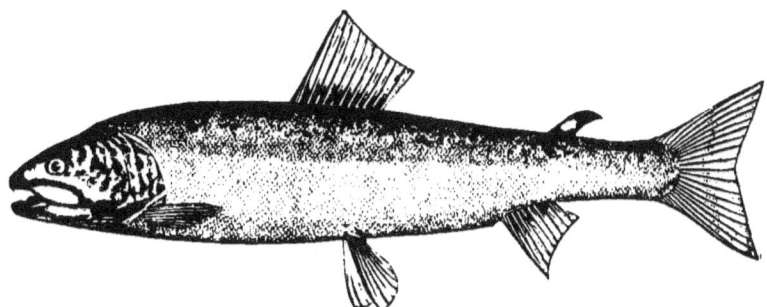

Fig. 160. —SALMO GENIVITTATUS (HECKEL AND KNER).

its breadth nearly half the length. The eye is twice its diameter from the snout, five times its diameter from the hinder opercular margin, and the breadth of the convex frontal region is more than twice the orbital diameter. The lunate fold at the front of the orbit is small. The mouth is large, and the strong, broad, maxillary bone reaches far behind the eye. The dentition is strong; the pre-maxillary teeth are larger than those in the maxillary bone; the vomerine plate carries six teeth in a transverse series in front, and a parallel double row posteriorly.

The pre-operculum is rounded; the lower edge of the operculum is directed backward. It is higher than the sub-operculum, which is wider than the operculum. The number of branchiostegal rays is ten on the right, eleven on the left. The last gill-arch carries about six short, thick, blunt rake-teeth.

The distance from the snout to the occiput is one-third of the distance from the snout to the dorsal fin; and the distance between the adipose fin and the dorsal is equal to the distance between the dorsal and the occiput. There is a slight indentation of the profile behind the eyes, but it is less marked than in

*S. dentex.* The ventral contour is rather less arched than the dorsal profile. The part of the tail behind the anal fin is conspicuously long.

The dorsal fin has eleven jointed rays, the longest of which are equal to those of the anal fin. The adipose fin is large and long. The ventral fins are under the last third of the base of the dorsal; their longest rays are scarcely shorter than those of the anal. The longest rays of the pectoral are equal to the terminal rays of the caudal. The caudal fin is but little notched; it has seventeen jointed rays, two unjointed terminal rays, and five or six truncated rays on each side.

The scales are about the same size as those in *S. obtusirostris.* At the beginning of the dorsal fin there are twenty scales above the lateral line. The perforated scales in the lateral line are the smallest of all. The scales become small on the belly, and are a little larger on the sides of the tail. The lateral line runs nearly horizontally in the middle of the height in the hinder two-thirds of the body. Ten distinct pores may be traced on the branch of the cephalic canal, which runs along the lower jaw.

The colour of this species is peculiar. The whole of the head is covered with large brownish-black patches, which are drawn out longitudinally, and give an irregular banded appearance, comparable to the watering on silk. The free edges of the scales are silvery, but there is a band of black pigment, where they cover each other laterally, giving the fish a delicately crossbarred aspect; but there are no spots on the body, caudal, or other fins.

## Salmo labrax (Pallas).

### D. 10, A. 9.

The flesh of this fish is esteemed as food, and becomes red in cooking. It is common in the Crimea, being often speared at the entrance to the harbour of Sevastopol, and is taken with spears or nets about Kherson. It has very much the aspect of *Salmo fario,* but there are only ten rays in the dorsal fin, and nine in the anal. The pectoral fins are bluish, and the dorsal has irregular black spots.

## CHAPTER IX.

### FRESH-WATER FISHES OF THE ORDER PHYSOSTOMI (*continued*).

FAMILY SALMONIDÆ (*concluded*) : - SUB-GENUS CHARR: Salmo salvelinus — S. umbla — S. alpinus — S. nivalis- S. killinensis—S. perisii - S. colii—S. willughbii -S. rutilus- S. carbonarius—S. grayi- S. bucho—S. lossos- GENUS LUCIOTRUTTA : L. leucichthys--GENUS OSMERUS: Smelt—GENUS COREGONUS: Species with the upper jaw prolonged- Species with the snout obliquely truncated- Species with the snout vertically truncated -Powan Species with the mandible longer than the snout—Pollan—Vendace —GENUS THYMALLUS : Grayling -Thymallus microlepis.

## SUB-GENUS : **Salvelini, or Charr.**

THE Salvelini is a group or sub-genus of the Salmon tribe, popularly known as Charr, and widely distributed in Europe, Northern Asia, and North America. The species of Charr are united together, because their vomerine teeth are limited to the head of the vomer. This character, insignificant enough in itself, is useful in facilitating the study of a large tribe of variable fishes which require classification. The members of this group, like Trout, vary with almost every sheet of water, so that local races and varieties have been developed, which many writers term species. These mutations of form do not confuse our conception of a species; because they demonstrate a gradual modification of characters, with varied geographical distribution, which enables us to recognise the steps by which circumstances of existence have changed structures which, if examined in one locality only, seem to be stable.

We would greatly reduce the number of species of Charr, although they are founded on characters which in other groups might characterise fishes well; but the variability of specific character in almost every one of the species adopted in the British Museum Catalogue favours their union into larger species, such as were in favour with Louis Agassiz, Yarrell, and the earlier naturalists. Thus we find the number of fin-rays presenting but little difference in a large number of so-called species of Charr. The number of vertebræ is almost as constant. The colouring is on the same general plan in them all. And even the pyloric appendages in the majority of these fishes vary in number between small limits.

The number of scales in the lateral line, however, is one of the most variable characters in the geographical races, and we are disposed to associate *Salmo salvelinus*, *S. umbla*, *S. alpinus*, *S. nivalis*, *S. killinensis*, *S. perisii*, *S. willughbii*, and *S. colii* as types of *Salmo salvelinus*, in which the variation is cumulative in the direction of increased size of the scales, as expressed by their diminished numbers in the lateral line, in the several forms enumerated, in which the numbers are 220, 200, 195 to 200, 190, 180, 170, 165, and 160. Some of the species found in other localities fill in gaps in this gradation. *Salmo hucho*, however, may be considered a distinct species, since it differs from the ordinary Charr in having no median teeth on the hyoid bone, a character in which it resembles the *Salmo fontinalis* of the northern parts of North America, a fish which has recently been successfully introduced into British lakes and rivers.

But while we advocate the institution of larger species for closely-allied fishes, it seems indispensable that the constituent varieties of those species should be kept distinct, and treated as though they were species in process of development, still showing their connection with the parent stock.

## Salmo salvelinus (LINNÆUS).

D. 12—13, A. 13, P. 13, V. 9, C. 6 17/6—7. Sq. lat. 220.

The Alpine lakes of Austria and Bavaria yield a fish which is known in those countries as the *Salbling*. It has an elongated form of body, though scarcely so elongated as the *Huch*; but the proportions vary considerably

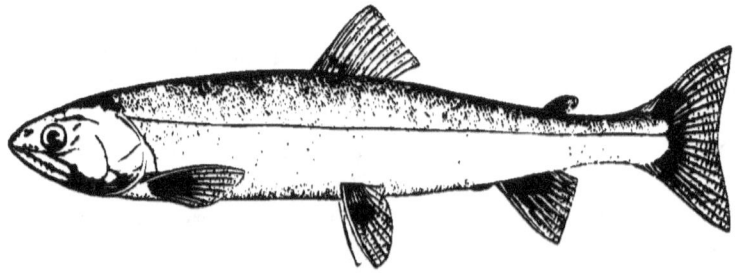

Fig. 161.—SALMO SALVELINUS (LINNÆUS).

with sex, age, and localities, so that the species exhibits a multitude of varieties. The young have a very short snout. The length of the head is sometimes one-sixth of the length, and sometimes the fish is only four and a half times as long as the head; hence the height of the body may equal the length of the head, or be greater or less. Usually, the height of the head is two-thirds of its length, and its thickness half the head-length. The greatest height of the body and its thickness are always nearly equal to each other. The head is five and a half to six times as long as the eye. The eye is less than twice its diameter from the snout, and more than twice its diameter from the other eye, so that the frontal interspace is broad. The gape of the mouth is wide; the maxillary bone usually reaches back for fully half the length of the head. Both jaws are equal, or, if there is any inequality, the lower is the shorter (Fig. 161).

The teeth are in a single row in the pre-maxillary, maxillary, palatine bones, and lower jaw. Those in the maxillary bone are comparatively small. The tongue is not toothed at the edges, but a double row of teeth, parted by a deep furrow, extends along the middle. The hyoid bone has a simple longitudinal row of teeth. The dentition of the vomerine plate is very variable; generally there are six or seven teeth arranged in a nearly equal-sided triangle, with its apex directed backwards, or there may be a simple transverse row, or two longitudinal rows, with nine teeth in each.

The operculum is rounded. There are usually ten branchiostegal rays on the right side, and eleven on the left, though there are occasionally ten on both sides. The rake-teeth of the gill-arches are thin, long, and finely denticulated.

The dorsal profile is much more convex than the ventral contour. The dorsal fin commences rather in front of the middle of the length. It is about two-thirds of the height of the body. It is truncated, but the truncation scarcely reduces the height of the last rays one-half. The adipose fin is opposite to the end of the anal fin. The latter is as high as the dorsal, but more truncate behind. The ventrals are opposite the hinder part of the dorsal. They are well developed, their longest rays being equal to those of the anal. The terminal rays of the forked caudal are a little longer than the pectoral fin, and are from one-seventh to one-sixth of the length of the fish.

The scales are very minute, delicate, and firmly attached to the skin. They are smallest and most elongated on the throat and belly, and largest on the sides, where they are more circular.

The lateral line is straight, and nearer to the back than the abdomen. It comprises from one hundred and twenty-four to one hundred and thirty simple pores, although there are generally over two hundred and twenty scales in the lateral line, but only every second scale is perforated. Among the cephalic canals the lower jaw has the pores most distinct.

The number of pyloric appendages is from thirty to forty. All are moderately long, and none exceed the length of those forming the first circle.

The colour is almost as variable as that of the Common Trout. In the living fish at spawning-time the upper part of the head and back are brownish-green, the sides of the body lighter, and the lower jaw is yellowish-white. The under side passes through all shades of orange to vermilion, from the throat to behind the ventral fins, the colour being most brilliant in front of the ventrals, getting paler behind, and passing at last into yellowish-white. Blackish points, spots, and stripes, are seen on the branchiostegal rays, and the sides are ornamented with rounder and smaller spots, which present a variety of colours, from deep red to white; the majority of these spots are near the lateral line, or below it. The dorsal fin is greenish-yellow or brownish, with cloudy spots. The pectorals and ventrals are vermilion-red. The first ray of the ventral and of the anal fin is pure white, while the anal is pale red, except in the middle, where it is dark grey or black, like the pectoral. The caudal fin is greenish-brown, becoming faintly red in the lower lobe. The iris is reddish-yellow, but more or less pale. Out of spawning-time the colours are less intense, especially on the abdomen, and differ with the locality and water. They are paler, too, in the lower-lying lakes, even though they are deep; but in the lakes among the high mountains the colours are more intense, as in the Altau See, Grundel See, and Alm See. The colours are most beautiful and brilliant in the Gosau See and the Tappenkar See, near St. Johann, in Pongau. Among them are found some individuals which are quite pale, such as Fitzinger's White Salbling, var. *carnea*, with pale-violet back, rose-red sides, yellowish-white belly, and pale-red fins.

The size of the Salbling varies. In some lakes, such as the Mond See and Langbath See, the fishes are rarely heavier than half a pound, or longer than eight or nine inches; and the size is no larger in the variety *carnea* from the Traun See, and in the fish known as Schwarzreuterl, from the Langbath See, Upper König See, or Bartholomäus See. The usual size in most lakes is a foot long, with a weight of half a pound to a pound. Fishes two feet long, weighing five to six pounds, are much rarer, but the species is occasionally taken weighing eighteen to twenty pounds in the Fuschler See and the Hinter See, near Berchtesgaden. The size is diminishing from over-fishing.

The variations with sex are remarkable. The head is considerably longer in the male than in the female, not only in the adult, but in the young. In the male the total length of the fish is four and two-thirds as much as the length of the head, but in the female it is five and a half times as long as the head. The male has a wider mouth, longer maxillary bone, and longer rays to all the fins. The colour is similar in both sexes, but the shallower the water the paler the colour. The varieties in the different lakes are constant.

The Salbling never leaves the clear mountain waters, and usually frequents considerable depths. In the Austrian Alps it is found as high as six thousand feet above the sea, as in the Grün See.

These fishes feed on univalve fresh-water shells, and on various small fishes. The usual time for spawning is December, though some spawn as early as November, and others not till January. They increase rapidly, but their growth is slower than that of Trout, which often occur in the same lakes. They do not associate with the Trout. They are hardy, and can be removed from one lake to another, often with advantage. Thus Salbling taken from the Grundel See, on the Elm, were put into a lake four thousand feet up the mountain, when they rapidly gained from three to five pounds' weight, and became larger than the native fishes of the same species.

The flesh of the Salbling is held in extraordinary esteem. It is sometimes red, sometimes white, the colour of the flesh varying with the season of the year, the lake, or even with the water in which fishes have been temporarily placed. The excellence of the flesh varies in a distance of two or three miles. On this account the species has not been thought suitable for artificial cultivation. Like so many other fishes, it is captured easily at spawning-time, when its usual vigilance is relaxed. It is taken in Upper Austria with a net, known as the *Segen*, which has a sack-shaped form, and is furnished with two lateral expansions, or wings, so that it can be drawn through the water to shore by two boats, each manned by two men. In this way a plentiful supply is obtained. The smaller fishes, weighing less than half a pound, are thrown back into the lake. Von Siebold says sterile individuals are common in the Bavarian lakes, where they weigh about three pounds or three pounds and a half.

This species has sixty-four vertebrae, and thirty-six pyloric appendages to the intestine.

## Salmo umbla (Linnæus).

D. 12, A. 12—13, P. 14, V. 9, C. 19. Scales: lat. line 200.

A well-marked variety of *Salmo salvelinus* is found in the Swiss lakes Constance, Neuchâtel, and Geneva, which is known to the Germans as *Röthel*, or *Rothforelle*, and to the French as *Ombre Chevalier*, sometimes written *Omble-chevalier*. The body is higher, the scales are larger, the teeth in the pre-maxillary and maxillary bones are stronger than in *S. salvelinus*, and the belly is never red. The entire fish is about four and a half times as long as the head, and about five times as long as the greatest height of the body in front of the dorsal fin (Fig. 162).

The snout is always more pointed than in the Salbling, and the profile of the back is similar to the ventral profile for the first half of the length. Exclusive of the furrow for the eyelid, the eye is more than its own diameter from the point of the snout, and three times its diameter from the margin of the operculum. The frontal interspace is narrower than in the Salbling, being in width only one and a half times the diameter of the eye. The jaws are equal in length; the mouth is wide, while the maxillary bone extends behind the eye. All the tooth-bearing elements of the jaws have large teeth of uniform size, which are stronger than in the Salbling. Those on the palatine bone, which form a single row, and those on the tongue in a double row, are as strong as the others. The hyoid bone has a row of feeble teeth; but the smallest teeth are on the vomer. They usually number seven or eight, and may be grouped as variously as in *S. salvelinus*, though often placed in a simple transverse row. The pre-operculum and operculum are rounded and their margins are nearly parallel. There are eleven branchiostegal rays on the right side, and twelve on the left.

The dorsal fin begins in front of the middle of the body, and is nearly as high as long. The anal fin is lower and shorter. The small adipose fin is

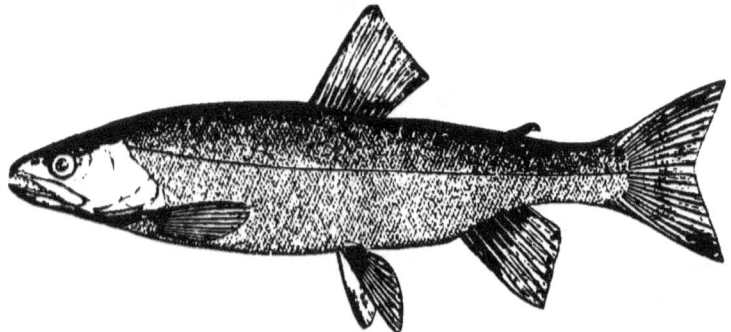

Fig. 162.—SALMO UMBLA (LINNÆUS).

opposite the termination of the anal. The ventral fins are opposite the middle of the dorsal, and have longer rays than the anal. The terminal rays of the forked caudal fin are a little longer than the pointed pectoral fin.

The scales furnish marked evidence of affinity with *S. salvelinus*, for along the lateral line there are similarly one hundred and twenty to one hundred and twenty-two scales, perforated with mucus pores; but in the neighbouring rows there are twenty to twenty-five fewer scales in a row than in that species. The cephalic canals are distinct, and the sub-orbital branch opens with six or seven large pores.

The number of pyloric appendages is similarly about thirty-six, and there are sixty-five vertebræ instead of sixty-four.

The colour at the back is bluish-grey, sometimes without, sometimes with, small round yellowish spots. The lower parts of the body are whitish and silvery, but sometimes slightly tinged with yellow or red. In old males the operculum and sides of the belly have a smudged appearance, as though marked with a stick of charcoal. The dorsal fin is marked with isolated black spots; like the anal and caudal, its colour is blue, though lighter than the back. The iris is silvery.

This fish usually weighs less than a pound; but two or three pounds' weight is not rare. It is not so frequently taken as other Salmon in Lake Constance, because it frequents deep water. It is commoner in the Lake of Geneva, where it also affects great depths, except at spawning-time. It there grows to a larger size, twelve pounds' weight being not unusual, while specimens of twenty-five to thirty pounds are said to have been taken.

The flesh is reddish, and esteemed. The fish spawns in January and February, but never leaves its native lake. According to Blanchard, it spawns in October and November. Its eggs are semi-transparent, and yellow.

The elder Agassiz united the species of *Salmo*, which Linnæus had named *umbla*, *salvelinus*, *alpinus*, and *salmarinus*, as different conditions of the same species, a view which Professor Nilsson also adopted after examination of the original specimens. And if we have rather followed authors like Heckel and Kner, and Günther, who keep them distinct, it is not because they possess any claims to specific distinction, but because all species consist of varieties, and may give rise to varieties which are limited in geographical range. *Salmo umbla* is, therefore, to be regarded only as a western form of *Salmo salvelinus*, and both Charr only differ as varieties from forms met with in Britain and Scandinavia.

## Salmo alpinus (LINNÆUS).

D. 13, A. 12, P. 13, V. 10.   Scales: lat. line 195 to 200.

This Charr may be regarded as a northern variety of *Salmo salvelinus*. It is characteristic of Scandinavia, ranging up to Lapland, occurs in Iceland, in the Orkneys, and in, at least, the northern counties of Scotland. The number of vertebræ has been observed to vary between fifty-nine in the Orkney fishes, and sixty-two in those of Scandinavia. The length of the body and head, the form of the operculum, and other characters, are variable.

The back is dark green and the under-part of the abdomen is light red. There are small round spots on the sides, which vary in tint between red and green. In Scandinavia it is limited to the higher lakes, and is said to be found in those which are frozen through the greater part of the year.

It spawns in September on stony ground, when the abdomen becomes a dark red. The eggs are as large as those of the Trout. It is taken with nets; is usually between nine and seventeen inches long, but is said to grow to a weight of fifteen pounds. The flesh is red, and accounted delicious.

Dr. Günther describes the body as being slightly compressed and elongated, five or six times as long as high. In mature specimens the head is about as long as the body is high, but in the young the relative length of the head is greater. The diameter of the eye is one-sixth of the length of the head. The snout is compressed and conical, half as long again as the orbital diameter. The jaws are equal in front. The maxillary bone extends but little beyond the hinder margin of the orbit. It contains twenty to twenty-two teeth of moderate size. There are six teeth in each pre-maxillary. There are nineteen teeth on each palatine bone, six pairs on the tongue, three pairs on the vomer, in two longitudinal series, and fifteen on each mandible. The width of the inter-orbital space does not exceed twice the orbital diameter. The operculum, rounded obtusely behind, is sometimes as long as high, sometimes with the length only two-thirds of the height.

The origin of the dorsal fin is nearer the snout than the base of the caudal; it is truncated, and its base is a third longer than its last ray. The adipose fin is distant fully twice the length of its base from the dorsal. The anal fin is midway between the root of the caudal and the ventral; it has a shorter base than the dorsal. The caudal fin is forked, with pointed lobes. The ventral is below the middle of the dorsal. The pectoral is longer than the rays of the dorsal. It covers less than half the distance between its root, and that of the ventral.

The number of coecal appendages is thirty-six to forty-two.

## Salmo nivalis (GÜNTHER).

D. 14, A. 13. Scales: lat. line 190.

Farther north, in the rivers and lakes of Iceland is a Charr, closely allied to *Salmo alpinus*, so that it might be regarded as a representative of that form in Iceland. It is known to reach a length of twenty-one inches. Dr. Günther describes the body as slightly compressed and elongated. The greatest depth is equal to the length of the head, and is one-fifth of the length of the fish. The maxillary bone extends beyond the orbit in the adult. The teeth are rather small. The breadth of the frontal interspace is less than twice the orbital diameter.

The pectoral fin measures more than half the distance between its base and the root of the ventral, and this elongation of the pectoral fin is one of its best distinctive characters. The longest ray of the dorsal is equal to that of the pectoral. The lower parts of the body are of a deep orange colour, and the lower fins have the anterior margin orange-coloured or white. This species has sixty-two vertebrae, and forty-one pyloric appendages.

## Salmo killinensis (Günther).

D. 14—15, A. 13, P. 13, V. 9. Scales: lat. line 180.

This fish is described by Dr. Günther from Loch Killin, in Inverness-shire, from adult examples ten to fifteen inches in length. The head and body are thick, and but slightly compressed. The greatest depth of the body is equal to the length of the head, and measures two-ninths of the total length. The maxillary bone scarcely extends beyond the hind margin of the orbit. The lower jaw is rather shorter than the upper; the snout is blunt. The diameter of the eye is less than the length of the snout, and half the width of the interorbital space. The teeth are very small. The sub-operculum is very short and high.

This species is remarkable for the excessive development of the fins, the pectoral being scarcely shorter than the head, and the dorsal, which is high and long, is a little shorter than the pectoral. The ventral fin extends nearly to the vent. The caudal is broad.

The head and upper parts of the body and fins are brownish-black; the lower parts are orange-coloured, and there are very small light spots on the sides, which are not very conspicuous. The anterior margins of the abdominal fins are white or orange-coloured, as in the Icelandic Charr, *Salmo nivalis*, of which this fish may be regarded as a Scottish representative.

The number of vertebræ is the same (sixty-two), and the pyloric appendages vary between forty-four and fifty-two.

## Salmo perisii (Günther).

D. 13—14, A. 12—11, P. 12—13, V. 9. Scales: lat. line 170.

The lakes of North Wales, especially the Lake of Llanberis, yield a Charr, known as the Torgoch, which we may regard as the Welsh variety of *Salmo salvelinus*. Adult specimens have a length of nine inches. It has been carefully described by Dr. Günther, whose description brings out its close resemblance to the Charr of Windermere. It has the same compressed elongated body. The eye is similarly one-fifth of the length of the head, but the mandible is slightly curved upward so as to reach the upper jaw.

The anterior nostril is round, open, and surrounded by a membrane, which becomes a small flap posteriorly, that covers the smaller oblong posterior nostril. This character is one of the most distinctive peculiarities of the Welsh Charr.

The maxillary bone has the same posterior extension, with nineteen to twenty-one medium-sized teeth. There are six or seven teeth in each pre-maxillary, seventeen in each mandible, seven on the vomer, in a **V**-shape, fifteen on each palatine bone, and five pairs on the tongue. Only three branchiostegal rays are exposed in a side view. The lower branch of the outer branchial arch has thirteen straight lanceolate gill-rakers.

The fins present but little variation from the other types, though the base of the pectoral is overlapped by the opercular apparatus.

The scales are similar to those of *S. willughbii*, and the perforated series of the lateral line similarly does not correspond to the transverse series. The colour is essentially the same; but the sides show numerous orange-red spots. The fins have the same white anterior margin, red colour on the abdomen, and black on the back and tail, though both dorsal and caudal fins have lighter margins.

### Salmo colii (Günther).

D. 14, A. 12, P. 13, V. 9. Scales: lat. line 160.

This is a small fish, seven to eight inches long, described by Dr. Günther, from the Irish loughs, Dan and Esk. It has the same form as other Charr, in which the depth of the body is one-fifth of the total length of the fish. The eye is similarly one-fifth of the length of the head. The frontal interspace is half as wide again as the eye, very slightly convex, with a faint median ridge. The maxillary bone has thirteen to seventeen small teeth, and all the other teeth are small. There are four to six in the pre-maxillary, fifteen each in the mandible and palatine bones, three on the vomer, and four pairs on the tongue.

The base of the dorsal fin is longer than its last ray. The base of the pectoral fin is not overlapped by the operculum; the fin is shorter than the head. The back is bluish-black, with silvery sides, dotted over with light salmon-coloured markings; the belly is reddish. The fins are black, though there is a reddish tinge in the pectoral and ventral, and the ventral and anal have a white margin. The number of vertebræ is sixty-three; and the forty-two pyloric appendages are extremely short. It is perhaps most nearly related to the Windermere Charr.

### Salmo willughbii (Günther).

D. 12—13, A. 12, P. 13—14, V. 19. Scales: lat. line 165.

The Charr of Windermere also occurs in Cumberland and in Loch Bruiach, in Scotland. The Lancashire fishes are eleven inches long, with fifty-nine vertebræ, and thirty-two to thirty-five pyloric appendages; but in Scotland some variation is perceptible, for the vertebræ are sixty-one to sixty-two, the pyloric appendages thirty-nine to forty-four; while the fishes are seven to eight inches long. The number of scales in the lateral line is fewer, but otherwise the fish is so closely allied to *S. killinensis* and *S. nivalis*, that it may be regarded as another link in the chain of variation from *Salmo salvelinus*.

The head and body, according to Dr. Günther, are compressed. The fish is four times as long as high. The orbital diameter is one-fifth of the length of the head; the length of the sub-conical snout exceeds the diameter of the eye, and the breadth of the convex frontal interspace is less than twice the diameter of the eye. This area has a prominent ridge on the middle, with two series of pores. The nostrils are just in front of the eye. The maxillary bone carries twenty to twenty-one teeth of medium size, and extends as far back as the posterior orbital border.

There are four teeth in the pre-maxillary bone, seventeen in each mandible, two pairs on the vomer, fifteen on each palatine bone, and four pairs on the tongue. Nearly all the branchiostegal rays are exposed in a side view of the fish. The lower branch of the outermost branchial arch carries eleven lanceolate slightly-curved gill-rakers.

The dorsal fin commences midway between the snout and the base of the caudal. The length of its base is equal to the height of the last ray. The ventral fin is below the hinder part of the dorsal, and is two-thirds as long as the pectoral. The scales are ovate, minute on the back, and hidden in the skin. They are largest in the lateral line. As in *S. salvelinus*, the back is a dark green, and the belly deep red. The sides of the head are silvery, finely dotted with black below. The lower fins are red, with white anterior margins, and blackish in the middle.

## Salmo rutilus (Nilsson).

A small Charr, about twelve inches long, is found in the lakes of south-western Norway, which has been thus named. We only know that the body is five and a third times as long as the head, and that it has a pointed snout, with the lower jaw sometimes prominent.

The colour is brownish above, yellowish at the sides, orange, or red below, while the sides have small light spots. The red ventral and anal fins have white anterior margins. These characters are insufficient to distinguish the fish from the British Charr, and we can only suspect that this is another slight variation from the type so well represented in Great Britain.

## Salmo carbonarius (STRÖNN).

Mr. Lloyd tells us of a Charr, found only in Lake Sigdal, in Norway, which the people distinguish as *Gautesfisk*, characterised by its small eyes, blunt white snout, and inflated belly, which led Nilsson to distinguish it as *S. ventricosus*. It reaches a length of a foot, and is taken with the hook in winter. Dr. Günther regards it as identical with the fish found in the wooded lakes of Western Norway, which the people name the *Kellmund*.

The latter has a soft, white, insipid flesh, and is taken in the summer with a bait of live frogs; but it lives in deep water, and does not come to the surface, even to spawn. The only difference from *S. ventricosus* is apparently the absence of the abdominal inflation. The head and upper parts are black, the sides grey, with small round white spots; the under parts are white, and the fins black, though the ventral and anal fins have white margins. Nothing is known of the scales, internal anatomy, or other characters.

## Salmo grayi (GÜNTHER).

D. 13—14, A. 12, P. 13—14, V. 9. Scales: lat. line 125.

Lough Melvin, in the north-west of Ireland, between Ulster and Connaught, yields a peculiar Charr, with sixty vertebræ, and thirty-seven pyloric appendages, first recognised and described by Dr. Günther. He finds it to be eleven inches long, with the head and body compressed, and the depth at the origin of the dorsal fin one-fourth of the total length.

As in the English Charr, the snout is slightly compressed, sub-conical, and longer than the diameter of the eye, which is one-fifth of the length of the head. The frontal space between the eyes, as in the Charr of Windermere, is convex, with a median ridge, two series of pores, and similar width. The nostrils are a little more distant from the eye. The maxillary bone has sixteen small teeth, with those at the back rudimentary. All the other teeth are small, and slightly fewer than in the English Charr, being four in the premaxillary, twelve in the mandible, two to four on the vomer, fifteen on the palatine bone, and four pairs on the tongue. The two or three outer branchiostegal rays are exposed laterally, as in the Welsh Charr. The lower branch of the outer branchial arch has only nine straight lanceolate gill-rakers, instead of thirteen, as in that fish.

The dorsal fin commences a little farther forward, and its base is a little longer. The base of the pectoral fin is not overlapped by the opercular apparatus, so that in this character it resembles the English rather than the Welsh fish. The pectoral fin is as long, or longer than the head, and the ventral is three-quarters of this length.

The scales are larger than in other British forms. Between the dorsal fin and lateral line they are nearly square, with the hind margin rounded, and those of the lateral line are not larger than the others. Even the small scales on the back are distinctly formed. Those on the sides have a silvery centre, and a blackish margin. The head and back are bluish-black; the sides and belly are silvery, inclining below to a reddish tinge. There are light orange dots on the sides. The fins are blackish, with a white margin to the ventral.

## Salmo hucho (LINNÆUS).

D. 13, A. 12, P. 17, V. 9—10, C. 19.   Scales: lat. line 180.

The River Charr of the Danube occurs nowhere else. It is sometimes known as *Rothfisch*, but more frequently as *Huch*. The Hungarians name it *Galócza* (Fig. 163).

Its form is elongated, slender, and somewhat cylindrical. The round body and elongated form readily separate it from other Salmon. The height of the

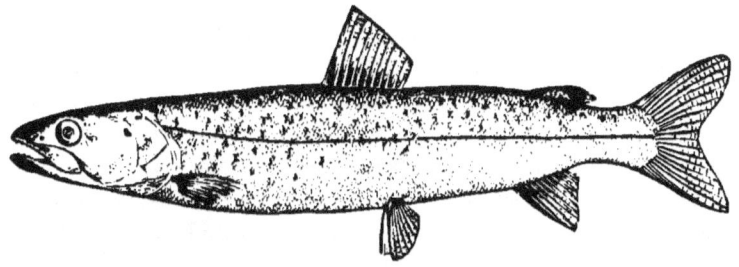

Fig. 163.—SALMO HUCHO (LINNÆUS).

body is only half to two-thirds of the length of the head; and in the young fish the entire length is four and a half times the length of the head. The head is nearly as broad as high; the eye is close to the frontal profile, and even in young specimens its diameter is only one-seventh of the length of the head; It is twice the orbital diameter from the snout, and the same distance from the other eye; though in young fish the breadth of the inter-orbital space is less. The half-moon-shaped eyelid is less developed than in the May Trout and the Salmon. The small nares are rather near to the eye. The gape of the mouth is wide; the maxillary bone extends behind the eye, and the blunt-pointed snout extends slightly beyond the closed jaws. The teeth are in single rows; they curve backward, and are developed on all the elements of the jaws. They are smallest on the maxillary bone, but numerous. The vomerine bones are strong, and have a simple curved transverse row of four to six teeth; flanked on each side with a stronger row of teeth on each of the palatine bones. The tongue is bordered on each side with a row of six to eight curved teeth. The thick mucus skin of the mouth forms projecting folds or corrugations between the rows of teeth; and in some of these depressions the new teeth can be seen developing. There is no median series of teeth along the hyoid bone.

The edge of the pre-operculum is far removed from the eye. The operculum is triangular, with its extremity covered by a broad sub-operculum, which reaches farther back. The gill-aperture is very wide. The number of branchiostegal rays is ten on the right, and eleven on the left. Four branchial arches are strongly developed, with very long, slender, finely-denticulated rake-teeth on the concave side. The pseudobranchiæ are large and comb-shaped.

The frontal profile rises from the snout in a flat arch, and then runs nearly horizontally along the back, but the ventral profile forms a regular flat arch to the vent. The dorsal fin may commence rather behind the middle of the body. It is nearly as long as high, and truncated. The adipose fin is opposite the hinder part of the anal, and is rather long and high. The anal fin is as high as the dorsal, but more truncated behind. The ventral fins are shortest, and opposite the termination of the dorsal. The longest rays of the pectoral, which is pointed, exceed those of the dorsal and anal. The caudal fin is forked.

The scales are much smaller than in either the May Trout or the Salmon Trout. They are elongated, and everywhere of the same size. On the lateral line Heckel and Kner counted two hundred, Günther one hundred and eighty. The lateral line is nearer to the dorsal than the ventral profile, and slightly curved in its anterior half. The number of pyloric appendages, which is two hundred, distinguishes this fish from all other Salmon. They are short, nearly equal in length, and are arranged on a large portion of the intestine, so as to overlap, like the scales of a fir cone, except in the median line. There are two convolutions to the intestine.

The colour of the upper part of the head and back is a dark-brownish green, passing into violet, which shades into reddish-grey, while the throat and belly have a clear silvery lustre. The head and body are covered with very small dark-grey or black spots, between which larger round spots of black are seen on the skull, temporal region, and operculum, and on the back towards the lateral line. Higher up, and farther back, they gradually assume a half-moon shape. The dorsal fin has a few black spots at its base, and a dark border above, and a similar border characterises the caudal fin. The other fins are pale yellow or pale red and unspotted. In old age the black spots on the body vanish. The iris is grey-green, with a clear yellow ring round the pupil.

This is the largest and heaviest fish of the Salmon tribe found in the Danube, if not the largest Salmon known. It attains a length of four to six feet, and a weight of forty to sixty pounds, though fishes of one hundred pounds are sometimes taken. It is met with throughout the Austrian empire, chiefly in the various tributaries of the Danube, which flow from the Alps, more especially the Inn, Salzach, Ager, Traun, Ens, Steyr, and Traisen, but is always found in the Save and Drave. It occurs in the Schiul in Transylvania, and in the Alt and Maros, and in the Danube from Bavaria downwards. It is rare in the Ammer See and Chiem See. It ascends the tributary streams to spawn; and only large waterfalls, like the Traun Fall, or high weirs, form obstructions over which it is unable to leap. Spawn matures when the fish weighs four pounds. It spawns in April and May, and is reputed to excavate furrows in the ground with its tail, like other large fishes of the Salmon tribe; and Heckel and Kner state that while thus occupied its attention is so absorbed as to be undisturbed by a boat passing over it.

Its growth is rapid, but the fish is not tenacious of life. Its flesh is whitish, and well-flavoured, but not so much esteemed as the May Trout or Austrian Salmon Trout. It is captured with nets, but is also taken with the rod. It is frequently speared as it lies still at the bottom, with the head up stream, or shot with a ball.

It flourishes in ponds which have a good current flowing through them, and would be well suited for pond cultivation were it not so voracious. When put in ponds it should not exceed a pound in weight, and if then fed with a plentiful diet of Gudgeon, Rudd, Carp, Roach, and Dace, it increases two pounds a year. It is liable to a fungoid skin disease, which first appears on the margins of the fins, and spreads up to the head, and down to the tail.

## Salmo lossos (GÜNTHER).

D. 11, A. 10—11, P. 12, V. 9.

Dr. Günther founds this species upon a fish known in Russia as *Lossos*, which is described by Pallas as *Salmo hucho*. It is common in the rivers flowing into the Baltic, and in the fish-ponds of St. Petersburg. It is found in the River Kama, but is rare in the Volga, and it occurs in the Caspian Sea.

It is remarkable for the small number of its pyloric appendages, which may be as few as twenty-four. It reaches a length of two feet six inches.

It has a conical head, blunt snout, thick compressed body, with convex sides. The teeth are pointed, recurved, and longer in the lower jaw than in the upper. The teeth on the palatine bones are smaller than those on the tongue. The scales are small, inclined to be silvery, with scattered spots. The belly is white and unspotted. The head, which is blotched with brown, is about two-ninths the length of the body.

### Genus: **Luciotrutta** (Günther).

This genus includes two species of migratory Trout, one of which lives in the river Mackenzie, and the other in the Caspian Sea, the Volga, and rivers of Russia.

The type is characterised by the mandible projecting far beyond the upper jaw. The mouth has a wide cleft, and the long maxillary bone is prolonged far backward. The teeth are remarkable for their minute size; they are villiform, and occur in bands on the vomer, palatine bones, and tongue. The maxillary is toothless. The gill-rakers are rough on the inner side. There are many pyloric appendages.

### Luciotrutta leucichthys (Güldenstädt).

D. 15, A. 14, V. 11.

This fish is known only from the descriptions of Güldenstädt, Pallas, and Lepechin. It goes up the Volga periodically, and appears to similarly frequent the great rivers of Siberia, ascending them from the Arctic Ocean. It is usually about three feet long, and resembles the Salmon in form and size. Both the maxillary and mandible are toothless. The cleft of the mouth is wide, the tongue triangular, free, and somewhat rough. The dentition on the bones of the palate is so fine that it is more easily felt than seen. The eyes are large and lateral. The nares are midway between the eye and the blunt snout. It is artificially cultivated in Russia.

### Genus: **Osmerus** (Artedi).

The Smelts are closely related to Salmon on the one hand, and to Coregonus on the other. The genus differs from both types in its internal anatomy, the stomach having a moderately long blind sac and very few short pyloric appendages. The eggs are small, like those of Coregonus, but the distinctive characters are in the dentition.

The pre-maxillary and maxillary teeth are small, especially the outer row. There is a transverse series of large teeth on the small vomerine bones, and conical teeth on the pterygoid and palatine bones; while the tongue has strong fang-like teeth in front, and small teeth behind. The opening of the mouth is wide, and the maxillary bone so long that it reaches almost to the hinder margin of the orbit. The pseudobranchiæ are rudimentary. The scales of the body have no lustre, are transversely oval, and of medium size. Three species of Smelt have been recognised, one limited to the coasts and fresh waters of Northern and Central Europe, another, scarcely distinguished by smaller scales, represents it on the Atlantic side of the United States, while a third, common in the Bay of San Francisco, is limited to the Pacific coast of North America. The allied *Thaleichthys* is an oily fish, that is capable of being burned as a candle, and is known as the Candle Fish along the Pacific coast.

## Osmerus eperlanus (LINNÆUS).—The Smelt.

D. 11, A. 13—16, P. 8—11, V. Scales: lat. line 60—62, transverse $\frac{7}{11}$.

The name Smelt is said to be derived from the odour which this fish exhales when freshly taken from the water, an odour which is not always the same, but is usually suggestive of cucumber, though Pennant also compares it to violets. In Germany the perfume would appear to be stronger, since Benecke says the Smelt has a frightful smell of putrid cucumbers. The odour would seem to arise from the mucus secretion which lubricates the fish, since it is readily imparted to the hand, and soon disappears as the skin of the fish is dried. The French name it *Éperlan*. The Dutch name *Spierling* appears to have been introduced by the Dutch engineers into the East of England, and survives in the local English names, Spirling and Sparling, by which the fish is sometimes known. Although generally distinguished in Germany as the *Stint*, it has many provincial names, among which Pennant mentioned *Stinkfisch*, and he quotes Linnæus as saying that in spring, when the peasants come to buy it, all the streets of Upsala are filled with its foul odour, and he adds that agues are then prevalent. In Sweden it is known as *Slom*. Not only is it found in the countries enumerated, but penetrates southward into Central Germany, and in Russia is found in the White Sea and Baltic, and the lakes which are connected with those waters.

Smelts ascend rivers to spawn, and according to Yarrell, live in fresh water from August to May. Pennant remarks that they appear in certain rivers a long time before they spawn, being taken in great abundance in November, December, and January, in the Thames and Dee, but in other rivers not till February. They spawn in March and April, after which they return to the salt water. They never come into the Mersey as long as there is any snow-water in the river. Smelts are not commonly captured in the English Channel, though taken at Brighton. They ascend the Seine. They are abundant in the Medway, plentiful at Boston in Lincolnshire, Norwich, and King's Lynn, and make their way into the Bedford river, Hundred-Foot, and other artificial waterways of the Bedford Level which supply Cambridge with this fish. They are also abundant at Ulverstone on the West coast, and other localities in Lancashire.

Smelts are common in the interior of Jutland, and in the great lakes and rivers of the middle part of Scandinavia, where the large fishes, six to eight inches long, keep by themselves in schools, and go by the name of *Slom*, while the smaller variety, two to four inches long, is found in distinct shoals, and goes by the name of *Nors*.

Mr. Lloyd says the Smelt is of a dull disposition, slow in its movements, and prefers large lakes with sandy bottoms. It is seldom found singly, and lives in deep water during the greater part of the year.

Great prejudice exists against the fish both in Scandinavia and Prussia, on account of its odour. Many believe that the odour makes it unwholesome as food, and it is reputed as insipid and of disagreeable flavour, but this is in the land of the Salmon. Fishermen declare that it drives away other fishes from the fishing grounds. Mr. Lubbock speaks of the Roach and Dace fleeing from the Smelt in Norfolk, as it ascends the rivers. Some French writers have regarded the odour as a protection from enemies, and in certain localities the smell is said to be so penetrating that anything dipped in the water where the fishes abound becomes impregnated with it. We cannot but think that national prejudice is here somewhat unjust to the Smelt, since in France Blanchard finds only a beautiful odour of violets, and the small variety called *Nors* is largely taken in Scandinavia for bait, and is reputed the best bait for other fishes. We may, perhaps, think that the fishes are more discriminating than the people; though the nature and intensity of the perfume probably depend on the insects on which it feeds. Nevertheless, it is an important article of food for the poor, and is either dried in the sun or salted down for winter use in Sweden. The poorer classes along the Haffs of Prussia and other parts of the southern coast of the Baltic Sea almost live upon it. In the Kurische Haff, where their abundance is almost incredible, fishes sometimes reach a length of a foot, and they have been used for feeding cattle, for manure, and for the manufacture of oil and guano.

In England the Smelt is regarded when freshly taken as one of the most choice and delicately-flavoured fishes that comes to table; it is always served fried.

In form of body this fish is elongated from four and a half to five and a half times as long as the head; and is six to seven times as long as deep. The back is rounded, and the abdomen somewhat compressed. The ventral and dorsal contours are very slightly arched. The lower jaw extends far in front of the upper, so that the chin forms the extremity of the head, and the gape of the mouth is deeper than wide; the teeth are dense and slender on both the pre-maxillary and maxillary bones. The large teeth at the extremity of the vomer and on the tongue curve inward. The size of the eye is variable; it may be little less than one-quarter or less than one-sixth of the length of the head. The breadth of the frontal interspace is about twice the orbital diameter.

The dorsal fin begins midway between the extremity of the head and the base of the caudal fin, or a little farther back. The adipose fin is midway between the end of the dorsal fin and the root of the caudal, and opposite to the middle of the anal, which has a long base. The caudal fin is deeply forked.

The lateral line is not traced beyond the first eight or ten scales. The scales are delicate and easily fall off; they show concentric lines of growth with the nucleus far back, as in an oyster-shell. Their absence of lustre gives the fish a semi-transparent waxy aspect, and in a living state the vertebral column, ribs, and viscera may be seen through it. The colour is ashy-green or bluish on the back, and dull yellow below, sometimes pinkish. A faint silvery band extends down the side. The ashy tone is largely due to diffused points of black pigment. The iris, lower jaw, and sides of the head are silvery. The dorsal, pectoral, and caudal fins are grey with pigment spots; the other fins are colourless.

Smelts live on Whitebait, worms, small shrimps, and insects; and are eaten by the Zander and Perch. They spawn in spring, and sometimes change the spawning ground. In Sweden fishes may spawn before the ice breaks up, and the chosen localities in Lake Wener are at the mouths of streams which flow into the lake. According to Eckström, the spawning takes place at night; and the fishes are reputed to leave the spawning-grounds at dawn and return at sundown. It is a peculiarity of this fish that the spawning, which lasts from eight to fourteen days, takes place by preference in stormy weather, and it is on this account, Eckström tells us, a snow-storm in the spawning-season is termed by the Swedes a Smelt-blast.

Mr. Lloyd states, on the information of a friend resident near the north shore of Lake Wener, that Smelts are occasionally taken there of nearly a pound in weight. We give this remarkable assertion not without a suspicion that some other fish has been mistaken for the Smelt, for Mr. Lloyd saw no specimens larger than those which are common in England.

The fishing is generally carried on during the spawning-season in the night-time, when large fires are kindled on the shores to attract the fishes out of the deeper water, to within the sweep of the net. The net used is attached to an iron ring six feet in diameter.

There is no doubt that Smelts are capable of being naturalised in ponds. Yarrell mentions that Colonel Meynell kept them successfully for four years in a three-acre pond which had no communication with the sea; and Sir T. Maryon Wilson introduced Smelts from Rochester to a pond at Serles, near Lewes, where they bred and flourished.

Professor Huxley has drawn attention to the remarkable condition of the ovaries in the Smelt. They are divided into two, and are placed one behind the other, so that the left ovary is in front, and the right ovary behind. These ovaries have large oviducts which extend below the swim-bladder, one long and the other short, on each side of the intestine. The common genital outlet in which they terminate is situate between the vent and the urinary aperture; while the ovary agrees in structure with that of the Salmon tribe, the oviducts constitute a singular difference.

## Genus: **Coregonus** (Artedi).

The genus Coregonus has the body covered with scales of moderate size, which easily fall off. The cleft of the mouth is small. The maxillary bone is broad and short, and does not extend behind the orbit. The teeth are extremely minute and deciduous, and when present in the adult fish are usually limited to the tongue. The dorsal fin is higher than long, and the caudal is deeply forked. The air-bladder is very large, undivided, and opens into the throat by a short duct.

The stomach is horse-shoe-shaped. There are about one hundred and fifty pyloric appendages of nearly uniform length and thickness, attached to the slender intestine, behind the portal region. The intestine is often completely filled with Entomostraca. The eggs are small.

Coregonus is found in the temperate and northern parts of Europe, Asia, and North America, in fresh waters; and many species periodically go down to the sea. In North America it is abundant in the lakes and fresh waters, and known everywhere as White Fish. At spawning-time several rows of scales on the sides of the body develop tubercles, like those of Cyprinoid fishes. Individuals of the same size herd together in schools. The species are all valued for food, and most are esteemed for their delicate flavour; they are as numerous as in the genus Salmo. The geographical range is limited in some types, while other types are as widely distributed as some kinds of Salmon. The species are often difficult to separate, and are best distinguished, according to Dr. Günther, by the shape of the snout, its relative length, the development of the maxillary bone, form of the supplementary maxillary bone, length of the mandible, height of body and tail, position of the dorsal fin, and number of scales and vertebræ; but these characters often define species on small differences.

Most species have the upper jaw the longer, though many have the lower jaw longer than the snout, and differences in the length of snout are easily recognised, and we have followed Dr. Günther in grouping the species by this characteristic.

(1.) *Species with the upper jaw prolonged.*

## Coregonus oxyrhynchus (LINNÆUS).

D. 14, A. 14—15. Scales: lat. line 75—81, transverse $\frac{9-10}{12}$.

This fish frequents the coasts, and enters the fresh waters of Belgium, Holland, Germany, Denmark, and Sweden. It is known in Belgium as *Hautin*, is common in Antwerp, and sent thence to Paris, where it is sold in the fish-markets as *L'outil*. In Holland it is plentiful, known to fishermen as the *Houting*, and also supplies the French markets. In Germany it is popularly known as *Schnäpel*. It is very rare in France, but is said sometimes to be caught in the Doubs. It ascends the Rhine as far as Strassburg, is found in the Maas and Scheldt, Weser, Elbe, and the rivers flowing into the North Sea and Baltic. It thrives as well in the sea as in lakes and rivers. Mr. Day records it as a British species from Lincolnshire, Chichester, and the Medway, taken with Smelts.

*C. oxyrhynchus* is easily distinguished from all other species by its long conical fleshy snout. The head is compressed and small. It is two-ninths or one-fifth of the total length, exclusive of the caudal fin, and the height of the body slightly exceeds the length of the head. The maxillary bone extends below the fatty eyelid, and is more than one-quarter of the length of the head. The mouth is at the lower and hinder extremity of the soft black snout. Behind the head the back is moderately curved, but enough to give the aspect of a high shoulder. The scales are more circular than in other species, with denser concentric striæ. The dorsal fin is in the middle of the back. The pectoral is as long as the head, exclusive of the snout.

The fish measures from sixteen to eighteen inches; individuals of twenty inches, with a weight of two pounds, are recorded in Denmark, while in Sweden the weight may be eight to twelve pounds.

It migrates and spawns at the end of October or beginning of November. As in other species, rows of tubercles then appear on the scales of the male; two rows above the lateral line, and three below. After spawning, when these growths have fallen off, the scales show longitudinal bands in both sexes.

This species has fifty-eight vertebræ.

## Coregonus lloydii (Günther).

D. 14, A. 15. Scales: lat. line 90, transverse $\frac{11}{10}$.

The *Nabb-sik* of Sweden was first regarded as a distinct species by Dr. Günther. It has thirty-nine vertebræ in the thorax, and twenty-three in the tail. The length is about twenty-two inches. It is distinguished by smaller scales, and longer and blunter snout from *Coregonus oxyrhynchus*, which it somewhat resembles. The height of the body is greater than the length of the head, the former being one-fourth and the latter one-fifth of the length of the fish, exclusive of the caudal fin. The upper jaw is produced into a short, fleshy cone, and the profile of the snout makes a less acute angle than in *C. oxyrhynchus*. The maxillary bone extends to the anterior border of the fatty eyelid, and is, therefore, relatively shorter than in the former species. The supplementary bone of the maxillary is long and narrow. Behind the head the back is strongly arched. The pectoral fin is as long as the head, without the snout.

In this species there are fully two hundred pyloric appendages to the intestine.

It is found in Lake Wener and other localities. This species is very closely allied to *C. oxyrhynchus*, of which it may prove to be a thick-nosed variety.

(2.) *Species with the snout obliquely truncated.*

## Coregonus lavaretus (Linnæus).

D. 14, A. 15. Scales: lat. line 90—94, transverse $\frac{11}{12}$.

Many writers have described this fish, which is extremely variable, under different names, such as *C. maræna*, *C. fera*, and we owe the definition of characters which establish its geographical range, as indeed is the case with the whole Salmon tribe, to the patience and genius of Dr. Günther. It is a lake fish, widely distributed in the great lakes of Switzerland, the Tyrol, and Sweden; Ladoga, Onega, and Peipus in Russia; Madua See in Pomerania, and it occurs in Mecklenburg. It frequents the Baltic, and is found in the Haffs on the Prussian coast. It is everywhere valued for food (Fig. 164).

In Sweden it is known as *Sik*, or *Knebb sik*, or *Helge sik*, which in Russia becomes *Sig*, a name applied indifferently to many species, for in that country as many as thirty-five varieties have been named. In Northern Germany it is *Maräne* or *Selmäpel*, in Bavaria it is *Bodenrenke*, and in the lakes of Austria *Kröpfling* and *Kindling*, and in Lake Constance *Sandfelchen* or *Weissfelchen*.

Its size varies with the locality, being only fourteen to fifteen inches long, with a weight of half a pound, in the Austrian lakes, while in Lake Constance

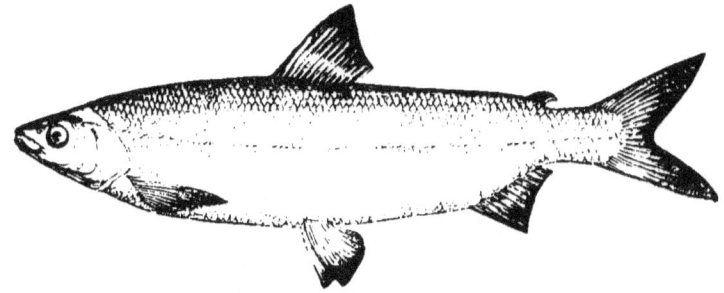

Fig. 164. —COREGONUS LAVARETUS (LINNÆUS).

and the Prussian lakes it reaches a length of two feet or more, and a weight of four to six pounds.

The greatest height of the body exceeds the length of the head, and is twice the thickness. The fish is five times as long as high; the head is larger than in *C. wartmanni*. The upper jaw is obtuse, longer than the lower, with the snout obliquely truncated, so that the thick blunt nose projects forward as in *Chondrostoma*. The mouth is large and the maxillary bone extends below the front part of the fatty eyelid. It may be from one-third to little more than one-fourth the length of the head. The supplementary maxillary bone is semicircular, broad, and short. The eye is nearly one-fifth of the length of the head, and is separated by its own diameter from the snout. It is more than its own diameter from the other eye. The half-moon-shaped fold in front of the eye is rather large but variable, and the pupil extends towards it, as in other species of the genus. A wide mucus-canal developed behind the eye, becomes narrow in the suborbital region, and is often prolonged forward below the nares in Swiss and other southern specimens. The opercular bones are strong and extend backward and upward, so that the head appears prolonged in those directions. The dorsal profile forms a flat curve similar to the ventral outline, and is nearly straight between the dorsal and adipose fins.

The fins show no distinctive characters from *C. wartmanni*, except that the edge of the dorsal is more curved, the pectoral is longer and more pointed, the small adipose fin pointed, the caudal fin deeply forked, and its terminal rays just exceed the length of the head. The base of the tail is thicker and higher; the lateral and cephalic canals are similar. There are sixty to sixty-one vertebrae, and the colour is similar but not so deep, and there is less black pigment on the body.

At spawning-time the fins are pale red, often finely dotted with black, though never edged with black. But in South Germany black edging is found; in Prussia it occurs in the variety named *C. maræna*. The species spawns in the end of November or December in shallow, gravelly places; spawning, according to the Lake Constance fishermen, happens a fortnight earlier than in the *C. wartmanni*; though Jurine says it spawns in February. The eggs number from 20,000 to 50,000, and measure from two and a half to three millimètres in diameter.

At ordinary times the fish lives in great depths, and when drawn up in nets the air-bladder expands and inflates the body, so as to suggest the idea of the fish having a full crop. It lives on worms, insects, and small mollusca. The flesh is white and firm, greatly esteemed when fresh, and, according to Benecke, has then an agreeable odour of cucumber.

## Coregonus lapponicus (Günther).

D. 15, A. 16.  Scales: lat. line 94—100, transverse $\frac{11}{9}$

This fish reaches a length of sixteen inches; the total length without the caudal fin is about five times the length of the head and four times the height of the body. The diameter of the eye is one-fifth of the length of the head and two-thirds of the length of the snout. The head is small, of moderate height, with the snout obliquely truncated and the upper jaw projecting beyond the lower. The frontal space between the orbits is rather convex, and in width one and a third times the diameter of the eye. The maxillary bone barely extends below the beginning of the fatty eyelid; the supplementary bone of the maxillary is short, broad, and semicircular. The pectoral fin is longer than the head without the snout. The commencement of the dorsal fin is nearer to the back of the head than to the adipose fin. This fish is closely allied to *Coregonus lavaretus*, but has smaller and shorter jaws. It is found only in Lapland, and is recorded from the river Munio. Although Cuvier and Valenciennes indicated a Coregonus from Lapland, under the name of *C. sikus*, it is not possible to recognise it from their description.

### Coregonus gracilis (Günther).

D. 14, A. 15—16. Scales: lat. line 91—100, transverse $\frac{11}{12}$

This Swedish species is about fourteen inches long. In form of head it most resembles *Coregonus hiemalis*. The body increases in depth with age, and in the adult the height and the length of the head are respectively one-fourth and one-fifth of the total length, without the caudal fin. The head is five times as long as the eye. It is relatively small, much compressed, with the snout pointed and rounded rather than obliquely truncate, still with the upper jaw projecting beyond the lower. The interorbital space is convex and less than twice the diameter of the eye. The maxillary bone, which measures one-fourth of the length of the head, is not strong, and extends back only to the fatty eyelid. The supplementary maxillary bone is short and semicircular. The tail is rather elongated; the dorsal fin is nearer to the back of the head than to the adipose fin; the pectoral fin increases a little with age.

### Coregonus widegreni (Malmgren).

D. 14, A. 16—17. Scales: lat. line 90, transverse $\frac{10}{11}$

A Coregonus, having the usual proportions of body, has been thus named, and is found in the lakes of Sweden and Finland. Its snout is obliquely truncated in front, and the upper jaw is considerably longer than the lower, so that it belongs to the type of *C. lavaretus*, though the snout is not quite so long. The maxillary bone extends a little farther backward, but the supplementary maxillary is of similar form and size, and we have no doubt of the propriety of regarding this form as a variety of that species.

(3.) *Species with the snout vertically truncated.*

### Coregonus wartmanni (Bloch).

D. 14—15, A. 15—16, V. 12, P. 15—17, C. 19. Sq. lat. 80—92, transverse $\frac{10}{11}$

This fish (Fig. 165) is moderately elongated and compressed, with a small pointed head and jaws of equal length, the upper reaching back as far as the front of the orbit. The greatest height of the body in front of the dorsal fin is less than one-fifth of the entire length, and the height always exceeds the length of the head, which is one-sixth of the entire length of the fish. The diameter of the somewhat large eye is one-fifth of the length of the head; it is separated by its own diameter from the snout; and the width of the frontal interspace between the eyes is one and a half times the orbital diameter. There is a half-moon-shaped fatty border in front of the orbit, but it is smaller than in most other Salmonidæ. The nares are small and double, midway between the eyes and snout. The snout is vertically truncate; the mouth is terminal and the broad upper jaw partly covers the mandible. The mouth is toothless, but there are short pointed teeth on the tongue. The frontal profile rises in a moderate curve, and the dorsal profile is highest at the beginning of the dorsal fin. The ventral profile is a more flattened curve in its middle part.

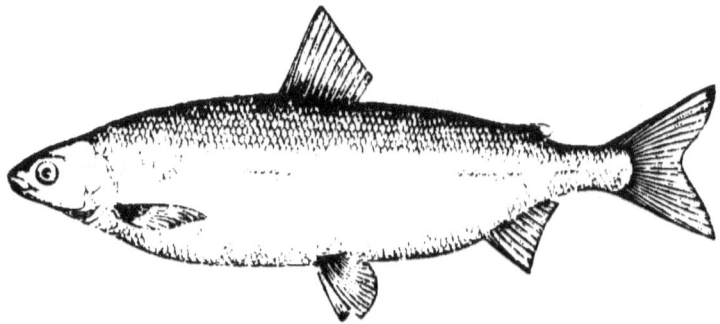

Fig. 165.—COREGONUS WARTMANNI (BLOCH).

The gill-openings are wide; they open below in the middle of the throat, and extend laterally as high as the middle of the eye. The rake-teeth of the anterior gill-arch are long, finely notched, and crowded together, while those of the fourth gill-arch are pointed and not more than ten in number. There are nine branchiostegal rays. The pseudobranchiæ are comb-like and well developed.

The dorsal fin commences in front of the middle of the body. It is truncated behind, the longest rays are two-thirds of the height of the body, and the last ray is one-third of the length of the fifth and sixth.

The anal fin is placed far back, and resembles the dorsal, but is only two-thirds as high. It is opposite to the small and low adipose fin. The ventral fins are below the hinder part of the dorsal, as long as the pectorals, and have the free margin rounded. The pectoral fins are pointed behind; the caudal fin is deeply forked, evenly lobed, longer than the head, with seven to eight short truncated rays on each side of the tail.

The scales are nearly circular, soft, finely marked with concentric stripes, entire at the margin, and overlapping nearly half their diameter. They are smallest on the back, throat, and bases of the pectoral and caudal fins. The spine-like scale over the base of the ventral is long, and ends in a membranous point. The lateral mucus-canal runs in a nearly straight line above the middle of the height of the side, and opens with simple pores. The suborbital branch of the cephalic canal is strongly developed, and is sunk so deeply in the bone that the suborbital ring appears to be double.

The colour of the upper part of the head and back, as far down as the lateral line, is blue and silvery. The sides of the head, and the abdomen are silvery. The lateral line is spotted with black. All the fins are yellowish-white with broad bluish-black edging, which is especially marked in the ventral and anal fins. The iris is silvery. In Lake Constance the fish is sometimes dark in colour. In young individuals the fins are colourless. The common length of this fish is from fifteen to eighteen inches, when it usually weighs from one and a half to two pounds; but the length varies from nine to twenty-eight inches, and it passes under different names among fishermen, according to its size. It has fifty-nine to sixty vertebrae.

This species stocks certain lakes almost to the exclusion of other fishes, and is particularly common in the Atter See, Gmünden See, and Fuschler See, in Austria, and is found in most of the large lakes in Austria, Bavaria, and Switzerland, especially in Lake Constance. According to Blanchard and Moreau, it is found in Lake Bourget, and other places in the east of France. It usually lives at a considerable depth, and in cold weather descends deeper, often going down to a hundred fathoms.

Spawning-time is somewhat uncertain, taking place in spring, summer, and autumn, according to different observers; according to Von Siebold, in November and December; February and March, according to Heckel and Kner. Most fishes then come into shallow water, but this species keeps at a depth of ten fathoms. The spawning lasts for about fourteen days or three weeks, when the fishes collect in immense numbers, and many are so pressed together that the scales on their sides are rubbed off and float for some distance over the surface of the water above the spawning ground. It does not spawn till ten to twelve inches long.

This species lives on worms, insect-larvae, the spawn of other fishes, and many kinds of animal life, as well as vegetable substances. It does not eat when spawning. It increases rapidly; but is not tenacious of life, dying as soon as taken out of the water.

These fishes are greatly eaten by Trout. They are taken with large nets sunk deep in the lakes at spawning-time, and their capture constitutes an important local trade. They are eaten fresh on Lake Constance and other

Swiss lakes, and are an esteemed delicacy. In some localities they are salted, smoked, or baked, and packed for export.

This species is known in Upper Austria as *Rheinanken*; in Bavaria as *Renke*, in the Tyrol as *Renken*, and in Lake Constance as *Felchen*, though the names vary with age even in the same locality; and many of the great lakes have independent systems of nomenclature.

## Coregonus clupeoides (LACÉPÈDE).—The Powan.

D. 14—15, A. 13—16, V. 11—12, P. 17, C. 19.
Scales: lat. line 73—90, transverse 9/11.

The Powan belongs to the group of species with the snout vertically truncated, and is confined to the lakes of Great Britain. It commonly weighs three to four pounds, and may reach a length of as much as sixteen inches. It is known in Loch Lomond as the *Powan*, and is usually termed the Freshwater Herring, which it is not unlike in appearance. It abounds in Ullswater and the lakes of Cumberland to a height of two thousand six hundred feet above the sea, and is known in the Lake district as the *Schelly*. It is also found in the lakes of Wales, especially Bala Lake, where it is said to have been abundant before Pike were put into the lake at the beginning of this century; it is known as the *Gwiniad*. It is gregarious, and, according to Pennant, approaches the shore in spring and summer in immense shoals, so that as many as eight hundred have been taken in a single sweep of the net in one of the Westmoreland lakes; and Pennant says the Rev. Mr. Farrish, of Carlisle, wrote that he was assured by an Ullswater fisherman that one summer he took between seven and eight thousand at a single draught.

It sometimes wanders from Bala Lake, but is recorded by Pennant in the Dee only at Llandrillo, six miles down the stream. This excellent observer states that it spawns in December in Llyntegid, but since his time no observations on its habits appear to have been made. Some of his specimens weighed between three and four pounds. It dies very soon after it is captured, and requires to be eaten at once, though it is often preserved with salt. It is said to be insipid as food; but though Pennant may have thought so, probably, when contrasting it with Trout or Salmon, we have pleasant recollections of the delicate Powan of Loch Lomond as a fish to be eaten again in September.

Günther compares it to *Coregonus wartmanni*. The head is one-fifth of the length of the fish, but the body is deeper than the head is long. In shape the head is sub-triangular, rather truncated at the snout, with the jaws nearly equal, though the upper is slightly the longer. The maxillary bone extends below the front margin of the eye, and the supplementary maxillary is broad and semicircular.

There are a few minute teeth on the tongue and jaws, though from their small size they may be easily worn away. The eye is rather large, one-quarter of the length of the head, and separated from the snout by its own diameter. As in other species of the genus, the nostrils are close together, midway between the eye and the snout, and rather near the profile of the head. The dorsal and abdominal profiles are moderately convex, as in the Herring. The lateral line is in the middle of the side. The dorsal fin is midway between the snout and the base of the adipose fin; it is higher than its own base is long. The ventral fin is below the middle of the first dorsal, and the elongated scale at its base is one-third of the length of the fin. The pectoral fin is inserted low down on the throat and is more pointed than the ventral. The anal fin is nearer the caudal than the ventral; it is as deep as its base is long. In colour, the back and upper part of the head are dark blue; the sides are paler, often with a tinge of yellow; the belly and under side silvery. All the fins are dull bluish-black, darkest at the margin. There are one hundred and twenty pyloric appendages to the intestine; thirty-eight vertebræ in the thorax, and twenty in the tail.

## Coregonus hiemalis (Jurine).

D. 14, A. 15. Scales: lat. line 76—90, transverse $\frac{9}{11}$

In the Lake of Geneva is a Coregonus known as *la Gravenshe*, the same fish in Lake Constance is *Kropffelchen*. It occurs generally in the lakes of Switzerland and South Germany, such as the Ammer See; it is also known as *Kilch* and *Kilchen* (Fig. 166).

This species is somewhat smaller than the other German forms of Coregonus, ranging from eight to eleven inches in length. It is chiefly caught in the spring and at the close of summer, at spawning-time. It lives in considerable herds at great depths, mostly upon the bottom, where it feeds on small mollusca. It is less prized as food than the other species which occur with it.

*C. hiemalis* has a shorter body than the other Continental representatives of the genus. The body is four and a half times as long as high (Fig. 166). The length of the head is less than the height of the body; it is two-ninths of the length of the fish without the caudal fin. The snout is rather short, as long as the eye, obliquely truncated so that the upper jaw projects beyond the mandible. The breadth of the frontal interspace between the eyes is one and a third times the orbital diameter, and the distance of the hinder edge of the operculum behind the orbit is twice the orbital diameter. The fold of skin in front of the eye is fairly large, and the narrow pupil is pointed towards it. The usual mucus-canal is developed behind and below the eye. The maxillary bone extends back below the front margin of the orbit and is one-fourth of the length of the head. The supplementary bone on the maxillary is rather narrow. The tongue has five pointed curved teeth. There are eight branchiostegal rays. The pseudobranchiæ are large, but the fringing is short. The gill-rakers on the last branchial arch are fine, pointed, rather short, and about nine in number. The frontal and dorsal profiles form a continuous arch.

The dorsal fin commences in front of the middle of the body; it is as

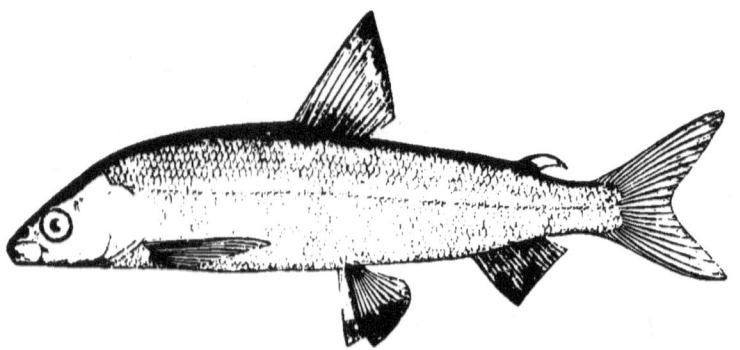

Fig. 166.—COREGONUS HIEMALIS (JURINE).

long as the caudal. The ventral and pectoral fins are also equal; and, as in other species, the anal has the shortest rays. The adipose fin is broad and blunt. The colour is always pale; on the back it is brownish-yellow or grey. The upper part of the head is yellowish-white, and the sides and operculum are silvery. The pectoral fins are colourless and transparent, but all the other fins are bordered with black; the iris is silvery. The flesh is firm and well-flavoured.

This fish spawns during September and October. It commonly lives in a depth of about forty fathoms. When captured the body becomes distended by the reduction of the atmospheric pressure on the gas in the air-bladder, which is so inflated as often to burst the abdominal walls; and as the torn condition is unsightly, the fish is but little caught. Sometimes, when flying before a pursuing Pike, or otherwise rising to the surface attracted by insects on the water, it falls back dead from the pressure of the expanded air on the blood-vessels.

## Coregonus maxillaris (Günther).

D. 13, A. (14)—16.  Scales: lat. line 87—(96), transverse $\frac{11}{11}$

Species which resemble each other in the vertically truncate form of snout do not present very striking differences in other parts of the body. Dr. Günther admits the close correspondence of this form with *C. wartmanni*, but distinguishes it as having the body much more elevated. But as it is found only in Lake Wener, it is quite possible that this may be a representative variety rather than a distinct species. Excluding the caudal fin, the height of the body, as measured by Günther, is two-sevenths, and the length of the head two-ninths of the total length. The eye is one-fifth of the length of the head, and shorter than the snout. The lower jaw is a little shorter than the upper. The space between the eyes is flat. The maxillary bone of the upper jaw is strong, and the supplementary bone upon it is semicircular. It is known in Scandinavia, according to Mr. Lloyd, as the *hof-sik*. It is eighteen inches long.

We are inclined to regard some other of Dr. Günther's species also in the light of northern representatives of *C. wartmanni*, just as the *C. clupeoides* is a British representative of the same species. And though we give short notes on the following types, it is with a view of showing the nature of the diversity which a species may assume in Lake Wener in which they occur.

## Coregonus humilis (Günther).

D. 13, A. 15.  Scales: lat. line 90—96, transverse $\frac{10}{11}$.

This variety grows to the same size as *C. maxillaris*, and is similarly found in Lake Wener. It is referred to by Mr. Lloyd as *Martinsmess-sik*. Excluding the caudal fin, the fish is four times as long as high, and the head is two-ninths of the length. The eye is one-fifth of the length of the head, and two-thirds of the length of the snout. The head is flattened, and in the male there are numerous small black dots on the head and dorsal fin.

There are sixty vertebræ.

## Coregonus megalops (Widegren).

D. 14, A. 16, V. 11.  Scales: lat. line 92.

A Coregonus is found in the lakes of Northern Sweden which may be distinguished from those of Lake Wener. The height of the body is nearly equal to the length of the head. The snout is truncated, shorter than the diameter of the eye, which measures little more than one-third of the length of the head. The maxillary bone extends back to below the front margin of the eye. The jaws are equal in length.

## Coregonus nilssoni (CUVIER AND VAL.).

D. 14—15, A. 15. Scales: lat. line 85—88.

This Coregonus has a small head and elongated body; the body, without the caudal fin, being six times as long as the head, while the depth of the body is no more than the head length.

The truncated snout is rather larger than the eye, the jaws are equal, and the diameter of the eye is from one-fourth to one-fifth of the length of the head.

It is known from Lake Ring, near Lund, in Sweden, where it is named *Blåsik*, and attains a length of ten to twelve inches. It is said to occur in other lakes in Sweden and Lapland.

(4.) *Species with the mandible longer than the snout.*

## Coregonus albula (LINNÆUS).

D. 12—13, A. 15—16, P. 15—16, V. 12, C. 19.
Scales: lat. line 75—84, transverse 9/10.

*Coregonus albula* belongs to a division of the genus in which the mandible is longer than the upper jaw, into a shallow notch of which it fits. The head approximates in aspect to that of the Herring. There are five teeth on the tongue. The maxillary bone does not reach back to the eye. The eye is as long, or nearly as long, as the snout, and about one-fourth of the length of the head. The iris is silvery. The fins present no distinctive peculiarities in form or position. The lateral line descends over the pectoral fins and then runs in a straight line in the middle of the body. The dorsal outline is rather more arched than the ventral contour. The colour is greenish-blue above; the sides and belly are silvery. The dorsal, caudal, and adipose fins are grey; the other fins are colourless. There are fifty-seven to fifty-eight vertebræ in the vertebral column.

This species is found in the north of Europe, from the Ural region in Russia, to Mecklenburg, Scandinavia, Poland, and Silesia. It never takes the fly for trout, and is chiefly captured in nets.

Dr. Günther regards the species as made up of four varieties.

There is, first, the variety *a*, distinguished as *C. albula*, which has fifty-seven vertebræ, and is found in Lake Wener. The eye is as long as the snout, fully one-fourth of the length of the head. The pectoral fin is as long as the head without the snout. In males the height of the body is one-sixth of the total length, exclusive of the caudal fin, or rather more. In females the height is rather more than one-fifth of that length, while the length of the head varies between one-fifth and two-ninths. It is adult at a length of six inches.

The second variety, *β*, called *C. norvegica*, has the eye rather shorter than the snout, and a little less than one-quarter of the length of the head. The pectoral fin is rather longer than the head without the snout, and is one-fifth of the length of the fish without the caudal fin. The length of the head is two-ninths, and the height of the body rather more than one-fifth, of the total length, exclusive of the caudal fin. It is characteristic of Norway, is adult at a length of seven inches, and also has fifty-seven vertebræ.

The third variety, *γ*, named *C. marænula*, has the eye as long as the snout, one-fourth of the length of the head. The pectoral fin is as long as the head without the snout, but though these proportions resemble those of the variety *C. albula*, the fin is two-elevenths of the length exclusive of the caudal, in *C. marænula*, and two-thirteenths of that length in *C. albula*. The head is one-fourth of the length, and the fish, without the caudal fin, is four and two-third times as long as high. It is characteristic of North Germany, is adult at a length of eight inches, though exceptionally reaching a length of a foot. It has fifty-eight vertebræ.

Dr. Günther's fourth variety, *δ*, named *finnica*, is known only from the Gulf of Finland, where it is seven inches long. The head is two-ninths of the total length without the caudal fin.

Probably most species admit of being separated in this way into constituent varieties, and we have given this classification as the best means of illustrating the variability of the type under different geographical conditions.

Except at spawning-time, which lasts a fortnight, it frequents deep water; but begins to travel in September and October, and, reaching shallow places, drops about ten thousand large eggs, with noisy and lively movements, during November and December. The female in Sweden is said to rub herself against stones to get rid of the eggs.

The flesh is greatly esteemed for food. It is eaten fresh, but is sometimes smoked. It is known in Germany as *Moranke* or *Kleine Marane*. In Sweden it is *Siklöja*, *Smasik*, or *Sik*. In Norway it is *Lakesild*, *Skadd*, or *Wemme*.

## Coregonus vimba (Linnæus).

D. 12, A. 14—16.  Scales: lat. line 80, transverse $\frac{9}{10}$.

Popularly distinguished under the name *Sik-wimma*, this variety of Coregonus is widely distributed in Sweden. It is most closely allied to *C. albula*, with which it agrees in fin rays, arrangement of the scales, and number of the vertebræ. There are similarly seven longitudinal rows of scales between the lateral line and the root of the pectoral fin. The diameter of the eye is in both cases equal to the length of the snout, and about one-quarter of the length of the head, and we are unable from Dr. Günther's diagnosis to recognise any sufficient character to separate it as a species. Mr. Lloyd states that the adipose fin is said to be slightly serrated.

The habits of the different kinds of *Sik* are similar. They live in deep water; but in spring approach the shore, following the Smelt to the spawning-ground, for the purpose of feeding on the spawn. In the autumn they return to the shore again for their own spawning. The least blow on the head, according to Mr. Lloyd, causes death. The fish is reputed very cunning, and the Swedes have a saying of a sly person that he is " as cunning as a *Sik*." The flesh is eaten fresh or salted, and is occasionally smoked. During the long winter in Lapland it furnishes a considerable part of the food of the people.

## Coregonus pollan (Thompson).—The Pollan.

D. 13—14, A. 12—13, V. 12, P. 15—16, C. 23.
Scales: lat. line 80—86, transverse 9, 11.

The Pollan is known only from the loughs of Ireland, where it is commonly about six inches long, though occasionally twice this length. Its habits are not dissimilar to those of the Powan. It approaches the shores in large shoals in spring and summer in search of food, and even far on into the autumn. It is commonly fished for in the afternoon, probably for the evening meal. It is rarely taken with the fly, from which we may infer that it is not insect-eating, but feeds on small mollusca. Mr. Thompson records that seventeen thousand two hundred and twenty were taken in a single draught in Lough Neagh in one September, and were sold on the spot for £23 6s. 8d. They are largely consumed in Belfast, where the cry of " Fresh Pollan ! " is more common than that of " Fresh Herring ! " It is taken to neighbouring places by rail, but requires to be eaten speedily after capture.

Its favourite spawning ground is a hard rocky bottom. The female fish is rather the larger. The stomach is found to contain the small bivalve shell *Pisidium*, the fry of Stickleback, Entomostraca, and *Limaea*, besides larvæ of insects, and the freshwater shrimp, *Gammarus*.

The species varies a little in proportions in the different lakes. The length of the head is equal to the depth of the body, which is about one-sixth of the total length of the fish. The eye is one-quarter of the length of the head. The jaws are equal in length; both may have a few delicate teeth, and the teeth are chiefly seen on the tongue.

The lateral line descends a little at the opercular region, and then runs straight to the tail. There are eight longitudinal series of scales between the lateral line and the ventral fin. The colour of the upper part of the body is of a dark blue, the sides and abdomen are silvery. The dorsal, caudal, and anal fins are dark, with finely-diffused pigment. Nearly all that is known of this species is due to its original discoverer.

There are sixty to sixty-one vertebræ.

## Coregonus vandesius (RICHARDSON).—The Vendace.

D. 11, A. 13, V. 11.  Scales: lat. line 68–71, transverse $\frac{8}{10}$

The name Vendace, according to Pennant, is derived from the French *Vandoise* (Dace) which the Scottish fish resembles in the whiteness of its scales; and in Scotland the tradition runs that it was introduced in the time of Mary Queen of Scots. As the fish does not occur on the Continent, it is more probable that the Continental taste of the Frenchmen about her Court for freshwater fishes may have brought it into notice, and provided it with a name by which it is now identified. It is found in the lochs near Lochmaben in Dumfriesshire, and also in Windermere and Bassenthwaite, and comes down the Annan to the Solway Firth.

Sir William Jardine states that the constitution of the fish is so delicate that it is quite unable to bear transport. In habit it nearly resembles the Powan; they both swim in large shoals and are taken only with the net. They make their way against the direction of the wind, and on dull days are readily captured near the shore of the loch, but during warm and clear weather retire to deep water, better able apparently to accommodate themselves to the increased pressure than to the increased temperature. The Vendace feeds on Entomostraca and other minute forms of life. It spawns in October and November, and then often jumps out of the water into the air. The females are larger and more numerous than the males.

The flavour is compared to that of the Smelt, and the fish is regarded as a delicacy, probably, Jardine remarks, on account of the difficulty of always obtaining a supply; though Yarrell, writing from practical experience, says he considers the fish quite entitled to its high character for excellence.

Without the caudal fin the fish is four times as long as high. The total length is five and a quarter times that of the head, and the diameter of the eye is about one-third of the length of the head, so that it is longer than the snout. The lower jaw is the longer. The small mouth contains a few minute teeth on the tongue; the ventral and dorsal profiles are similar. The lateral line is in the middle of the side; the longest rays of the dorsal fin are twice the length of its base. The adipose fin is near the caudal; the ventral fin begins under the front part of the dorsal. The fins are all conspicuously long, though the pectoral is shorter than the head.

The colour of the upper part of the body is a dull greenish or bluish tint, while the sides and lower part of the body are silvery, though the colour varies with the nature of the bottom. The dorsal fin takes the colour of the back, and there is a paler colour of the same tint in the fins of the lower part of the body. There are fifty-six vertebræ in this species, which is two fewer than in the Powan.

In Russia Dr. Grimm enumerates many other species of Coregonus, such as *C. nelma* (Pallas) in Lake Koobin, in the Dwina and rivers of Siberia; *C. polkur* (Pallas) and *C. omul* (Lepechin), both occur in the Mezen and Petchora; *C. peled* in the Petchora only; *C. fera* (Jurine); *C. Baerii* (Kessler), in Lakes Ladoga and Onega; *C. maræna* (Bloch) in Lake Peipus and the lakes in Poland; and *C. lscholmugensis* (Danilewski) found only in Lake Onega.

## Genus: **Thymallus** (Cuvier).

The genus Thymallus is found in the temperate parts of the northern hemisphere, occurring in North America, Northern Asia, and Europe. Its most distinctive characters are the great length and size of the dorsal fin, which includes from thirteen to twenty-three rays; and secondly, the mouth and maxillary bones are small and furnished with teeth, which, though small, are developed on the jawbones, the head of the vomer and palatine bones, and are absent from the tongue. The pyloric appendages to the intestine are less numerous than in Coregonus or Salmo. The stomach is horse-shoe-shaped; the air-bladder is very large.

The only European examples are the widely distributed *Thymallus vulgaris*, its variety *Thymallus æliani*, which is limited to Lago Maggiore, and the Dalmatian species, *T. microlepis*. Two species are limited to North America: the *Poisson bleu* of the Canadians, or *Hewlook-powak* of the Eskimo, found north of the Mackenzie River, and the other, *T. tricolor*, in Lake Michigan.

## Thymallus vulgaris (LINN.EUS).—The Grayling.

D. 20—23, A. 13—16, P. 16, V. 10—11, C. 19.

Scales: lat. line 75—88, transverse $\frac{9}{14}$.

In the rivers of England, *Thymallus vulgaris* (Fig. 167) is known as the Grayling; in France it is *l'Ombre*; in Upper Italy, *Temolo*; and in Germany, *die Esche* or *der Asch*, and *Harr* in Scandinavia. The name of the genus dates apparently from the time of Ælian, who found the fish in the Ticino and Adige.

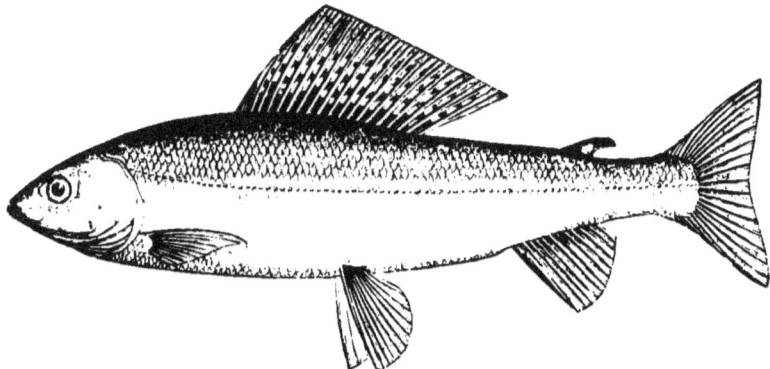

Fig. 167.—THYMALLUS VULGARIS (NILSSON).

The name is attributed to the thyme-like odour which the fish exhales. Walton says, "some think he feeds on water-thyme, and smells of it on being first taken out of the water," but Pennant observes that he never could distinguish the imaginary scent. Lloyd attributes the odour to the fish eating insects which have a strong odour, so that it may sometimes be absent.

It is a voracious fish, which Pennant states will eagerly take a bait, rising readily to the fly, though flies are probably an autumn luxury, for it is not till October that it takes the fly freely, and then, when the taste is once acquired, speedily gets into fine condition, and feasts on flies through November; but in the days when insect life is over and the work of spawning is done, the Grayling, out of condition, has to feed on what the streams may provide, and its taste then lies in the direction of univalve mollusca, small worms, young fishes, small fishes like Minnows and Gudgeons, spawn, and larvæ of insects.

When one remembers the great refracting power of water it seems remarkable that the Grayling should often jump out of the water and succeed in catching insects in the air, but the swiftness of their movements has been compared to the passing of a shadow.

As a rule the fish is solitary, preferring clear shallow streams with pools, having a bottom of sand or gravel. At spawning-time it mates, and they then swim in couples. The pair excavate holes with their tails, in which the female deposits the eggs, which are four millimètres in diameter, and orange-coloured, though Blanchard found them to be white and opalescent in France. As soon as the male has covered them with the milt, they are protected with small stones, and abandoned. The spawning often takes place early in March, but sometimes in April and May; though, according to Canestrini, the species spawns in Upper Italy between January and April; and Lloyd mentions May or the beginning of June in Scandinavia; the time depending upon temperature, and in Continental waters being governed by the melting of the ice in spring. In spawning thus early it differs from the Salmon. The young brood is usually hatched out in fourteen days, commonly in June; the process being more rapid than in other members of the Salmon tribe.

The young grow quickly, and under favourable circumstances may reach a weight of a pound or a pound and a half in two years, when they begin to spawn. Many individuals, according to Von Siebold, are sterile; such specimens are paler in colour and have smaller fins. On progress to maturity they encounter many dangers, and are by no means tenacious of life. Where Salmon abound they feed on Grayling; and water birds are discriminating enough to prefer a Grayling, when he can be caught. The fish is not easily naturalised anywhere. Sir Humphry Davy failed in getting it to live in brackish water, and it is only by a system of slow change of the water that they are introduced into ponds in Germany.

The flesh would seem to have been even more prized in ancient times than now, and numerous laws in the Codex Austriacus testify to the care with which its delicate life was protected by the earlier rulers of Austria. It had a reputation for being easily digested, and its fat, under the name of *Oleum æschiæ*, was once much used in medicine for its reputed healing properties.

Grayling are commonly caught with the net and weir-basket, though taken with the line even when baited with artificial flies. In Upper Austria, Heckel and Kner record the practice of an ingenious if unsportsmanlike mode of fishing, which consists in the application of the decoy such as was formerly in general use for the capture of birds. At spawning-time a female fish is taken just ready to deposit the eggs. A thread is tied firmly to the dorsal fin of the fish, and fixed at the other end to a small stake driven in the bottom of the brook. A net is then placed on the bottom, and as soon as the males come near the net is quickly raised.

The Grayling is not recorded from Ireland, has only recently been introduced into Scotland, and is most abundant in the limestone country of the Pennine chain, in the Chalk streams of the South of England, and Triassic sands of the Central districts. The Severn, Dee, Trent and its tributaries, and the tributaries of the Yorkshire Ouse, are all favourite Grayling waters. Pennant says the largest fish he ever heard of, taken near Ludlow in the Wye, was above half a yard long and weighed four pounds, six ounces. But from time to time specimens only a little inferior to this in size are taken in the South of England. Specimens weighing four pounds have been got in Denmark, and a weight of five pounds is recorded in Northern Scandinavia. The Grayling is found also in Lapland, where the gastric juice is said to be used in making cheese from the milk of the reindeer. In France and Germany it measures from a foot to fifteen inches. It occurs only in the East of France, the Lake of Geneva, the Auvergne, and some tributaries of the Rhône. In Switzerland it is characteristic of Lake Constance and the other lakes. It occurs in the mountain lakes of Austria, Hungary, and Transylvania, and their tributary streams, where it is sometimes two feet long, though commonly a foot and a half. It is found in Lombardy, Piedmont, Venice, and Istria. In Russia it is confined to the small rivers, and upper parts of the large rivers which flow into the Arctic Ocean, White Sea, Baltic, Black Sea, and Caspian.

The shape of the Grayling is eminently elegant; the body is five times as long as high, and six times as long as the head, though in some individuals the head may be a little larger, and the body a little higher. The body is twice as high as thick, but the head is only one-third higher than broad. The eyes are as far from each other as from the snout. Their diameter, including the adipose margin, is one-quarter of the length of the head. The dorsal profile is more arched than the ventral outline, and the convexity is more developed in the front of the body than in the genus Coregonus. The double nares are nearer to the eye than to the snout, which is blunt and somewhat broad, and projects a little over the lower jaw, so that the aspect of the mouth is inferior.

The pre-maxillary and maxillary bones are small, the latter extends back as far as the anterior margin of the pupil; though the mouth is small. The delicate teeth are largest in the pre-maxillary bone; they form a simple row in the maxillary bone, but are in groups on the vomer and palatine bones. The termination of the tongue is a round free point. The gill-openings are very wide; the number of branchiostegal rays, according to Günther, is seven to eight. Heckel and Kner state that the number of branchiostegal rays is sometimes ten on each side, sometimes nine on the right and ten on the left, though a St. Petersburg example shows eleven on the left and ten on the right, an unsymmetrical arrangement very common in the genus. The gill-rakers are pointed, and compressed at the base. Those of the last branchial arch are nine or ten in number, short, conically pointed, and not denticulated. The opercular shield is remarkable for the small size of the operculum, and the broad sub-operculum, which is placed high up.

The dorsal fin commences in the second third of the length, with a short ray. The rays increase in length to the first jointed ray, but the height of the fin differs with age, and it is only in the fourth to the fifth year that the fin acquires its full development, and then when laid back it touches the adipose fin with its last ray. The adipose fin, which is common to all the Salmon tribe, is highly suggestive, as probably a consequence of fatty degeneration, in which case it may perhaps represent a second fin of jointed rays which has entirely disappeared in these fishes, but may have existed in some ancestral type; nevertheless, that there is no necessary connection between a fatty fin and a bony skeleton to support it is shown by the development of a fatty fin on the backs of Dolphins and other cetacea. The anal fin is opposite to the adipose fin; it has commonly four unjointed and eleven jointed rays; it is the shortest of all the fins, and strongly truncated behind. The ventral fins are opposite the middle of the dorsal; they are broader than the pectoral, but rather shorter. The caudal fin is as long as the head; its forking is very variable, and it is evenly-lobed.

The lateral line descends at the back of the head, and then runs straight to the tail a little above the middle of the body; it opens with simple pores in a smaller row of scales than those on each side of it. The mucus-canals of the head are similar to those in Coregonus. They are well marked in the suborbital bone, open with a row of pores over the eye, and there are three or four pores on each branch of the lower jaw. The scales are adherent, larger than in the Salmon. They have fine concentric striping, are generally convex at the free margin, with a few short delicate rays. The posterior border is notched and truncate, so that the scale is much higher than long. The scales are smaller on the back than on the sides, but smallest on the abdomen and throat; they become gradually larger round the ventral fin.

The scales on the sides commonly have the aspect of being arranged in parallel lines, and the name Grayling has sometimes been supposed to be a corruption of "grey lines," due to this appearance. The head and fins are free from scales, but the caudal is covered at its longest ray for half its length with small delicate scales. Naked patches occur on the throat on both sides of the densely-scaled median line. These patches are larger in the young than in the old fish, and probably have no importance in classification, though Valenciennes founded a species, *Thymallus gymnothorax*, on fishes which had the belly naked and a hundred scales in the lateral line.

The colour varies considerably with the condition of the fish, season of the year, and age. The back is generally greenish-brown, becoming grey on the sides, while the belly is silvery. The head is brown above, yellow at the sides, with spots of black, which also occur in the fore part of the body, especially over the lateral line, and brownish-grey longitudinal stripes run in the direction of the rows of scales. The ventral and anal fins are violet, often with brown transverse bands. The pectoral is yellow, becoming red at spawning-time. The dorsal and caudal fins are bordered with black; they are generally red, and sometimes blue, and the dorsal commonly shows dark-brown spots or bands extending horizontally. The caudal may also be decorated. In the young fish the fins are transparent, and even the dorsal may be free from spots. Some Swedish writers state that during the spawning season the usual white colour of the belly becomes red.

The number of pyloric appendages is small, varying from nineteen to twenty-four. The last is conspicuously thicker than the others. The walls of the stomach are very thick. The liver is not lobed. The air-bladder is very long, and united to the abdominal walls; it extends back behind the vent, terminates in a point in front, and opens by a pneumatic duct into the alimentary canal. Two kidneys extend along the length of the abdomen. There are thirty-nine vertebræ in the thoracic region, and twenty-two in the tail, of which eight extend into the fan of the caudal fin. There are thirty-six pairs of ribs, and the accessory ribs are well-developed. Günther mentions that in the Neckar the stomach is infested with a parasite, *Ascaris capsularia*, and with *Cucullanus salaris*.

The Grayling is eminently a lover of clear, rapid streams, and, like the Trout, is more commonly met with in mountain brooks and small rivers than where they empty themselves into lakes, a point of natural history which Tennyson has recorded in his " Brook " :—

> " With here and there a lusty Trout,
> And here and there a Grayling."

In the Lago Maggiore a *Thymallus* occurs, which was named by Cuvier and Valenciennes *T. æliani*, but its chief difference from *T. vulgaris* is in having only seventeen rays in the dorsal fin, which is three fewer than in *T. vulgaris*. It may be regarded as a local lake variety.

## Thymallus microlepis (STEINDACHNER).

D. 13—14, A. 12. Scales: lateral line $\dfrac{\text{above } 19-22}{110}$
$\text{below } 17-19$

The body is elongated, and the tail compressed, so as to form a sort of edge, which disappears on the fore-part of the back. The height of the body slightly exceeds the length of the head.

The height of the head is two-thirds of its length. The snout is somewhat truncate, and projects a little over the mouth. The cleft of the mouth is short, and its angles are below the middle of the eye. There is a row of small pointed teeth in the pre-maxillary and maxillary bones, and in the mandible. The lateral rows of teeth on the tongue are much larger. There are two long rows of vomerine teeth. The eye varies from a quarter to less than a fifth the length of the head. The frontal region of the head is nearly flat. There are eleven branchiostegal rays. The gill-aperture is wide. The hinder border of the operculum is very little curved. The lower border of the pre-operculum is more rounded.

The dorsal profile rises in a moderate arch to the dorsal fin; the ventral contour is flatter. The dorsal fin is in advance of the middle of the body. The insertion of the ventral fin is in the middle of the length, and below the hinder part of the dorsal. The dorsal fin is as high as its base is long, and about half as long again as the head. The anal fin is two-thirds the length of the dorsal. Both these fins have their posterior free margins slightly concave. They are most elevated at the fifth ray, which is jointed and branched. The caudal fin is forked, with pointed lobes, which are only a little shorter than the length of the head; its base is covered with scales. The pectoral fin, which has fourteen rays, is longer than the ventral, but shorter than the head. The ventral fin contains ten rays; over its base is a rather long spur-like scale, which is covered with smaller scales in its front part. The adipose fin is above the hinder half of the anal.

The scales are smaller than in the Common Thymallus, but quite as firm in texture, and more nearly uniform in size. The body is completely covered with scales, without any of the naked patches which characterise the Grayling. The smallest scales are on the throat, and increase in size towards the ventral fin. The largest scales are above and below the lateral line.

The lateral line is nearly straight. It includes one hundred and ten perforated scales, of which from four to eight are upon the base of the caudal fin.

The colour of the back is greenish-brown. The sides and ventral region are a brilliant silver-white. The greater part of the middle of the sides of the body is marked with grey or black pigment spots, among which, in varying number, are scattered flecks of orange-red. The dorsal fin is yellowish-white, and has its middle part adorned with reddish stripes in the direction of the rays, while a blackish border margins its anterior and superior edges. The pectoral, ventral, and anal fins are orange. Blackish spots extend along the free edge of the caudal, and similar spots are sometimes seen on the dorsal and middle of the anal. The length is from seven to ten inches.

The swim-bladder is long, thin, and pointed at the ends.

There are upwards of thirty pyloric appendages.

This species is described by Steindachner from the mountain streams of Vergoraz, in Dalmatia. It is distinguished by the small number of dorsal rays, the greater number of rows of scales, and the dentition of the tongue.

## CHAPTER X.

#### FRESH-WATER FISHES OF THE ORDER PHYSOSTOMI (*concluded*).

FAMILY ESOCIDÆ GENUS Esox: Pike.—FAMILY UMBRIDÆ—GENUS UMBRA—Umbra Krameri—FAMILY CYPRINODONTIDÆ—GENUS CYPRINODON : Cyprinodon calaritanus - C. ibericus GENUS FUNDULUS Fundulus hispanicus—FAMILY MURÆNIDÆ—GENUS ANGUILLA : Common Eel—Broad-nosed Eel—Dalmatian Eel.

## FAMILY: ESOCIDÆ.

### GENUS: **Esox** (ARTEDI).

THE Pike genus is the only type of its family, and is a physostomous fish, not very distantly removed from the Salmon tribe, and some intermediate genera occur in the rivers of Africa, Australia, and North America. Among the more strikingly distinctive characters are, the absence in Pikes of the adipose fin, the situation of the dorsal fin, in the position of the adipose fin of the Salmon, the glandular condition of the pseudobranchiæ, and the absence of pyloric appendages to the intestine. It resembles the Salmon tribe in having the upper jaw formed by the pre-maxillary bones in front, and the maxillary bones at the sides, but the maxillary bone is toothless. The teeth on the mandible vary in size, and those on the pre-maxillary, vomer, palatine, and hyoid bones are sickle-shaped. The caudal fin is forked, the body is long, and the snout is long, broad, and depressed, with the mandible exceeding the length of the upper jaw. The genus Esox is limited to fresh water, and is distributed through Europe, Asia, and North America. It may be regarded, however, as rather an American than an Old World type, since all the seven species mentioned by Günther are found in the United States.

## Esox lucius (LINNÆUS).—The Pike.

The Pike (Fig. 168) is universally distributed in Britain and in Europe. It is one of the older inhabitants of England, its remains occurring plentifully in the peat of the Fens, with those of extinct mammals and birds. When young the fish is known as a Jack, when a little larger it is often termed a Pickerel, and the old fishes were formerly called Luce, a designation common in the time of Chaucer, as the familiar lines in the "Canterbury Tales" show :—

> "Full many a fatte partricke had he in mewe,
> And many a Breme and many a Luce in stewe."

In France it is known as *Brochet*, and by a figure of speech it is sometimes spoken of in that country, as in England, as the "Fresh-water Shark." It is absent from some Departments in the South of France, and has not been recorded in Spain.

The Dutch term it *Greepfisch* or *Snoek*. In Germany it is *Hecht*, and is absent only from the highest mountain lakes, occurring indifferently in all waters, though, as in Britain, preferring such as do not flow rapidly. In Italy it is *Luccio*; in Hungary the provincial name is *Csuka*; in Roumania, *Stuke*; in Bohemia, *Stika*; and in Poland, *Szczupak*. It, however, extends farther south, and the British Museum collection contains examples from the Lake of St. Stefanos, in Turkey. It is found in all the rivers of Russia in Europe, and in Siberia, with the exception of the Crimea and the Trans-Caucasus region. In

Fig. 168.—ESOX LUCIUS (LINNÆUS).

Sweden it is found throughout the peninsula, extending even beyond the limits of the birch-tree in Lapland. It is known to the Swedes as *Gädda*. It is never met with in salt water, though found in the lagoons of Venice; but it diminishes in size and number as it approaches the open sea in Sweden.

The aspect of the Pike is very striking, and quite unmistakable. The broad, low, compressed, and remarkably long head, with the backward position of the dorsal fin, distinguishes it from every other fish. The head may be one-third of the length of the body, but is usually somewhat less; it is three times as long as broad. The fish is six times as long as high, and the thickness of the body may exceed half its height. The sides are flat, and the transverse section is somewhat four-sided. The head is usually about six times as long as the eye, though in old fishes it may be eight times as long as the eye. The eye is nearly in the middle of the length of the head, and close to the frontal profile. The eyes are rarely separated by more than their own diameter, and this distance diminishes with age. The nares are in front of the eye. The mouth is nearly horizontal, the lower jaw projects in front of the small pre-maxillary, but the maxillary bones extend back for nearly half the length of the head. The teeth on the mandible are recurved; small teeth occur between the larger ones. They are arranged in a single row, usually five or six on each side, and are conspicuously strong. Many rows of small, sickle-shaped teeth cover the pre-maxillary, the hinder teeth being the largest. The palatine bones carry many rows of these sickle-shaped teeth; they are curved inwards, and are longest on the inner side. The middle of the band of vomerine teeth is more than half as wide as the band of palatine teeth. The number of branchiostegal rays is fourteen or fifteen. They are denticulated on the inner side. The dorsal fin commences at about a head-length from the base of the caudal; its longest rays are equal to those of the anal fin, which is opposite to it, or placed a little farther back still. The ventral fins are as long as the pectoral, and are rather behind the middle of the body. The terminal rays of the forked caudal fin are half as long as the head.

From the back of the head to the anal and dorsal fins the ventral and dorsal profiles are nearly straight and sub-parallel, though the ventral outline more commonly forms a curve. Behind the fins the depth of the tail decreases rapidly. There are no scales on the upper part of the head and snout, but the cheeks and upper part of the operculum are covered with delicate scales. The largest scales are on the sides, but they are scarcely equal to one-third of the orbital diameter. The scales have concentric striae; they are smallest on the breast and belly, and have a few large festoons on the free border. Some of those on the back and sides are divided at the base by a deep median longitudinal notch, as though they possessed mucus-canals, like the scales of the lateral line; but there is room for further study of the scales of the Pike. Small scales cover the bases of both lobes of the caudal fin. The lateral line runs parallel to the back and at one-third of the height from it. It is not so distinctly marked as are the cephalic canals, which have large pores. Twelve of these pores can be traced on the upper part of the head, and five large pores on the branch to the lower jaw. The sub-orbital branch is smallest.

The colour of the back is black, the sides are grey with yellow spots, which often become irregular transverse bands. The head is marbled like the body. The belly is white, but dotted with black. The colour varies with age, season of the year, and locality. At spawning-time the grey colour of the sides becomes a beautiful green, the pale yellow flecks deepen to a golden-yellow, and the gills are deep red. The pectoral and ventral fins are reddish; the dorsal, anal, and caudal are brownish and flecked with black. The olive-green variety has been termed the Grass Pike, and the pure yellow variety with large blackish bands is known in Germany as the Pike King. In size the Pike is scarcely inferior to the Salmon.

Frank Buckland enjoyed special opportunities for seeing large specimens, and among the finest which came under his notice were several forty-six inches long and thirty-five to thirty-six pounds in weight. Though still larger fishes are reputed to swim, Buckland warns us that, "to put it in plain language, more lies have been told about the Pike than about any other fish in the world."

In Austria it is frequently from twenty to thirty pounds in weight, and Heckel and Kner say that it grows in the Atter See to a weight of forty to forty-eight pounds. Yarrell tells us of a Pike from Loch Lomond which weighed seventy-nine pounds, and Mr. Thompson in his "Natural History of Ireland" mentions one taken in County Clare which weighed seventy-eight pounds, but I am unable to find personal testimony to larger fishes than those seen by Buckland. In Sweden the reputation of a Pike four feet long and eighty pounds in weight reached Mr. Lloyd, but the monster appeared a long time ago, and though dragged five times up to the gunwale of the punt, was never captured.

There is reason to suppose that the large fishes of forty-six inches were not more than fifteen years old, and the legends of fishes with the rings bearing ancient dates have not that quality of veracity which is required by science; for although the skeleton of the great fish said to be seventeen feet long, and reputed to be 267 years old, was preserved in the Cathedral of Mannheim, its bones furnished the unexpected gloss on the old story that it had been manufactured out of smaller fishes.

The Pike grows rapidly. Sometimes it weighs only half a pound in the first year, and two pounds in the second year, but a two-year-old fish may with exceptional feeding weigh six or seven pounds. It is the greediest of predaceous fishes, and the boldest robber in the waters. It is by no means exclusively a feeder on fish, but is as omnivorous as a pig.

Much of the food of a Pike may consist of frogs, leeches, weeds, Trout, Carp, or other fishes; but young wild-ducks, water-hens, coots, and water-rats do not come amiss, and when hungry it will swallow anything that comes within reach of its jaws. Mr. Jesse states that a Pike of five pounds consumes thirty Gudgeon a week, which is an average of four to five a day. It can, however, eat more, since when five Roach were thrown to this Pike in succession, four of them, each four inches long, were swallowed at once, while the fifth disappeared in about a quarter of an hour. Tales are told of Pike seizing the limbs of bathers, and attempting to capture the snouts of mules and donkeys which had gone down to the water to drink. But there seems to be no doubt about the story told by Mr. Jesse, of a man who, endeavouring to capture a large Pike, which had got into a shallow creek, received severe wounds on the arms in the combat.

Pike are often put into lakes and ponds to keep down the number of Trout; and, on the authority of Mons. Weern, Lloyd states that the Pike in a pond are mostly stationary, but the Trout are in constant motion. On one occasion Mons. Weern saw a Pike of seven to eight pounds' weight seize a Trout fully as large as itself across the body. The Trout made desperate but ineffectual efforts to get free, and after two hours became exhausted. The Pike then commenced gorging its prey, beginning with the head, and it was not till the expiration of three whole days that it had succeeded in completely swallowing its victim. The process of digestion must have continued much longer, as the fish had a very swollen appearance for a week afterwards, and was hardly able to move from the spot, even when poked with a stick. But Pike readily eat each other, though there is usually a slight advantage in size on the part of the attacking fish. They begin this practice when less than an inch long, and probably continue it throughout life, since we hear of one Pike of nine pounds with its head and throat firmly embedded in the jaws of another Pike of ten pounds, and Dr. Burton tells how the lad who took Pike thus united, wondered to see "a muckle fish wi' twa tails."

Near the coast of Sweden the sea-eagle, the osprey, kite, and probably other birds of prey, swoop down on Pike which may be basking at the surface of the water. If the fish is a small one the bird flies away with its booty to the eyrie; but often bird and fish are equally matched in strength, and then desperate conflicts ensue, which end in the death of both, for after the bird has struck its talons into the flesh of the Pike they cannot be disengaged. If the Pike is more powerful than the bird, the latter is borne down to the bottom, and drowned, and then succeeds, if true, one of the most curious incidents of natural history. For, incredible as the story seems, Eckström, the Rev. M. Möller, and other writers, state that the flesh of the Pike heals with the talons of the bird in its back, while the bird becomes converted into a skeleton, which is carried about by the Pike. One skeleton, which had long been exhibited by a Pike in Lake Wetter, had acquired a greenish tinge, and was regarded by the fishermen as a harbinger of misfortune. Mr. Lloyd tells us of another skeleton carried on the back of a Pike in Lake Fryksdal, which was known to the fishermen for some time as the Sjötroll, or Water-sprite, and they fled from it in fear. It is said to have appeared like the horns of an elk, or reindeer, moving rapidly on the water; but at last Lieut. J. Lekander put a shot in the Pike which carried it, and solved the mystery, by proving the Water-sprite to be the skeleton of a sea-eagle.

The flesh of the Pike when spitchcocked is excellent, and of a bright white colour, but is thought to have the choicest flavour when taken from streams, like those of the English Fen-land, where Smelts abound.

In Ireland the Pike spawns as early as February, but it is a month or two later in England. In North Germany it spawns from February to April, but in South Germany spawning takes place from March to May. The young spawn earlier than the old fishes, and in Scandinavia there are three successive spawnings, which correspond with the disappearance of ice, the pairing of frogs, and the unfolding of the leaf on trees; and these broods are known to the people, according to Mr. Lloyd, as ice Pike, frog Pike, and blossom Pike. The female is always larger than the male, and is sometimes attended by four males. The males are said to be more numerous than the females. At spawning-time the female remains quiet, and the males are reputed to rub themselves against her body till she deposits the yellow, somewhat large eggs, which are three millimètres in diameter. The number of eggs varies with the fish. In a Pike weighing twenty-eight pounds there were 292,000 eggs, weighing one pound five ounces, but in a fish weighing thirty-two pounds the eggs weighed five pounds, and numbered 595,000, so that in the large fish they weighed nearly four times as much, but were only twice as numerous. Pike first breed when they are three years old. According to Benecke, the eggs are hatched in about fourteen days, and the young have a large umbilical sac. In Sweden the period of incubation is from twenty-five to thirty days. Except when spawning, the Pike lives a solitary life.

The same species of Pike is plentiful in Manitoba, and occurs in the great Canadian lakes, and in the United States. Pike are taken with nets of all kinds, especially in the winter, but they are captured with the rod and line at all times from June till February.

The number of thoracic vertebræ varies from forty-one to forty-three; the caudal vertebræ number from twenty to twenty-one. The stomach is not well defined from the intestine. The intestine has two convolutions, and is about one-quarter longer than the body. The liver is on the left side, and is undivided. The air-bladder is simple, and loosely attached to the walls of the abdominal cavity throughout its length; a short pneumatic canal communicates with the pharynx. The kidneys are at first slender, but unite behind into a thick mass. The peritoneum is pearly-white.

The arms of the city of Luçon are three silver Pike.

## Family: UMBRIDÆ.

### Genus: **Umbra** (Kramer).

The genus Umbra includes small fresh-water fishes which are allied to Pike on the one side and to the Garfish tribe on the other. The pre-maxillary bones form the front, while the maxillary bones form the sides of the jaws. There are villiform teeth in the jaws, on the vomer, and on the palatine bones. The eye is small. The stomach is a simple expansion of the intestine, and there are no pyloric appendages. Both the head and the oblong body are covered with cycloid scales, on which there are no radiating striæ. The lateral line is not conspicuous. The dorsal fin is opposite the ventral, or a little behind it. The anal fin is short, and the caudal rounded.

Only two species are known: one limited to the South of Europe, the other to the United States.

### Umbra krameri (Müller).

D. 15—16, A. 7—8, V. 6, P. 13. Scales: lat. line 33—35, transverse $\frac{5}{7}$

The fish thus named is popularly known in Austria as the *Dogfish*, and in Hungary as the *Rihahal* (Fig. 169). In these countries it frequents stagnant

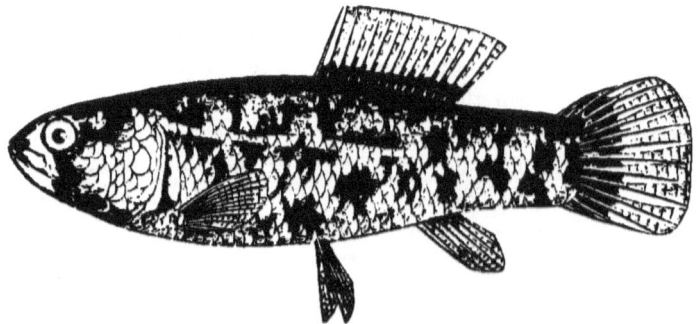

Fig. 169.—Umbra krameri (Fitzinger).

waters and grows to a length of three or four inches, and is especially found in the neighbourhood of the Neusiedler See, Moosbrunn near Vienna, and Teufelsbach near Pesth, and in streams which flow into the Plattensee. It is also said to occur in the neighbourhood of Odessa.

The scaled head gives this fish its most distinctive character. It has the body four and a half to five times as long as deep, but is deeper than usual behind the dorsal fin, so that its appearance is rather strong than elegant. The head is nearly as long as the body is deep, and twice the thickness of the body. The head is longer than high and twice as long as broad, though the thickness increases with age. The diameter of the eye is one-quarter the length of the head. The eyes are separated by their own diameter and are the same distance behind the extremity of the snout. The small double nares are nearer to the extremity of the snout than to the eye. The maxillary bone extends under the middle of the eye and forms part of the margin of the moderately large mouth. The lower jaw is rather the longer and is pointed in the middle. There are no teeth on the maxillary bone or on the tongue, but the pre-maxillary, mandible, vomer, and palatine bones have fine-pointed teeth in bands. The opercular apparatus is large, with a rounded margin and large gill-aperture. There are five branchiostegal rays on the right and six on the left. The pseudobranchiæ are invisible and glandular. The first gill-ray is so slender and fine, according to Heckel and Kner, that it might be easily overlooked; the hindermost ray is very wide and flat. Both the belly and back are broad and rounded;

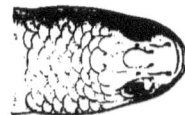

Fig. 170. — HEAD OF UMBRA KRAMERI, SEEN FROM ABOVE.

the ventral profile is rather the more convex of the two, and the depth of the tail exceeds the thickness of the body. The somewhat long dorsal fin begins behind the middle of the body; its branched rays are equal in length. The anal fin has a short base and is below the hinder part of the dorsal; it is rounded. The ventral fins, which are narrow, are just in front of the beginning of the dorsal and reach back to the vent. The pectoral fin is as long as the ventral, and its free border is similarly rounded. The middle rays of the caudal fin are the longest, making the outline of the fin rounded; it is not quite so long as the head. The fins constitute characters as distinctive as anything in the structure of this fish.

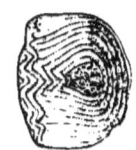

Fig. 171.—SCALE OF UMBRA KRAMERI.

Except the point of the snout and jaws, the whole head is covered with scales, so that it passes insensibly into the back (Fig. 170). On the body the scales are large and nearly circular, but the attached basal margin is somewhat truncated (Fig. 171). The scales cover each other like roofing-slates, are soft, not firmly attached, with very fine concentric lines and no radiating rays.

The lateral line is marked only by a paler band, descending from the neck down the side of the body, and shows no visible pores. It is roughly parallel to the back and nearer to it than to the ventral margin.

The cephalic canals are distinct, and are traced by their pores along the pre-operculum round the eyes and on the lower jaw. The bony elements of the sub-orbital ring are strongly developed.

The large stomach extends back in the abdominal cavity to the ventral fins, and is surrounded by the moderately large liver. The liver opens by a wide duct into the pyloric region. The intestine is slender, and makes a forward curve under the liver, and then curving again, passes into the large intestine, which runs straight to the vent. The air-bladder extends the entire length of the abdominal cavity, and opens by a short pneumatic canal into the lower and hinder wall of the gullet. The ovaries of the female form long sacs, which are small in front and covered with black pigment; the eggs are small. The kidneys are small in front and expand in the region of the vent to thick glandular masses. The small long urinary bladder and the oviduct open behind the vent. The male has in this region a projecting papilla.

The colour is reddish-brown, becoming blackish-brown on the back, and paler on the belly. The lateral line is indicated by a pale-yellow or copper-red line. The head and body are spotted and dotted irregularly with dark-brown pigment. The dorsal and caudal fins are brownish, the former has some spots; the other fins are pale. The male sometimes has a red line along the middle of the belly from the throat to the vent.

Males are rarer than females, and have a more elongated body. They are smaller fishes, often scarcely two inches long, while the females measure three inches, or three inches and a half.

This *Hundsfisch* lives in company with the River Bullhead, Crucian Carp, and Roach, in the marshes adjoining the Hungarian lakes, but it prefers muddy bottoms of deep pools. It is rare for more than five or six to be taken at a time, as it is shy, active, and buries itself in the mud.

When swimming, the pectoral and ventral fins move alternately like the feet of a dog in running, and the dorsal fin at the same time moves with a wave-like motion like that seen in the Sea-horse tribe, Hippocampus and Syngnathus. This condition is apparently due to the muscles which are attached to the individual rays being capable of independent action. Even when the fish is otherwise still, the last three or four rays of the dorsal fin oscillate as in Hippocampus. The animal is sometimes at rest in a horizontal, sometimes in a vertical position, and may remain for hours together with the head directed either upward or downward, then with a sudden rapid movement of the tail it comes from deep water to the surface and swallows air, which is ejected in large bubbles through the gill-openings, and it then breathes slowly for some time. Heckel and Kner kept this fish in captivity for a year and a half, and fed it with small pieces of raw meat, which were never touched till they came to rest at the bottom, though at last it learned to take food from the hand. The fishes spawn in confinement; the eggs are as large as millet-seed. The species is reputed poisonous by the fishermen, who have a superstitious belief that it is unfortunate to catch Dogfish. Formerly they were frequently brought to the Vienna market, and are still seen occasionally scattered among the great masses of *Misgurnus fossilis* brought from the Neusiedler See.

## Family: CYPRINODONTIDÆ.
### Genus: **Cyprinodon** (Lacépède).

Cyprinodon is the type of a family of fresh-water fishes, which are for the most part viviparous, and present marked differences of aspect and ornament in the two sexes. The head as well as the body is covered with scales. The pre-maxillaries form the margin of the upper jaw. The mandible is short, with the bones on each side firmly blended. The teeth have an incisor character, are notched, of moderate size, and arranged in a single series. The scales are rather large; the dorsal and anal fins are larger in the male than in the female.

The species of Cyprinodon extend along the Mediterranean region, occur in the Dead Sea, the Jordan, the valley of the Tigris, and in Persia and Abyssinia; other species are found in the United States.

### Cyprinodon calaritanus (Bonelli).

D. 9—10, A. 10—11, V. 6, P. 16, C. 14. Scales: lat. line 26, trans. 9 10.

In Italy this fish is known as the *Nono*. It is common about the lagoons of Venice, and ascends some of the streams which flow into them, but has been taken of a larger size at Treviso, in the Sile, than in Venice itself. It is rarely found in the sea. The largest specimens measure less than three inches.

It is an interesting species, as having been found by Canon Tristram in the hot springs of Sidiahkbar in the Sahara, and it is also found at Susa and Tunis.

The males have nine or ten very distinct silvery cross-bars; each bar is about as wide as a scale. The caudal fin shows an indistinct band on its hinder half and the front rays of the dorsal fin are black.

The female is silvery on the sides and has a number of black vertical bars on the sides, which do not extend to the back or belly; but neither the silver bars of the male nor the black lines of the female are developed in the young. The colour of the fish is yellowish or pale olive-green.

The body, without the caudal fin, is three times as long as high, and four times as long as the head. The head is broad, with a blunt snub snout; the lower jaw is more prominent than the upper, and the cleft of the mouth is nearly vertical. The eye is more than one quarter of the length of the head, and is equal to the length of the snout. The frontal interspace is nearly twice as wide as the orbital diameter. The dorsal and anal fins are high in males, and would reach to the caudal fin if laid back, though the dorsal commences in about the middle of the body at the eleventh scale of the lateral line, while in females it is somewhat farther back, though still corresponding to the same scale. The first anal ray is below the fifth ray of the dorsal; the tail is compressed, and the caudal fin is truncated.

Canestrini says that cats and small dogs which eat these fishes die poisoned.

Dr. Günther recognises a second Italian species, *Cyprinodon fasciatus*, known from males only, which has only eight or nine rays in the anal fin. It is recorded from brackish waters at Venice, and in Sardinia. But neither Canestrini nor Giglioli accepts the species, which seems to us to be only a variety of *Cyprinodon calaritanus*.

## Cyprinodon ibericus (Val.).

D. 10—12, A. 9—10. Scales: lat. line 23—24, transverse 9.

In this species the females are larger than the males, and of different aspect. The female is about one inch and three-quarters long, and the male measures less than an inch and a half. The female is ornamented on the sides of the body with two to four rows of longitudinal round black spots, which become more marked and may occasionally assume the aspect of bands towards the caudal region. In the anterior part of the tail there are sometimes one or two ill-defined transverse bands. The dorsal and anal fins, which are nearly opposite to each other, are sometimes slightly spotted with brown; the dorsal fin contains ten to twelve rays. The anal fin has usually ten rays, though they are occasionally reduced to nine. The length of the body is from three and a half to four times the greatest height; the head is about one-fourth the entire length of the fish.

The males are adorned with from twelve to sixteen transverse bands of a bright silver colour, which are at first short and limited to the middle of the side, above the base of the pectoral fin, but subsequently encircle the body. They are separated from each other by broad brown transverse bands. The caudal fin is slightly rounded, as in the female, and is ornamented with from three to five dark brown transverse bands. The anal and dorsal fins are marked with black spots, which are arranged in longitudinal bands, and these occasionally become confluent in lines. Generally the spots are limited to the hinder half of the anal fin. The proportions of the body are nearly the same as those of the female. The diameter of the eye is about one-third of the length of the head. In the greatest length there are twenty-three to twenty-four scales in a row; and in the greatest height nine scales. In the lower jaw there are fourteen to sixteen, and in the pre-maxillary bone sixteen to eighteen teeth, each of which has three yellow-brown points. The abdominal wall is black. The intestine, which has three marked folds, is nearly twice as long as the body.

This species lives chiefly on small mollusca. It spawns at the end of April and during May. It is most abundant where the Vega canal enters lake Albufera in Valencia; and Dr. Steindachner, to whom we are indebted for these observations, also found it in the canals of the plains of Murcia. The species is known to the fishermen as *Pececillo*. It is very nearly allied to *Cyprinodon calaritanus*, of the head of the Adriatic.

## Genus: Fundulus (Lacépède).

*Fundulus* is a Cyprinodont genus, which is represented by many species in Central America, the West Indies, northern part of South America, and the south of the United States. In the Old World a species is found on the east coast of Africa, and another in Spain. The bones of the mandible are firmly united, and the upper jaw is protractile. The teeth are conical, and adapted to the insectivorous habits of the fish. The dorsal fin is opposite the anal.

## Fundulus hispanicus (Val.).

In this species the males and females differ in the size and shape of the body. In both sexes the colour is the same and the back greenish, dotted with golden-brown, while the belly is golden-yellow. At the base of the scales, on the upper part of the head, there are brown spots, which become more numerous farther back, and are arranged into four or five longitudinal rows in the upper part of the body. The caudal region is also marked with delicate spots, which are arranged to give the aspect of a net-work. The cheeks and operculum are spotted with black.

In the female there is a median blue-grey band, which extends along the side from the orbital border to the caudal fin. The dorsal and caudal fins are very faintly and partially spotted with brown. In the female the anal fin is the larger; anal, dorsal, and caudal are all slightly rounded. The greatest height of the body is over the ventral fin, and is about equal to the length of the head. The diameter of the eye is four-fifteenths of the length of the head.

In the males the blue-grey longitudinal band is absent, and the hinder part of the body is covered with black spots, and crossed with nine to twelve transverse brown bands.

The anal fin is longer than in the female, but less deep, and is of the same height as the dorsal fin. Toward their posterior extremities both fins become pointed. Both these fins and the caudal fin are densely covered with black spots, which, however, are absent from the outer parts of all three fins. The unpaired fins at their free edges are margined with black, which is also seen on the extremity of the ventral fin and occasionally on the pectorals. The height of the body is equal to one and two-thirds its breadth; and is about one-fourth of the length of the fish.

There are twenty teeth in the lower jaw in both sexes; the upper edge of the lower jaw is margined with black. In the pre-maxillary there are twelve to fourteen large recurved teeth in front, with a row of very small teeth behind. There are twenty-nine scales in a line between the head and caudal fin. Over the ventral fin there are eight to nine scales in a transverse row. The scales are relatively large and are marked with many concentric rings. The upper side of the head and forward part of the back are broad and flat, and the body reaches its greatest height over the ventral fin. The alimentary canal forms two simple folds, and has more delicate walls than in *Cyprinodon ibericus*. The male measures $2\frac{3}{10}$ inches in length, while the female has a length of $2\frac{8}{10}$ inches.

The habits are similar to those of Cyprinodon. It is recorded by Dr. Steindachner from among the vegetation in the canals of Valencia, about Seville, in Catalonia, and in the Albufera lake.

## Family: MURÆNIDÆ.

### Genus: **Anguilla** (Cuvier).

Eels have an elongated cylindrical body, which is spotted with rudimentary scales embedded in the skin. There are no ventral fins in any of these fishes; and they were hence placed, in an older classification, in a group named Apodes. The fins of the Eels are essentially those of immature fishes. For Alexander Agassiz has shown in the young condition of many fishes the existence of a fin extending from the dorsal surface round the tail to the ventral surface. And this condition persists through life in Eels, so that neither the dorsal nor anal fins are separate from the caudal. The upper jaw is formed by the pre-maxillary bones in front and the maxillary bones at the sides. The jaws are equal in length, and carry small teeth in bands. The gills open by narrow slits at the base of the pectoral fins.

The species of the genus were formerly made on the form of the snout, size of the eyes, width of the bands of teeth, and other characters which are now known to be very variable, and Dr. Günther finds more valuable characters for classification in the position of the dorsal fin, in advance of, or behind, the vent, in the teeth of the mandible being in one or in two series, in their relative size, in the relative length of the tail to the body, the nature of the lips, the cleft of the mouth, &c.

The larger number of species of the genus are found in the south-east of Asia, the south-east of Africa, the Malay Archipelago, and Australia, the West Indies, and the United States. Probably only one species occurs in Europe.

Eels differ from other migratory fishes in habit, for while they for the most part come into rivers to spawn, and descend to the sea to grow, Eels grow in the rivers and spawn in the sea.

## Anguilla vulgaris (Turton).

The Eel (Fig. 172) is a very variable fish, known in Germany as *Aal*; as *Auhor* in Bohemia, and *Wegorz* in Poland; to the Swedes it is *Al*; it is *Anguille* in France, and *Anguilla* in Italy; it is *Anguilas* in Spain, and *Enguia* or *Eiroz* in Portugal. The Eels of the Nile were famous in the earliest ages, and are said to have been worshipped, and Dr. Badham pointedly remarks that the divine honours paid to the race by the sons of Ham could not preserve them from the jaws of gluttonous Greeks. A gentleman of ancient Greece is reported to remark, "You Egyptians worship the Eel as a deity; your idol is my idol too; I adore him in a dish." The tastes of the luxurious Romans for Eels have descended to the lazzaroni of modern Italy, who, in Naples at least, feast on Eels, if possible, as the choicest Christmas fare.

In England they are said to have given names to Ellesmere on the Mersey, and Elmore on the Severn; not to mention Ely, which has always been famous for its great abundance of eels, though modern drainage of the fen-land has greatly limited the areas over which they are now distributed. The size varies with the locality, and in the Eastern counties they are taken from time to time of from four to ten pounds' weight. Buckland met with no larger samples, though Yarrell saw the skins at Cambridge of Eels taken at Wisbech, which weighed twenty-three and twenty-seven pounds respectively.

The head of the Eel is from one-eighth to one-ninth of the length of the fish. The vent is four times the length of the head behind the extremity of the snout. The breadth of the head is from two-fifths to a third of its length. The height of the head is one-half its length. The greatest height of the body exceeds its thickness by about one-third, so that the transverse section of the fish is vertically ovate. The cleft of the mouth is between one-fourth and one-third of the length of the head, and in the broad-nosed variety its length is equal to the width of the head in front of the eye. The eye, which is rather small, is covered with transparent skin, as in the *Misgurnus fossilis*. It is placed above the angle of the mouth, and separated from the other eye by a frontal interspace, which measures from once to twice the orbital diameter. The distance

Fig. 172. ANGUILLA VULGARIS (TURTON).

of the eye from the snout varies with the bluntness of the nose, but is never more than twice the diameter of the eye. The lips are thick and fleshy; the lower lip may project slightly in front of the lower jaw. On the upper lip, near the end of the snout, there are two short open pores, projecting almost like aborted barbels. The nares are simple, oval, and near to the upper border of the eye. The premaxillary, maxillary, vomer, and mandible, are covered with small teeth of uniform size, in bands. The lancet-shaped tongue ends in a free point. The gill-aperture is a half-moon-shaped cleft in front of the base of the pectoral fin, which is ovate. There are ten branchiostegal rays, which are long and slender, and united to the skin which covers the head. Four gill arches are connected with the small operculum.

The dorsal fin commences at a distance behind the snout of two-and-a-half times the length of the head, and extends down nearly two-thirds of the total length of the body of the Eel. The rays at first are very short, being only about one-quarter of the height of the body, but become much higher as they pass backward, and, at the end of the tail are twice as high as the depth of the tail. The dorsal fin passes insensibly into the caudal, and this similarly passes into the anal, which is as much developed as the dorsal, and extends forward to the vent, which is three-and-a-half times the length of the head from the snout. The rays are so soft, curved, and thickly covered with skin, that they are counted with difficulty; but, according to Benecke, there are 1,100 rays in the combined dorsal, caudal, and anal fin. There are eighteen or nineteen rays in the pectoral fin.

The scales of the Eel are peculiar. They are extremely delicate, and are arranged obliquely in two directions, at right angles to each other, so that they give the skin an irregular, puckered appearance. They are small, elongated, and transparent, formed of a single layer of cells, which are arranged in concentric rows. The largest scales have a length of about two lines. They nowhere overlap, but form a sort of network, with scaleless interspaces. There are smaller scales on the head, and encircling the bases of the fins. The lateral mucus canal is so large, and its walls so thick, that it is visible to the naked eye in transverse section. It opens with simple pores. The cephalic canals are not less well developed; the branch which runs over the eye terminates, as already remarked, in a short barbel pore behind the upper lip. There is a suborbital branch, and there are seven or eight pores on the branch upon the lower jaw.

The air-bladder is large, and one-fifth the length of the fish; its anterior blunt end is behind the stomach, and the pointed posterior end reaches back to the vent. And there is a pneumatic canal which rises between two heart-shaped bodies, and becoming narrower anteriorly, opens into the dorsal side of the œsophagus. The urinary bladder is rather large, and more triangular than pear-shaped. Its opening is close behind the vent.

The colour of the back, down to the middle of the side, is dark green. The upper part of the head is darker, but shades into brown. The belly, throat, and lower jaw are white, sometimes with a silvery lustre, sometimes yellowish. The dorsal and caudal fins and hinder part of the anal fin have a brownish-green colour, darker than the back. The pectoral fin is brownish-black with a black border. The iris is of pale gold, almost silvery, but the colours vary with the locality and season of the year. The pectoral fin and anterior part of the anal are often pale, with a slightly reddish border. The anal, like the dorsal, may have a yellowish-green border.

The Eel ordinarily lives in deep water; it is common on muddy bottoms, but prefers clear water. It is a voracious fish, and cases are known of Eels eating each other, as well as snakes, water-rats, young water-fowl, and river crayfish, but from the small size of the mouth their usual food is the fry of other fishes, fish-spawn, worms, insect-larvæ, and decomposing animal substances. Probably the Eel feeds on fresh-water mollusca, but takes its prey at night when other fishes are at rest, while by day it hides itself in holes, or buries itself in mud. It is said often to leave the water to seek food on the grass and on cultivated fields. It certainly may sometimes leave the water, but we are aware of no evidence concerning the reason for this habit. Yarrell suggests that Eels hunt frogs, and it is quite possible that, observing the tadpoles leave the water as they become frogs, the Eels may follow them; or, as worms leave the ground at night, they may not improbably hunt worms. It has also been suggested that they quit unsuitable waters to find others which please them better.

The Eel is proverbially tenacious of life. It can remain for a long time out of water, perhaps because the narrow gill-opening enables the gill-chamber to retain some moisture. It is susceptible to changes of temperature, and is absent from the northern part of Scandinavia, apparently owing to the cold of winter. As the cold season comes on, Eels which have not gone down to the sea, everywhere bury themselves, and again become active in the spring. Yarrell quotes Mr. Young as stating that on moving a gravel bank in the River Shin in Sutherland, in October, he found it swarming with young Eels at a depth of six to fifteen inches. Mr. Young concluded that the Eels had spawned in the gravel, but it seems more probable that the cold weather had driven them under ground to enjoy the radiated heat of the earth.

Desmarest, of the Jardin des Plantes, had an Eel which had been domesticated for thirty-seven years. From 1838 till 1853 it was kept in a large earthen pan in a room, the water being changed once a week. The vessel was too small for it to rest otherwise than coiled. It was subsequently placed in summer in a large zinc tank holding twenty buckets of water, changed every fortnight; but returned in winter to the smaller vessel. It was fed on beef sliced into worm-like forms, which were seized upon while falling, but were never eaten after they had reached the bottom. It refused worms and little fishes, and would take neither bread nor vegetable substances. It took food freely from April till October, and during the winter abstained from all nourishment. During the hot season it fed only once a week. It was perfectly tame, recognised those who fed it, and never bit any one. It swam only in the morning and evening; but when the temperature grew hotter, its movements were more active. In the month of May it became more quiet even than in the winter, and then deposited small white bodies which were regarded as eggs. It subsequently became greatly agitated, and on two occasions threw itself out of its tank, and was found on the gravel, where it remained without movement, nearly dead. On another occasion it was frozen in winter, but by thawing with tepid water was liberated. Yarrell speaks of an Eel that was frozen and buried in the snow, and then after four days placed in water and slowly thawed, when it soon recovered.

The reproduction of the Eel is still imperfectly understood, and from the time of Aristotle has been a mystery which invited the wildest conjectures. As Eels were seen at times to come out of the mud, they were supposed to be bred in the mud. The viviparous Blenny (*Zoarces viviparus*), popularly known in Germany and Italy as the Eel-mother, has been supposed to be the parent of Eels. In many parts of England village people still believe that Eels are generated out of hair from a stallion's tail, cut into short lengths. The eggs were discovered nearly simultaneously about 1780 by Mondini of Bologna, and O. F. Müller; but it was not till 1838 that Rathke described the ovaries fully. They are two yellowish or pale-red band-like organs, as wide as a finger, which extend along both sides of the vertebral column for nearly the whole length of the visceral cavity. They are broadest in the middle and become narrower behind, where they are closer together. They have numerous transverse folds, which are narrow, but somewhat unequal in size, and the folded appearance is lost posteriorly. There is no oviduct, and the eggs are discharged into the cavity of the body, and make their exit by two apertures, covered by a membranous fold, between the vent and the urinary opening. The ovaries contain so much fat, that, although very conspicuous, their true nature is easily overlooked; and the eggs, being immature, are commonly only one-tenth of a millimètre in diameter, and therefore only to be discovered by the microscope. Each egg is surrounded by a net-work of fat cells; but in a young fish of twenty centimètres there is much less fat, and the eggs are more easily found. The number of eggs may be estimated at many millions. The male Eel was discovered by Syrski in 1873. It is smaller than the female, being rarely over fifteen inches long. It is found only in brackish water, and in the sea. The male contains, in place of the ovaries, a milt of different shape. This consists of two tubes which run down both sides of the abdominal cavity, and in the middle of their length have an immense number of small vesicular expansions. It is as difficult to discover ripe spermatozoa in these organs, as to find eggs in the ovaries of the female. The male fish has a more pointed head, with large eyes, lower dorsal fin, darker back with stronger metallic lustre, and a white belly.

There are some early statements that Eels of the most minute size have been taken in ponds where those ponds are unconnected with streams; which seems to leave the question open whether there may not in some localities be land-locked races of Eels.

The Eel is stated to spawn once in its life and then to die. Benecke quotes some observations made by Dr. Jacoby at Trieste, in which it is affirmed that the dead bodies of female Eels, with the ovaries empty, are seen in great numbers floating off the mouths of rivers; but confirmation of so interesting an observation is desirable.

The eggs steadily increase in size during the later part of the year. In August their diameter is ·09 of a millimètre; in September ·10 of a millimètre; in October ·16 of a millimètre, and in November from ·18 to ·23 of a millimètre. Nevertheless, fishes were found later, in December and January, in the rivers and Haffs on the Prussian coast, with eggs having a diameter of only ·03 to ·09 millimètre. Hence it appears that the Eels which were going to spawn had already left for deeper water, and that all Eels do not spawn every year.

The young come up from the sea in myriads, travelling in North Germany at night in March and April, or in some streams as late as May. Old fishes have never been observed returning with them, so that the old Eels which run down to the sea on stormy nights in the autumn, are lost to the fisherman. Crespon records seeing, near the mouth of the Rhône, a mass of these young Eels, each about two inches long, united into a huge globular lump. It constantly rose and descended in the water; and the young gradually became detached into a kind of rope, and ascended the river, keeping to the banks, and entering every creek and tributary. The procession lasted for fifteen days.

Dr. Ehlers saw a similar migration in the Elbe. The young Eels, three to four inches long, kept so close to the bank that they followed its every outline, in a band about a foot wide and of unknown depth, which was observed to pass without intermission for nearly two days. These great swarms, however, are not observed every year. In Britain the Eel-swarms, in Sutherland and the more northern rivers, come up the rivers in April or May, and the similar Eel-fare in the Thames is usually at the same period.

When the young are observed in May they rarely reach a length of four inches, but when they go down stream in October the length is nine and a half to ten inches.

In the rivers and lagoons on the coast of Italy the fry of the Eel, according to Professor Giglioli, swarm during December, January, and February, in countless myriads. They are vermiform and semi-transparent, are known as *Lèche*, and give occasion for active fishing.

In some localities, the Elvers, as the Eel fry are termed, are caught, scoured, cooked, and made into cakes, which are sometimes, in the west of England, cut in slices, and fried. Before the middle of this century the New River Water Company in London occasionally distributed Elvers with its water, so that it was no unusual circumstance for a number to be found in a cistern, and for water-pipes to be occasionally stopped by Eels, which had found their way into them, though this accident has since been guarded against by improved filtration. With the slightest assistance they ascend waterfalls, and the Irish proprietors have for a long time constructed ladders of straw bands to aid their ascent.

In rivers Eels are commonly taken at weirs with wicker baskets, known as Eel-bucks, which have a sugar-loaf form. Nets are also used, set across the stream on dark nights, when the Eels run. When captured with a line the baited hooks are sunk to the bottom. The fishes are also largely taken by spearing, especially in the winter, when they frequently hybernate, and the Eel-spear is put through the shallow water into the mud; or occasionally they are dug out when the disappearance of frost on the banks indicates their presence. The most celebrated fisheries on the Continent are those of Comacchio, near Venice, from whence they are sent all over Italy. At Elbœuf, on the Seine, Eels are taken in great numbers, and the fame of Narbonne, in the south of France, is not less for its Eels than for its honey. The English market is largely supplied from Holland. They are rarely eaten in Scotland, and are not eaten by the Jews.

The flesh is highly prized, firm, and can be cooked in a variety of ways. Eel pies were formerly in general demand, and Eel-pie Island, in the Thames, is still known for the commodity from which it derives its name.

Small Eels have always been an important item of food with the poorer classes when stewed, while there are few limits to the attractions which Eels may not offer in the hands of an experienced cook.

Eels have many enemies, and are preyed upon by carnivorous fishes, water-birds, and mammals which frequent streams. They are occasionally liable to epidemics, and die in vast numbers.

Their distribution in Europe is exceptional, being absent from the Danube and south-east of Europe, the Black Sea, the Caspian, and rivers which flow into them, though they are met with in Italy, and, according to Steindachner, in the Minho, Douro, Tagus, and near Madrid in Spain.

In a species so widely distributed, ranging through the north of Africa, and, according to Günther, found in North America, larger variation might have been expected in the number of vertebræ than has been observed.

In the body the vertebræ number from forty-five to forty-six, and in the tail from sixty-eight to seventy-one.

Pennant distinguished an Eel under the name of Grig, which has been long known as a distinct species, *Anguilla latirostris*. Its distribution is world-wide, being found not only throughout Europe, but recorded in the British Museum Catalogue from the Nile, China, New Zealand, and the West Indies. It has no peculiarities of habit, and is not known to exceed five pounds in weight; but although the proportions are somewhat different from those of the Common Eel, it may be doubted whether this broad-nosed form is more than an extreme variation of the common type.

The head is flattened, broad, and rounded behind; the eye large; the lips are fleshy and broad; the angle of the mouth is behind the back of the eye; and the tail is much longer than the body.

## Anguilla eurystoma (HECKEL AND KNER).

There is an Eel found in Dalmatia, growing to a length of twenty-two inches, which is imperfectly known, but is remarkable for the great length of the head; for while in the Common Eel the body is usually from eight to nine times as long as the head, in this Dalmatian variety the body is only a little over six times as long as the head. The head is nine and a half times as long as high, and the cleft of the mouth is equal to the height or greatest breadth

Fig. 173.—ANGUILLA EURYSTOMA (HECKEL AND KNER).

of the head. The head is ten times as long as the diameter of the eye; and the eye is only twice its diameter from the snout. The snout is rather broad, and is nearly one-sixth of the length of the head.

The lower jaw is rather more projecting than in the Common Eel. The teeth are stronger, and the bands on the vomer are longer and broader. The gill-opening is larger, and it stretches deeper under the pectoral fin on the throat. The pectoral fins are short, but the longest of the twenty-one rays scarcely exceeds the length of the mouth. The dorsal fin commences one-third of the length of the fish behind the snout, and is of nearly uniform height. The vent is in advance of the middle of the length. The anal fin is similar to the dorsal. The tail is strongly compressed, and gradually decreases in height. The lips, dentition, barbel-pores, nares, skin over the eyes, lateral and cephalic canals, and colour, are similar to those of *Anguilla vulgaris*. The scales also are similar, only they are larger.

## CHAPTER XI.

### FRESH-WATER FISHES OF THE ORDER GANOIDEI.

FAMILY ACIPENSERIDÆ –GENUS ACIPENSER: Acipenser glaber—Sterlet– A. gmelini—A. stellatus—A. schypa– A. guldenstadtii– A. naccarii– A. nardoi– A. heckelii– A. nasus–– The Sturgeon—A. huso.

## FAMILY: ACIPENSERIDÆ.

### GENUS: Acipenser (ARTEDI).

THE Sturgeons are a small group of fishes, forming a distinct division of the class. The body is elongated, and almost cylindrical, tapering conically to the tail. They have the tail fashioned on the heterocercal plan, in which the rays are chiefly developed on the lower lobe of the caudal fin, and the vertebral column is prolonged into the upper lobe. The skeleton is cartilaginous. The skin is commonly armoured with bony bucklers, which, in this genus do not unite into a continuous armour, but occur in five longitudinal rows, leaving the skin between them exposed, or defended only with small scattered bony scales.

The head has the snout produced far in front of the mouth, which is situate on the under side. Between the mouth and the extremity of the snout are four barbels in a transverse series.

The cartilaginous framework of the head is covered with armour. The shields are more or less quadrate. The elongated snout is formed by a prolongation of the vomerine element below, and the nasal and ethmoid elements above. The nostrils are double, and in front of the eye. The mouth is protractile, toothless, transverse, relatively small, and lies in a cartilaginous protuberance, formed of three elements.

Its anterior edge is margined by thick lips, which are often rudimentary at the angle of the mouth.

The eyes are placed laterally, behind the nares, and commonly differ in size on the two sides of the head. The gill-cover appears to correspond partly to the operculum of bony fishes, and partly to the sub-orbital arch. The gills are arranged like those of bony fishes. The gill membranes at the throat are confluent, and united to the isthmus between them. There are no branchiostegal rays. The true gills are four in number, comb-like, with their points free, and there are two accessory gills. On the upper edge of the operculum there is a small spiracle.

The five rows of shields, which defend the body, are arranged in the following way: first, a median row in the middle of the back; secondly, a lateral row above the middle of each side; and thirdly, a row on each ventral angle, which extends back to the ventral fin. As each of these scutes has a sharp edge or keel running along its middle, they impart to the fish a somewhat pentagonal aspect. Between these rows of plates the skin is not entirely naked, but is partly covered with smaller shields or bony protuberances of different forms and sizes, and is smooth where these are not present. But towards the end of the tail the bony scutes are flat, small, quadrate in form, and closely packed. They extend upon the upper lobe of the caudal fin. There are two large scutes on the under side of the head, angular posteriorly, which are placed behind the gill-aperture. All the smaller scutes are extremely variable, differing in form and size with the age of the fish; and occasionally some of the larger scutes are lost. The fins are seven in number, and are formed of closely-placed, compressed, and jointed rays, which are usually curved, and finely denticulated on both sides. The first ray of the pectoral fin is strong and bony; the denticulations on the rays are directed backwards, and are very unequally developed in different species, and are often mere asperities. The dorsal fin is placed far back; the anal is opposite to its posterior part. The ventral fins are close in front of the vent and much smaller than the pectoral fins, which are rather powerful and long.

The vertebral column extends to the extremity of the upper lobe of the caudal fin, within which it is directed upward with a strong curve. The upper lobe of the fin is much longer than the lower, and makes some approach to the scythe-like form sometimes seen in the tails of Sharks. The intestine is short; the fleshy stomach is divided by a pyloric valve from the small intestine. The large intestine terminates in a spiral valve, like that which characterises the intestines of Sharks and other fishes of Günther's group, Palæichthyes. The liver is large and firm; it forms two principal lobes and many accessory lobes, and it surrounds the biliary duct. The pancreas is subdivided into numerous pyloric appendages, and opens into the intestine by a wide outlet near to the biliary duct. The kidneys are narrow in front, and wider posteriorly, and extend behind the vent, where the urinary duct opens in a depression, which also includes the apertures of the reproductive organs.

The ovaries and milt both extend through nearly the whole length of the ventral cavity. The ovaries are not in connection with an oviduct, and open freely into the ventral cavity by a funnel-shaped process. The milt is an agglomeration of small rounded bodies like currants. The air-bladder, which lies behind the stomach, is a large long or oval sac, and opens by a pneumatic canal into the dorsal wall of the œsophagus.

The bulbus arteriosus has two rows of valves at its commencement, and a third row at its termination. The large aorta extends in a channel on the under side of the cartilaginous vertebral column.

Günther remarks that the geographical distribution of Sturgeons is nearly identical with that of Salmon. They are scattered over the whole of the northern regions of the earth. Some species are confined to fresh water, but they mostly live in the sea, and go up the rivers into lakes. In some streams they are found in incredible multitudes, and remain in them for months. Towards winter they seek deep holes in the rivers, or more commonly the mouths of rivers and inlets of the sea, where they crowd together and pass the winter in a hybernating condition. According to Lepechin, they burrow their heads into the mud, so that their tails stick up straight in the water like a fence. Sturgeon are among the largest of freshwater fishes; and they occasionally reach a length, in the larger species, of eighteen feet. They are very voracious, and live chiefly on soft-bodied animals, worms, spawn, and bottom-feeding fishes; but many also take water-birds, which are swallowed whole. Sturgeons increase very rapidly, but the numbers in the rivers of western Europe have greatly diminished. Even in Hungary they were formerly an important source of income to the fishermen, and were commonly taken of seven or eight hundred pounds' weight, while occasionally fishes were captured twice as heavy. Forty or fifty years ago it was rare to see a fish in the Vienna market of less than one hundred pounds, and now only very small ones are caught. The diminution in size is due partly to the capture of spawning fishes, and partly to excessive fishing, which has reduced the value of the fisheries, just as the Salmon fishing of the Tweed and other British rivers was injured in former years.

In Britain the Sturgeon is only an occasional visitor; and by far the larger number of individuals as well as species is found in the south-east of Russia. Several species are North American, and are met with on the Pacific coast and in Californian rivers, on the Atlantic coast, in the Mississippi, and in some of the great lakes. Certain of the species are common to Europe and the North American continent, while there is a peculiar genus limited to the Mississippi and its tributaries, which has no spiracles, and has the tail completely enveloped in bony armour.

Sturgeon fishing is carried on in different ways in summer and winter. In autumn, the Sturgeons which hybernate bury their noses in the mud on the bottom of a river. About January the fishermen assemble, and at a given signal, get to work with poles upward of sixty feet long, each having six to ten prongs at the end. Holes are made in the ice. This operation disturbs the fishes, and this instrument put down through the holes often strikes one of them, as it swims away. Many fishermen have the luck to take ten large Sturgeon in a day; but sometimes the fisheries barely pay their expenses, for many days may pass without a fish being speared. This mode of fishing is pursued on a large scale by the Cossacks of the River Ural. The first fish is given to the Church, and the others are sent away on sledges as fast as possible. Frequently travelling fish-merchants buy up the fishes and prepare the roes and flesh for market.

In summer, fishing villages are established near the mouths of the great Russian rivers. Russian or Greek merchants hire pieces of the banks, and construct the necessary buildings for storing salt, and for sleeping accommodation for the fishermen, from twelve to twenty of whom share a hut. Mills are set up for grinding the salt, boats are provided, and people of many races are engaged in the work. The men live on fish, usually with mutton on Sundays. An elevated outlook is set up on the bank, in which a man watches the approach of the shoal, and is often able by its characteristic movements to distinguish the species from an immense distance, as the fishes move up the river. At Rubinsk, on the Volga, in the Russian government of Yaroslav, the fisheries in spring and summer draw together a hundred thousand people, who work continuously for the season, and return to their homes in winter.

Fifteen thousand fishes have been taken at one fishery in a single day; and when the fishing has been intermitted for a day, the fishes have sometimes completely blocked a river 28 feet deep and 360 feet wide, so that the backs of the uppermost appeared above the water.

## Acipenser glaber (MARSILIUS).

The Austrians term the *Acipenser glaber*, *Glattdick* or *Glatt-stor*, but it reaches Austria only through the Danube, its home being in the Black Sea and Sea of Azov, from which it ascends all the rivers flowing into these waters. It is not a large species, and no specimens have been taken in Hungary weighing more than sixty pounds, and fishes of less than thirty pounds are not often seen (Fig. 174).

In this species the entire length of the fish is about five and a half times the length of the head. The snout is short and rounded, broad, and rather thick, and the profile rises rapidly from it to the first dorsal shield at the back of the head, where the body attains its greatest height. The armour on the skull, which only sheathes the thick cephalic cartilage, corresponds in a remarkable way with the bones which usually enclose the brain, and consists of flat, long, rather small shields, which are furrowed with radiating grooves placed close together (Fig. 175). At the back of the head, towards the median line, are two small unequal spike-shaped epiotic shields, which have their broad bases joined to the first dorsal shield. A median supra-occipital shield, somewhat Y-shaped,

Fig. 174.—ACIPENSER GLABER (MARSILIUS).

extends forward from the epiotic shields like an inter-parietal element, and divides the two largest scutes of the head from each other at their posterior end. These large plates are the parietal scutes. They exceed half the length of the

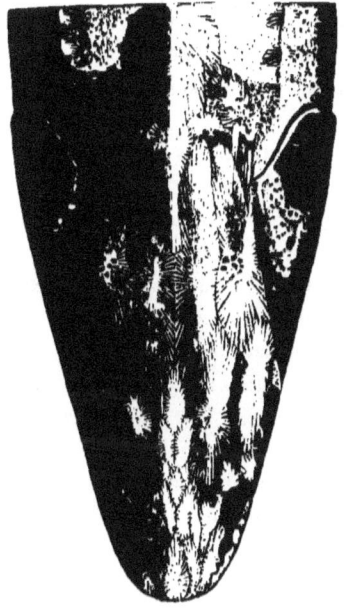

Fig. 175.— UPPER SURFACE OF HEAD OF ACIPENSER GLABER, SHOWING DERMAL SCUTES.

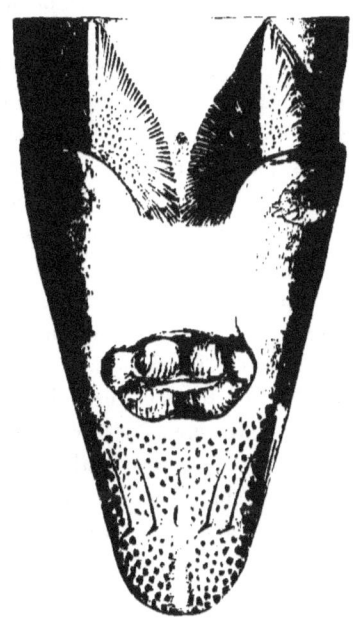

Fig. 176.— UNDER SIDE OF HEAD OF ACIPENSER GLABER, SHOWING MOUTH AND BARBELS, AND PORES ON THE UNDER SIDE OF THE ROSTRUM.

head, and are longer than in any other species. Flanking them laterally, are the similarly long squamosal shields, which are widest in front. Another pair of shields stretches forward from the angle between the parietal and squamosal. These are the frontal shields. They are not in contact in the middle line, because a number of small plates come between them, and extend backward between the anterior ends of the parietal shields. These small ossifications, of which there may be about eight, represent the ethmoid bone, which in birds often occupies a large area anterior to the frontal bones. External to the frontal plate, and coming above and behind the eye, so as to abut against the squamosal, is the long post-frontal shield. At the anterior margin of the frontal, above the nares, is the pre-frontal shield, and there is a small nasal scute near the nasal margin.

There are a few other scutes in front, extending to the extremity of the snout, which would be regarded as pre-maxillary and maxillary plates, if it were not that the jaws are far removed from the extremity of the rostrum, so that they must probably be regarded as distinct in origin from the bony element so named in the skulls of other vertebrates. Behind the epiotic, there are shields which connect the back of the skull with the pectoral arch. The under side of the head is naked, but the skin is in places rough with ridges, though in front of the gill-opening the throat is defended with one long shield on each side. The continuity of the armour of the head is broken only by numerous large pores, which open through the soft skin, and run in the usual directions of the cephalic canals, and give off similar branches to the hinder part of the head, and the supra-orbital and infra-orbital areas. The latter branches run to the end of the snout; but numerous pores, irregularly placed, extend beyond them, and cover the entire under surface of the head, as far back as the upper lip (Fig. 176). The eyes are small, a little elongated, and protected by a strong projecting, scutal margin. They are nearer to the end of the snout than to the mouth. The breadth of the frontal interspace between them is equal to six and a half times the orbital diameter. The nares are oval, and small, but the lower aperture is larger than the upper one. The barbels, which are nearer to the snout than to the mouth, are rather long, and reach back nearly to the upper lip. They are round at their bases, are rather compressed for half their length, and thence are distantly fringed. The lips form a thick cushion round the wide mouth; they are deeply notched in the middle, but not divided.

The body is a little higher than wide; and its diameter is one-eighth of the total length. All the shields of the five longitudinal rows upon it, are rather distant from each other. Only the scutes of the median dorsal row are well developed; the rows on the sides are small, and those on the ventral margins are very small. The number of dorsal shields varies from twelve to sixteen. The first dorsal shield is largest and highest, and the remainder of the dorsal series gradually decrease in size. Except the last, which is lozenge-shaped, these scutes all have a heart-shaped base, and a blunt keel, which ends in a slight hook. Their radiated ornament is feebly developed. The lateral shields, according to Heckel, increase in number from thirty-five in the young to sixty in the adult. They are lozenge-shaped, have only a slight ridge, and decrease in size posteriorly. The lateral line is seen between them as a row of small, round, bony scales, which overlap each other, like the slates on a roof. The ventral shields are only twelve to fifteen in number. They are most distinct in front, but are far apart, and are scarcely perceptible behind. There are no shields on the tail, behind the vent and the dorsal fin. On the anterior part of the belly behind the gill-openings, the surface is covered and defended by two large shields, longitudinally keeled, pointed behind, and beautifully sculptured with radiating rays. The skin between the rows of scutes is thickly covered with small bony scutes, which are denticulated posteriorly, but become smaller towards the belly, and lose their denticulations.

The pectoral fin consists of a very strong bony ray, and thirty jointed soft rays, of which the longest is equal to the thickness of the body. The length of the base of the dorsal fin exceeds its height; its end is opposite to the anal fin. The height of the anal fin exceeds the length of its base, and it is longer than the ventral fin. All these fins are truncated behind. In the caudal fin the long upper lobe has thirty-nine rays, and the lower has sixty-five jointed rays, and sixteen unjointed rays. There is an unsymmetrical bony shield in front of the lower lobe of the caudal, in front of the anal, and in front of the dorsal fins.

The colour of the back is reddish-grey, paler towards the sides. The belly and lateral shields are dirty white. The barbels are white. The iris is yellow.

The young in this species are distinguished by the form of the snout, which is relatively longer, more pointed, and more curved upward, while the median ridge on its under side is more strongly developed. Hence the relative length of the head is then different, while the barbels are farther from the extremity of the snout than in the adult. Moreover, the growth of the shields is not proportionate to the growth of the body, so that in the young fish the shields are nearer together in all the rows, while those of the back partly overlap each other, and all have the keels sharper and hooks stronger. The caudal fin is then less developed, and the lower lobe is scarcely one-third as long as the upper. The pectoral fins, on the other hand, are remarkably developed in the young, and are nearly as long as the head.

*Acipenser glaber* grows to a length of six or seven feet, and may attain a weight of sixty pounds, but specimens of more than three or four feet length are not often captured.

# Acipenser ruthenus (Lin.).—The Sterlet.

The Sterlet is found in the Black Sea, the Sea of Azov, and Caspian, and all the rivers which flow into them, though it is absent from the south of the Caspian, and only stray specimens are occasionally taken in the River Kur. It is also abundant in the rivers of Siberia, which fall into the Arctic Ocean, and especially affects the Irtish. According to Dr. Grimm, the Sterlet made its way about forty years ago through canals into the Northern Dwina, which flows into the White Sea by Archangel, and here, under the cold conditions which are so necessary for it, it has not only settled down and become abundant, but acquired a short blunt snout and an arched back, and has developed a flavour which makes it more prized in St. Petersburg than the Sterlet from the

Fig. 177.—ACIPENSER RUTHENUS (LINNÆUS).

Volga. Indeed, in that river it deteriorates in flavour the farther south it is taken. It ascends the Danube regularly as far as Vienna, is not rare at Linz, and is often got at Passau and in Bavarian waters. It rarely reaches a length of three feet, and in the Danube is seldom over two feet. Fishes weighing over seven pounds are scarce, and the demand has been so great that in the Volga and Dnieper, Sterlet only four or five inches long are now captured. The flesh and roe of this species are more esteemed than those of the larger Sturgeons (Fig. 177).

The fish is four-and-a-half times as long as its head, and there is very little difference between the width of the back of the head and the fore part of the body. The body is highest directly under the third dorsal shield, and this measurement is half as long as the head. The small snout is rather flatly compressed, and three times as long as the mouth aperture is wide. The upper part of the head is covered with irregular radiated shields, which have raised central points and correspond in position and relation to each other with the bones which in most vertebrates roof-in the head.

The forms of the shield are altogether characteristic, and the strongly elevated central point gives the head a curious tuberculated appearance. The supra-occipital shield has a long blunt triangular form; the epiotic ossifications are rather longer and quite as wide. The parietal ossifications terminate well behind the eye, where the notch between them is occupied by the small divided ethmoid. The margins of the squamosal scutes are parallel to those of the parietal scutes, external to which they are placed. The frontal scutes are small, the post-frontal is notched out above the eye, and there are no distinct pre-frontal or nasal ossifications. The small ethmoid shields cover the rostrum to the extremity of the snout, but they are so closely united together as to have the aspect

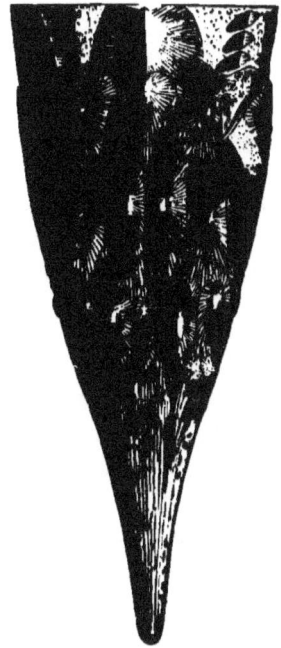
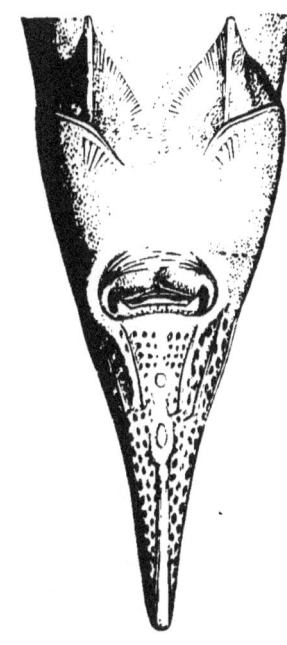

Fig. 178. ACIPENSER RUTHENUS, SHOWING THE CRANIAL SCUTES ON THE UPPER SIDE OF THE HEAD.

Fig. 179. - ACIPENSER RUTHENUS, SHOWING THE UNDER SIDE OF THE HEAD, MOUTH, BARBELS, PORES, ETC.

of a single bone (Fig. 178). The cephalic mucus-canal subdivides in the usual way, and opens by numerous pores which cover the under side of the rostrum, except along the median line, in which there are three tubercles, two of which lie in front of the barbels. The barbels are nearer to the mouth than to the extremity of the snout; their ends are fringed on the inner side (Fig. 179).

The mouth is not exceptionally large, has the upper lip small, and only slightly notched; the lower lip is broader and always divided in the middle, but both halves are in contact. The eyes are small, of equal size, nearly round, and a little in advance of the position of the mouth. The diameter of the frontal interspace is three-and-a-half times the orbital diameter. The opercular shield is small.

The shields on the body are similar in character to those on the other species, though the scutes of the lateral rows are more closely packed together. The number of dorsal shields varies from eleven to fourteen; they have the base broad and nearly triangular, but the last is lozenge-shaped. The first shield is the largest, but the fourth to sixth rise highest. All have a strong smooth median ridge, which is prolonged backward into a rather long pointed hook. They decrease in size towards the dorsal fin, and between the last scute and the dorsal fin there are usually many flat irregular bony plates.

The lateral scutes are sixty to seventy in number, and have each a long oblique lozenge shape. They are sharply keeled in the middle and striped. The number of ventral shields varies from ten to eighteen, they are small, nearly triangular, and ridged, but are more distant from each other than the lateral shields. There are no shields between the dorsal and caudal fins, but three or four flat oval small scutes are found between the vent and the anal fin. Between the rows of shields the skin is covered with small bony scales, which are uniform in size, and become denticulated at their hinder border, but on the belly they become merely rough points.

The fins are strong; the pectorals are longer than the diameter of the body, the bony ray reaches nearly to the extremity of the fin. The dorsal fin is longer than its height, and the height of the anal and ventral fins exceeds their respective lengths. The caudal fin is powerfully developed, and the jointed rays are remarkably long.

The colour of the back is greyish-brown or yellowish-brown. The shields are dirty white. The fins are grey, though the ventral and anal are rather red. The iris is yellow. Albino varieties are occasionally found.

The females have a flatter forehead, and the snout is rather thinner, longer, and more curved.

In young fishes the snout is relatively less elongated than in other species, but the pectoral fin is more elongated, being more than twice the height of the body.

This species spawns in May and June, and yields the best caviare. It lives about six or seven years.

According to Dr. Grimm, it sells at a lower price at Astrakhan than any other Sturgeon, the wholesale price being from two to two-and-a-half roubles the pood of thirty-six pounds, for fishes measuring fourteen inches or more; and one to one-and-a-half roubles the pood for fishes of less than fourteen inches. But when salted the larger Sterlet command a rouble (two shillings and tenpence) the pood additional. It is frequently pickled.

## Acipenser ruthenus. Variety gmelini (F. & H.).

We have already spoken of the liability to variation of the *Acipenser ruthenus*, and Dr. Günther has classed the *Acipenser gmelini* as a further illustration of the capacity of the species for change of character, but the variety is so marked in many structural details that we have considered it necessary to briefly describe it (Fig. 180).

It is distinguished by its broad triangular rostrum or snout from the Sterlet, which it resembles in form of body, and in the thick compressed lozenge-shaped lateral shields. The head measures one-fifth of the entire length, the height of the body is rather more than its thickness, and this is

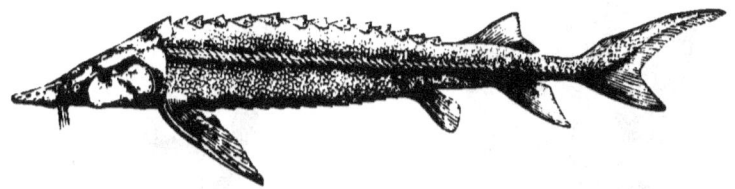

Fig. 180.—ACIPENSER RUTHENUS, VAR. GMELINI (FITZINGER AND HECKEL).

about one-eighth of the total length. The arched forehead rises somewhat abruptly from the snout, and the elevation continues to the back of the head. The greatest height of body is measured from the first dorsal shield to the base of the pectoral fin. According to Heckel and Kner's figure (Fig. 181), there is considerable difference in the form and proportions of the cephalic scutes. First, the supra-occipital scute is small and entirely separated from the epiotic scute in contact with the long parietal scutes. Secondly, the parietal scutes are not divided in front because no distinct ethmoid scutes are developed. Thirdly, the post-frontal scutes, which are external to the frontal, and notched out for the orbit, are pointed behind, and extend back as far as the frontal scutes with their points external to the squamosal scutes, instead of penetrating into those ossifications. There are no distinct pre-frontal or nasal ossifications, and the frontal plates in front are not sharply distinguished from the compact ossifications which extend over the snout. All the sculpturing on the shields is finer than in *A. ruthenus*.

The pores on the under side of the head are similar to those in the Sterlet, as are also the rows of pores in the cephalic canals. The middle bony ridge on the under side of the rostrum decreases gradually in width from its extremity towards the mouth, and is completely broken four times, making four round wart-like tubercles, of which the three anterior are in advance of the barbels. The barbels are rather compressed, and fringed at their ends. The upper lip is small, and slightly notched, and the thickened under lip is quite divided in the middle (Fig. 182).

The eyes are relatively large, and the breadth of the frontal interspace is

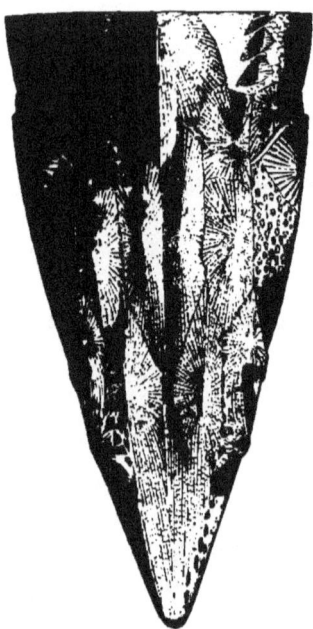

Fig. 181.—ACIPENSER GMELINI, HEAD SEEN FROM ABOVE.

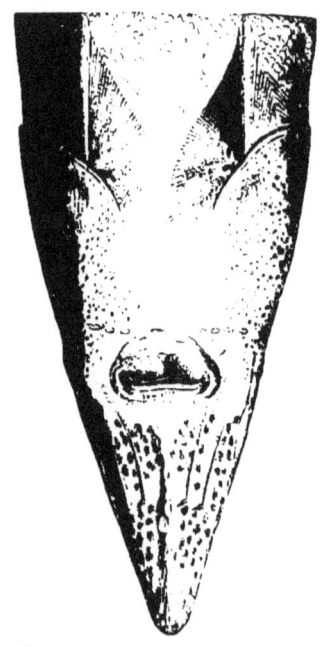

Fig. 182.—ACIPENSER GMELINI, HEAD SEEN FROM BELOW.

three times the orbital diameter. The lower nasal aperture is larger than the upper.

There are fourteen dorsal shields, each with a wide heart-shaped or triangular base, except the last, which, though long, is the smallest. Each has a smooth, sharp medium edge, which ends in a short pointed hook. The lateral row consists of sixty-two narrow shields, which are feebly keeled, have a long lozenge shape, and are notched posteriorly with comb-like denticulations. The ventral shields number thirteen to fifteen; they are very small, but keeled.

There are two or three little star-shaped flat scutes between the anal fin and the vent, and the skin between the rows of shields is covered with denticulated bony scales, like those of the Sterlet, only larger.

The strongly-developed pectoral fin is nearly as long as the head. The anal fin is higher than the dorsal; the ventrals are shortest. The lower lobe of the caudal is half the length of the upper lobe, which is nearly as long as the head; it terminates in a rounded point, as in the Sterlet. The colour is brown above, the belly and shields are yellowish-white. The iris is yellow.

This form is characteristic of the Black Sea, but is nowhere common. It ascends the Danube as far as Pesth, or, occasionally, to Vienna. It spawns in spring.

## Acipenser stellatus (PALLAS).

This species is known in Russia as the *Serruga*, in Bavaria as the *Sterahausen*, and in Austria as the *Scherg*. It is found in the Black Sea and the Sea of Azov and their rivers, and is met with in the Caspian and the Rivers Ural and Kur. Though it grows to the same length as the common Sturgeon, it is a much lighter fish; and at Astrakhan its weight, according to Dr. Grimm, does not exceed thirty pounds, though exceptional fishes of twice that weight are sometimes taken. When four feet long it weighs about twenty pounds; but

Fig. 183.—ACIPENSER STELLATUS (PALLAS).

examples as small as eight pounds are caught in Hungary. It is regarded on the evidence of chemical analysis as the most valuable food fish of the Sturgeon family, and is much esteemed. It begins to ascend the rivers early in March, and great shoals often ascend the Danube for several weeks, but rarely get beyond Komorn, though it goes up the Theiss to Tokay. It has been seen in the Isar. It spawns in May and June, and, after spawning, most of the fishes descend again to the sea, though some remain in the rivers, and they are caught all the year round in the Danube. It feeds on worms, spawn, and mud which contains the smaller forms of life. Hungarian fishermen state that it lives from fifteen to twenty years (Fig. 183).

*Acipenser stellatus* has a more slender body than any other Sturgeon, and the snout is more slender and elongated. The head is as wide as the body, but its diameter is less than half the length of the head. The fish is about four-and-a-half times as long as its head, and the snout alone is nearly one-sixth of the total length. Owing to the elongation of the snout, the upper surface of the head is concave in front, and rises in a moderate convex curve over the skull to the back. As in the other species, the back of the head is covered by bony shields, which are elongated in the direction of the head length, have elevated centres, and are marked by radiating ribs. The first dorsal shield is triangular, and its anterior point only touches the small wedge-shaped supra-occipital shield, which is contained between the hinder ends of the parietal shields. Neither the supra-occipital, parietal, nor first dorsal scutes touch the epiotic shield, which is in contact with the hinder margin of the squamosal. The squamosal, parietal, and supra-occipital plates all terminate in the same

Fig. 184. ACIPENSER STELLATUS, HEAD SEEN FROM ABOVE.

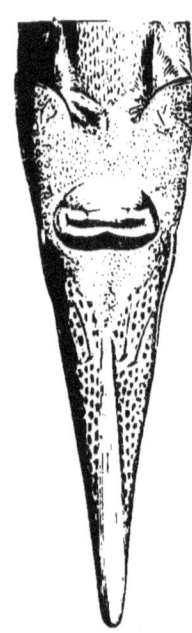

Fig. 185.—ACIPENSER STELLATUS, HEAD SEEN FROM BENEATH.

transverse line. The parietal shields are long, and extend forward to the anterior third of the orbit. The frontal shields are in the usual position, with the post-frontal behind the eye, and the pre-frontal in contact with it, in front of the eye, so as to exclude the frontal scute from the orbital border. There is a broad and deep furrow or depression on the head between the parietal and frontal plates (Fig. 184).

The nares consist of two large contiguous apertures adjacent to the eye and immediately in front of the pre-frontal scute. The rostrum, or snout, is covered with a number of elongated plates, some of which are partly marked with radiating grooves, between which pores open from the cephalic canal. On the under side of the snout there is a long, raised, smooth area, the hinder termination of which becomes narrower, and extends beyond the barbels. The inner pair of barbels is slightly posterior to the outer pair. They do not extend nearly so far back as the mouth (Fig. 185).

The whole under part of the snout, on each side of the median ridge, is covered with large pores for as far back as the barbels reach.

The somewhat large mouth is situate below and rather behind the eyes. It is somewhat transverse, and is margined in front by a small cartilaginous fold of the lips, which is slightly notched laterally. The eyes are small and oval; the right eye is about one-seventh larger than the left, and the breadth of the frontal interspace is equal to five times its diameter. The upper nasal aperture is much smaller than the lower narine.

The shields of the longitudinal rows are well developed, those of the dorsal row are strong, have a heart- or lozenge-shaped base, and a strongly-raised point, which is not the hindermost part of the keel. They vary in number from twelve to sixteen. The first is less firmly united to the back of the head than usual. The two following are the least elevated. The sixth to eighth are highest—all have the characteristic radial ribbing. The lateral shields are lozenge-shaped, with serrated edges, and are often separated by interspaces equal to the length of the base of the shield. They vary in number from thirty to forty.

The ventral shields are ten to twelve in number, they are lozenge-shaped, and similar to the dorsal shields, only much smaller. There are one to three similar shields between the anal fin and vent, but there are none of these small shields behind the dorsal fin. The shields at the sides of the head, which support the shoulder girdle, are regarded by Professor Parker as supra-temporal. They are large and triangular, coarsely radiated, with a slight keel. Between the rows of shields the skin is thickly covered with small denticulated bony scales, which are irregularly placed. They are most dense behind the pectoral fin, and are largest between the dorsal and lateral shields, where they have a star shape, and notched border. On the belly their shape and arrangement are much more regular, and in the middle line, from the throat to the vent, there is usually a sharply-defined median row, especially in the male.

There is frequently a similar row of star-like scales on each side of the ventral shields. All the fins are somewhat small. The pectoral is only one-fifth longer than the diameter of the body. The anal, as usual, is opposite the hinder part of the dorsal. The upper lobe of the caudal is slender and scythe-shaped. The upper caudal lobe is much larger than the lower lobe, and both are pointed.

The colour of the back is a reddish-brown, or, sometimes, bluish-black. The sides and belly are white, but the under side of the snout is flesh-coloured. The shields are dirty white. The iris is yellow or silvery.

In the young the relative length of the head is, as usual, much greater, being, in specimens a foot and a half long, more than one-quarter of the length of the fish. The operculum is better developed; the lower lobe of the pectoral fin is relatively smaller, and all the shields are close together. The females have the snout shorter, the bony ray in the pectoral fin weaker, and the dorsal and lateral shields smaller, though the terminal hook is rather longer.

## Acipenser schypa (GÜLDENSTÄDT).

There is some difference of opinion as to the claim of this fish to specific distinction from *Acipenser güldenstädtii*, and certainly the resemblance in external characters is sufficiently close to justify Dr. Günther in uniting them; but there is some difference in the geographical distribution, and a not inconsiderable difference in the shields of the head, which makes it convenient to

Fig. 186.—ACIPENSER GÜLDENSTÄDTII, VAR. SCHYPA.

keep them distinct, though nothing but an examination of a large series of specimens, such as is possible only in their native home, could test the value of the characters by which we shall define the species (Fig. 186).

Dr. O. Grimm states that the distribution of the *Schyp*, as he terms this species, which is known in Austria as the *Dick*, is the same as that of *Acipenser stellatus*, except that it is also found in the Sea of Aral, where no other Sturgeon occurs. It is rare in the Black Sea and Sea of Azov, though it ascends the Danube as far as Komorn. The great centres of the fishery are in the Sea of Aral and on the rivers Ural and Kur, where it reaches a weight of about sixty pounds. It is not so large in the Danube, and those taken in Hungary vary from twenty-four to forty pounds in weight.

It is a short-nosed species, in which the head is one-sixth of the total length. The breadth between the opercular plates is equal to that of the fore-part of the body, and this measurement is the same as the greatest height of the body, which is over the base of the pectoral fin. The broad rounded snout rises obliquely to the back of the head, where it is overlapped by the first dorsal shield, which, in its position, has quite the aspect of belonging to the cephalic series (Fig. 187).

This shield is nearly circular, has a strongly-elevated keel, terminating backward in a hook, which is the hindermost part of the plate. It is contained laterally between the large epiotic plates, behind which about one-half the scute projects. The supra-occipital shield rather recalls the form of the same element in *Acipenser glaber*, but wants the posterior V-shaped notch, and is not so compressed laterally. Its hinder expansion extends back behind the parietal shields, which it divides for nearly half their length, and in front of it is a small distinct ossification, which still further separates the parietals. The

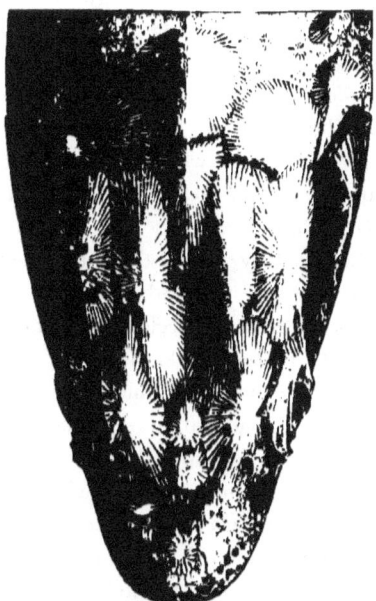
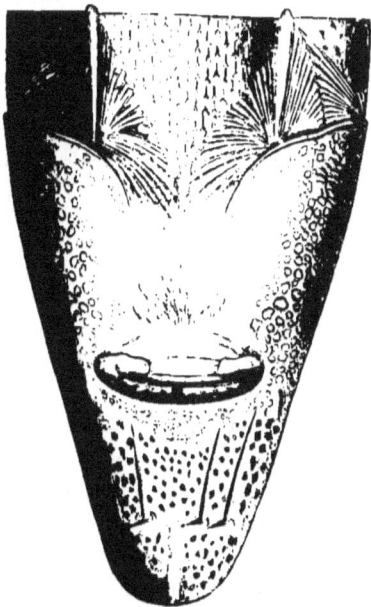

Fig. 187.—ACIPENSER SCHYPA, HEAD SEEN FROM ABOVE.

Fig. 188.—ACIPENSER SCHYPA, HEAD SEEN FROM BELOW.

squamosal is broad, wider than the parietal, and about two-thirds of its length. The frontal plates are somewhat irregular in shape, and do not meet in the median line. The upper nasal aperture is in contact with the outer margin of the frontal scute, while the ethmoid ossifications, three in number, are between the frontals, but do not penetrate far between the parietals. The post-frontal scute is external to the frontal, and forms the upper and hinder border to the orbit. Other shields, some of large size and rayed, cover the space between the nares and the extremity of the snout. The plates are not all in contact, and there are marked interspaces dividing the supra-occipital and parietal from the first dorsal and epiotic, between the frontal and post-frontal, and, to a less extent, between the parietal and squamosal. Pores are developed between the squamosal and frontal shields, and occur in rows between the small shields on the snout. On the under side of the snout (Fig. 188) the median ridge is small and smooth, and reaches back only to the barbels, which are nearer to the extremity of the snout than to the mouth. The outer barbels are the longer. The mouth-aperture is wide, bounded anteriorly by a small upper lip, which is entire, and is produced only for a short distance beyond the angles of the mouth, so as to leave the entire centre of the lower jaw free. The eyes are of moderate size, equal, and ovate. The breadth of the frontal interspace is six times the orbital diameter. The lower and larger narines are smaller than the diameter of the eye. The shields on the body are strong, but distant from each other. The number of dorsal shields is about ten to eleven, their bases are sometimes round, sometimes heart-shaped; the keel is smooth, strongly elevated, and forms a curved point or hook, which extends backward. The first dorsal shield is not firmly attached to the supra-occipital shield; the third shield is the highest. Those which succeed the third are broader than long, and become smaller as they extend down the back. They are regularly rayed. The lateral shields vary in number from thirty to thirty-three; they are lozenge-shaped, finely rayed and keeled, but towards the tail the keel is prolonged into a hook. The ventral shields also each have a keel ending in a hook; they are more ovate, and seven to nine in number. One or two similar shields lie between the vent and anal fin; and there are often a few smaller bony scales behind the dorsal fin, on both sides of the body. The skin is smooth and shining between the rows of shields, but there are star-shaped bony scales thickly scattered over the surface. They vary in size and shape. Some of the larger are elevated into a spine, and they form an irregular row between the dorsal and lateral shields. There are small pores on the under side of the snout. The skin is smooth about the mouth, eyes, nares, gill-aperture, and the region of the pectoral, ventral, and anal fins. All the fins are pointed. The pectoral fin is longer than the diameter of the body. One unjointed soft ray succeeds its bony ray. The dorsal fin is very strongly notched out posteriorly. The upper lobe of the caudal fin is not conspicuously long.

The colour is blackish-grey above, and yellowish-white below; the shields are dirty white.

In the young the head is relatively large, and is about one-quarter of the entire length; and then the barbels are midway between the mouth and the extremity of the snout. The upper lip is notched. The lower lobe of the caudal fin is less developed. The length both of the snout and of the hooks on the shields varies with age, as in all other species of the genus.

## Acipenser güldenstädtii (BRANDT AND RATZEBURG).

This species (Fig. 189) is known to the Austrians as *Wardick*, or *Warlück*, and in Hungary is termed *Tok*. Dr. Grimm speaks of it as the Eastern Sturgeon. He states that its distribution is similar to that of the Sterlet, except that it keeps nearer to the mouths of rivers, and goes farther into the sea. It is found all over the Caspian, and is caught in the Kur and rivers of Persia, and in Siberian rivers. It lives farther north than the Sterlet. Heckel and Kner state that it is found throughout the year in the Danube, and ascends all its larger tributaries. It rarely gets beyond Pressburg, and is taken only occasionally at Vienna. It lives on worms and mud contain-

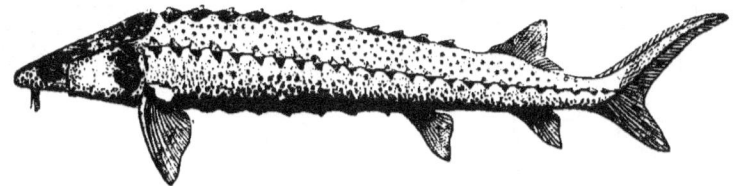

Fig. 189.—ACIPENSER GÜLDENSTÄDTII (BRANDT).

ing organic matter. It spawns in the Danube in May and June, and is reputed by Hungarian fishermen to live for fifteen to twenty years. In Russia this species is known to reach a length of 270 centimètres, but the usual size in Astrakhan does not exceed 180 centimètres. Heckel and Kner state that fishes ten to twelve feet long weigh 150 to 160 pounds. In Hungary it is rarely taken weighing less than thirty pounds. Its flesh is much valued for food. It yields excellent caviare and isinglass.

The body is as wide as high, and it is usually eight-and-a-half times as long as thick; but the fish is only five-and-three-quarter times as long as the head. The outline of the snout and frontal region is very like that in the *A. schypa*; but in *A. güldenstädtii* the distance from the mouth to the snout is much less. The shields which cover the head are large, broad, and marked with strong irregular rays (Fig. 190). They are often separated by a narrow interspace, so as not to be perfectly in contact. Thus the first dorsal scute, which is sub-circular, does not touch either the supra-occipital or the epiotic scutes; but between these shields, and extending into the large interspace between the epiotic, parietal, and squamosal shields, is a number of small, densely-packed scales. The supra-occipital shield is oblong, inclining to lanceolate, being narrower in front, where it penetrates for half its length between the parietal shields. The squamosal shields are broad, extend back a little farther than the parietal, and are shorter than these shields, but longer than the frontal shields, which converge anteriorly, but do not meet, being divided by the ethmoid plate, which has small ossifications round it, and penetrates a little way between the extremities of the parietal scutes. The post-frontal scute extends quite as far back as the frontal, and forms the upper and hinder border of the orbit. The extremity of the snout is covered with small, star-shaped, irregular plates and bony processes, and shows two rows of black-

Fig. 190.—ACIPENSER GÜLDENSTÄDTII,
HEAD SEEN FROM ABOVE.

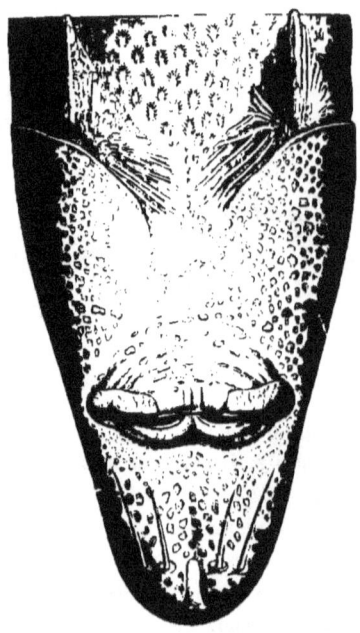

Fig. 191.—ACIPENSER GÜLDENSTÄDTII,
HEAD SEEN FROM BELOW.

coloured pores. The under side of the snout is covered with large pores, which have somewhat the aspect of the meshes of a net, except in the median line.

There is a short median ridge in front, feebly developed, and extending only as far back as the barbels (Fig. 191). Its farther course is indicated by a furrow. The barbels are much nearer to the extremity of the snout than in *A. schypa*, and, when laid back, do not reach near to the mouth. The mouth is large, and occupies nearly the entire width of the under side of the head. The small upper lip is not only deeply notched, but completely divided, though the two halves are in contact. The entire middle of the lower jaw is destitute of lip. The eyes are large, nearly round, the left rather smaller than the right, which is one-fourth of the width of the frontal interspace between the orbits. The hinder border of the orbit is vertically over the anterior edge of the mouth. The diameter of the lower and larger narine is equal to that of the right eye.

The body shields in all the five rows are rather distant from each other. The dorsal shields number twelve or thirteen, they have round or heart-shaped bases; the keel rises gently for half the length of the shield, and then terminates abruptly, usually in a recurved hook, but sometimes in a straight point, which in old individuals may be worn off. The first shield is slightly the largest, and the succeeding shields become smaller and relatively shorter as they approach towards the dorsal fin. There are no scutes behind the dorsal fin. All the scutes are coarsely and irregularly ribbed, with the ribbing confused towards the centre.

The number of shields along the sides is different on the right and left of the body, as in other species, and the number varies from twenty-four to thirty-six. They have a lozenge shape, with flat keels, and are largest in the middle. Bony granules occur in the interspaces between them, and indicate the course of the lateral line. The ventral shields are nine to ten in number, round, and elevated to a central point.

Between the vent and anal fin there is one large shield, and between the anal fin and the caudal fin there is a small shield on each side. The skin between the principal rows of scutes is smooth and shining, with bony scales of different sizes, more or less thickly scattered over its surface, showing a raised star-shape or a round denticulated form. A few larger hooked scales form an indefinite single or double row between the dorsal and lateral rows of shields. A similar row of much larger scales, about eight in number, runs from the region of the pectoral fin some distance towards the ventral fin. These shields are much larger than the lateral shields.

Compared with *A. schypa*, the pectoral fins in this species are relatively broader, and their length is scarcely equal to the height of the body; a long unjointed soft ray similarly succeeds the strong bony ray in both species. Both lobes of the caudal fin are more strongly developed than in *A. schypa*; the lower lobe is broader, less pointed, and is half as long as the upper lobe.

The colour of the back is a bluish ash-grey. Below the lateral line it is white. The shields are dirty white. The barbels are white with blackish points. The iris is silvery.

In the young, seven or eight inches long, the length of the snout is twice the diameter of the mouth; and the snout is then thin, pointed, and much curved. The upper side has long keeled shields. The median ridge on the under side of the snout ends in a strong hook. The shields which support the operculum are less developed, the lower lobe of the caudal fin is absent, and many of the later rays of the fin, which afterwards become jointed, are still undivided. The pectoral, dorsal, and caudal fins are black at their bases, and yellowish-white at their edges.

The females have the snout rather blunter and longer, and the males are distinguished by their olive-brown colour.

## Acipenser naccarii (BONAPARTE).

Small Sturgeons are met with in the Adriatic which belong to this species. They reach a length of three feet, but are more frequently about twenty inches long. *Acipenser naccarii* goes up the river Po in May, and though never abundant, supplies the markets of Milan and Pavia. It is found in the autumn in the lagoons of Venice, where it is generally small. The Italians have no special name for it, and it is termed *Storione*, in common with the other species. Its colour is blackish-brown above; the shields and under side are dirty white.

At the base of the pectoral fin the body is higher than broad, the height being one-ninth of the total length. The length of the fish is four and three-quarter times the length of the head. Seen from above, the outline of the head is lanceolate (Fig. 192).

All the shields covering the head are finely rayed, regular, and closely packed together. The temporal shields are short and narrow, rather smaller than the epiotic shields, and much smaller than the frontal. The parietal shields are long, notched behind to receive the broad heart-shaped supra-occipital, and tapering in front, where they are notched to receive the small anterior division of the ethmoid which separates the frontal shields. The post-frontal forms the upper border of the orbit. In front of the frontal are the pre-frontal, nasal, and other ossifications. The anterior end of the snout is only partly covered with small irregular little plates, and shows a row of large pores on each side running from in front of the nares forward to the end of the snout. These pores are connected with the cephalic canals. The first dorsal shield is broad. The mouth is large, but its breadth is not so great as the interspace between the eyes. It is placed in the middle of the length of the head. The distance between the outer barbels is equal to their distance from the point of the snout, but this breadth is less than the interspace between the anterior nares. The barbels are nearly equally long, and do not reach back to the mouth.

The number of dorsal shields varies from eleven to fourteen. The first is slightly the largest, and its anterior extremity is received in the notch at the back of the supra-occipital shield. The succeeding four or five shields are as broad as long, but towards the dorsal fin they become relatively longer, and have strong keels, each prolonged into a backwardly directed spine. Behind the dorsal fin is a double row of three or four small shields, or there may be a large long one and two smaller. The lateral line is covered by forty long lozenge-shaped shields, twice as deep as long, with straight spines. There are ten shields in the ventral row, which extends as far as the ventral fin, and behind it there is a further series of ten, which may number nine or eleven on one side.

There are two keeled shields between the ventral and anal fins, the more

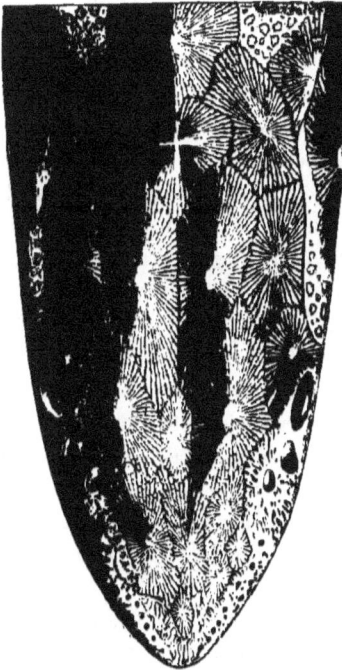

Fig. 192.—UPPER SURFACE OF HEAD OF ACIPENSER NACCARII.

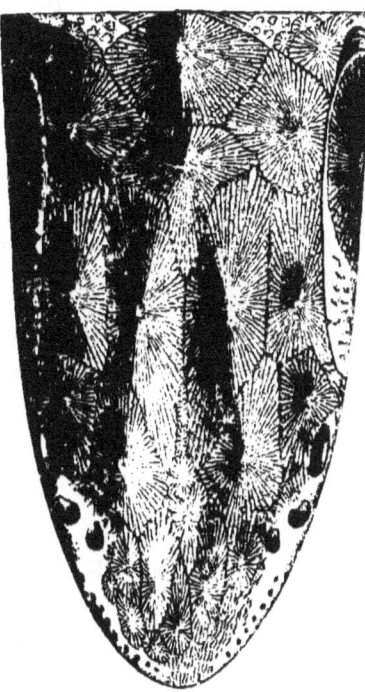

Fig. 193.—UPPER SURFACE OF HEAD OF ACIPENSER NACCARII, VAR. SARDOI.

anterior of which is the largest of all the shields. There is generally a long shield, without a keel, behind the anal fin. Between the rows of shields the skin is covered with starlike scales which vary in size; anterior to the pectoral fin they have a regular lozenge shape. The bony ray of the pectoral fin is strong, and longer than the breadth of the head over the squamosal bones. There are thirty-two to thirty-five truncate rays on the upper margin of the upper lobe of the caudal fin. There is a small rayed shield in front of the base of each of the ventral fins.

## Acipenser naccarii; variety nardoi (HECKEL).

A variety of the Adriatic Sturgeon, which does not differ in size or habit, and is known only from the river Po and the lagoons of Venice, is characterised by its yellow colour. With this are associated certain differences in the shape of the head and forms of the cephalic shields. The relation between the height of the body and its length is the same as in *A. naccarii*, but its thickness is greater, being equal to the height; and since the body is five and a third times as long as the head it follows that the head is relatively shorter. The outline of the snout is less pointed than in the type of the species, and the length of the snout is equal to the breadth between the nasal apertures. The shields on the upper part of the head have a strong general resemblance to those already described, but there are some conspicuous differences (Fig. 193).

The first dorsal shield is broadly pentagonal; the epiotic shield is broad and sub-quadrate, instead of oblong; the supra-occipital terminates laterally in sharp angles, and its parietal sides are notched out angularly by the parietals, so that the median portion extends farther between the parietals. The parietal shields are long, and as they reach as far forward as the anterior margins of the eyes they come nearer to the extremity of the snout, because the eyes are placed farther forward, in this form than in *A. naccarii*.

The squamosal shields are broad and participate in the elongation. The post-frontal are more decidedly triangular; the frontal more decidedly lozenge-shaped, while the ethmoid ossifications are small, three or four in number, and penetrate far between the parietals, though not reaching quite so far back as do the frontal shields. In front of these are pre-frontal, nasal, and other small ossifications. There is a row of pores near to the anterior border of the snout, which is otherwise covered with an osseous crust. The mouth is placed as in *A. naccarii*, and the lower lip is not developed in the middle. The outer barbels reach back to the mouth.

The dorsal shields are thirteen or fourteen in number. The first is the broadest, and the succeeding three or four are also broader than long, but otherwise present no distinctive peculiarities. Two smaller median shields are developed behind the dorsal fin. The lateral shields, forty in number, are twice as broad as long and keeled in the usual way. There are ten ventral shields, which are rounded and keeled. Two similar shields, rarely three, of which the hinder is commonly the larger, are developed in front of the anal fin. A small shield, without a keel, is placed behind the anal fin. The star-shaped bony scales are developed over the skin, as in *A. naccarii*, and the lozenge-shaped scales on the throat are densely packed in rows, but are embedded in the skin. The strong bony ray of the pectoral fin is longer than in *A. naccarii*.

## Acipenser heckelii (Fitzinger).

This short-nosed Sturgeon, with a rounded snout, is found in the Adriatic sea and river Po. It is more frequently met with than *A. naccarii*, and is known in Venice as *Copese*, but nothing is definitely ascertained of its habits or life-history. Dr. Günther regards it as a variety of *A. güldenstädtii*, but the differences from that type are strongly marked, although there is a close general resemblance. It varies in length from eighteen inches to four feet and a half (Fig. 194).

The colour of the head is black, and the upper part of the body is blackish. The lower part of the body and shields are dirty white.

The head measures one-fifth of the entire length. The fish is eight and

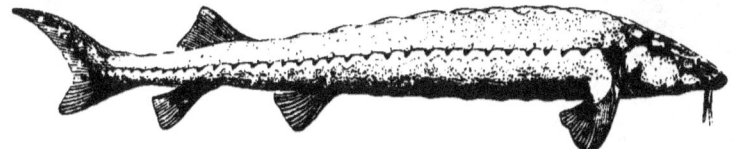

Fig. 194. ACIPENSER HECKELII (FITZINGER).

two-thirds times as long as high. The height and breadth are equal. The form of the head is similar to that of *A. güldenstädtii*, but shorter, and the contour of the snout, seen from above, is semicircular.

All the shields of the head are comparatively flat, finely rayed, granular, and closely packed together. The supra-occipital is narrow in front, expanded laterally in a broad V-shape, and notched behind to receive an anterior process of the first dorsal shield (Fig. 195).

The epiotic shields are longer than wide, but broad and sub-hexagonal; each abuts against the first dorsal, the supra-occipital touches the parietal by a narrow margin, and penetrates into the squamosal with the anterior angle. The parietal shields are long, and entirely separated from each other by a series of ethmoid ossifications, the posterior of which divides the extremity of the supra-occipital shield; the squamosal shields are broad and short. Their centres are well behind the centres of the parietal shields, and in a line with the anterior termination of the supra-occipital shield. The frontal shields are broad, but do not extend to the orbit, which is margined above and behind in the usual way by the broad post-frontal shield, which extends farther back than the frontal. The nasal and other shields lie in front of the remarkable ethmoid ossifications.

The eyes are rather large, round, equal in diameter, and entirely seen in a view of the skull from above. The frontal interspace which separates them measures five times the orbital diameter. The nares are large, rather distant, and the lower narine is nearly equal to the diameter of the eye. There is a small row of pores on the anterior margin of the snout, over which numerous small ossifications extend.

The mouth lies immediately below the orbits in the middle of the under side of the head, and its width is nearly equal to the width of the frontal

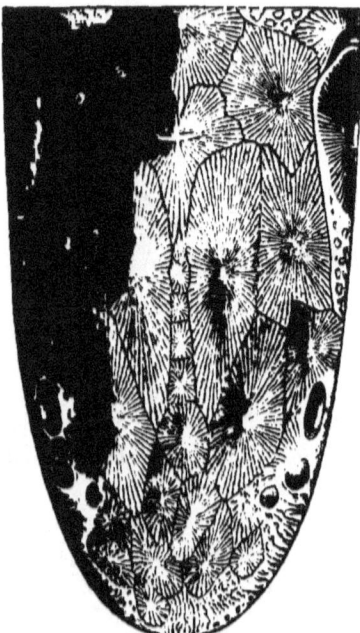 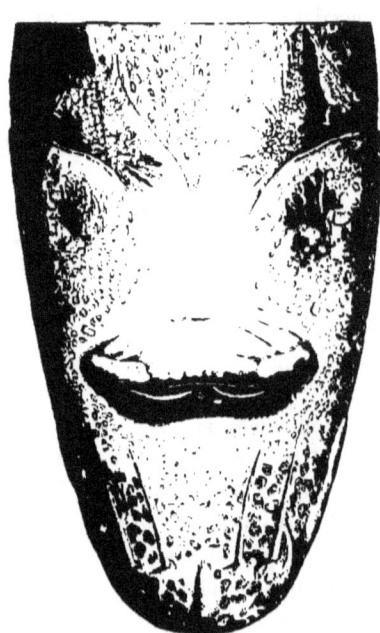

Fig. 195.—HEAD OF ACIPENSER HECKELII, SEEN FROM ABOVE.

Fig. 196.—HEAD OF ACIPENSER HECKELII, SEEN FROM BELOW.

interspace. The hinder lip is very imperfectly developed. The outer pair of barbels is the longer, and when laid back reaches nearly to the mouth. The barbels are near to the anterior end of the snout; in the median line in front of them is a slight ridge, and the under side of the head, both behind and in front, is reticulated with large pores (Fig. 196).

There are twelve slightly elevated dorsal shields, which have no conspicuous spine, and are only slightly keeled. The first is broadest and flattest. At first the shields are broader than long, but after the third the length and breadth are equal, and then they become narrower. The margins of their bases are serrated. There are five shields behind the dorsal fin, of which the middle plate is the largest. The lateral row, which is not placed so high on the side as in some other forms, consists of from thirty-two to thirty-four lozenge-shaped shields, which are small, elongated vertically, and distant from each other; they are keeled longitudinally, and the keel rises into a spine.

The ventral shields, eight or nine in number, are round and have similar keels. There are two shields in front of the anal fin, and two more behind. These are unequal in size; the two most distant from the fin are the larger. Between the rows of shields the skin is densely filled with small star-shaped scales, which become lozenge-shaped on the breast, where they are arranged in lines.

The pectoral fin is short and rounded; its bony ray is no longer than the width of the skull over the squamosal bones. The ventral, dorsal, and anal fins are obliquely truncate, and the dorsal is as high as the anal.

The caudal fin is not well developed, but its upper lobe is as long as the head.

## Acipenser nasus (HECKEL).

This is another Adriatic Sturgeon, found in the River Po and the lagoons of Venice, and brought, according to Heckel, to the Venice fish-market, with other species, in autumn.

It appears to be rare, since Heckel possessed but one specimen, twenty-six inches long, and the species is not represented in the collection of the British Museum.

The colour of the body is brown above, with the head paler.

The proportions of the body are very similar to those of *A. heckelii*, the breadth being equal to the height, and this measurement is about one-ninth of the length of the fish, while the head is one-fifth of the total length. The snout contracts regularly in front of the eyes to a narrower lanceolate form than in the other species; and the part of the head which lies in front of the anterior narines would form nearly an equal-sided triangle. The profile, seen from the side, rises concavely to the back of the head, which is convex transversely. The shields of the head are chiefly remarkable for the immense size of the ethmoid, which reaches from the extremity of the snout back between the frontals, and its extremity just penetrates the points of the parietals. The condition of this plate would, therefore, justify us in regarding all the small ossifications upon the snout in other species as divisions of the ethmoid (Fig. 197).

The first dorsal shield is remarkably broad, being only a little narrower than the transverse width of the two parietals. The supra-occipital is T-shaped, tapering in front, and convex behind, receiving a slight tooth from the first dorsal shield in the middle of the posterior border. The epiotic shields are longer than wide; the squamosals do not extend as far backward as the parietals, and their centres are slightly anterior to those of the latter, while the centres of the epiotics and supra-occipital are in a transverse line. The frontal shields are large and extend a little farther back than the post-frontal scutes. The pre-frontal shields are strongly developed, and form the anterior borders of the orbits. The nasal shields are anterior to the frontal. The eyes are partly

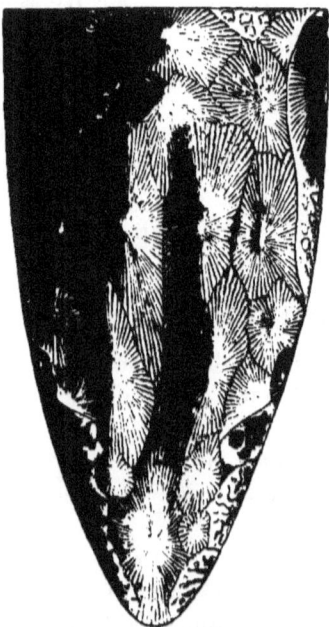

Fig. 197.—HEAD OF ACIPENSER NASUS, SEEN FROM ABOVE.

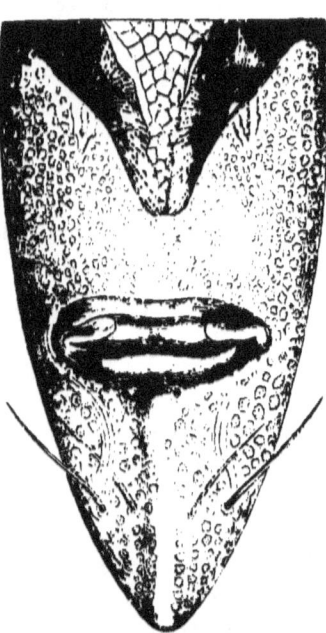

Fig. 198.—HEAD OF ACIPENSER NASUS, SEEN FROM BELOW.

covered by the pre-frontal and post-frontal plates. A median longitudinal depression extends along the head between the frontal and parietal shields.

The under side of the head shows the mouth, as usual, in the middle of the head-length; it is nearly as wide as the frontal interspace between the orbits. The middle portion of the lower lip is not developed. A strong elevated median ridge extends from the extremity of the snout backward to the upper lip, widening somewhat behind. The barbels are rather in front of the middle distance, between the mouth and the end of the snout. The outer pair is the longer; they reach back to the upper lip (Fig. 198).

There are thirteen well-developed shields on the back. They are broader than long, but become longer after the seventh, and the keel is then sharper, and ends in a point.

There are three shields behind the dorsal fin, two side by side, and the third, which is oval, is behind them. There are forty-one lateral shields, which have a lozenge shape, with serrated margins, and are transversely keeled. There are ten ventral shields; posteriorly they become somewhat larger, and more pointed at the ends. Two shields, similar to the ventral series, are formed between the ventral and anal fins. There is one very long and slender shield behind the anal fin.

The whole of the skin is covered with rough star-shaped scales, which vary in size. On the throat they are quadrate or lozenge-shaped, and closely packed together, so as to present the aspect of a tesselated pavement, though Heckel and Kner compare them to the ganoid scales of the American Bony Pike (*Lepidosteus*).

The pectoral fin is as long as the body is high; the height of the dorsal and anal fins is equal. Both lobes of the caudal fin are pointed, but the upper lobe is greatly elongated, and is much longer than the head.

## Acipenser sturio (LINNAEUS).—The Sturgeon.

The Sturgeon which is commonly taken in British rivers, is known in France as *l'Esturgeon*, though often called by fishermen *l'Estiious*. In Germany the popular name is *Stor*, and in Scandinavia *Storge*, or *Storjer*. In Italy it is *Storione commune*.

It is very widely distributed, and frequents the whole Atlantic coast of North America, as well as the Atlantic coast of Europe. From this distribution it is known to the Russians as the Western Sturgeon, to distinguish it from the Eastern Sturgeon (*A. güldenstädtii*). It is absent from the Caspian, and is not common in the Black Sea, from which it ascends some of the rivers, though not recorded by Heckel and Kner in the Danube.

According to Canestrini, it enters most of the rivers of Italy in March, April, and May, and is more common in the river Po than the other Sturgeons of the Adriatic. Steindachner found it during the spring months, in the Spanish peninsula, in the Douro and Tagus. It is common in all the great rivers of France, the Rhône, Garonne, Loire, and Seine. It ascends the Rhine, and has been taken at Speyer, and, more rarely, at Basel. It is found in all the rivers of North Germany, and goes up the Elbe into Bohemia, and the Vistula into Galicia. In the Baltic it usually lives in deep water, from which it rises in the spring to enter the German, Russian, and Scandinavian rivers. In Britain it is only an irregular visitor, but is not rare in our fish-markets, especially in London during the summer season.

As might be expected from the wide geographical distribution and diverse habitat of this species, it is extremely variable, and its synonymy is, on this account, difficult to disentangle. Indeed, any one who should critically compare the figures of the species given by modern naturalists might well hesitate to regard them as representations of the same kind of fish; but, like other Sturgeons, this type varies with age, and the long pointed snout of the young, which, as usual, curves upward, becomes replaced by a broader, shorter snout, in which the curvature is lost. The body is, at first, pentagonal in section, but the surfaces between the rows of scutes afterwards become filled out and rounded. The lower lobe of the caudal fin, which is at first absent, develops with age. The unjointed rays of the pectoral fin subsequently become jointed. Certain differences may be observed in the shields of the body and the head, as may be seen by comparing the figure of a fresh head (Fig. 200) with the

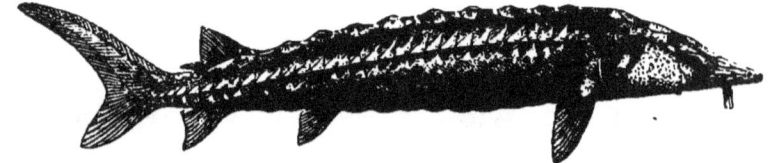

Fig. 199.—ACIPENSER STURIO (LINNÆUS).

dried head figured by Yarrell, differences sufficiently obvious to have induced Sir John Richardson to describe the British Sturgeon as a distinct species. Probably no work would throw more light upon the nature of species than a thorough investigation of the variability of *Acipenser sturio*.

In form of body (Fig. 199), and length and shape of snout, this species most resembles *A. gmelini*. The height of the body over the pectoral fins exceeds the thickness by one-third, so that the section is vertically ovate, and the aspect of the fish slightly compressed. The head is as wide as the body is high, and this measurement is one-eighth of the total length. The head is two-ninths of the length of the fish, but in the young it is one-quarter of the length. The extremity of the snout seen from above is triangular. The profile rises with a gentle convexity to the back of the head.

The shields which cover the head are rather roughly granular than radiated, and in close contact (Fig. 200). The first dorsal shield is only slightly wider than the supra-occipital. The latter is **T**-shaped, with both the transverse portions and anterior stem broad. An anterior denticle from the keel of the first dorsal scute penetrates a little into the supra-occipital. The epiotic shields are parallel and long. They reach a little farther forward than the supra-occipital. Their centres are nearly in a line with that of the supra-occipital. The squamosal shields are small, smaller than the epiotic, and their centres are in a transverse line with those of the long parietals.

The parietal shields are more in contact than usual, the narrow ethmoid ossification in front only penetrating slightly between them, while it assists (with small anterior ethmoid ossifications) in dividing from each other the long and large frontal scutes which are only a little shorter than the parietal scutes. The post-frontal scute is narrow, and extends behind and over the orbit for the

Fig. 200. — HEAD OF ACIPENSER STURIO, SEEN FROM ABOVE.

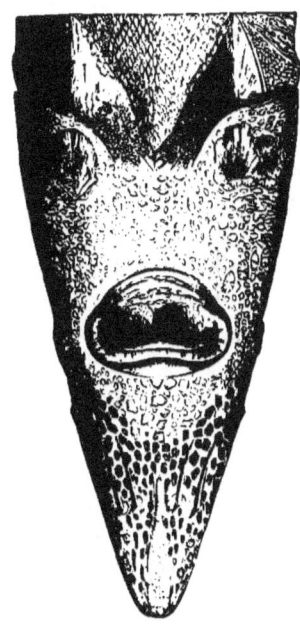
Fig. 201. HEAD OF ACIPENSER STURIO, SEEN FROM BELOW.

eye. It unites with the pre-frontal scute in front. The snout is covered with a number of closely packed elongated ossifications, which, however, leave a naked margin, which shows pores. The operculum is large, flat, and granular. The eyes are rather large, ovate, equal, and look laterally, and the breadth of the frontal interspace is less than four times the diameter of the eye. The nares are somewhat in advance of the eye, and the lower narine is longer, and not so large as the upper rounder one. On the under side of the head the mouth is a little nearer to the gill-aperture than to the snout; it is placed, as usual, between the eyes, but its breadth is less than the frontal interspace. The lip is expanded into two large pads on the lower jaw, but they are separated by a median depression. There are the usual four barbels, placed in a transverse row in advance of the middle of the space between the mouth and extremity of the snout. They are rather short. They are divided by a median ridge, which does not extend far behind them, but widens in front, where it is covered by rough shields. When the snout becomes broad and thick, the barbels are midway between the eye and the snout.

The bony armour on the body is well developed. The number of dorsal shields in the median line of the back is usually eleven, but may be thirteen. Each dorsal shield has a more or less distinct keel, which is more developed in the hinder part of the body than in front, but the form and size of the shields varies with age, though all have their upper surfaces roughly granular and radiate, and ornament is seen only at the margin. The lateral shields have the vertically elongated lozenge-shaped form, and usually number from twenty-nine to thirty-one, though Dr. Günther finds in some young specimens twenty-six or twenty-seven, and mentions an example from the Black Sea with thirty-four shields. The anterior part of each scale has a process which is embedded in the skin, and is overlapped by the scale in front of it. The keels on the scutes become sharper and higher towards the tail. The ventral shields are nine to ten in number, and similar in shape to the dorsal shields, the first two are smaller than the others. There are three shields behind the ventral fin, the two smaller ones form a pair in front of the single large shield. There are three or four longer shields behind the anal fin. There are three to five pairs of shields behind the dorsal fin, followed by a single larger one, which extends on to the upper lobe of the caudal fin.

The skin is covered with small bony scales which have almost the character of shagreen. They become larger towards the head, and are sometimes arranged in oblique rows (Figs. 199, 201).

The pectoral fin is inserted low down at the back of the head. The strong bony ray is only three-quarters of the length of the succeeding jointed rays. The dorsal fin, rather in front of the anal fin, is angularly notched, so as to be nearly divided into a large anterior part and a small posterior lobe. The upper lobe of the caudal fin does not exceed the length of the head; the lower lobe is moderately developed.

The colour of the back is usually a dull reddish-brown, but varies to a blue or yellowish-grey; the belly is white, inclining to silvery; the shields are grey. The iris is yellow.

The usual size of the fish in Austria and Italy is from five to six feet, but it may reach a length of eighteen feet, and a weight of two hundred to five hundred pounds; the size is not larger in France or North Germany. Pennant speaks of one taken in the North Sea which weighed four hundred and sixty pounds, and a fish eleven feet four inches long, cast for Buckland, weighed five hundred and forty-one pounds.

The food of this species consists of many kinds of invertebrata, especially worms, small crustacea, small mollusca, and mud containing organic matter. Sometimes the intestine is found completely filled with sand which has been swallowed with food.

The species spawns in the south of Europe in April or May, rarely in March; but in the north of Europe the spawning is commonly in May or June. Fishes have been taken in British rivers with eggs developed in January and May, and with some of them as large as small peas at the end of June, while others were then as small as small-shot. Benecke describes the eggs as two millimètres in diameter, and numbering many millions. Mr. B. Parfitt states that the ovaries weighed eighteen to twenty pounds in a fish seven feet long, captured at Exmouth. The young are hatched in five days. Sturgeon-breeding is carried on by artificial impregnation of the eggs in Schleswig-Holstein. When hatched, the young only require to be watched in the river for from ten to fourteen days, and soon go down to the sea.

Pennant remarks, and the statement is repeated by writers in many countries, that the Sturgeon makes no resistance either when caught in a net or hooked, but is drawn out of the water like a lifeless lump. It, however, lives out of water for some time, and a specimen is referred to by Thompson as having lived out of water thirty-six hours before it was killed.

In England the Sturgeon is a royal fish, belonging, by Act of Parliament, of the reign of Edward II., to the sovereign, except where it has been granted by charter to certain corporations, as at Boston in Lincolnshire. When it came into the Thames and passed above London Bridge it was claimed by the Lord Mayor, but the last Sturgeon which appears to have possessed the endurance necessary to undergo this perilous adventure is said to have been captured in 1832. It was customary for the Lord Mayor in former times to present the Sturgeon to the sovereign. The regard in which it is held as food varies in different countries. The flesh is commonly white with yellow fat, and often a faint pink tinge. Towards the end of the tail it becomes red. It is firm, and varies in flavour in different parts of the body. A good cook is said to be able to serve it as beef, mutton, pork, or poultry. The last two are obtained from the fore part of the body. The middle of the tail yields a flesh which has the character of veal, and sometimes adds to this the flavour of lobster. The smaller part of the tail closely resembles mutton. Although an excellent fish it never commands a high price in the London market, usually selling retail at from fourpence to eightpence a pound, according to the locality in which it is sold.

In earlier days the Sturgeon appears to have been much more common in American rivers than now, and formerly as many as three hundred a day have been hooked out of some of the rivers of Virginia in May, June, and July.

*Acipenser latirostris*, of Parnell, is a variety of the Common Sturgeon, distinguished by the thickness as well as the breadth of the snout. The specimens originally described were nearly eight feet long, and weighed eight stone. The back and sides are light grey inclining to olive, and the belly is dirty white. Dr. Parnell remarks that it differs from the Common Sturgeon in having the tip of the snout much broader than the mouth, in having the barbels much nearer to the end of the snout than the mouth, and in the slight elevation of the keel of the dorsal plates. According to Yarrell, there are thirteen dorsal shields, and four between the dorsal and caudal fins. There are thirty-two lateral shields and fifteen in each ventral row, with three or four behind the ventral fins.

## Acipenser huso (LINNÆUS).

This is the largest of all the Sturgeons. It is found in the Black Sea, the Sea of Azov, and the Caspian Sea, and enters the rivers which drain into those waters. It is much rarer in the Mediterranean, but is sometimes taken in the

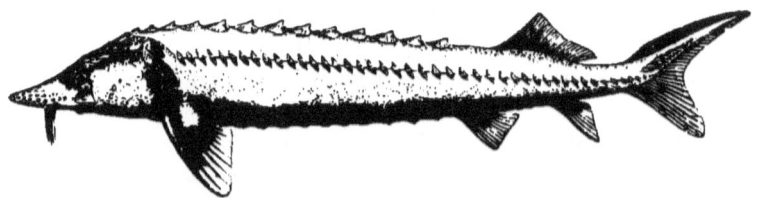

Fig. 202.—ACIPENSER HUSO.

rivers of Italy. In Russia it is known as the *Beluga*, and in the valley of the Danube its popular name is *Hausen*, though the Hungarians also call it *Viza*. Now it is rarely seen above Pressburg.

It was formerly much more abundant, and attained a far larger size in the Danube than it ever grows to at the present day; but its great size tempted capture. Fishes twenty-four feet long were common, and it was slaughtered in Hungary in thousands. At the present day it varies from one hundred to nine hundred pounds in weight, and occasionally reaches twelve to fifteen hundred pounds. In Russia very heavy fishes are sometimes caught. Grimm tells us of a Sturgeon taken in the river Ural in 1847, which weighed 1,600 pounds. One was caught at Saratov in 1869, which weighed 2,760 pounds, while in 1813 a Beluga taken at Sarepta weighed 3,200 pounds. Fishermen report having sometimes caught Belugas so large that they were unable to drag them from the river; but large specimens like these are altogether exceptional. *Acipenser huso* often reaches a weight of 800 to 1,200 pounds, but the average size in the Astrakhan market is from 360 to 400 pounds, though smaller ones, down to a weight of 120 pounds, are often seen. Owing to its large size — and the females are larger than the males — the weight of the roe is immense. Pallas states that the roe weighed 800 pounds in a fish of 2,800 pounds, but this is an unusually large proportion, for Dr. Grimm mentions that in *Acipenser stellatus* the roe is commonly one-tenth of the weight, one-quarter of the weight in *A. schypa*, and one-fifth in *A. güldenstädtii*. It is captured not only for its flesh, 10,800,000 pounds' weight of which is annually exported from Astrakhan, but also for the roe, which in common with that of other species is made into caviare; and for the isinglass which is made from the air-bladder.

The Beluga migrates like the other Sturgeons, and the wanderings commence with the beginning of spring. It crowds into the mouths of the rivers in large herds before the ice has disappeared, and it is no rare circumstance for the fishes to be injured, or even to have their snouts broken off by the drifting ice. Observers differ as to whether it regularly descends to the sea, and just as it would appear to remain in the sea for some time, or, at least, until it attains a weight of twenty pounds — for smaller individuals never occur in the Danube — so there is evidence to show that, in the Volga, at any rate, it sometimes passes the winter in a hybernating condition in the rivers. It is a sluggish and timid fish, fleeing from other fishes, and especially from the pursuit of the Sterlet. Yet, owing to its large size, a blow from its tail would easily capsize a small boat. When captured it makes a grunting noise, and it soon becomes powerless. It seeks soft muddy places in the river, and often remains inactive, as though asleep. During the day it sometimes swims on the surface, with its head projecting in the water, and sometimes remains at the bottom, with its snout buried in the sand or mud. It is very voracious, and feeds on a variety of animal and vegetable substances. Small fishes, especially of the Carp tribe, are eaten by it, and water-birds have occasionally been found in its stomach.

The proportions of the body are not very different from those of other species, the thickness being equal to the height, and about one-eighth of the total length (Fig. 202).

The head is rather narrower than the body, and tapers interiorly in a blunt cone. The length of the snout scarcely exceeds the width of the mouth; its extremity, is curved slightly upward, and is free from shields, so that the cartilage, of which it consists, is semi-transparent.

The shields upon the head are more feebly developed than any other species, are only nine in number, and have serrated margins. The first dorsal shield is entirely separated from, and behind the head. The supra-occipital shield is half-moon-shaped, with the convexity directed forward. The epiotic shields are long and narrow, and diverge backward from the sides of the supra-occipital. The parietal shields are small, much smaller than the squamosal, which are prolonged backward much farther, and extend quite as far forward. The frontal shields extend between the outer border of the squamosal shields and the

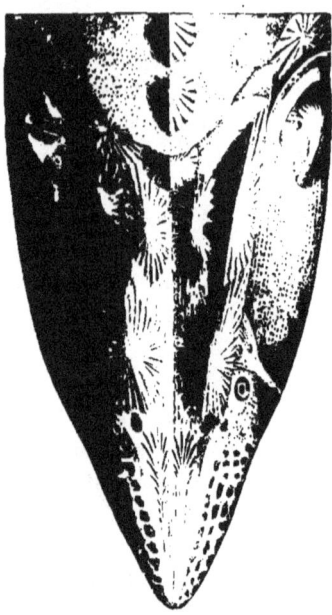

Fig. 203.—HEAD OF ACIPENSER HUSO, SEEN FROM ABOVE.

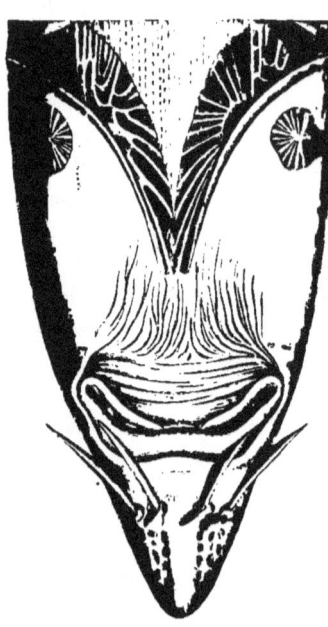

Fig. 204.—HEAD OF ACIPENSER HUSO, SEEN FROM BELOW.

nares, passing just above the eyes, and converging a little forward. They do not meet in the median line, or touch the parietal shields; and the middle line of the skull is occupied by a furrow (Fig. 203).

The eyes are small, round, of equal size, and placed vertically over the mouth. They look outward and upward, are far from the outline of the head when seen from above, and are separated by eight times the orbital diameter. The narines are small and nearly equal in size. The outer sides of the snout are covered with pores, and similar pores are seen on its under side divided by a short broad median ridge. The mouth is far in advance of the middle of the length of the under side of the head, and occupies nearly its entire breadth.

The cartilaginous jaws are surrounded by a fleshy lip, which is absent from the middle of the lower jaw. The upper lip is thickest in the middle (Fig. 204).

The barbels are very near the mouth, and when laid back reach beyond its middle portion; they are compressed at the base, expanded in the middle, and smooth and pointed at the extremities.

There are the usual five rows of shields, but they are not strongly developed, so that the body has less of a pentangular aspect than in most of the other species. There are generally twelve or thirteen dorsal shields. Their surfaces are coarsely rayed and rough. They are smallest in front; the sixth to eighth are the largest and highest, while the hindmost have the longest bases, and their spines are directed backward. There are two or four small shields in front of the dorsal fin, arranged in pairs. The lateral shields are forty to forty-five in number. They have a lozenge shape, are very small, but largest in the middle; they are nearly flat in front, and towards the tail the keel is but little elevated; their surfaces are pitted, and occasionally show obscure radiation. The lateral line can be traced between them.

The ten to twelve ventral shields are of median size, thicker than the lateral shields, but less developed than in other species. Their bases are star-shaped in front and round behind; they are radiated, and have a keel, which extends backward. There are one to three round and rather flat shields. The skin is thickly covered with small spinous bony granules, which form parallel rows. They form irregular groups on the throat, as well as behind the pectoral and ventral fins. On the anterior part of the sides, towards the lateral shields, they become star-shaped.

The rounded pectoral fin is rather shorter than the height of the body. The bony ray is weak and short. The dorsal fin is twice as long as high, and truncate behind. The upper lobe of the caudal fin is shorter than the head; the lower lobe is rounded, and is relatively well developed.

The colour of the snout is yellowish-white. The back is dark ash-grey. The under side of the body and the shields are dirty white. The iris is silvery, with a yellow tinge.

The differences which appear with age are more remarkable than in other species. As the fishes get older the shields do not increase in size so rapidly as in other species; and they gradually allow intervals of naked skin to appear between them, and become thinner. The median ridge on the under side of the snout becomes flattened. The bony granules in the skin decrease in size, and in old individuals are only half as large as in young individuals. In a young specimen about a foot and a half long, all the shields of the head are in close contact. Their central points are strongly elevated, and directed backward. The dorsal shields are in contact, and have high, sharp keels and hooks; and the other shields are similarly strong in the young fish; and the younger the fish the more marked are these characters. As in several other species, the lower lobe of the caudal fin is rudimentary, and all the soft rays of the fins are undivided in young specimens.

Pallas states that hermaphrodites are not rare, in which the ovary is developed on one side and the milt on the other.

Hungarian fishermen state that the fishes come up the Danube in spring, from March to May, to spawn, and that there is a second migration in the autumn, from August to December. If, however, they spawn in the Danube, young fishes are never seen there.

In rivers in which the species is still common, its presence is indicated by peculiar movement of the water; and they are sufficiently large to break down considerable obstacles to their progress.

The fishery in the Danube is carried on in a variety of ways. Lines are stretched over the river, to which bright glistening hooks, without barbs, but well baited, are suspended, and sunk to different depths, so as to intercept the fishes like a curtain. As the Sturgeon come up they strike the festoons of hooks with their noses, and turn about till they get entangled, when the fisherman seeing the strain on the line, knows where a fish may be found. In the more rapid parts of the Danube, piles are driven into the river so as to leave channels, through which the fishes must pass. When swimming on the top of the water they are harpooned; and they are also shot with ball. When wounded, a rope is passed through the gill-aperture, and the fish is drawn to land. They are also taken with long wide-meshed nets, the opposite ends of which are secured by fishermen in two boats. When a *huso* strikes the net he turns back, and is pursued by the fishermen until driven into shallow water, when it often throws itself on to the bank. In the Volga this species is frequently taken with baited hooks suspended to lines stretched over the river, as in the Danube. It is more useful than the other Sturgeons, on account of its size, but its flesh, isinglass, and roe are not so much valued as those of some of the smaller species. The wholesale price of the fresh Beluga at Astrakhan varied, in 1882, from three roubles, eighty copecks, to four roubles, thirty-five copecks a pood of thirty-six English pounds, equivalent to about fourpence a pound. It is sold either alive or frozen or salted. It is also dried in the sun (then termed *balik*), and often smoked.

The method of preparation of caviare from the roe is essentially the same for all species. The ovaries from which the caviare is to be made are first of all beaten with rods to loosen the eggs from the tissue in which they are contained; and are separated from this membrane by being passed through a sieve. They are then salted according to the taste of the markets, packed in barrels, and exported. The commonest kind is the pressed caviare. From this only the larger and stronger portions of connective tissue are removed; it is then dried in the sun, and trodden with the foot into small kegs. A superior quality is prepared by salting the roes in long troughs, then partially drying them on nets or sieves, through which they are afterwards pressed into kegs. The best caviare is prepared by granulating the roe in linen sacks. The sacks with the eggs in them are then laid in brine, and afterwards hung up for some time to dry, when the caviare is pressed into small barrels.

The air-bladder is converted into isinglass. It is first washed, then turned inside out, and dried. The next step is to remove the internal layer, which is easily detached. The bladder is again moistened, hung in the shade, and subsequently cut into strips which are stretched on the bark of a tree to dry.

The isinglass yielded by the Sterlet is considered to be finer than that yielded by *A. huso*.

The total annual export of caviare from Astrakhan amounts to 3,240,000 pounds' weight. The annual produce of isinglass is 124,000 pounds' weight, and the spinal cartilage, which when dried is known as "vyaziga," is of the same weight. The annual value of the Sturgeon fishery in European Russia is stated by M. Danilosky at 8,000,000 roubles, of which the flesh in its various forms is valued at 5,000,000, the caviare at 2,250,000, the isinglass at 600,000 roubles, and the vyaziga at 100,000 roubles. There can be no doubt but that the Sturgeon fishery is greatly decreasing in Russia from over-fishing, just as it has declined in other countries. Von Baer estimated the produce thirty years ago at nearly one million poods more than it is now. Yet in some localities, like the River Kur, the yield is well sustained; but, though just as many fish are taken, they are of smaller size, and with the smaller size of the fish the quantity of useful products diminishes, so that the yield of caviare and of isinglass is steadily declining.

Thus, according to Sokoloff, in the middle of this century the roe of the Sturgeon (probably meaning *A. ruthenus*) was about one-nineteenth of the total weight of the fish, in the third quarter of the century the percentage of roe had diminished to one twenty-seventh, and in 1881 it was as low as one-forty-fourth, so that the River Kur will probably soon cease to yield caviare at all.

## CHAPTER XII.

#### FRESH-WATER FISHES OF THE SUB-CLASS CYCLOSTOMATA.

GENUS PETROMYZON: Lamprey — Lampern — Mud Lamprey, or Pride.

### GENUS: **Petromyzon** (ARTEDI).

THE fishes included in this genus belong to a distinct Division of the Class Fishes; and are placed in an Order named *Marsipobranchii*. The body is Eel-like or worm-like in form, and undergoes a remarkable metamorphosis. The skin has neither scales, nor scutes, nor lateral line, though there are a few scattered pores upon the head. The eyes are feebly developed in the larval stage of existence, though they subsequently acquire a more perfect and larger condition. There is but one nasal aperture, in the middle line of the upper part of the head. The mouth is suctorial; and when open is circular and seen to be filled with a large number of horny denticles, yellowish-brown in colour, and arranged in rows. When closed the mouth is a longitudinal furrow. The animal adheres by suction to various substances and to fishes, and thus has a semi-parasitic character. There are consequently no true jaws; and the modifications of the head due to its suctorial habit, have led Professor W. K. Parker to compare it with the tadpole of the frog, and even to regard it as more nearly related to such a type than to fishes or other vertebrata. The skeleton is cartilaginous and remains in an embryonic condition. The notochord not being divided, the vertebræ are not separated from each other, though their arches are distinctly developed. The skull is cartilaginous; but the brain is formed in the way usual in fishes, and gives off the usual nerves. There are seven branchial openings for the gills, and the gill sacs are supported by a complicated cartilaginous framework, in which the branched lateral arches are connected together at the sides, so as to surround the gill-openings; and they are also connected below in the median line, forming a structure termed the extra-branchial basket, which terminates behind in a transverse convex cartilage which is defined as extra-pericardial.

The alimentary canal is straight and simple, and terminates in a spiral valve. The heart contained in a pericardium has its several parts divided by valves, and resembles the heart of a fish of the ordinary type. The reproductive organs are single, there being but one ovary or one milt in each individual, though Yarrell describes two pores through which the spawn is discharged. There is no air-bladder. The kidneys are well developed, and their ducts open into the abdominal pore.

The fish has no limbs; but there are two dorsal fins in which rays are developed. Petromyzon lives in rivers, where it spawns, and then goes down to the sea. The eggs are small.

Dr. Günther enumerates four genera of true Lampreys, distinguishing Petromyzon as having the maxillary tooth bicuspid, Ichthyomyzon as having the maxillary tooth tricuspid, Mordacia as having tricuspid maxillary teeth, and Geotria as having four lobes on the maxillary plate. Petromyzon is the only genus found in Europe, but it also extends to North America and West Africa, and one species is recorded from Buenos Ayres.

## Petromyzon marinus (LINNÆUS).—The Lamprey.

The Sea Lamprey is known in France as *Lamproie*, in South Germany as *Seelamprete*, in Austria it is termed *Pricke*, in Italy *Lampreda*, in North Germany the popular name is *Meeraenaange*, and in Sweden *Hafs-nejnoga*.

Thus widely distributed along the coasts and rivers of Europe (excepting in the region of the Black Sea), it ranges southward to the west coast of Africa, and is found in North America, from Nova Scotia southward.

It is usually met with on the bottom, but swims with a snake-like movement, and is sometimes transported adhering to the bodies of other fishes. It is often taken at sea, and has been known to attack Gurnards, Mackerel, Cod, and many other fishes; and Couch mentions many examples of fish in which the flesh had healed after holes had been thus rasped in the body. It has occasionally fastened on to bathers, and is said to be attracted by the tar of boats. Pennant speaks of one that attached itself to a stone weighing twelve pounds, from which it was separated with difficulty; and the popular name, Lamprey, has been regarded as signifying stone-sucker, but we rather agree with Badham, in deriving it from *lang*, long, and *prey*, *prick*, or *pride*, local names for the smaller River Lamprey.

Day states that the fishery in the Severn at Worcester begins in February and lasts till May. The Lamprey is taken in the Thames in May and June; but Jardine tells us that it ascends the Scottish rivers about the end of June, and remains in them till the beginning of August. On the Continent it ascends the rivers in spring, both in North Germany and in Italy; but the period at which it enters rivers varies with the stream. Although it may attack other fishes, and sometimes eat flesh from their bodies, it will feed upon any kind of soft animal food. It is usually from twenty to thirty inches long, but larger specimens are sometimes found; and in the Elbe and Havel it is known to reach a weight of five or six pounds (Fig. 205).

The body is cylindrical, and more elongated than that of any other species of the genus. The distance from the extremity of the sucking lip to the first gill-arch is about one-ninth of the total length; and the greatest height of the forepart of the body is more than half this measurement. The relatively small eye is only twice its diameter from the first gill-aperture. The eyes are separated by five times the orbital diameter. The simple nasal aperture is a long round opening, placed between the eyes; it is surrounded with white spots, and is the extremity of a blind tube, which does not penetrate the palate, as in its marine ally, the Hag (*Myxine glutinosa*). The mouth is built upon

Fig. 205.—PETROMYZON MARINUS (LINNÆUS).

a great circular cartilage, which closely resembles the incomplete but similar cartilage in the Tadpole, and shows no sign of division into two parts; the external margin of the mouth consists of a girdle of short, slender, branched cirrhi, or little tentacles. Within this band are the teeth, which are arranged in rows. In the middle of the disc, behind the entrance to the throat, is a curved plate, with seven to eight strong sharp denticles (Fig. 206). Opposite this, in front, are two teeth, sometimes stronger than the three other lateral teeth on each side, which complete the innermost circle. External to these are eight rows of small denticles extending backward from the great mandibular tooth, which rows are nearly parallel to each other, but diverge a little laterally. The rows of teeth anterior to these are larger, six in number on each side, and curved outward and backward, but the two anterior rows on each side do not reach the central circle of teeth. On the tongue, which works backward and forward, there are three strong denticulated horny teeth. All the teeth are hollow, and easily fall off from the epithelium to which they are attached, but they are renewed,

Fig 206.—HEAD OF PETROMYZON MARINUS, SEEN FROM BELOW.

and the new teeth are already formed when the old ones are lost. The mouth shuts laterally instead of vertically.

The seven gill-apertures are scarcely separated from each other by the width of an orbital diameter, and each is margined in front by a membranous fold, and defended posteriorly by papillæ; and there are two valves behind, which meet like folding doors. These gill sacs open into a longitudinal canal, which is below the œsophagus; it terminates posteriorly above the pericardium, and opens in front into the gullet by a valve like a glottis. An artery runs between each pair of gill sacs, so that each sac receives branches from two arteries, like the gills of osseous fishes. The water for respiration may either be taken by the mouth, or by the gill-apertures when the animal is attached. Dumeril describes a gland between the buccal disc and the respiratory cage, which is enveloped in a strong muscle, and contains a yellow, or yellowish-brown, thick, acid fluid. The excretory duct opens into the mouth. It is regarded by some observers as the salivary gland.

The first dorsal fin begins behind the middle of the body; its base exceeds the length of the head. It is divided from the second dorsal by a distinct interspace. The second dorsal has a long triangular form; it increases in height to a point, and is then as high as the tail beneath it is deep; the height then declines towards the caudal fin, with which it is confluent. The caudal fin is small, rounded at its extremity, and extends symmetrically on both sides of the termination of the vertebral column. In the position of the anal fin there is sometimes a membranous ridge prolonging the caudal fin to the vent. The vent is one-quarter of the length of the fish from the extremity of the tail, and has a short longitudinal opening, which shows the urogenital papilla.

The cephalic mucus-canals are well-developed. There is a row of pores above and behind each eye, and another row below the eye, which is prolonged to the lip, and there are other rows of pores over the gill-apertures, becoming irregular on the back and sides; there are also rows on the throat. The body is always very slimy.

The colour of the back and sides varies with locality and season, but is usually grey or some tint of yellow, spotted or clouded with brown or black.

At spawning-time in the spring, a crest, comparable to that of the water salamander, is developed on the back, between the neck and the first dorsal fin, and on the under side of the body, between the vent and the caudal fin. The ovaries are very large. The fishes excavate furrows on the river bed, by removing stones, which are lifted by suctorial action of the mouth; and the male and female are said by Jardine to moor themselves to stones while spawning. According to Yarrell, the spawn escapes in both sexes by a membranous sheath, which is prolonged beyond the apertures of the body. It is deposited in the excavated grooves and covered with sand. After spawning the fish is exhausted, and goes down to the sea. Its flesh is well-flavoured; but some peoples have a prejudice against it. It is not commonly eaten by the inhabitants of Scotland, Ireland, or Cornwall. Pennant considered it best in March, April, and May, when the flesh is firmer than at other times.

## Petromyzon fluviatilis (LINNÆUS).—The Lampern.

The River Lamprey, or Lampern of English writers, is known in Germany as *Flussneunauge*. In Austria it is not always distinguished from the *P. marinus*, being popularly known either as *Pricke* or *Neunauge*. The Italians term it *Lampredone*, and the Swedes *Nejnoga*. Its distribution is very similar to that of the Sea Lamprey, but, besides ranging over the other parts of Europe, it is also distributed in Russia; and, in North America, ranges up to Alaska. A variety of the species is found in Japan.

In English rivers, especially the Severn, Trent, and Thames, the Lampern

Fig. 207.—PETROMYZON FLUVIATILIS (LINNÆUS).

was formerly a not unimportant article of local commerce. Pennant states that in his time vast quantities were taken about Mortlake; and that the Dutch purchased them for bait for the Cod fishery, taking 450,000 in a season, at forty shillings a thousand; and he states that about 100,000 a year were sent to Harwich to be used as bait. The number in the Thames is now much diminished. Buckland agrees in regarding Lamperns as the best possible bait for Cod; and states that a considerable trade is carried on in the Trent, in catching them as bait for the deep-sea fisheries. The Lampern fishery goes on from the end of August till March; 3,000 have been taken at Newark in one night. When captured they are sent in wicker baskets to the fishing ports on the east coast, but must be kept moving while in transit to avoid suffocation. Their value for bait is partly owing to the fish being very tenacious of life. The Dutch fishermen manage to keep them alive for weeks, and use them in the Turbot fishery.

The Lampern feeds on worms, insects, and their larvæ, small fishes, and chance animal substances. In England it commonly reaches a length of twelve or fifteen inches; it is sometimes a few inches longer in Continental countries. It has much the same aspect and Eel-like shape as the preceding species, and is compressed at the tail (Fig. 207).

The head is smaller than the anterior part of the body; the greatest thickness between the gill-openings is two-thirds of the height. The length of the head to the last branchial aperture, is fully one-fifth of the length of the fish. The mouth is terminal and nearly circular, so placed as to look obliquely downward and forward. It is surrounded by a short fringe of tentacles, within which there is a row of very small teeth, which are scarcely visible, and easily fall off. The arrangement of teeth in the oral disc is very different from that in the Sea Lamprey, and the form of the opening of the gullet, which contains the tongue, is quadrate rather than circular (Fig. 208).

Fig. 208.—PETROMYZON FLUVIATILIS, HEAD SEEN FROM BELOW.

At the back of the gullet there is a large plate similar to that in the Sea Lamprey, only much less curved, and it similarly contains about eight horny denticles. The corresponding plate on the opposite side or front of the mouth similarly contains two teeth, but instead of being close together, as in the Sea Lamprey, they are wide apart, at the anterior corners of the mouth. There are no teeth at all behind the posterior or mandibular tooth-plate; and in front of the maxillary tooth-plate there is only a crescent of about five little teeth scarcely larger than those of the outer circle to which they are parallel. But there are three tooth-plates arranged in a longitudinal row on each side of the mouth. The middle plate on each side has three denticles arranged transversely. The upper plates have two denticles, each transversely prolonging the line of the maxillary teeth, and the lower plates have also two denticles prolonging the transverse line of the mandibular bar. The tongue within the mouth has two teeth which are transverse, parallel to each other, and carry many denticles.

The eye has the same lateral position as in the Sea Lamprey, but varies in size from one-ninth to one-twelfth of the length of the head, reckoning to the last branchial aperture. It is separated from the other eye by twice its diameter, and in the young fish the distance is less. It is seven times its diameter from the last gill-aperture. Between the eyes on the upper side of the head is the single nasal aperture which extends forward as a short tube. The gill-apertures are arranged in a longitudinal series, as in the Sea Lamprey, with a valve in front and a serrated margin behind. They are about the diameter of the eye from each other.

The dorsal fins are similar to those of the Sea Lamprey. The first dorsal begins about the middle of the length of the fish; it is short, and forms a long low triangle, slightly rounded. Its base is one-seventh of the entire length, and it is not so high as the second dorsal which begins at some little distance behind it. The fin rises rapidly to a point and similarly declines posteriorly; but, before disappearing, merges in the caudal fin, which extends round the extremity of the vertebral column.

The anal fin is usually but slightly developed; it is inseparable from the caudal; and, as a ridge, extends forward to the vent. The dorsal fins have the aspect of adipose fins, but contain branched cartilaginous rays. The vent lies in the last fourth of the length; the perforated urogenital papilla projects through it.

The lateral line is not visible. The cephalic canals are indicated by a row of pores behind and over the eye, and extend forward to the upper lip.

The colour of the back is usually steel-blue, but olive-green in some localities; the sides are yellowish and the belly is silvery-white. The fins are violet. The iris is golden-yellow with a few dark specks.

The males rarely differ in aspect from the females, but are of rather smaller size. The milt of the male lies in the median line of the body, and is transversely divided into a number of little laminæ; it ends in a point beneath the anterior lobe of the liver. The ovary of the female stretches the entire length of the ventral cavity, terminating opposite the last gill-opening. The intestine is straight, and, as in the Sea Lamprey, has no trace of a stomach. The liver consists of long pointed lobes which lie on the right side. The kidneys in the female are only half as long as the ovaries, but in the male they are as long as the milt.

The Lampern is limited to low-lying country, and is found in lakes, rivers, and brooks, and marshes, but spawns only where the water runs rapidly over stony bottoms. The spawning takes place in March and April, and then the skin acquires a bright metallic lustre. The eggs are small, one millimètre in diameter, opaque, and of a greyish-yellow colour. After spawning the Lampern usually dies. The development of the spawn depends upon the weather; and Benecke, who has studied this fish in North Germany, states that in some years only very small broods appear. The young of this species were discovered by August Müller in 1854, and found to possess distinctive characters, and they have since been bred at Cambridge by artificial fecundation of the eggs by Balfour and Salvin. The young have no dorsal fin; the caudal extends much farther forward. The upper lip is expanded into a horse-shoe-shaped hood. The liver possesses a gall-bladder which disappears in the adult Lampern, and the otolites of the auditory capsules are similarly lost in the adult fish. The fishes are not usually met with, but Benecke states that they may be dug out of the mud every year when an arm of the River Alle dries up; and he believes that their metamorphosis into Lamperns takes place in the sea.

River Lampreys are captured in Germany with nets or baskets from December to Easter; and in the rivers flowing into the Baltic are taken both in the spring and autumn. The Lampern is esteemed as food, and Buckland considered that there is no finer fish. In Germany it is regarded as difficult of digestion, and is usually pickled with spices, potted, smoked, or salted; and when fresh is stewed or made into pies, as in this country.

Kessler has described the *Petromyzon wagneri* from the Caspian, from which it ascends all rivers which flow into that water.

## Petromyzon branchialis (LINNÆUS).—The Pride.

This Lamprey is rarely more than eight or ten inches long, and about as thick as a swan's quill. It extends throughout the British Islands, and in the southern part of England is locally known as the Pride. From its habit, first observed by Pennant, of frequenting the mud rather than concealing itself under stones, Couch named this little fish the Mud Lamprey. It is distributed throughout the rivers of Europe, from Sweden to Spain, Italy and Russia, and ranges into the Black and Caspian Seas. The same species is found in the western part of North America. It is valued as bait on account of the toughness of its flesh.

In aspect and character, the Pride, or Sandpiper, closely resembles the Lampern (Fig. 209). The head is a little smaller. The external fringe to the oral disc is better developed, but the outer circle of teeth is not developed. The mandibular plate is semicircular as in the Sea Lamprey; it carries twelve blunt teeth which are equal in size. At the anterior corners of this plate, in the middle of the sides of the mouth, there is a plate with three teeth in a transverse row (Fig. 210). As in *P. fluviatilis*, there is a smaller plate with two teeth anterior to this, and a plate with one tooth posterior to it, so that the lateral teeth are formed on the same plan as in *P. fluviatilis*; and there are similarly two teeth distant from each other at the anterior corners of the mouth. The teeth are not always uniformly developed. The eyes are smaller than in the Lampern, separated from each other by the diameter of the orbit, and they are a little nearer to the mouth. The gill-apertures are separated from the eye and from each other by distances equal to the diameter of the eye. The single nasal opening is less conspicuous and round.

The first dorsal fin may or may not be connected with the second dorsal.

but there is usually an interspace between them; their height varies, and the second dorsal is sometimes higher than the body beneath it. The anal fin is more developed in the female than in the male, being in the latter a mere ridge which prolongs the line of the caudal. The vent is nearer to the head than in the Lampern. The skin of the whole body shows annular transverse markings. The colour does not differ from that of the Lampern, except that the back is more frequently olive-green.

This species spawns in March or April. The eggs are yellowish-brown, one millimètre in diameter. After spawning, the old fishes probably die.

Fig. 209.—PETROMYZON BRANCHIALIS (LINNÆUS).

August Müller traced the development of the young, which are hatched in eighteen days, and found that they were identical with the fish which had been named *Ammocœtes branchialis*. At first there are eight branchial slits, but the first aperture is soon obliterated. The young bury themselves in the mud; and grow slowly.

Fig. 210.—PETROMYZON BRANCHIALIS, HEAD SEEN FROM BELOW.

They are as well known in the north of Germany as the Pride, but though they are eaten in some places, are generally used only as bait. In Austria the fish is known as *Uhlea*, and in Germany as *Querder*.

In the larval state the head is remarkably small, and the length to the first gill-aperture is no more than the height of the body, and is one-fifteenth of the total length, or a little less. The eye is extremely small, and is hidden in the thick skin of the head so as to be scarcely visible, and the eyes are separated by twice the orbital diameter. The nasal aperture is triangular, and has an elevated margin. The upper lip forms two-thirds of a circle, and reaches forward in front of the lower lip. There is no trace of the fringe to the lip which is so characteristic of the mature animal. The inner surface of the upper lip is covered with papillæ, behind which there is a fringe of more or less branched barbels at the entrance to the throat. There are no teeth. The gill-apertures are small, half-moon-shaped, and lie in a longitudinal furrow. The whole body is marked with transverse rings. The dorsal fins are low and have fine rays. The caudal fin is prolonged to the vent as a membranous ridge. The vent is a long groove. The skin of the abdomen is transparent. The largest specimens measure seven inches. It is very hardy, and its habits are like those of the earth-worm. Its ovaries are well developed.

The fish remains in this state for from three to four years (Fig. 211). And then the metamorphosis is completed in about ten days, during which the oral disc with its tentacles and teeth are formed, the barbel-like processes

Fig. 211.—LARVAL FORM OF PETROMYZON BRANCHIALIS, FORMERLY KNOWN AS AMMOCŒTES.

in the mouth disappear, the eyes develop, the respiratory tube is completed, the intestine and whole body shorten, and the reproductive organs become opaque. The bright silver colour is put on, and the fins grow and become yellowish-white. With these modifications the habits of the fish change, and now having become adult it seeks clear water, uses its suctorial disc, and no longer limits its diet to microscopic forms of food.

# GENERAL INDEX.

*(For Index to Species, see p. 440; and for Index to Common Names, see p. 443).*

Abramidina, 210 — 244; characteristics of family, 210; distribution of, *ib.*
Abramis, 210—231; characters of genus, 210—211; distribution of, *ib.*
—— ballerus, description of, 218, 219; fin formula, 218; illustration of (Fig. 121), 218; colour of, 219; size of, *ib.*; habits of, *ib.*; distribution of, *ib.*; vertebræ of, *ib.*
—— (blicca) björkna, 223—226; fin formula, 223; distribution of, 223, 225, 226; description of, 223—226; illustration of (Fig. 124), 223; size of, 224; spawning of, *ib.*; colour of, 225; pharyngeal teeth of (Fig. 125), 225; bastard form of, 226.
—— bipunctatus, characters of, 226—228; illustration of (Fig. 127), 227; Blanchard on variations in colour of, *ib.*; habits of, 227, 228; distribution of, 228; diseases of, *ib.*; southern representative of, 236.
—— brama, 211—215; fin formula, 211; description of, 211, 212; illustration of (Fig. 116), 210; pharyngeal teeth of (Fig. 117), 211; size of, 212; habitat of, *ib.*; food of, *ib.*; habits of, 212, 213; as a food fish, 213; spawning of, *ib.*; mode of capture of, *ib.*; distribution of, *ib.*; var. *retula* (Fig. 118), 214; var. *gehini*, 215.
—— elongatus, description of, 217, 218; illustration of (Fig. 120), 217; distribution of, 218.
—— fasciatus, description of, 228, 229; fin formula, 228; distribution of, 228; colour of, 229.
—— laskyr (Fig. 126), 225, 226; distribution of, 226.
—— leuckartii, hybrid between Leuciscus rutilus and A. brama, characters of, 221, 222; illustration of (Fig. 123), 222; distribution of, *ib.*; Siebold on, *ib.*
—— persa, 218.
—— sapa, description of, 219, 220; fin formula, 219; illustration of (Fig. 122), 220; colour of, *ib.*; size of, 222; distribution of, *ib.*; spawning-time of, *ib.*
—— tenellus, 218.
—— vimba, description of, 215, 216; fin formula, 215; illustration of (Fig. 119), *ib.*; colour of, 216; size of, *ib.*; habits of, *ib.*; spawning period, *ib.*; distribution of, *ib.*
Acanthopterygii gobiiformes of Dr. Günther, 57.
Acanthopterygii, 19, 23—49.
Acerina, 33—38; characters of genus, 33; distribution of genus, 33.
—— cernua, 33—36; fin formula, 33; size, 33; description of, 33—35; illustration of (Fig. 12) 34; external characters, 33—35; internal characters, 35; colour, *ib.*; size, *ib.*; distribution of, *ib.*; habits of, *ib.*; food of, 36; spawning of, *ib.*; capture of, *ib.*; Buckland on, *ib.*
Acerina rossica, 38; fin formula, 38.
—— schrætzer, 36—38; description of, 36—38; illustration of (Fig. 13), 37; weight of, 38; habits of, 38; distribution of, 38.
Acipenseridæ, 22, 381.
Acipenser, distinctive characters of genus, 381, 382; distribution of, 383; habits of, *ib.*; fishing for, *ib.*, 384.
—— glaber, description of, 384—387; illustration of (Fig. 174), 385; head of (Figs. 175, 176), *ib.*; colour of, 387; distinctive peculiarities of young of, *ib.*; size of, *ib.*
—— güldenstädtii, 399—402; local names of, 399; distribution of, *ib.*; illustration of (Fig. 189), *ib.*; size of, *ib.*; description of, 399—401; head of (Figs. 190, 191), 400; colour of, *ib.*; distinctive characters of young, 401, 402; distinctive external sexual characters, 402.
—— heckelii, distribution of, 400; size of, *ib.*; colour of, *ib.*; description of, 405—407; illustration of (Fig. 194), 405; head of (Figs. 195, 196), 406.
—— huso, 414—419; home of, 414; local names of, *ib.*; illustration of (Fig. 202), *ib.*; size and weight of, *ib.*, 415; migrations of, 415; description of, 415—417; head of (Figs. 203, 205), 416; hermaphrodite forms of, 418; spawning of, *ib.*; method of capture of, in Danube, *ib.*; preparation of caviare from roe of, 418, 419; preparation of isinglass from air-bladder of, 419; commercial value of, *ib.*
—— latirostris of Parnell, 414.
—— naccarii, description of, 402—404; distribution of, 402; head of (Fig. 192), 403.
—— nardoi, characters of, 404; head of (Fig. 192), 403.
—— nasus, description of, 407—409; home of, 407; colour of, *ib.*; head of (Figs. 197, 198), 408.
—— ruthenus, 388—390; distribution of, 388; size of, *ib.*; description of, 388—390; illustration of (Fig. 177), 388; cranial scutes on upper side of head (Fig. 178), 389; under side of head, mouth, barbels, pores, &c. (Fig. 179), *ib.*; exceptional largeness of mouth, *ib.*; colour of, 390; distinctive characters of young of, *ib.*; spawning of, *ib.*; market value of, *ib.*
—— ruthenus, variety *gmelini*, distinctive

characters of, 391; illustration of Fig. 180, ib.; head of Figs. 181, 182), 392; distribution of, 393.

Acipenser schypa, 396—398; distribution of, 396; description of, 397, 398; illustration of (Fig. 186), 396; head of (Figs. 187, 188), 397; colour of, 398; distinctive characters of 293; young, ib.

—— stellatus, description of, 393—396; distribution of, 393; illustration of (Fig. 183), ib.; value as a food fish, ib.; head of (Figs. 184, 185), 394; colour of, 395; young of, 396.

—— sturio, 409—414; local names of, 409; wide distribution of, ib.; variability of type, 410; illustration of (Fig. 199), ib.; description of, 410—412; head of (Figs. 200, 201), 411; colour of, 412; size of, ib; food of, ib.; spawning of, 413; a royal fish, ib.; flesh of, ib.

—— sturio, var. *latirostris* of Parnell, 414.

Acipenseridæ, 381—419.

Agonostoma, 69.

Alburnus, 232—239; characters of genus, 232; distribution of, ib.

—— alburnellus, description of, 236; fin formula, ib.; illustration of (Fig. 133), ib.; distribution of, ib. Var. *fracchia* (Fig. 134), 236, 237.

—— breviceps, 235.

—— chalcoides, description of, 239; fin formula, ib.; distribution of, ib.; use of, for food, ib.

—— fabræi (Blanchard), 235.

—— fracchia, 236.

—— lacustris, 235.

—— lucidus, 232—235; fin formula, 232; distribution of, ib.; description of, 232, 233; illustration of (Fig. 131), 233; colour of, ib.; habits of, ib.; spawning of, ib. et seq.; parasites of, 234; imitation pearls prepared from scales of, ib.; var. *lacustris*, 235; var. *breviceps* (Fig. 132), 235.

—— mento, 237—239; fin formula, 237; description of, ib. et seq.; illustration of (Fig. 135), 238; size and colour of, ib.; Heckel and Kner on mode of deposition of spawn by, 239; as a food fish, ib.

—— mirandella (Blanchard), 235.

Anacanthini, 2, 19, 82—89.

Anatomy of fish, *see* Fish Structure.

Anguilla, distinctive characters of genus, 372, 373; distribution of, 373.

—— eurystoma, 380; home of, ib.; description of, ib.; illustration of (Fig. 173), ib.

—— latirostris, 379.

—— vulgaris, 373—380; local names of, 373; size of, 373, 374; illustration of (Fig. 172), 374; peculiar scales of, 375; internal anatomy of, ib.; habits of, ib., 376; voracity of, ib.; tenacity of life in, 376; domestication of, ib.; reproduction in, ib.; eggs of, 378; migration of young, ib.; method of capture, 379; flesh of, ib.; enemies of, ib.; distribution of, ib.

Aspius, 229—231, characters of genus, 229; distribution of, ib.

—— rapax, description of, 230, 231; illustration of (Fig. 128), 230; pharyngeal teeth of (Fig. 129), 231; habitat of, ib.; food of, ib.;

deposition of spawn upon stones, ib.; mode of capture, ib.; distribution of, ib.

Aspro, 44—49; generic characters, 44.

—— streber, 48, 49; distribution of, 48; spawning period, 48; size of, 49; habits of, 49.

—— vulgaris, 46—48; characters of, 46—48; distribution of, 47, 48; illustration of Fig. 17), 47; characters of spine, ib.; eggs of, 48; habitat of, 48; superstitions concerning, 48.

—— zingel, 44—46; description of, 44—46; illustration of (Fig. 16), 45; colour of, 46; spines and scales as protecting organs, ib.; size of, ib.; flesh of, ib.; distribution of, ib.; period of spawning, ib.

Atherina boyeri, 66.

—— hepsetus, 66.

—— lacustris, 67, 68; fin formula, 67; description of, 67, 68; distribution of, 68.

—— mocho, 66.

—— presbyter, 66.

Atherinidæ, 19, 66—68; characters of family, 66.

Aulopyge, 125—128; description of, 125; limited distribution of, ib.

—— hügelii, description of, 125—128; illustration of female (Fig. 51), 126; of male (Fig. 54), 127; under side of head of female (Fig. 52), 126; lateral line and skin, showing pigment spots (Fig. 53), ib.; remarkable condition of anal aperture in female, ib.; variations in size of sexes, 127; modifications of internal anatomy, 127, 128; pharyngeal teeth of (Fig. 55), 128.

Barbus, 113—125; diversity of distribution of genus, 113, 114; characters of, 113, 114; Günther on limitation of species of, 113, 114; diagnosis of genus, 114; division of into three sections, 114.

—— bocagei, description of, 122, 123; distribution of, 123; bastard form of, 125.

—— caninus, 119, 120; fin formula, 119; description of, 119, 120; illustration of (Fig. 49), 119; distribution of, 120.

—— chalybeatus, description of, 124, 125; Steindachner on hybrids of, 125.

—— comiza, 121, 122; fin formula, 121; description of, 121, 122; Steindachner on crosses of, 122, 125.

—— equus (Fig. 48), 118.

—— petenyi, 120—121; fin formula, 120; description of, 120, 121; illustration of (Fig. 50), 120; distribution of, 121.

—— plebejus, 118—119; fin formula, 118; distribution of, 118; description of, 118, 119; illustration of (Fig. 47), 118; variety *equus* (Fig. 48), 118.

—— sclateri, description of, 123, 124; fin formula, 123; home of, ib.

—— vulgaris, 114—117; fin formula, 114; description of, 115, 116; illustration of (Fig. 46), 115; pharyngeal teeth of (Fig. 45), 114; size of, 116; habits of, ib.; food of, ib.; spawning of, 116, 117; reputed poisonous character of roe, 117; flesh of, ib.; prejudice of Thames and other anglers against, ib.; distribution of, ib.

Blennidæ, 19, 64—66; characters of family, 63.
Blennius alpestris (of Professor Blanchard), var. of *B. vulgaris*, 65, 66.
—— vulgaris, 64—66; fin formula, 64; description of, 64, 65; illustration of (Fig. 27), 65; external sexual characters in, *ib.*; distribution, *ib.*
Blicopsis abramorutilus, bastard of Abramis blicca and Leuciscus rutilus, 226; description of, *ib.*; distribution of, *ib.*

Carassius, 104–113; characters of genus, 104.
—— auratus, 110–112; fin formula, 110; home of, *ib.*; domestication of, *ib.*; description of, 110, 111; Dr. Günther on variations of race of, 111; Badham on, 111, 112; Buckland on, 111, 112; treatment of, in captivity, 112; as a food fish, *ib.*
—— bucephalus, 112, 113; fin formula, 112; home of, *ib.*; description of, 112, 113.
—— gibelio, 107.
—— humilis, 110.
—— oblongus, 109.
—— vulgaris, 104–110; fin formula, 104; characters of, 104—106; illustration of (Fig. 40), 105; colour and weight, 106; habits, *ib.*; spawning of, *ib.*; flesh of, *ib.*; capture of, *ib.*; distribution of, 107; varieties of, 107—110; var. *gibelio* (Fig. 42), 107—109; pharyngeal teeth of (Fig. 43), 108; var. *oblongus* (Fig. 44), 109, 110.
Carpio kollarii, a hybrid between C. carpio and C. carassius, 102—104; illustration of (Fig. 38), 103; pharyngeal teeth of (Fig. 39), *ib.*
Chondrostoma, 193—206; characters of genus, 193; its distribution, *ib.*
—— auratus, 196.
—— caerulescens, 196.
—— dremeri, 197.
—— genei, distribution of, 199; description of, 199, 200; fin formula, 199; illustration of (Fig. 109), *ib.*; head of (Fig. 110), *ib.*; colour of, 200.
—— knerii, description of, 202, 203; fin formula of, 202; home of, *ib.*; illustration of (Fig. 112), *ib.*; head of (Fig. 113), 203; colour of, *ib.*
—— nasus, 193—197; fin formula, 193; distribution of, 193, 194, 196, 197; description of, 194, 195; illustration of (Fig. 104), 194; head of (Fig. 105), *ib.*; Dr. Günther on a variety of, 195; use of, as food, *ib.*; spawning of, 195, 196; parasites of, 196; var. *auratus*, *ib.*; var. *caerulescens*, *ib.*; pharyngeal teeth of (Fig. 106), 197; var. *dremei*, *ib.*
—— meigii, 125, 203.
—— phoxinus, description of, 197, 198; fin formula, 197; illustration of (Fig. 107), 198; head of (Fig. 108), *ib.*; distribution of, *ib.*
—— polylepis, 122; description of, 204, 205; fin formula, 204; distribution of, *ib.*; hybrid of (with Leuciscus arcasii), 206.
—— rhodanensis (Blanchard), 197.
—— rysela, description of, 203, 204; fin formula, 203; variability of, *ib.*; distribution of, 204.
—— soëtta, 200–202; fin formula of, 200; distribution of, 200; description of, 200, 201; illustration of (Fig. 111), 201; colour of, *ib.*; size of, *ib.*; habits of, 202; capture of, *ib.*
Chondrostoma willkommii, 122, 125; description of, 205, 206; fin formula, 205; distribution of, *ib.*
Clupea, 256—264; characters of genus, 256; distribution of, *ib.*
—— alosa, 257—261; fin formula, 257; distribution of, *ib.*; local names of, *ib.*; illustration of (Fig. 144), *ib.*; stories concerning, by Badham, Buckland, and others, 258; description of, 258—261; food of, 260, 261; habits of, 261; size and colour of, *ib.*
—— caspia, 264.
—— finta, description of, 261, 262; fin formula of, 261; habits of, 262; local and general distribution of, 262.
—— harengus, 262, 263; Pennant on, 262; whitebait discovered to be the young of herring by Günther, 263; characters of, *ib.*; how captured, *ib.*; food of, *ib.*
—— pontica, 263, 264; fin formula, 263; distribution of, *ib.*; use as food, 263; description of, 263, 264.
Clupeidæ, 21, 256—264.
Cobitis, 252—255; characters of genus, 252; distribution of, *ib.*
—— elongata, 254.
—— larvata, 255.
—— tænia, 252—255; illustration of (Fig. 142), 252; fin formula, *ib.*; description of, 253, 254; respiratory function of intestine, 254; var. *elongata* (Fig. 143), 254; var. *larvata*, 255.
Cobitidinæ, 245—255; characters of, 245; distribution of, *ib.*
Coregonus, 336–354; characters of genus, 336; distribution of, *ib.*, 337; Russian species of, 352.
—— albula, fin formula, 348; description of, 348, 349; distribution of, 348; var. *a, albula*, 349; var. *β, norwegica*, 349; var. *γ, maraenula*, 349; var. *δ, finnica*.
—— clupeoides, fin formula, 344; description of, 344, 345; distribution of, 344; habits of, *ib.*; flesh of, *ib.*; weight of, *ib.*
—— finnica, 349.
—— gracilis, fin formula, 341; description of, *ib.*
—— hiemalis, fin formula, 345; local names of, *ib.*; description of, *ib.*, 346; habits of, 345; food of, *ib.*; illustration of (Fig. 166), 346; spawning of, *ib.*
—— humilis, fin formula, 347; description of, *ib.*
—— lapponicus, fin formula, 340; description of, *ib.*, 341.
—— lavaretus, fin formula, 338; distribution of, 339; local names of, *ib.*; description of, 339, 340; illustration of (Fig. 164), 339; spawning of, 340; habits of, *ib.*; food of, *ib.*; flesh of, *ib.*
—— lloydii, fin formula, 338; description of, *ib.*; distribution of, *ib.*
—— maraenula, 349.
—— maxillaris, fin formula, 347; description of, *ib.*
—— megalops, fin formula, 347; distribution of, *ib.*; description of, 348.

Coregonus nilssoni, fin formula, 348; description of, *ib.*; distribution of, *ib.*
—— norwegica, 349.
—— oxyrhynchus, fin formula, 337; distribution of, *ib.*; length of snout, *ib.*; description of, *ib.*; size of, 338; migration and spawning of, *ib.*
—— pollan, fin formula, 350; home of, *ib.*; description of, 351; size of, *ib.*
—— vandesius, fin formula, 351; distribution of, *ib.*; Sir Wm. Jardine on the constitution of, *ib.*; habits of, *ib.*; spawning of, *ib.*; description of, 352; colour of, *ib.*
—— vimba, fin formula, 350; description of, *ib.*; habits of, *ib.*; usefulness as a food fish in Lapland, *ib.*
—— wartmanni, fin formula, 341; description of, 341—343; illustration of Fig. 165), 342; colour of, 343; abundance of in certain lakes. *ib.*; spawning of, *ib.*; food of, *ib.*; capture of, 343, 345; local names of, 345.
—— widegreni, fin formula of, 341; description of, *ib.*
Cottidæ, 19, 49—57.
Cottus, 49—57; characters of genus, 49.
—— gobio, 50—55; fin formula, 50; distribution of, *ib.*; illustration of (Fig. 18), *ib.*; illustration of head of Fig. 19, 50; absence of scales upon, 51; colour, *ib.*; internal characters of, *ib.*; distribution of, 52; habits of, *ib.*; time of spawning of, 52; nest and eggs of, *ib.*, 53; mode of capture, 53; flesh of, *ib.*; habits of fry, *ib.*; origin of common name (Miller's Thumb), *ib.*, 54. Varieties:—i. —*ferrugineus*, 54; illustration of (Fig. 22); illustration of head of (Fig. 23), 55; ii. —*microstomus*, 53, 54; illustration of (Fig. 20); 53; illustration of head of (Fig. 21), 54.
—— poecilopus, 55, 56; fin formula, 55; illustration of (Fig. 24), *ib.*; description of, *ib.*, 56; distribution of, 56.
—— scorpius, 56, 57; fin formula, 56; description of, *ib.*, 57; spawning of, 57; habitat of, *ib.*; distribution of, *ib.*
Cyclostomata, 22, 420.
Cyprinina, 94; characters of family, *ib.*; subdivisions of, *ib.*; distribution of, *ib.*
Cyprinodon calaritanus, fin formula, 369; distribution of, *ib.*; occurrence in hot springs, *ib.*; sexual characters of, *ib.*; description of, 370; poisonous property of, *ib.*
—— characters of genus, 369; distribution of, *ib.*
—— fasciatus, 370.
—— ibericus, distinctive sexual characters of, *ib.*; food of, 371; spawning of, *ib.*; distribution of, *ib.*
Cyprinidæ, 20, 94.
Cyprinodontidæ, 22, 369—372.
Cyprinus acuminatus, 99, 101; illustration of (Fig. 33), 100.
—— carpio, 95—102; fin formula, 95; illustration of (Fig. 32), 95; description of, 95—98; rapidity of growth of, 97; variations in size of, *ib.*; habits and food of, *ib.*; hybernation of, 98; spawning of, *ib.*; eggs of, *ib.*; longevity of, *ib.*; tenacity of life, *ib.*; flesh of, *ib.*, 99; Thames water

specimens of, *ib.*; process of "felting," *ib.*; diseases of, *ib.*; varieties of, 99—102.
Cyprinus hungaricus, 101, 102; illustration of (Fig. 35), 101; pharyngeal teeth of (Fig. 36), *ib.*
—— leuciscus (*Linn.*), 164.
—— nordmanni, 102.
—— regina (Fig. 37), 102.

Esocidæ, 22, 360—365; distinctive characters of family of, 360; distribution of, *ib.*
Esox, 360—365; characters and distribution of genus, 360.
—— lucius, 360—365; local names of, 361; illustration of (Fig. 168), *ib.*; distribution of, *ib.*; description of, 361, 362; colour of, 362; weight of, 363; rapidity of growth of, *ib.*; food of, *ib.*; tales concerning, *ib.*; voracity of, 364; attacked by birds of prey, *ib.*; spawning of, 365; capture of, *ib.*

Fish-Growth, 12, 13; metamorphosis in life history of lampreys, *ib.*; size depending greatly upon nature of food, *ib.*; stimulation of growth by insectivorous food, 13; growth of perch, 27.
Fish-Structure, air-bladder in (Fig. 7), 11; absence of in lamprey, *ib.*; position of in other fishes, *ib.*; connection of with intestine in early stages of fish life, *ib.*; in the Cyprinina, 94.
—— digestive organs, 11; dissection showing internal anatomy of carp (Fig. 8), *ib.*; liver and pancreas, *ib.*; pyloric appendages, *ib.*; liver, 12; gall-bladder, *ib.*; spleen, *ib.*
—— fins, description of, 4, 5; explanation of fin formulæ, 5; points to be noted in describing, *ib.*; use of, *ib.*
—— gill-arches (and operculum), 7; constitution of gill cover, 8; connections of, *ib.*; pseudobranchiæ, function of, *ib.*; gill-rakers, *ib.*; spiracles, *ib.*; respiratory sacs in lamprey, *ib.*; inter-operculum (Fig. 4), *ib.*
—— mouth in fishes, position of, 7; jaws of, *ib.*; mode of opening, *ib.*; suctorial in lamprey, *ib.*; lips, *ib.*; teeth (Fig. 3), *ib.*; dental formula, *ib.*; remarkable development of teeth in certain cases, *ib.*; variability in form and arrangement of teeth in different families, *ib.*
—— reproductive organs, 12; oviparous nature of all European fresh-water fishes, *ib.*; ovaries, *ib.*; milt, *ib.*; oviducts, *ib.*; discharge of eggs, *ib.*; development of brilliant colours and tubercles in breeding season, *ib.*; dissection of carp to show position of (Fig. 9), *ib.*
—— scales, description of two types, *cycloid* and *ctenoid* (Figs. 1 and 2), 5, 6; simplicity of in eel, 6; overlapping of, *ib.*; variation in size of, in different parts of the body, *ib.*; absence of, from fins, *ib.*; lateral line of scales, *ib.*; perforations in, as an outlet for mucus, *ib.*; presence of scutes instead of, in sturgeons, *ib.*; arrangement of, *ib.*; number, and how to express in form of a formula, *ib.*; periodical shedding of, in some fishes, *ib.*

FISH-STRUCTURE, sensory organs in fishes, *eye*,
6; eyelids in some fishes, 7; *nostrils*, peculiarity
of in lampreys, *ib.*; position of, *ib.*;
sensitive tentacles, *ib.*
—— skeleton (Fig. 4), 8; skull (Fig. 5), 9;
distinctive characters of skull of fish, *ib.*;
branchial apparatus, *ib.*; correspondence in
position of scutes in sturgeon's head with
bones in skull of bony fishes, 10; description
and relative position of bones forming skull
of fish, *ib.*; vertebral column, *ib.*; vertebræ
(Fig. 6), *ib.*; fin bones, 10, 11.
FISHES, classification of, 2, 19; distinctive characters
of groups, 2, 19; essential diagnosing
characters of orders, 3; genera, 19, *ib.*; species,
*ib.*; difficulties in distinguishing between
species and varieties, 3; influence of environments
upon form and establishment of local
races, 3, 4.
—— geographical distribution, 13; table of
genera, 14—18.
Fundulus, 371, 372; distinctive generic characters
and distribution, 371.
—— hispanicus, characters of, 371; external
distinctive sexual characters of, 371, 372;
size of, 372; habits of, *ib.*; distribution of,
*ib.*

Gadidæ, 19, 82.
Ganoidei, 22, 381.
Gasterosteidæ, 19, 73—81.
Gasterosteus, 73—81; characters of genus, 73.
Gasterosteus aculeatus, 73—80; fin formula, 73;
characters of, 73—76; illustration of (Fig. 28),
74; size, 76; voracity of, *ib*; used as manure,
76, 77; parasite of, 76; distribution of, *ib.*;
pugnacity of male, 77; formation of nest by
male, *ib.*; deposition of eggs, 77, 78; fertilisation
of eggs, *ib.*; hatching, 78; watchfulness
and attention of male over young, *ib.*; description
of eggs of, *ib.*; Continental names of, 79;
illustration of var. *brachycentrus* (Fig. 29),
79. Varieties: 76, 79, 80. i. *Gasterosteus
gymnurus*, 76; ii. *G. semiarmatus*, *ib.*; iii.
*G. semi-loricatus*, *ib.*; iv. *G. trachurus*, *ib.*;
v. *G. noveboracensis*, *ib.*
—— argyropomus, 79.
—— brachycentrus, 79.
—— breviceps, 81.
— burgundianus, 81.
— gymnurus, 76.
— lævis, 81.
— lotharingus, 81.
— noveboracensis, 76.
— occidentalis, 81.
—— pungitius, 80—81; fin formula, 80; description
of, 80; colour of, 81; fishing season
of, in Sweden, 81; mode of capture of, 81;
eggs of, 81. Varieties. i. *G. breviceps*, 81;
ii. *G. burgundianus*, *ib.*; iii. *G. lævis*, *ib.*;
iv. *G. lotharingus*, *ib.*
— semiarmatus, 76.
— semiloricatus, 76.
— spinulosus, 80.
— trachurus, 76.
Gobiidæ, 19, 57—63; characters of family, 57,
58.

Gobio, 128—132; characters and distribution of
genus, 128.
—— fluviatilis, description of, 61; 128—131;
illustration of (Fig. 56), 129; of head of (Fig.
57), *ib.*; pharyngeal teeth of (Fig. 58), 130;
variety of, *ib.*; habits and habitat of, *ib.*;
spawning of, *ib.*; mode of capture of, *ib.*; distribution
of, *ib.*
—— uranoscopus, description of, 131—132;
illustration of (Fig. 59), 131; head of (Fig.
60), *ib.*; distribution of, 132.
Gobius, 57—63; characters of genus, 58.
Gobius avernensis, 60.
—— burncisteri, 61.
—— constructor, 61.
—— eckströmii of Dr. Günther, 63.
— — fluviatilis, 61.
—— gymnotrachelus, 61.
—— kessleri, 61.
—— lacteus, 61.
—— lugens, 61.
—— martensii, 58—61; fin formula, 58; description
of, *ib.*— 60; illustration of (Fig. 25),
59; scales of (Fig. 26), *ib.*; peculiarities of
colour in, 59, 60; reproductive organs in, 60;
flesh of, *ib.*; distribution of, *ib.*, 61.
Varieties of: i. *G. avernensis*, 60; ii. *G.
panizzæ*, *ib.*; iii. *G. punctatissimus*, 61.
—— macropus, 61.
—— marmoratus, 61.
—— melanostomus, 61.
—— minutus, 62, 63.
— — panizzæ, 60.
— — punctatissimus, 60.
—— semilunaris, 61—63; fin formula, 61;
description of, *ib.*, 62; distribution of, 62.

Hybrids of carp, 102; of barbus, 122; of barbus
and chondrostoma, 124, 125; of chondrostoma
and leuciscus, 206; of alburnus and leuciscus,
237, 239; of salmo, 283, 308.

Labrax lupus, description of, 29; fin formula,
*ib.*; distribution of, *ib.*; inordinate greediness
of, *ib.*; as a food fish, *ib.*, 30; size of,
30; weight of, *ib.*; colour of, *ib.*
Latrunculus pellucidus, life period limited to
one year, 58.
Leucaspius, 240—242; characters of genus, 246;
distribution of, *ib.*
—— delineatus, 240—242; fin formula, 240;
characters of, *ib.*, 241; illustrations of (Figs.
136, 137), *ib.*; colour of, *ib. et seq.*; Von
Siebold on the pharyngeal teeth of, 242; size
of, *ib.*; distribution of, *ib.*
Leuciscus, 133—185; Dr. Günther on, 133;
critical remarks upon characters of genus, *ib.*
— — adspersus, 144—145; fin formula, 144;
home of, *ib.*; description of, 144, 145; illustration
of (Fig. 72), 145.
—— alburnoides, description of, 180—181; fin
formula, 180; distribution of, 181.
—— albus, 160.
—— arcasii, description of, 178—179; fin
formula, 179; distribution of, *ib.*; local
names of, *ib.*; hybrid of (with *chondrostoma
polylepis*), 206.

Leuciscus aula, 142—144; fin formula, 142; description of, *ib.* 143 : illustration of (Fig. 70), 142 ; pharyngeal bones of (Fig. 71), 143 ; size of, *ib.*; distribution of, 143 ; Günther, Heckel, and others on varieties of, 143, 144 ; var. *rubella*, 143 ; illustration of, Frontis. ; var. *basak*, 144 ; illustration of, Frontis.
—— arrigonis, Description of, 181, 182 ; fin formula, 181 ; spawning period of, 182 ; eggs of, *ib.* ; distribution of, *ib.*
—— basak, 144.
—— bearnensis, 163.
—— borysthenicus, 141.
—— cavedanus, 158.
—— cephalus, 155—160 ; fin formula, 155 ; distribution of, 155; description of, *ib.*, 157 ; Dr. Günther on, 156 ; illustration of (Fig. 82), 157 ; pharyngeal teeth of (Fig. 81), *ib.*; habitat of, *ib.*; food of, *ib.* ; Buckland on, 158 ; mode of catching its prey, *ib.*; baits for capture of, *ib.* ; Heckel on diseases of, *ib.* ; spawn of, *ib.* ; variability of species, *ib.*; var. *cavedanus* (Fig. 83), 158, 159 ; var. *albus* (Fig. 84), 160.
—— chalybæus, 163 ; illustration of, Frontis.
—— dergle, 136.
—— erythrophthalmus, 134—138 ; fin formula, 134 ; distribution of, 134, 138 ; characters of, 134, 135 ; illustration of (Fig. 61), 134 ; pharyngeal teeth of (Fig. 62), 135 ; variation in colour with locality, 135 ; size of, 136 ; habits and habitat of, *ib.* ; food of, *ib.*; spawning of, *ib.*; var. *dergle* (Fig. 63), 136, 137 ; var. *scardafa* (Figs. 65, 66), 137, 138 ; var. *plotizza* (Fig. 64), *ib.*; var. *macrophthalmus* (Fig. 67), *ib.*
—— friesii, description of, 153—155 ; distribution of, 153 ; illustration of (Fig. 79), 154 ; head of (Fig. 80), *ib.* ; colour of, *ib.*; development of tubercles at spawning season, 155 ; pharyngeal teeth of, *ib.*
—— heckelii, description of, 183 ; fin formula, *ib.*; colour of, *ib.*; distribution of, *ib.*
—— hispanicus, description of, 179 ; fin formula, *ib.* ; habitat of, *ib.*; distribution of, *ib.*; size of, *ib.*
—— idus, 139—142 ; local names of, 139 ; description of, *ib.*, 140 ; illustration of (Fig. 68), 139 ; pharyngeal teeth of (Fig. 69), *ib.* ; colour of, 140, 141 ; size of, 140 ; habits of, *ib.*; distribution of, 140, 141 ; var. *miniatus*, 141 ; var. *lapponicus*, *ib.*
—— illyricus, description of, 164, 165 ; fin formula, 164 ; illustration of (Fig. 88), *ib.* ; colour of, 165; size of, *ib.*; large scale of (Fig. 89), *ib.*
—— lancastriensis, 162.
—— lapponicus, 141.
—— lemmingii, 182, 183 ; fin formula, 182 ; colour of, 183 ; distribution of, *ib.*
—— macrolepidotus, description of, 181 ; fin formula, *ib.*
—— macrophthalmus, 138.
—— microlepis, description of, 169, 170 ; fin formula, 169 ; illustration of (Fig. 93), 170 ; distribution of, *ib.*
—— miniatus, 141.
—— monticellus, 172—175 ; fin formula, 172 ; description of, *ib.*, 174 ; illustration of (Fig. 95), 172 ; pharyngeal teeth of (Fig. 96), *ib.*; food of, 173 ; spawning of, 174 ; flesh of, *ib.* ; var. *sarigayi*, 174, 175.

Leuciscus pausingeri, illustration of (Fig. 74), 149.
—— phoxinus, 175—178 ; description of, 175, 177 ; fin formula, 175 ; local names of, *ib.* ; size of, *ib.* ; illustration of (Fig. 98), 176 ; pharyngeal teeth of (Fig. 99), *ib.*; variations in colour of, *ib.*; external sexual differences, 177 ; hatching of, natural and artificial, *ib.* ; habits of, *ib.* ; parasites of, 178 ; as articles of food, *ib.* ; distribution of, 178.
—— pigus, 150—153 ; fin formula, 150 ; illustration of (Fig. 75), *ib.*; pharyngeal bones of (Fig. 76), *ib.*; description of, 150, 153 ; head of (Fig. 77), 151 ; development of spots upon males at spawning-time, 152 ; seasonal colours of, 152, 153 ; vertebræ of, 153 ; capture of, 153 ; grossness of food of, *ib.* ; var. *virgo* (Fig. 78), 151.
—— roseus, 153.
—— pictus, description of, 165, 166 ; fin formula, 165 ; Günther on, *ib.*
—— plotizza, 138.
—— polylepis, description of, 185 ; fin formula, *ib.* ; distribution of, *ib.* ; period of spawning, *ib.*
—— pyrenaicus, description of, 184 ; fin formula, *ib.* ; distribution of, *ib.* ; Günther on, 184, 185 ; spawning of, 184.
—— rodens, 163.
—— rostratus, 164.
—— rubella, 143.
—— rutilus, 145—149 ; fin formula, 145 ; description of, 145—147 ; illustration of (Fig. 73) 146 ; size of, 147 ; Buckland on, *ib.* ; wide distribution of, *ib.*; Yarrell on, *ib.*; Grimm on, *ib.* ; commercial value of, 148 ; spawn of, *ib.*; origin of common name of, *ib.*; Lund on, *ib.* ; eggs of, *ib.*; hybernation of, *ib.* ; distribution of, *ib.*; Heckel and Kner on varieties of, 149 ; var. *pausingeri*, *ib.*
—— savignyi, 174.
—— scardafa, 137.
—— svallize, description of, 166, 167 ; fin formula, 166 ; distribution of, 166, 167.
—— tenellus, description of, 171 ; fin formula, *ib.* ; illustration of (Fig. 94). *ib.* ; colour of, *ib.*
—— turskyi, description of, 168, 169 ; fin formula, 168 ; Canestrini on limits of species, *ib.* ; illustration of (Fig. 92), 169.
—— ukliva, description of, 167, 168 ; fin formula, 167 ; distribution of, *ib.* ; size of, *ib.* ; illustration of (Fig. 91), *ib.* ; colour of, 168 ; hybrid of, 237.
—— virgo, 151.
—— vulgaris, 160—164 ; fin formula of, 160 ; local names of, *ib.*; description of, 160, 161 ; illustration of (Fig. 85), 161 ; colour of, *ib.*; habitat of, 162 ; food of, *ib.*; utility as a food fish, *ib.*; spawning period of, *ib.* ; capture of, *ib.* ; var. *lancastriensis*, 162 ; head of (Fig. 86), 163 ; var. *bearnensis*, *ib.* ; var. *chalybæus*, 163 ; var. *rodens*, *ib.*; var. *rostratus* (Fig. 87), *ib.*; var. *Cyprinus leuciscus* (Linnæus), 164.

Lota, 82—84 ; characters of genus, 82.

Lota vulgaris, 82—85; characters of, 82—84; illustration of (Fig. 30), 83; size of, 84; distribution and habitat of, 84, 85; spawning of, 85; monstrous form of, *ib.*; distribution of, *ib.*

Lucioperca, 38—44; characters of genus, 38.
—— sandra, 39; fin formula, *ib.*; description of, 39, 41; variation in size of scales of, 40; illustration of (Fig. 14), *ib.*; colour of, 41; weight of, *ib.*; distribution of, *ib.*; local names of, *ib.*; habitat and habits of, *ib.*; time and mode of spawning, 42; use of roe as food, *ib.*; mode of capture of, *ib.*; value as a food-fish, 42; artificial culture of, *ib.*; voracity of, *ib.*; diseases of, *ib.*
—— volgensis, 43, 44; description of, 43; distribution of, *ib.*; size of, *ib.*; illustration of (Fig. 15), *ib.*

Laciotrutta, 332; characters of genus, *ib.*
—— leucichthys, fin formula of, 332; distribution of, *ib.*; description of, *ib.*

Misgurnus, 245—248; characters of, 245.
—— fossilis, 245—248; fin formula, 245; description of, 245—247; illustration of (Fig. 140), 246; colour of, 247; size of, *ib.*; Bauham on habits of, *ib. et seq.*; flesh of, 248; food of, *ib.*; eggs of, *ib.*; distribution of, *ib.*

Mugil, 68—73; characters of genus, 68, 69; Dr. Günther on mode of feeding of, 68; distribution of, 69.
—— auratus, 71.
—— capito, 69, 70; fin formula, 69; distribution, 69, 70; description of, *ib.*; food of, 70; spawning of, *ib.*; capture of, *ib.*
—— cephalus, 70, 71; fin formula, 70; distribution of, *ib.*; description of, 71; weight of, *ib.*
—— chelo, 71, 72; fin formula, 71; distribution of, *ib.*; description of, 71, 72.
—— saliens, 71.
—— septentrionalis, 72, 73; fin formula, 72; description and distribution, 72, 73.

Mugilidae, 19, 68—73.
Muraenidae, 22, 372—380.

Nemachilus, 248—251; characters of, 248; distribution of, *ib.*
—— barbatulus, 249—251; fin formula, 249; origin of common name, *ib.*; local names of, *ib.*; illustration of (Fig. 141), *ib.*; characters of, *ib. et seq.*; colour of, *ib.*; habitat of, *ib.*; habits of, *ib.*; food of, *ib.*; smallness of air-bladder of, 251; spawning of, *ib.*; distribution of, *ib.*

Osmerus, 332—336; characters of genus, 332, 333; distribution of, 333.
—— eperlanus, 333—336; fin formula, 333; origin of popular name, *ib.*; local names of, *ib.*; distribution of, *ib.*; Yarrell on spawning of, *ib.*, 334; rivers, etc., it abundantly affects, 334, habits of, *ib.*; prejudices against, *ib.*; size of, 335; description of, *ib.*, 336;

capture of, 336; naturalisation of, *ib.*; Huxley on ovaries of, *ib.*

Paraphoxinus, 186—188; characters of genus, 186.
—— alepidotus, description of, 186, 187; illustration of (Fig. 100), 186; pharyngeal teeth of (Fig. 101), 187; absence of scales upon, *ib.*; colour of, *ib.*; distribution of, *ib.*
—— croaticus, description of, 187, 188; fin formula, 187; distribution of, 188; local names of, *ib.*

Pelecus, 242—244; characters of, 242; distribution of, *ib.*
—— cultratus, description of, 243, 244; fin formula, 243; illustration of (Fig. 138), *ib.*; pharyngeal teeth of (Fig. 139), 244; size of, *ib.*; colour of, *ib.*; distribution of, *ib.*

Perca, distinctive characters of genus, 23; distribution of, *ib.*
—— flavescens, 25.
—— fluviatilis, 23—28; fin formula, 23; description of, 23—25; illustration of (Fig. 10), 24; colour of, *ib.*; varieties of, 25, 26; size of, 26; habits of, *ib.*; habitat of, *ib.*; food of, *ib.*; modes of capturing, 27; slowness of growth in, *ib.*; weight of, 27; flesh of, *ib.*; as a food-fish, *ib.*; manufacture of isinglass from skin, *ib.*; deformities in, 28; diseases of, *ib.*; distribution of, *ib.*; origin of common name, *ib.*
—— italica, 26.
—— vulgaris, 26.

Percarina, 30; characters of genus, *ib.*; distribution of, *ib.*
—— demidoffii, description of, 31, 32; fin formula, 31; illustration of (Fig. 11), *ib.*; colour of, 32; distribution of, *ib.*

Percidae, 19, 23—32.

Petromyzon, 420, 429; distinctive characters of genus, 420, 421; Günther on genera of, 421.
—— branchialis, description of, 427—429; distribution of, 427; illustrations of, adult (Figs. 209, 210), 428; larval form of (Fig. 211), 429; spawning of, 428, local names of, *ib.*; distinctive characters of larvae, 428, 429; colour of, 429; food of, *ib.*
—— fluviatilis, 424—427; local names of, 424; distribution of, *ib.*; illustration of (Fig. 207), *ib.*; value for bait, *ib.*; food of, *ib.*, 425; description of, 425, 426; head of (Fig. 208), 425; colour of, 426; habitat of, *ib.*; spawning of, *ib.*; distinctive characters of young, 426, 427; method of capture in Germany, 427.
—— marinus, 421—424; local names of, 421; distribution of, *ib.*; habits and habitat of, origin of popular name of, *ib.*; food of, 421, 422; description of, 422, 423; illustration of Fig. 205), 422; head of (Fig. 206), *ib.*; colour of, 423; spawning of, *ib.*, 424.
—— wagneri, 427.

Petromyzontidae, 22, 420.
Physostomi, 2, 90—380.

Pleuronectes flesus, 85-87; fin formula, 85; distribution of, 85, 86; habits of, 86; adaptation of colour and locality, *ib.*; spawning of, *ib.*; eggs of, *ib.*; description of, 87.

Pleuronectes italicus, 87—89; fin formula, 87; distribution of, ib., 88; description of, ib.
Pleuronectidae, 19, 85.

Rhodeus, 207—210; characters and distribution of genus, 207.
—— amarus, 207—210; fin formula, 207; description of, 207, 209; illustration of (Fig. 114), 208; pharyngeal teeth of (Fig. 115), ib.; sexual and seasonal variations in colour, ib.; size of, 209; contents of stomach of, ib.; habitat of, 210; bitterness of flesh of, ib.; distribution of, ib.

Salmo, 265—332; characters of, 265, 267; Günther on, 266; influence of water and food upon colour, ib.; Günther on diagnosing specific characters of, ib. et seq.; distribution of, 267.
—— alpinus, fin formula of, 323; distribution of, ib.; description of, ib., 324.
—— argenteus, specific characters of, 278; description of, 283, 284; colour of, 284.
—— ausonii, 276.
—— autumnalis, specific characters of, 278; description of, 295.
—— bailloni, specific characters of, 278; fin formula, 314; description of, ib.; home of, ib.
—— brachypoma, specific characters of, 278; fin formula of, 309; home of, ib.; description of, ib.
—— cambricus, specific characters of, 278; fin formula, 282; distribution of, ib.; description of, ib., 283; food of, 283; hybrids of, ib.
—— carbonarius, home of, 328; description of, ib.
—— carpio, specific characters of, 278; fin formula, 296; illustration of (Fig. 151), ib.; description of, ib., 297; spawning of, 297; local names of, ib.
—— colii, fin formula, 326; home of, ib.; description of, ib.
—— dentex, specific characters of, 278; fin formula of, 298; powerful jaws of, ib.; description of, ib., 299; illustration of (Fig. 152), 298; colour of, 299; non-migration of, 300; flesh of, ib.; distribution of, ib.
—— fario, specific characters of, 278; fin formula, 288; critical remarks respecting varieties of, ib.; distribution of, ib., 289; description of, 289; habits of, ib., 290; food of, 290; flavour of flesh influenced by the nature of the stream, ib.; longevity of, ib.; var. ausonii, characters of, 276; distribution of, 290; description of, 290, 292; illustration of (Fig. 150), 291; sexual variation in dentition, ib.; variation of colour with food, 292; habitat of, ib.; size of, 293; habits of, ib.; food of, ib.; spawning of, ib.; eggs of, and their incubation, ib., 294; artificial breeding of, 294; diseases of, ib.; internal anatomy of, ib.
—— ferox, specific characters of, 278; fin formula of, 313; distribution of, ib.; description of, ib.; colour of, ib.; Mr. C. St. John on catching, ib., 314.
Salmo gallivensis, specific characters of, 278; fin formula, 291; description of, ib., 295; parr marks of, 295.
—— genivittatus, special characters of, 278; fin formula, 315; home of, ib.; illustration of (Fig. 160), ib.; description of, ib., 316; colour of, ib.
—— grayi, fin formula, 328; home of, ib.; description of, ib., 329.
—— hardinii, characters of, 278, 281; distribution of, 281; colour of, ib.
—— hucho, fin formula, 329; home of, ib.; illustration of (Fig. 163), ib.; description of, 329, 330; colour of, 330; size of, 331; spawning of, ib.; habitat of, ib.; diseases of, ib.
—— killinensis, fin formula, 325; description of, ib.; excessive development of fins of, ib.
—— labrax, specific characters of, 278; home of, 316; account of, ib.
—— lacustris, specific characters of, 278; fin formula, 301; description of, ib., 305; illustration of (Fig. 156), 304; vomer of (Fig. 157), ib.; voracity of, 305; sterile forms of, ib.
—— lacustris; var. schiffermülleri, 305; illustration of (Fig. 158), ib.; vomer of (Fig. 159), ib.; description of, ib., 306; colour of, 306; weight of, ib.; habits of, ib.; spawning of, time of unknown, 307.
—— lemanus, specific characters of, 278; fin formula, 295; home of, ib.; description of, ib., 296.
—— levenensis, specific characters of, 278; fin formula, 307; home of, ib.; description of, ib., 308; parr of, ib.; hybrids of, ib.
—— lossos, fin formula, 331; distribution of, ib.; description of, 331, 332.
—— marsiglii, specific characters of, 278, 301; fin formula, 301; distribution of, ib.; description of, 302, 303; illustration of (Fig. 154), 302; vomer of (Fig. 155), ib.; colour of, 303, ib.; size of, ib.; food of, ib.; mode of hunting the prey, ib.; spawning of, ib.; habits of, 304; flesh of, ib.
—— microlepis, specific characters of, 278; fin formula, 288; description of, ib.
—— mistops, specific characters of, 278; fin formula, 285; description of, ib.
—— nigripinnis; specific characters of, 278; fin formula, 285; distribution of, ib.; description of, ib., 286; occurrence of sterile specimens of, 286.
—— nivalis, fin formula, 324; distribution of, ib.; description of, ib.
—— obtusirostris, specific characters of, 278; fin formula of, 286; distribution of, ib.; description of, ib., 287; illustration of (Fig. 147), 286; vomer of Fig. 148), 287; vomer, palatines, and maxillary arch of (Fig. 149), ib.; colour of, ib., 288; non-migration of, 288.
—— orcadensis, specific characters of, 278; fin formula, 308; races of, ib.; description of, 309.
—— perisii, fin formula, 325; distribution of, ib.; description of, ib., 326.
—— polyosteus, specific characters of, 278; fin formula, 309; home of, ib.; description of, ib.
—— rappii, specific characters of, 278; fin

formula of, 297; habitat of, *ib.*; description of, *ib.*, 298.
Salmo rutilus, home of, 327; colour of, *ib.*; specific characters of, *ib.*
—— salar, 267, 278; fin formula, 267; description of, 267—269; illustration of (Fig. 145), 267; head of (Fig. 146), 268; colour of, 269; Parr, Smolt, and Grilse, stages of, *ib.*; food of, *ib.*; cannibal propensities of, *ib.*; size of, 270; Buckland on age of, *ib.*; weight of, *ib.*; close season, *ib.*; length of, *ib.*; distribution of, *ib. et seq.*; salmon attractors, 271; habits of, *ib.*; spawning operations of, *ib. et seq.*; Pennant's account of spawning of, in Tweed, 273; other accounts of process, *ib.*; number of eggs of, *ib. et seq.*; hatching of eggs of, 274; habits of young, *ib.*; instinct of in finding native river, *ib.*; leaping powers of, 275; value of salmon fisheries, *ib. et seq.*; extermination of, in certain rivers, 276; artificial breeding of, *ib.*; hybrids of, 277; disease of, *ib. et seq.*
—— salvelinus, fin formula, 318; home of, *ib.*; illustration of (Fig. 161), *ib.*; description of, *ib.*, 319; colour of, 319, 320; size of, 320; sexual variations, *ib.*; food of, *ib.*; flesh of, 321; spawning of, *ib.*; how captured, *ib.*
—— schiffermülleri, 305.
—— spectabilis, specific characters of, 278; fin formula, 300; illustration of (Fig. 153), *ib.*; description of, 301; colour of, *ib.*; size of, *ib.*
—— stomachicus, characters of, 278; fin formula, 28; distribution of, *ib.*; origin of specific name, *ib.*; colour of, *ib.*; habitat of, 281; Day on, *ib.*
—— trutta, specific characters of, 278; fin formula, 310; distribution of, *ib.*, 312; habits of, *ib.*; abundance of in Tweed, *ib.*; food of, *ib.*; size of, 311; description of, *ib.*, 312; various fish identified with, *ib.*
—— —— umbla, fin formula, 321; distribution of, *ib.*; description of, 321, 322; illustration of, (Fig. 162), 322; colour of, *ib.*, 323; weight of, 343; flesh of, *ib.*
—— —— venernensis, specific characters of, 278; fin formula of, 284; distribution of, *ib.*; Lloyd on, *ib.*; colour of, *ib.*
—— willughbii, fin formula, 326; distribution of, *ib.*, 327; description of, 327.
Salmonidæ, 21, 265— 359; characters of family, 265; genera of, *ib.*
Salvelini, 317—332; characters of sub-genus, 317; distribution of, *ib.*; critical remarks concerning species of, *ib.*, 318.

Siluridæ, 19, 90—93; characters of family, 90.
Silurus glanis, 91—93; fin formula, 91; characters, 91, 92; illustration of (Fig. 31), 91; size and weight of, 92; habitat of, 92, 93; utility of barbels of, *ib.*; nature of food, *ib.*; superstitions relating to, *ib.*; spawning of, *ib.*; longevity of, *ib.*; flesh of, *ib.*; difficulty of capture of, *ib.*; distribution of, *ib.*; varieties of, *ib.*
Solea vulgaris, 88, 89; fin formula, 88; habits of, *ib.*; food of, *ib.*; description of body of, 89; distribution of, *ib.*
Squalius, 240.

Thymallus, 352—359; distribution of genus, 352; distinctive characters of, *ib.*
—— æliani, 357.
—— microlepis, fin formula, 358; description of, *ib.*; colour of, *ib.*; home of, 359.
—— vulgaris, 353—359; fin formula, 353; local names of, *ib.*; origin of generic name of, *ib.*; habits of, *ib.*; illustration of (Fig. 167), *ib.*; food of, 354; habitat of, *ib.*; spawning of, *ib.*; flesh of, *ib.*; method of capture of, *ib.*, 355; absence of from Ireland, 355; distribution of, *ib.*; description of, 355 —357; colour of, 357; varieties of, 352, 353.
Tinca, 188—193; characters of genus, 188.
—— vulgaris, 189—193; description of, 189 - 191; fin formula, 189; illustration of (Fig. 102), *ib.*; smallness of scales of, 190; pharyngeal teeth of (Fig. 103), *ib.*; slippery skin of, cause of, *ib.*; influence of habitat upon colour, 191; size of, *ib.*; tenacity of life in, *ib.*; Canestrini on males of, *ib.*; spawning of, 191, 192; peculiar habits of, 192; hybernation of, *ib.*; food of, *ib.*; flesh of, *ib.*; methods of serving for table, Badham on, *ib.*; on its medicinal uses, 193; distribution of, *ib.*

Umbra, characters of genus, 366.
—— krameri, 366—369; fin formula, 366; illustration of (Fig. 169), 366; head of (Fig. 170), 367; scale of (Fig. 171), *ib.*; distribution of, 366, 369; description of, 367, 368; internal anatomy of, 368; habits of, *ib.*; movements of fin when swimming, *ib.*; superstition concerning, 369.
Umbridæ, 22, 366—369.

# INDEX TO SPECIFIC NAMES.

abramorutilus (Bliccopsis), 226.
aculeatus (Gasterosteus), 73.
acuminatus Carpio), 99.
adspersus (Leuciscus), 144.
aeliani Thymallus), 357.
agassizii (Leuciscus), 173.
albula (Coregonus), 348; var. *norwegica*, 349;
   var. *maraenula*, 349; var. *finnica*, 349.
alburnoides Leuciscus), 180.
alburnellus (Alburnus), 236; var. *fracchia*, 236.
albus (Leuciscus), 159.
alepidotus (Paraphoxinus), 186.
alosa (Clupea), 257.
alpestris Blennius), 65.
alpinus (Salmo), 323.
amarus (Rhodeus), 207.
arcasii (Leuciscus), 179, 181.
argenteus Salmo , 283.
argyropomus (Gasterosteus), 79.
arrigonis (Leuciscus), 181.
aula (Leuciscus), 142; var. *rubella*, 144; var.
   *basak*, 144.
aurantius (Carassius), 110.
auratus (Chondrostoma), 196.
ausonii (Salmo), 290.
autumnalis (Salmo), 295.
avernensis Gobius), 60.

bailloni (Salmo), 314.
ballerus (Abramis), 218.
barbatulus (Nemachilus), 249.
basak (Leuciscus), 144.
bearnensis (Leuciscus), 163.
bipunctatus (Abramis), 226.
bjoerkna (Abramis), 223; var. *Laskyr*, 225.
bocagei (Barbus), 122.
borysthenicus (Leuciscus), 141.
boyeri (Atherina), 66.
brachycentrus (Gasterosteus), 79.
brachypoma (Salmo), 309.
brama (Abramis), 211; var. *vetula*, 215; var.
   *gehini*, 215.
branchialis (Petromyzon), 427.
breviceps (Alburnus), 235.
—— (Gasterosteus), 81.
bucephalus (Carassius), 112.
burgundianus (Gasterosteus), 81.
burmeisteri Gobius), 61.

caerulescens Chondrostoma), 193.
calaritanus Cyprinodon), 369; var. *fasciatus*, 370.
cambricus Salmo), 282.
caninus (Barbus), 119.
capito (Mugil), 69.
carbonarius (Salmo), 328.

carpio (Cyprinus), 95; var. *acuminatus*, 99; var.
   *hungaricus*, 100; var. *regina*, 100; var.
   *kollarii*, 101.
carpio (Salmo), 296.
cavedanus (Leuciscus), 158.
cephalus Leuciscus), 155; var. *cavedanus*, 158;
   var. *albus*, 159.
cephalus (Mugil), 70.
cernua (Acerina), 33.
chalcoides (Alburnus), 239.
chalybeatus (Barbus), 124.
chalybaeus (Leuciscus), 163.
chelo (Mugil), 71.
clupeoides (Coregonus), 344.
colii (Salmo), 326.
comiza (Barbus), 121.
constructor (Gobius , 61.
croaticus (Paraphoxinus), 187.
cultratus (Pelecus), 243.

delineatus (Leucaspius), 240.
demidoffii Percarina), 31.
dentex Salmo), 298.
dergle Leuciscus), 136.
dremaei (Chondrostoma), 197.

eckströmii (Gobius), 63.
elongata (Cobitis), 254.
elongatus (Abramis), 217.
eperlanus (Osmerus), 333.
eques (Barbus), 118.
erythrophthalmus (Leuciscus), 134; var. *dergle*,
   136; var. *scardafa*, 137; var. *plotizza*, 137;
   var. *macrophthalmus*, 138.
eurystoma (Anguilla), 380.

fabraei (Alburnus), 235.
furio (Salmo), 288; var. *ausonii*, 290.
fasciatus (Abramis), 228.
—— (Cyprinodon), 370.
ferox Salmo), 313.
ferrugineus (Cottus), 54.
finnica (Coregonus), 349.
finta (Clupea), 261.
flesus (Pleuronectes), 85.
fluviatilis (Gobio), 128.
   —— (Gobius), 60.
   —— (Perca), 23.
   —— (Petromyzon), 424.
fossilis (Misgurnus), 245.
fracchia (Alburnus), 236.
friesii (Leuciscus), 153.

gallivensis Salmo), 294.
gehini (Abramis), 215.

# INDEX TO SPECIFIC NAMES. 441

genei (Chondrostoma), 199.
genivittatus (Salmo), 315.
gibelio (Carassius), 107.
glaber (Acipenser), 384.
glanis (Silurus), 91.
gmelini (Acipenser), 391.
gobio (Cottus), 50; var. *microstomus*, 50; var. *ferrugineus*, 54.
gracilis (Coregonus), 341.
grayi (Salmo), 328.
guiraonis (Barbus), 120.
güldenstädtii (Acipenser), 399.
gymnostrachelus (Gobius), 61.
gymnurus (Gasterosteus), 76.

hardinii (Salmo), 281.
harengus (Clupea), 262.
heckelii (Acipenser), 405.
—— (Leuciscus), 183.
hepsetus (Atherina), 66.
hiemalis (Coregonus), 345.
hispanicus (Fundulus), 371.
—— (Leuciscus), 179.
hucho (Salmo), 329.
hugelii (Aulopyge), 125.
humilis (Carassius), 107.
—— (Coregonus), 347.
hungaricus (Carpio), 100.
huso (Acipenser), 414.

ibericus (Cyprinodon), 370.
idus (Leuciscus), 139; var. *orfus*, 20; var. *miniatus*, 141; var. *lapponicus*, 141.
illyricus (Leuciscus), 164.
italicus (Pleuronectes), 87.

kessleri (Gobius), 61.
killinensis (Salmo), 325.
knerii (Chondrostoma), 202; var. *meigii*, 203.
kollarii (Carpio), 101.
krameri (Umbra), 366.

labrax (Salmo), 316.
lacteus (Gobius), 61.
lacustris (Alburnus), 235.
—— (Atherina), 67.
—— (Salmo), 304.
laevis (Gasterosteus), 81.
lapponicus (Leuciscus), 141.
larvata (Cobitis), 255.
latirostris (Acipenser), 214.
—— (Anguilla), 379.
lavaretus (Coregonus), 338.
lemanus (Salmo), 295.
lemmingii (Leuciscus), 182.
lancastriensis (Leuciscus), 162.
lapponicus (Coregonus), 340.
laskyr (Abramis), 225.
leucichthys (Luciotrutta), 332.
leuciscus (Carassius), 164.
leuckartii (Abramis), 220.
lovenensis (Salmo), 307.
lloydii (Coregonus), 338.
lossos (Salmo), 331.
lotharingus (Gasterosteus), 81.
lucidus (Alburnus), 232; var. *lacustris*, 235; var. *breviceps*, 235; var. *fabraei*, 235; var. *mirandella*, 236.

lucius (Esox), 360.
lugens (Gobius), 61.
lupus (Labrax), 29.

macrolepidotus (Leuciscus), 181.
macrophthalmus (Leuciscus), 138.
maraenula (Coregonus), 349.
marinus (Petromyzon), 421.
marsiglii (Salmo), 301.
martensii, (Gobius), 58; var. *avernensis*, 60; var. *panizzae*, 60; var. *punctatissimus*, 60.
maxillaris (Coregonus), 347; var. *humilis*, 347; var. *megalops*, 347.
megalops (Coregonus), 347.
meigii (Chondrostoma), 203.
mento (Alburnus), 237.
microlepis (Leuciscus), 168, 169.
—— (Salmo), 288.
—— (Thymallus), 358.
microstomus (Cottus), 53, 54.
miniatus (Leuciscus), 141.
minutus (Gobius), 63.
mirandella (Alburnus), 236.
mistops (Salmo), 285.
mocho (Atherina), 66.
moles (Carassius), 107.
muticellus (Leuciscus), 170, 172.

naccarii (Acipenser), 402; var. *nardoi*, 404.
nardoi (Acipenser), 404.
nasus (Acipenser), 407.
—— (Chondrostoma), 193; var. *caerulescens*, 196; var. *auratus*, 196; var. *dromaei*, 197; var. *rhodanensis*, 197.
nigripinnis (Salmo), 285.
nilssoni (Coregonus), 348.
nivalis (Salmo), 324.
norwegica (Coregonus), 349.
noveboracensis (Gasterosteus), 76.

oblongus (Carassius), 109.
obtusirostris (Salmo), 286.
occidentalis (Gasterosteus), 81.
orcadensis (Salmo), 308.
oxyrhynchus (Coregonus), 337.

panizzae (Gobius), 60.
pausingeri (Leuciscus), 149.
perisii (Salmo), 325.
petenyi (Barbus), 120.
phoxinus (Chondrostoma), 197.
—— (Leuciscus), 175.
pictus (Leuciscus), 164.
pigus (Leuciscus), 150; var. *virgo*, 151; var. *roseus*, 155.
platessa (Pleuronectes), 88.
plebejus (Barbus), 118.
plotizza (Leuciscus), 137.
poecilopus (Cottus), 55.
pollan (Coregonus), 349.
polylepis (Chondrostoma), 206.
—— (Leuciscus), 185.
polyosteus (Salmo), 309.
pontica (Clupea), 263.
punctatissimus (Gobius), 60.
pungitius (Gasterosteus), 80.
pyrenaicus (Leuciscus), 184.

rapax (Aspius), 230.
rappii (Salmo), 297.
regina (Carpio), 100.
rhodanensis (Chondrostoma), 197.
rodens (Leuciscus), 163.
rossica (Acerina), 38.
rostratus (Leuciscus), 164 ; var. *chalybæus*, 163 ; var. *rodens*, 163 ; var. *rostratus*, 164.
rubella (Leuciscus), 144.
ruthenus (Acipenser), 388 ; var. *gmelini*, 391.
rutilus (Leuciscus), 145 ; var. *pausingeri*, 149.
—— (Salmo), 327.
rysela (Chondrostoma), 203.
—— (Leuciscus), 173.

salar (Salmo), 267.
salvelinus (Salmo), 318.
sandra (Lucioperca), 39.
sapa (Abramis), 219.
semiarmatus (Gasterosteus), 76.
semiloricatus (Gasterosteus), 76.
semilunatus (Gobius), 61.
septentrionalis (Mugil), 72.
scardafa (Leuciscus), 137.
schraetzer (Acerina), 36.
schypa (Acipenser), 396.
sclateri (Barbus), 123.
scorpius (Cottus), 56.
soëtta (Chondrostoma), 200.
spectabilis (Salmo), 300.
spinulosus (Gasterosteus), 80.
stellatus (Acipenser), 393.
stomachicus (Salmo), 280.
streber (Aspro), 48.
sturio (Acipenser), 409.
svallize (Leuciscus), 166.

tenellus (Leuciscus), 168, 171.
tenaea (Cobitis), 252 ; var. *elongata*, 254.
trachurus (Gasterosteus), 76.
trutta (Salmo), 310.
turskyi (Leuciscus), 168.

ukliva (Leuciscus), 167.
umbla (Salmo), 321.
uranoscopus (Gobio), 131.

vandesius (Coregonus), 351.
venernensis (Salmo), 284.
vetula (Abramis), 215.
vimba (Abramis), 215.
volgensis (Lucioperca), 43.
vulgaris (Anguilla), 373 ; var. *latirostris*, 379.
—— (Aspro), 46.
 (Barbus), 114.
—— (Blennius), 64 ; var. *alpestris*, 65.
—— (Carassius), 104 ; var. *moles*, 107 ; var. *gibelio*, 107 ; var. *humilis*, 107 ; var. *oblongus*, 109.
—— (Leuciscus), 160 ; var. *lancastriensis*, 162 ; var. *braunensis*, 163 ; var. *burdigalensis*, 163.
—— (Lota), 82.
—— (Solea), 88.
—— (Thymallus), 353 ; var. *æliani*, 357.
—— (Tinca), 189.

wartmanni (Coregonus), 341.
widegreni (Coregonus), 341.
willkommii (Chondrostoma), 205.
willughbii (Salmo), 326.

zingel (Aspro), 44.

# INDEX TO COMMON NAMES.

Aal-rutte (*Lota vulgaris*), 14, 82.
Ablette (*Alburnus lucidus*), 232.
Aborren (*Perca fluriatilis*), 23.
"Adventurer," 294.
Æsche (*Thymallus vulgaris*), 18, 353.
Alpine Trout, 292.
Apron, The (*Aspro vulgaris*).
Arundel Mullet, 73.

Bachforelle (*River-trout*), 16, 289.
Baggit, 271.
Barbeau (*Barbus vulgaris*), 14, 114.
Barbel (*Barbus vulgaris*), 114.
Barsch (*Perca fluriatilis*), 23.
Bass, The (*Labrax lupus*), 29.
Beluga, 415.
Bitterling (*Rhodeus amarus*), 210.
Black Sea Herring, 263.
Black-tail, 311.
Black Trout, 292.
Bleak, The (*Alburnus lucidus*), 232—235.
Blei (*Abramis brama*), 16, 211.
Blennies, 63—66.
Bordelière (*Abramis bjorkna*), 16, 226.
Bottom Trout of Lake Constance, 297.
Bream, The (*Abramis brama*), 211, 215.
Breamflat, 223.
Bullhead, The River (*Cottus gobio*), 50—55.
——, The Sea (*Cottus scorpius*), 56, 57.
Bull Trout, 312.
Burbot (*Lota vulgaris*), 82—85.

Candle-fish, 333.
Carp, common (*Cyprinus carpio*), 95—104; hybrid of, 102, 104.
——, The Bitter, 207.
—— Crucian, 104—110.
—— of Salonica, 104.
——, Prussian, 107.
Chabot de Rivière (*Cottus gobio*), 14, 50.
Charr, 317—332.
Chevaine (*Leuciscus cephalus*) 15, 155.
Chub, The (*Leuciscus cephalus*), 155—160.

Dace, The (*Leuciscus vulgaris*), 160—164.
Devil's-fish, 135.
Dobel (*Leuciscus cephalus*), 15, 155.
Dogfish (*Umbra krameri*), 366.

Eels, 372—380.
Eperlan (*Osmerus eperlanus*), 17, 333.
Epinoche (*Gasterosteus aculeatus*), 14, 73.
Father Lasher (*Cottus scorpius*), 56, 57.
Flounder (*Pleuronectes flesus*), 85—87.
Flussbarsch (*Perca fluriatilis*), 23.

Galway Sea Trout, 294.
Gangling, 141.

Gardon (*Leuciscus rutilus*), 15, 145.
Gersen (*Acerina cernua*), 33.
Gillaroo (*Salmo stomachicus*), 17, 280.
Gobies, 57—63.
Goby, spotted (*Gobius minutus*), 62.
Gold-fish (*Carassius auratus*), 110—112.
Gold Nose, 196.
Golden Carp, 141.
Golden Ide (*Leuciscus idus*), 141.
Golden Mackerel, 196.
Golden Tench, 141.
Gosen (*Lucioperca sandra*), 14, 39.
Goujon (*Gobio fluriatilis*), 15.
Graining, The, 162.
Grass Pike, 362.
Grayling (*Thymallus vulgaris*), 353.
Great Dalmatian Trout (*Salmo dentex*), 298.
Gremille (*Acerina cernua*), 33.
Grey Mullet, 69.
Grey Trout (*Salmo brachypoma*), 309.
Grilse, 269, 283.
Grundling (*Gobio fluriatilis*), 15, 128.
Gudgeon (*Gobio fluriatilis*), 128—130.

Hasling (*Leuciscus vulgaris*), 15, 160.
Hassnborren (*Labrax lupus*), 29.
Hecht (*Esox lucius*), 18, 360.
Hireling, 311.

Ide (*Leuciscus idus*), 139—142.
——, golden (*Leuciscus idus*), 141.

Jack, 360.

Karpfen (*Cyprinus carpio*), 14, 95.
Kaulbarsch (*Acerina cernua*), 35.
Kaulkopf (*Cottus gobio*), 14, 50.
Kelts, 269—278.
King Carp, 99.
Kippers, 269.

Lachs (*Salmo salar*), 16, 267.
Lake Trout (*Salmo lacustris*), 304.
Lampern (*Petromyzon fluriatilis*), 424.
Lamprey (*Petromyzon marinus*), 421—429.
Leather Carp, 99.
Loach, The (*Nemachilus barbatulus*), 249, 251.
Loach, The Spinous (*Cobitis taenia*) 252, 255.
Loch Leven Trout (*Salmo levenensis*), 307.
Loche d'Etang (*Misgurnus fossilis*), 16, 245.
Loche de Rivière (*Cobitis taenia*), 16, 252.
Loche Franche (*Nemacheilus barbatulus*), 16, 249.
Luce, 360.
Lupo (*Labrax lupus*), 29.

Maifisch (*Clupea alosa*), 16, 257.
Meerforelle (*Salmo trutta*), 17, 310.
Miller's Thumb (*Cottus gobio*), 50-55.
Minnow, The (*Leuciscus phoxinus*), 175—178.
Mountain Trout, 292.
Mouse Eater, The, 157.
Mud-lamprey (*Petromyzon branchialis*), 427.
Mullets, 68-73.
Myxus, 69.

Nasling (*Chondrostoma nasus*), 15, 193.
Nerfling *Leuciscus idus*), 15, 139.
Neunauge (*Petromyzon fluviatilis*), 18, 206.
Nosary (*Acerina rossica*), 38.

Parr, 269, 283, 285, 288, 293, 295, 308.
Perch, The (*Perca fluviatilis*), 23—28.
Perche (*Perca fluviatilis*), 23.
Perlfisch, 155.
Persico (*Perca fluviatilis*), 23.
Pfrille (*Leuciscus phoxinus*), 15, 175.
Pickerel, 360.
Pike (*Esox lucius*), 360—365.
——— King, 362.
——— Perch (*Lucioperca sandra*).
Pink, 271.
Pollan (*Coregonus pollan*), 350.
Pomeranian Bream (.*Abramis leukartii*), 222.
Pond Loach, 245.
Pope, The (.*Acerina cernua*), 33—36; " plugging a Pope," 36
Powan, 344.
Pride, The (*Petromyzon branchialis*), 427.

"Quixote," 294.

Red Eye, 134—138.
River Bullhead (*Cottus gobio*), 50—55.
River Trout, 288.
Roach, The (*Leuciscus rutilus*), 145-149.
Rothauge (*Leuciscus erythrophthalmus*), 15, 134.
Rudd (*Leuciscus erythrophthalmus*), 134—138.
Ruff, The (.*Acerina cernua*), 33—36.

Saddle Carp, 99.
Salbling, 320.
Salmon, The (*Salmo salar*), 267-278.
——— Peal, 271.

Salmon Trout (*Salmo trutta*), 310.
——— Trout of Austria (*Salmo marsiglii*), 301.
Sander (*Lucioperca sandra*), 14, 39.
Sandpiper (*Petromyzon branchialis*), 427.
Sand Smelt (.*Atherina presbyter*), 66.
Schlammbeissor (*Pond loach*), 245.
Schleihe (*Tinca vulgaris*), 15, 189.
Schrasen (.*Acerina schraetzer*), 14, 36.
Sea Bullhead (*Cottus scorpius*), 56, 57.
Sekrete (*Lucioperca sandra*), 14, 39.
Shad, The (*Clupea alosa*), 257—261.
Shedder, 271.
Smelts, 332-336.
Smolt, 269, 283.
Sole (*Solea vulgaris*), 88, 89.
Soodake (*Lucioperca sandra*), 14, 39.
" Soosh," 27.
Spitter, The, 194.
Spotted Goby (*Gobius minutus*).
Stachelfisch (*Gasterosteus aculeatus*), 14, 73.
Sterlet (.*Acipenser ruthenus*), 388.
Stichling (*Gasterosteus aculeatus*), 14, 73.
Stickleback (*Gasterosteus aculeatus*), 73-80.
———, four spined, 80.
———, ten spined, 80, 81.
Sturgeons, 381—419.

Telescope fish, 111.
Tench, The (*Tinca vulgaris*), 189-193.
Trout, 278—332. Table of principal European species of, 278; specific diagnosing characters of, 279; Mr. Day on, *ib.*; change of colour under different internal conditions, *ib.*, 280.
Trout of Lake Garda, 296.
Twaite Shad (*Clupea finta*), 261—262.

Vairon (*Leuciscus phoxinus*), 15, 175.
Vaudoise (*Leuciscus vulgaris*), 15, 160.
Vendace (*Coregonus vandesius*), 351.

Wels, 93.
White Bream (.*Abramis (blicca) bjorkna*), 223—226.
White Salmon of Pennant, 310.
Whitebait (*Clupea harengus*), 262, 263.
Whiting, 311.

Zander (*Lucioperca sandra*), 14, 39.

*Selections from Cassell & Company's Publications.*

## Illustrated, Fine Art, and other Volumes.

**Art, The Magazine of.** Yearly Volume. With about 500 choice Engravings from famous Paintings, and from Original Drawings by the First Artists of the day. An Original Etching forms the Frontispiece. 16s.

**Art Directory and Year-Book of the United States.** With Engravings. 7s. 6d.

**Along Alaska's Great River.** By FREDERICK SCHWATKA. Illustrated. 12s. 6d.

**American Academy Notes.** Illustrated. 2s. 6d.

**After London; or, Wild England.** By RICHARD JEFFERIES. *Cheap Edition*, 3s. 6d.

**Artist, Education of the.** By E. CHESNEAU. Translated by CLARA BELL. 5s.

**Bismarck, Prince.** By C. LOWE, M.A. 2 Vols., demy 8vo. With 2 Portraits. 24s.

**Bright, Rt. Hon. John, Life and Times of.** By W. ROBERTSON. 7s. 6d.

**British Ballads.** 275 Original Illustrations. Two Vols. Cloth, 7s. 6d. each.

**British Battles on Land and Sea.** By JAMES GRANT. With about 600 Illustrations. Three Vols., 4to, £1 7s.; Library Edition, £1 10s.

**British Battles, Recent.** Illustrated. 4to, 9s. Library Edition, 10s.

**Butterflies and Moths, European.** By W. F. KIRBY. With 61 Coloured Plates. Demy 4to, 35s.

**Canaries and Cage-Birds, The Illustrated Book of.** By W. A. BLAKSTON, W. SWAYSLAND, and A. F. WIENER. With 56 Fac-simile Coloured Plates, 35s.; half-morocco, £2 5s.

**Cassell's Family Magazine.** Yearly Vol. Illustrated. 9s.

**Cathedral Churches of England and Wales.** Descriptive, Historical, and Pictorial. With 150 Illustrations. 21s.

**Changing Year, The.** With Illustrations. 7s. 6d.

**Choice Dishes at Small Cost.** By A. G. PAYNE. 3s. 6d. *Cheap Edition*, 1s.

**Cities of the World:** their Origin, Progress, and Present Aspect. Three Vols. Illustrated. 7s. 6d. each.

**Civil Service, Guide to Employment in the.** *New and Enlarged Edition.* 3s. 6d.

**Civil Service. Guide to Female Employment in Government Offices.** Cloth, 1s.

**Clinical Manuals for Practitioners and Students of Medicine.** (*A List of Volumes forwarded post free on application to the Publishers.*)

**Cobden Club, Works published for the:—**
Local Government and Taxation in the United Kingdom. 5s.
The Depression in the West Indies. 6d.
The Three Panics. 1s.
Free Trade versus Fair Trade. 2s. 6d.
England under Free Trade. 3d.
Pleas for Protection examined. 6d.
Free Trade and English Commerce. By A. Mongredien. 6d.
The Trade Depression. 6d.
Crown Colonies. 1s.
Our Land Laws of the Past. 3d.
Popular Fallacies Regarding Trade. 6d.
Reciprocity Craze. 2d.
Western Farmer of America. 3d.
Transfer of Land by Registration. 6d.
Reform of the English Land System. 3d.

**Colonies and India, Our, How we Got Them, and Why we Keep Them.** By Prof. C. RANSOME. 1s.

**Columbus, Christopher, The Life and Voyages of.** By WASHINGTON IRVING. Three Vols. 7s. 6d.

**Cookery, Cassell's Dictionary of.** Containing about Nine Thousand Recipes, 7s. 6d.; Roxburgh, 10s. 6d.

**Cookery, A Year's.** By PHYLLIS BROWNE. Cloth gilt or oiled cloth, 3s. 6d.

**Cook Book, Catherine Owen's New.** 4s.

**Co-operators, Working Men: What they have Done, and What they are Doing.** By A. H. DYKE-ACLAND, M.P., and B. JONES.

*Selections from Cassell & Company's Publications.*

**Countries of the World, The.** By ROBERT BROWN, M.A., Ph.D., &c. Complete in Six Vols., with about 750 Illustrations. 4to, 7s. 6d. each.

**Cromwell, Oliver:** The Man and his Mission. By J. ALLANSON PICTON, M.P. cloth, 7s. 6d. ; morocco, cloth sides, 9s.

**Cyclopædia, Cassell's Concise.** With 12,000 subjects, brought down to the latest date. With about 600 Illustrations, 15s. ; Roxburgh, 18s.

**Dairy Farming.** By Prof. J. P. SHELDON. With 25 Fac-simile Coloured Plates, and numerous Wood Engravings. Cloth, 31s. 6d. ; half-morocco, 42s.

**Decisive Events in History.** By THOMAS ARCHER. With Sixteen Illustrations. Boards, 3s. 6d. ; cloth, 5s.

**Decorative Design, Principles of.** By CHRISTOPHER DRESSER, Ph.D. Illustrated. 5s.

**Deserted Village Series, The.** Consisting of *Éditions de luxe* of favourite poems by Standard Authors. Illustrated. Cloth gilt, 2s. 6d.; or Japanese morocco, 5s. each
Goldsmith's Deserted Village. | Wordsworth's Ode on Immortality
Milton's L'Allegro and Il Penseroso. | and Lines on Tintern Abbey.

**Dickens, Character Sketches from.** SECOND and THIRD SERIES. With Six Original Drawings in each, by FREDERICK BARNARD. In Portfolio, 21s. each.

**Diary of Two Parliaments.** By H. W. LUCY. The Disraeli Parliament, 12s. The Gladstone Parliament, 12s.

**Dog, The.** By IDSTONE. Illustrated. 2s. 6d.

**Dog, Illustrated Book of the.** By VERO SHAW, B.A. With 28 Coloured Plates. Cloth bevelled, 35s. ; half-morocco, 45s.

**Domestic Dictionary, The.** An Encyclopædia for the Household. Cloth, 7s. 6d.

**Doré's Adventures of Munchausen.** Illustrated by GUSTAVE DORÉ. 5s.

**Doré's Dante's Inferno.** Illustrated by GUSTAVE DORÉ. *Popular Edition*, 21s.

**Doré's Don Quixote.** With about 400 Illustrations by DORÉ. 15s.

**Doré's Fairy Tales Told Again.** With 24 Full-page Engravings by DORÉ. 5s.

**Doré Gallery, The.** *Popular Edition.* With 250 Illustrations by GUSTAVE DORÉ. 4to, 42s.

**Doré's Milton's Paradise Lost.** With Full-page Drawings by GUSTAVE DORÉ. 4to, 21s.

**Edinburgh, Old and New, Cassell's.** With 600 Illustrations. Three Vols., 9s. each ; library binding, £1 10s. the set.

**Educational Year-Book, The.** 6s.

**Egypt: Descriptive, Historical, and Picturesque.** By Prof. G. EBERS. Translated by CLARA BELL, with Notes by SAMUEL BIRCH, LL.D., &c. Two Vols. With 800 Original Engravings. Vol. I., £2 5s. ; Vol. II., £2 12s. 6d.

**Electrician's Pocket-Book, The.** By GORDON WIGAN, M.A. 5s.

**Encyclopædic Dictionary, The.** A New and Original Work of Reference to all the Words in the English Language. Nine Divisional Vols. now ready, 10s. 6d. each; or the Double Divisional Vols., half-morocco, 21s. each.

**Energy in Nature.** By WM. LANT CARPENTER, B.A., B.Sc. 80 Illustrations. 3s. 6d.

**England, Cassell's Illustrated History of.** With 2,000 Illustrations. Ten Vols., 4to, 9s. each.

**English History, The Dictionary of.** Cloth, 21s, ; Roxburgh, 25s.

**English Literature, Library of.** By Prof. HENRY MORLEY.
VOL. I.—SHORTER ENGLISH POEMS, 12s. 6d.
VOL. II. ILLUSTRATIONS OF ENGLISH RELIGION, 11s. 6d.
VOL. III.—ENGLISH PLAYS, 11s. 6d.
VOL. IV.—SHORTER WORKS IN ENGLISH PROSE, 11s. 6d.
VOL. V. SKETCHES OF LONGER WORKS IN ENGLISH VERSE AND PROSE, 11s. 6d.
Five Volumes handsomely bound in half-morocco, £3 5s.
Volumes I., II., III., and IV. of the Popular Edition are now ready, price 7s. 6d. each.

*Selections from Cassell & Company's Publications.*

**English Literature, The Dictionary of.** By W. DAVENPORT ADAMS. *Cheap Edition*, 7s. 6d.; Roxburgh, 10s. 6d.

**English Literature, The Story of.** By ANNA BUCKLAND. 5s.

**English Poetesses.** By ERIC S. ROBERTSON, M.A. 5s.

**Æsop's Fables.** With about 150 Illustrations by E. GRISET. Cloth, 7s. 6d.; gilt edges, 10s. 6d.

**Etching: Its Technical Processes, with Remarks on Collections and Collecting.** By S. K. KOEHLER. Illustrated with 30 Full-page Plates. Price £4 4s.

**Etiquette of Good Society.** 1s.; cloth, 1s. 6d.

**Eye, Ear, and Throat, The Management of the.** 3s. 6d.

**Family Physician, The.** By Eminent PHYSICIANS and SURGEONS. Cloth, 21s.; half morocco, 25s.

**Far, Far West, Life and Labour in the.** By W. HENRY BARNEBY. With Map of Route. Cloth, 16s.

**Fenn, G. Manville, Works by.** *Popular Editions.* Cloth boards, 2s. each.
Sweet Mace. | The Vicar's People.
Dutch the Diver; or, a Man's Mistake. | Cobweb's Father, and other Stories.
My Patients. | The Parson o' Dumford.
Poverty Corner.

**Ferns, European.** By JAMES BRITTEN, F.L.S. With 30 Fac-simile Coloured Plates by D. BLAIR, F.L.S. 21s.

**Field Naturalist's Handbook, The.** By the Rev. J. G. WOOD and THEODORE WOOD. 5s.

**Figuier's Popular Scientific Works.** With Several Hundred Illustrations in each. 3s. 6d. each.
The Human Race. | The Ocean World.
World Before the Deluge. | The Vegetable World.
Reptiles and Birds. | The Insect World.
Mammalia.

**Fine-Art Library, The.** Edited by JOHN SPARKES, Principal of the South Kensington Art Schools. Each Book contains about 100 Illustrations. 5s. each.
Tapestry. By Eugène Muntz. Translated by Miss L. J. Davis. | The Flemish School of Painting. By A. J. Wauters. Translated by Mrs. Henry Rossel.
Engraving. By Le Vicomte Henri Delaborde. Translated by R. A. M. Stevenson. | Greek Archæology. By Maxime Collignon. Translated by Dr J. H. Wright.
The English School of Painting. By E. Chesneau. Translated by L. N. Etherington. With an Introduction by Prof. Ruskin. | Artistic Anatomy. By Prof. Duval. Translated by F. E. Fenton.
The Education of the Artist. By Ernest Chesneau. Crown 8vo, 352 pages, cloth gilt, 5s. | The Dutch School of Painting. By Henry Havard. Translated by G. Powell.

**Fisheries of the World, The.** Illustrated. 4to, 9s.

**Five Pound Note, The, and other Stories.** By G. S. JEALOUS. 1s.

**Forging of the Anchor, The.** A Poem. By Sir SAMUEL FERGUSON, LL.D. With 20 Original Illustrations. Gilt edges, 5s.; or Japanese morocco padded, 6s.

**Fossil Reptiles, A History of British.** By Sir RICHARD OWEN, K.C.B., F.R.S., &c. With 268 Plates. In Four Vols., £12 12s.

**Four Years of Irish History** (1845-49). By Sir GAVAN DUFFY, K.C.M.G. 21s.

**Franco-German War, Cassell's History of the.** Two Vols. With 500 Illustrations. 9s. each.

**Fresh-Water Fishes of Europe, The.** By Prof. H. G. SEELEY, F.R.S. Cloth, 21s.

**Garden Flowers, Familiar.** FIRST, SECOND, THIRD, and FOURTH SERIES. By SHIRLEY HIBBERD. With Original Paintings by F. E. HULME, F.L.S. With 40 Full-page Coloured Plates in each. Cloth gilt, in cardboard box (or in morocco, cloth sides), 12s. 6d. each.

**Gardening, Cassell's Popular.** Illustrated. Complete in 4 Vols., 5s. each.

**Gladstone, Life of the Rt. Hon. W. E.** By BARNETT SMITH. With Portrait, 3s. 6d. *Jubilee Edition*, 1s.

**Gleanings from Popular Authors.** Two Vols. With Original Illustrations. 4to, 9s. each. Two Vols. in One, 15s.

*Selections from Cassell & Company's Publications.*

**Great Industries of Great Britain.** With 400 Illustrations. 3 Vols., 7s. 6d. each.
**Great Painters of Christendom, The, from Cimabue to Wilkie.** By JOHN FORBES-ROBERTSON. Illustrated throughout. *Popular Edition*, cloth gilt, 12s. 6d
**Great Western Railway, The Official Illustrated Guide to the.** With Illustrations, 1s.; cloth, 2s.
**Gulliver's Travels.** With 88 Engravings by MORTEN. *Cheap Edition.* 5s.
**Gun and its Development, The.** By W. W. GREENER. Illustrated. 10s. 6d.
**Health, The Book of.** By Eminent Physicians and Surgeons. Cloth, 21s.; half-morocco, 25s.
**Health, The Influence of Clothing on.** By F. TREVES, F.R.G.S. 2s.
**Heavens, The Story of the.** By Sir ROBERT STAWELL BALL, LL.D., F.R.S., F.R.A.S., Royal Astronomer of Ireland. With 16 *Separate Plates* printed by Chromo-Lithography, and 90 Wood Engravings. Demy 8vo, cloth, 31s. 6d.
**Helps to Belief.** A Series of Helpful Manuals on the Religious Difficulties of the Day. Edited by the Rev. T. TEIGNMOUTH SHORE, M.A. Price 1s. each.
Creation. By the Lord Bishop of Carlisle
Prayer. By the Rev. T. Teignmouth Shore, M.A.
The Resurrection. By the Lord Archbishop of York
The Divinity of Our Lord. By the Lord Bishop of Derry.
God. By the Rev. Prof. Monerie, M.A.
Miracles. By the Rev. Brownlow Maitland, M.A.
The Atonement. By the Lord Bishop of Peterborough
The Morality of the Old Testament. By the Rev Newman Smyth, D.D.

**Heroes of Britain in Peace and War.** In Two Vols., with 300 Original Illustrations. Cloth, 5s. each. In one Vol., library binding, 10s. 6d.
**Homes, Our, and How to Make them Healthy.** By Eminent Authorities. Illustrated. 15s.; half-morocco, 21s.
**Horse, the Book of the.** By SAMUEL SIDNEY. With 25 *fac-simile* Coloured Plates. Demy 4to, 31s. 6d.; half-morocco, £2 2s.
**Horses, the Simple Ailments of.** By W. F. Illustrated. 5s.
**Household Guide, Cassell's.** With Illustrations and Coloured Plates. *New and Revised Edition*, complete in Four Vols., 20s.
**How Women may Earn a Living.** By MERCY GROGAN. 1s.
**India, Cassell's History of.** By JAMES GRANT. With 400 Illustrations. 15s.
**India, the Coming Struggle for.** By Prof. ARMINIUS VAMBÉRY. 5s.
**India: the Land and the People.** By Sir JAMES CAIRD, K.C.B. 10s. 6d.
**In-door Amusements, Cards Games, and Fireside Fun, Cassell's.** 2s. 6d.
**Industrial Remuneration Conference.** The Report of. 2s. 6d.
**Insect Variety: its Propagation and Distribution.** By A. H. SWINTON. 7s. 6d.
**Invisible Life, Vignettes from.** By J. BADCOCK, F.R.M.S. Illustrated. 3s. 6d.
**Irish Parliament, The, What it Was, and What it Did.** By J. G. SWIFT MCNEILL, M.A. 1s.
**Italy.** By J. W. PROBYN. 7s. 6d.
**Kennel Guide, Practical.** By Dr. GORDON STABLES. Illustrated. 2s. 6d.
**Khiva, A Ride to.** By the late Col. FRED BURNABY. 1s. 6d.
**Ladies' Physician, The.** By a London Physician. 6s.
**Land Question, The.** By Prof. J. ELLIOT, M.R.A.C. 10s. 6d.
**Landscape Painting in Oils, A Course of Lessons in.** By A. F. GRACE. With Nine Reproductions in Colour. *Cheap Edition*, 25s.
**Law, About Going to.** By A. J. WILLIAMS, M.P. 2s. 6d.
**Liberal, Why I am a.** By ANDREW REID. 2s. 6d. *People's Edition*, 1s.

*Selections from Cassell & Company's Publications.*

**London & North-Western Railway Official Illustrated Guide.** 1s.; cloth, 2s.
**London, Greater.** By EDWARD WALFORD. Two Vols. With about 400 Illustrations. 9s. each.
**London, Old and New.** By WALTER THORNBURY and EDWARD WALFORD. Six Vols., each containing about 200 Illustrations and Maps. Cloth, 9s. each.
**London's Roll of Fame.** With Portraits and Illustrations. 12s. 6d.
**Longfellow, H. W., Choice Poems by.** Illustrated by his Son, ERNEST W. LONGFELLOW. 6s.
**Longfellow's Poetical Works.** Illustrated. £3 3s.
**Love's Extremes, At.** By MAURICE THOMPSON. 5s.
**Mechanics, The Practical Dictionary of.** Containing 15,000 Drawings. Four Vols. 21s. each.
**Medicine, Manuals for Students of.** (*A List forwarded post free on application.*)
**Microscope, The; and some of the Wonders it Reveals.** 1s.
**Midland Railway, The Official Illustrated Guide to the.** 1s.; cloth, 2s.
**Modern Artists, Some.** With highly-finished Engravings. 12s. 6d.
**Modern Europe, A History of.** By C. A. FYFFE, M.A. Vol. I., from 1792 to 1814. 12s.
**Morocco: Its People and Places.** By EDMONDO DE AMICIS. Translated by C. ROLLIN TILTON. With nearly 200 Original Illustrations. 7s. 6d.
**National Library, Cassell's.** In Weekly Volumes, each containing about 192 pages. Paper covers, 3d.; cloth, 6d. *A list sent post free on application.*
**National Portrait Gallery, The.** Each Volume containing 20 Portraits, printed in Chromo-Lithography. Four Vols., 12s. 6d. each; or in Two Double Vols., 21s. each.
**Natural History, Cassell's Concise.** By E. PERCEVAL WRIGHT, M.A., M.D., F.L.S. With several Hundred Ilustrations. 7s. 6d. Roxburgh, 10s. 6d.
**Natural History, Cassell's New.** Edited by Prof. P. MARTIN DUNCAN, M.B., F.R.S., F.G.S. With Contributions by Eminent Scientific Writers. Complete in Six Vols. With about 2,000 high-class Illustrations. Extra crown 4to, cloth, 9s. each.
**Natural History, Cassell's Popular.** With about 2,000 Engravings and Coloured Plates. Complete in Four Vols. Cloth gilt, 42s.
**Nature, Short Studies from.** Illustrated. 5s.
**Nimrod in the North; or, Hunting and Fishing Adventures in the Arctic Regions.** By FREDERICK SCHWATKA. Illustrated. 7s. 6d.
**Nursing for the Home and for the Hospital, A Handbook of.** By CATHERINE J. WOOD. *Cheap Edition.* 1s. 6d.; cloth, 2s.
**On the Equator.** By H. DE W. Illustrated with Photos. 3s. 6d.
**Our Own Country.** Six Vols. With 1,200 Illustrations. Cloth, 7s. 6d. each.
**Outdoor Sports and Indoor Amusements.** With nearly 1,000 Illustrations. 9s.
**ainting, Practical Guides to.** With Coloured Plates.
  **Animal Painting in Water Colours.** By F. Tayler. 5s.
  **China Painting.** By Florence Lewis. 5s.
  **Figure Painting in Water Colours.** By B. Macarthur and J. Moore. 7s. 6d.
  **Flower Painting in Water Colours.** First and Second Series. Cloth, 5s. each.
  **Neutral Tint, A Course of Painting in.** By R. P. Leitch. 5s.
  **Tree Painting in Water Colours.** By W. H. J. Boot. 5s.
  **Water Colour Painting Book.** By R. P. Leitch. 5s.
  **Landscape Painting in Oils.** By A. F. Grace. 25s.
  **Sketching from Nature.** By Aaron Penley. 15s.
**Paris, Cassell's Illustrated Guide to.** 1s.; cloth, 2s.
**Parliaments, A Diary of Two.** By H. W. LUCY, The Disraeli Parliament, 1874-1880. 12s. The Gladstone Parliament. 12s.
**Paxton's Flower Garden.** By Sir JOSEPH PAXTON and Prof. LINDLEY. Revised by THOMAS BAINES, F.R.H.S. Three Vols. With 100 Coloured Plates. £1 1s. each.

*Selections from Cassell & Company's Publications.*

**Peoples of the World, The.** By Dr. ROBERT BROWN. Vols. I. to V. now ready. With Illustrations. 7s. 6d. each.

**Perak and the Malays.** By Major FRED MCNAIR. Illustrated. 10s. 6d.

**Photography for Amateurs.** By T. C. HEPWORTH. Illustrated, 1s.; or cloth, 1s. 6d.

**Phrase and Fable, Dictionary of.** By the Rev. Dr. BREWER. *Cheap Edition*, *Enlarged*, cloth, 3s. 6d.; or with leather back, 4s. 6d.

**Pictures from English Literature.** With Full-page Illustrations. 5s.

**Pictures of Bird Life in Pen and Pencil.** Illustrated. 21s.

**Picturesque America.** Complete in Four Vols., with 48 Exquisite Steel Plates, and about 800 Original Wood Engravings. £2 2s. each.

**Picturesque Canada.** With about 600 Original Illustrations. Two Vols. £3 3s. each.

**Picturesque Europe.** Complete in Five Vols. Each containing 13 Exquisite Steel Plates, from Original Drawings, and nearly 200 Original Illustrations. £10 10s.; half-morocco, £15 15s.; morocco gilt, £26 5s. The POPULAR EDITION is published in Five Vols., 18s. each, of which Four Vols. are now ready.

**Pigeon Keeper, The Practical.** By LEWIS WRIGHT. Illustrated. 3s. 6d.

**Pigeons, The Book of.** By ROBERT FULTON. Edited and Arranged by LEWIS WRIGHT. With 50 Coloured Plates and numerous Wood Engravings. 31s. 6d.; half-morocco, £2 2s.

**Poems and Pictures.** With numerous Illustrations. 5s.

**Poets, Cassell's Miniature Library of the :—**

Burns. Two Vols. 2s. 6d.
Byron. Two Vols. 2s. 6d.
Hood. Two Vols. 2s. 6d.
Longfellow. Two Vols. 2s. 6d.
Milton. Two Vols. 2s. 6d.
Scott. Two Vols. 2s. 6d.
Sheridan and Goldsmith. 2 Vols. 2s. 6d.
Wordsworth. Two Vols. 2s. 6d.
Shakespeare. Twelve Vols., in box, 15s.

**Police Code, and Manual of the Criminal Law.** By C. E. HOWARD VINCENT, M.P., late Director of Criminal Investigations. 2s.

**Popular Library, Cassell's.** A Series of New and Original Works. Cloth, 1s. each.

The Russian Empire.
The Religious Revolution in the Sixteenth Century.
English Journalism.
The Huguenots.
Our Colonial Empire.
The Young Man in the Battle of Life.
John Wesley.
The Story of the English Jacobins.
Domestic Folk Lore.
The Rev. Rowland Hill: Preacher and Wit.
Boswell and Johnson: their Companions and Contemporaries.
The Scottish Covenanters.
History of the Free-Trade Movement in England.

**Poultry Keeper, The Practical.** By L. WRIGHT. With Coloured Plates and Illustrations. 3s. 6d.

**Poultry, The Book of.** By LEWIS WRIGHT. *Popular Edition*. With Illustrations on Wood, 10s. 6d.

**Poultry, The Illustrated Book of.** By L. WRIGHT. With Fifty Exquisite Coloured Plates, and numerous Wood Engravings. Cloth, 31s. 6d.; half-morocco, £2 2s.

**Rabbit Keeper, The Practical.** By CUNICULUS. Illustrated. 3s. 6d.

**Rainbow Series, Cassell's.** Consisting of New and Original Works of Romance and Adventure by Leading Writers. 192 pages, crown 8vo, price 1s. each.

As it was Written. By S. LUSKA.
Morgan's Horror. By G. MANVILLE FENN.
A Crimson Stain. By A BRADSHAW.

**Rays from the Realms of Nature.** By the Rev. J. NEIL, M.A. Illustrated. 2s. 6d.

*Selections from Cassell & Company's Publications.*

**Red Library of English and American Classics, The.** Stiff covers, 1s. each ; cloth, 2s each ; or half-calf, marbled edges, 5s. each.

Washington Irving's Sketch-Book.
Last Days of Palmyra.
Tales of the Borders.
Pride and Prejudice.
Last of the Mohicans.
Heart of Midlothian.
Last Days of Pompeii.
Yellowplush Papers.
Handy Andy.
Selected Plays.

American Humour.
Sketches by Boz.
Macaulay's Lays and Selected Essays.
Harry Lorrequer.
Old Curiosity Shop.
Rienzi
The Talisman.
Pickwick (Two Vols.)
Scarlet Letter.

**Romeo and Juliet.** *Edition de Luxe.* Illustrated with Twelve Superb Photogravures from Original Drawings by F. DICKSEE, A.R.A. £5 5s.

**Royal River, The: The Thames, from Source to Sea.** With Descriptive Text and a Series of beautiful Engravings. £2 2s.

**Russia.** By D. MACKENZIE WALLACE, M.A. 5s.

**Russo-Turkish War, Cassell's History of.** With about 500 Illustrations. Two Vols., 9s. each ; library binding, One Vol., 15s.

**Sandwith, Humphry.** A Memoir by his Nephew, T. HUMPHRY WARD. 7s. 6d.

**Saturday Journal, Cassell's.** Yearly Volume. 6s.

**Science for All.** Edited by Dr. ROBERT BROWN, M.A., F.L.S., &c. With 1,500 Illustrations. Five Vols., 9s. each.

**Sea, The: Its Stirring Story of Adventure, Peril, and Heroism.** By F. WHYMPER. With 400 Illustrations. Four Vols., 7s. 6d. each.

**Sent Back by the Angels.** And other Ballads of Home and Homely Life. By FREDERICK LANGBRIDGE, M.A. 4s. 6d.

**Shakspere, The Leopold.** With 400 Illustrations, and an Introduction by F. J. FURNIVALL. Small 4to, cloth, 6s. ; cloth gilt, 7s. 6d. ; half-morocco, 10s. 6d. ; full morocco, £1 1s.

**Shakspere, The Royal.** With Exquisite Steel Plates and Wood Engravings. Three Vols. 15s. each.

**Shakespeare, Cassell's Quarto Edition.** Edited by CHARLES and MARY COWDEN CLARKE, and containing about 600 Illustrations by H. C. SELOUS. Complete in Three Vols., cloth gilt, £3 3s.—Also published in Three separate Volumes, in cloth, viz. :--The COMEDIES, 21s.; The HISTORICAL PLAYS, 18s. 6d. ; The TRAGEDIES, 25s.

**Shakespearean Scenes and Characters.** Illustrative of Thirty Plays of Shakespeare. With Thirty Steel Plates and Ten Wood Engravings. The Text written by AUSTIN BRERETON. Royal 4to, 21s.

**Sketching from Nature in Water Colours.** By AARON PENLEY. With Illustrations in Chromo-Lithography. 15s.

**Skin and Hair, The Management of the.** By MALCOLM MORRIS, F.R.C.S. 2s.

**Smith, The Adventures and Discourses of Captain John.** By JOHN ASHTON. Illustrated. 5s.

**Sports and Pastimes, Cassell's Book of.** With more than 800 Illustrations and Coloured Frontispiece. 768 pages, 7s. 6d.

**Steam Engine, The Theory and Action of the: for Practical Men.** By W. H. NORTHCOTT, C. E. 3s. 6d.

**Stock Exchange Year-Book, The.** By THOMAS SKINNER. 10s. 6d.

**Stones of London, The.** By E. F. FLOWER. 6d.

**"Stories from Cassell's."** 6d. each ; cloth lettered, 9d. each.

My Aunt's Match-making.
Told by her Sister.
The Silver Lock.
A Great Mistake.
"Running Pilot."
The Mortgage Money.
Gourlay Brothers.

\*\*\* The above are also issued, Three Volumes in One, cloth, price 2s. each.

**Sunlight and Shade.** With numerous Exquisite Engravings. 7s. 6d.

**Telegraph Guide, The.** Illustrated. 1s.

**Tot Book for all Public Examinations.** By W. S. THOMSON, M.A. 1s.

**Trajan.** An American Novel. By H. F. KEENAN. 7s. 6d.

**Transformations of Insects, The.** By Prof. P. MARTIN DUNCAN, M.B., F.R.S. With 240 Illustrations. 6s.

*Selections from Cassell & Company's Publications.*

---

**Treatment, The Year-Book of.** A Critical Review for Practitioners of Medicine and Surgery. 5s.

**Twenty Photogravures of Pictures in the Salon of 1885**, by the leading French Artists. In Portfolio. Only a limited number of copies have been produced, terms for which can be obtained of all Booksellers.

**"Unicode":** The Universal Telegraphic Phrase Book. 2s. 6d.

**United States, Cassell's History of the.** By EDMUND OLLIER. With 600 Illustrations. Three Vols., 9s. each.

**United States, Constitutional History and Political Development of the.** By SIMON STERNE, of the New York Bar. 5s.

**Universal History, Cassell's Illustrated.** With nearly ONE THOUSAND ILLUSTRATIONS. Vol. I. Early and Greek History.—Vol. II. The Roman Period.—Vol. III. The Middle Ages.—Vol. IV. Modern History. 9s. each.

**Vicar of Wakefield and other Works** by OLIVER GOLDSMITH. Illustrated. 3s. 6d.

**Wealth Creation.** By A. MONGREDIEN. 5s.

**Westall, W., Novels by.** *Popular Editions.* Cloth, 2s. each.
    The Old Factory. | Red Ryvington.
    Ralph Norbreck's Trust.

**What Girls Can Do.** By PHYLLIS BROWNE. 2s. 6d.

**Wild Animals and Birds: their Haunts and Habits.** By Dr. ANDREW WILSON. Illustrated. 7s. 6d.

**Wild Birds, Familiar.** First and Second Series. By W. SWAYSLAND. With 40 Coloured Plates in each. 12s. 6d. each.

**Wild Flowers, Familiar.** By F. E. HULME, F.L.S., F.S.A. Five Series. With 40 Coloured Plates in each. 12s. 6d. each.

**Winter in India, A.** By the Rt. Hon. W. E. BAXTER, M.P. 5s.

**Wise Woman, The.** By GEORGE MACDONALD. 2s. 6d.

**Wood Magic:** A Fable. By RICHARD JEFFERIES. 6s.

**World of the Sea.** Translated from the French of MOQUIN TANDON, by the Very Rev. H. MARTYN HART, M.A. Illustrated. Cloth. 6s.

**World of Wit and Humour, The.** With 400 Illustrations. Cloth, 7s. 6d.; cloth gilt, gilt edges, 10s. 6d.

**World of Wonders, The.** Two Vols. With 400 Illustrations. 7s. 6d. each.

---

*MAGAZINES.*

*The Quiver, for Sunday Reading.* Monthly, 6d.

*Cassell's Family Magazine.* Monthly, 7d.

*"Little Folks" Magazine.* Monthly, 6d.

*The Magazine of Art.* Monthly, 1s.

*Cassell's Saturday Journal.* Weekly, 1d.; Monthly, 6d.

\*<sub>\*</sub>\* Full particulars of CASSELL & COMPANY'S **Monthly Serial Publications**, numbering upwards of 50 Works, will be found in CASSELL & COMPANY'S COMPLETE CATALOGUE, sent post free on application.

**Catalogues** of CASSELL & COMPANY'S PUBLICATIONS, which may be had at all Booksellers', or will be sent post free on application to the Publishers:—
    CASSELL'S COMPLETE CATALOGUE, containing particulars of One Thousand Volumes.
    CASSELL'S CLASSIFIED CATALOGUE, in which their Works are arranged according to price, from *Sixpence to Twenty-five Guineas.*
    CASSELL'S EDUCATIONAL CATALOGUE, containing particulars of CASSELL & COMPANY'S Educational Works and Students' Manuals.

CASSELL & COMPANY, LIMITED, *Ludgate Hill, London.*

*Selections from Cassell & Company's Publications.*

## Bibles and Religious Works.

**Bible, The Crown Illustrated.** With about 1,000 Original Illustrations. With References, &c. 1,248 pages, crown 4to, cloth, 7s. 6d.

**Bible, Cassell's Illustrated Family.** With 900 Illustrations. Leather, gilt edges, £2 10s.; full morocco, £3 10s.

**Bible Dictionary, Cassell's.** With nearly 600 Illustrations. 7s. 6d.

**Bible Educator, The.** Edited by the Very Rev. Dean PLUMPTRE, D.D., Wells. With Illustrations, Maps, &c. Four Vols., cloth, 6s. each.

**Bunyan's Pilgrim's Progress (Cassell's Illustrated).** Demy 4to. Illustrated throughout. 7s. 6d.

**Bunyan's Pilgrim's Progress.** With Illustrations. *Popular Edition*, 3s. 6d.

**Child's Life of Christ, The.** Complete in One Handsome Volume, with about 200 Original Illustrations. Demy 4to, gilt edges, 21s.

**Child's Bible, The.** With 200 Illustrations. Demy 4to, 830 pp. 143*rd Thousand. Cheap Edition*, 7s. 6d.

**Church at Home, The.** A Series of Short Sermons. By the Rt. Rev. ROWLEY HILL, D.D., Bishop of Sodor and Man. 5s.

**Commentary, The New Testament, for English Readers.** Edited by the Rt. Rev. C. J. ELLICOTT, D.L., Lord Bishop of Gloucester and Bristol. In Three Volumes, 21s. each.
Vol. I.—The Four Gospels.
Vol. II.—The Acts, Romans, Corinthians, Galatians.
Vol. III.—The remaining Books of the New Testament.

**Commentary, The Old Testament, for English Readers.** Edited by the Rt. Rev. C. J. ELLICOTT, D.D., Lord Bishop of Gloucester and Bristol. Complete in 5 Vols., 21s. each.
Vol. I.—Genesis to Numbers. | Vol. III.—Kings I. to Esther.
Vol. II.—Deuteronomy to Samuel II. | Vol. IV.—Job to Isaiah.
Vol. V.—Jeremiah to Malachi.

**Day-Dawn in Dark Places; or Wanderings and Work in Bechwanaland, South Africa.** By the Rev. JOHN MACKENZIE. Illustrated throughout. 3s. 6d.

**Difficulties of Belief, Some.** By the Rev. T. TEIGNMOUTH SHORE, M.A. *New and Cheap Edition.* 2s. 6d.

**Doré Bible.** With 230 Illustrations by GUSTAVE DORÉ. 2 Vols., cloth, £2 10s.; Persian morocco, £3 10s.; Original Edition, 2 Vols., cloth, £8.

**Early Days of Christianity, The.** By the Ven. Archdeacon FARRAR, D.D., F.R.S.
LIBRARY EDITION. Two Vols., 24s.; morocco, £2 2s.
POPULAR EDITION. Complete in One Volume, cloth, 6s.; cloth, gilt edges, 7s. 6d.; Persian morocco, 10s. 6d.; tree-calf, 15s.

**Family Prayer-Book, The.** Edited by Rev. Canon GARBETT, M.A., and Rev. S. MARTIN. Extra crown 4to, cloth, 5s.; morocco, 18s.

**Geikie, Cunningham, D.D., Works by:—**
Hours with the Bible. Six Vols. 6s. each. | Old Testament Characters. 6s.
Entering on Life. 3s. 6d. | The Life and Words of Christ. Two Vols., cloth, 9s. *Students' Edition*, Two Vols., 19s.
The Precious Promises. 2s. 6d. |
The English Reformation. 5s. |

**Glories of the Man of Sorrows, The.** Sermons preached at St. James's, Piccadilly. By the Rev. H. G. BONAVIA HUNT. 2s. 6d.

**Gospel of Grace, The.** By A. LINDESIE. Cloth, 3s. 6d.

"**Heart Chords.**" A Series of Works by Eminent Divines. Bound in cloth, red edges, One Shilling each.

**My Father.** By the Right Rev. Ashton Oxenden, late Bishop of Montreal.
**My Bible.** By the Rt. Rev. W. Boyd Carpenter, Bishop of Ripon.
**My Work for God.** By the Right Rev. Bishop Cotterill.
**My Object in Life.** By the Ven. Archdeacon Farrar, D.D.
**My Aspirations.** By the Rev. G. Matheson, D.D.
**My Emotional Life.** By the Rev. Preb. Chadwick, D.D.
**My Body.** By the Rev. Prof. W. G. Blaikie, D.D.
**My Soul.** By the Rev. P. B. Power, M.A.
**My Growth in Divine Life.** By the Rev. Prebendary Reynolds, M.A.
**My Hereafter.** By the Very Rev. Dean Bickersteth.
**My Walk with God.** By the Very Rev. Dean Montgomery.
**My Aids to the Divine Life.** By the Very Rev. Dean Boyle.
**My Sources of Strength.** By the Rev. E. E. Jenkins, M.A., Secretary of the Wesleyan Missionary Society.

**Life of Christ, The.** By the Ven. Archdeacon FARRAR, D.D., F.R.S., Chaplain-in-Ordinary to the Queen.
  ILLUSTRATED EDITION, with about 300 Original Illustrations. Extra crown 4to, cloth, gilt edges, 21s.; morocco antique, 42s.
  LIBRARY EDITION. Two Vols. Cloth, 24s.; morocco, 42s.
  BIJOU EDITION. Five Volumes, in box, 10s. 6d. the set.
  POPULAR EDITION, in One Vol. 8vo, cloth, 6s.; cloth, gilt edges, 7s. 6d.; Persian morocco, gilt edges, 10s. 6d.; tree-calf, 15s.

**Marriage Ring, The.** By WILLIAM LANDELS, D.D. Bound in white leatherette, gilt edges, in box, 6s.; morocco, 8s. 6d.

**Martyrs, Foxe's Book of.** With about 200 Illustrations. Imperial 8vo, 732 pages, cloth, 12s.; cloth gilt, gilt edges, 15s.

**Moses and Geology; or, The Harmony of the Bible with Science.** By SAMUEL KINNS, Ph.D., F.R.A.S. Illustrated. *Cheap Edition.* 6s.

**Music of the Bible, The.** By J. STAINER, M.A., Mus. Doc. 2s. 6d.

**Near and the Heavenly Horizons, The.** By the Countess DE GASPARIN. 1s.; cloth, 2s.

**Patriarchs, The.** By the late Rev. W. HANNA, D.D., and the Ven. Archdeacon NORRIS, B.D. 2s. 6d.

**Protestantism, The History of.** By the Rev. J. A. WYLIE, LL.D. Containing upwards of 600 Original Illustrations. Three Vols., 27s.; Library Edition, 30s.

**Quiver Yearly Volume, The.** With 250 high-class Illustrations. 7s. 6d.

**Revised Version—Commentary on the Revised Version of the New Testament.** By the Rev. W. G. HUMPHRY, B.D. 7s. 6d.

**Sacred Poems, The Book of.** Edited by the Rev. Canon BAYNES, M.A. With Illustrations. Cloth, gilt edges, 5s.

**St. George for England;** and other Sermons preached to Children. By the Rev. T. TEIGNMOUTH SHORE, M.A. 5s.

**St. Paul, The Life and Work of.** By the Ven. Archdeacon FARRAR, D.D., F.R.S., Chaplain-in-Ordinary to the Queen.
  LIBRARY EDITION. Two Vols., cloth, 24s.; morocco, 42s.
  ILLUSTRATED EDITION, complete in One Volume, with about 300 Illustrations, £1 1s.; morocco, £2 2s.
  POPULAR EDITION. One Volume, 8vo, cloth, 6s.; cloth, gilt edges, 7s. 6d.; Persian morocco, 10s. 6d.; tree-calf, 15s.

**Secular Life, The Gospel of the.** Sermons preached at Oxford. By the Hon. W. H. FREMANTLE, Canon of Canterbury. 5s.

**Sermons Preached at Westminster Abbey.** By ALFRED BARRY, D.D., L.C.L., Primate of Australia. 5s.

**Shall We Know One Another?** By the Rt. Rev. J. C. RYLE, D.D., Bishop of Liverpool. *New and Enlarged Edition.* Cloth limp, 1s.

**Simon Peter: His Life, Times, and Friends.** By E. HODDER. 5s.

**Voice of Time, The.** By JOHN STROUD. Cloth gilt, 1s.

*Selections from Cassell & Company's Publications.*

## Educational Works and Students' Manuals.

**Alphabet, Cassell's Pictorial.** Size, 35 inches by 42½ inches. Mounted on Linen, with rollers. 3s. 6d.

**Algebra, The Elements of.** By Prof. WALLACE, M.A., 1s.

**Arithmetics, The Modern School.** By GEORGE RICKS, B.Sc. Lond. With Test Cards. (*List on application.*)

**Book-Keeping.** By THEODORE JONES. FOR SCHOOLS, 2s.; or cloth, 3s. FOR THE MILLION, 2s.; or cloth, 3s. Books for Jones's System. Ruled Sets of, 2s.

**Commentary, The New Testament.** Edited by Bishop ELLICOTT. Handy Volume Edition. Suitable for School and general use.

| | | |
|---|---|---|
| St. Matthew. 3s. 6d. | Romans. 2s. 6d. | Titus, Philemon, Hebrews, and James. 3s. |
| St. Mark. 3s. | Corinthians I. and II. 3s. | Peter, Jude, and John. 3s. |
| St. Luke. 3s. 6d. | Galatians, Ephesians, and Philippians. 3s. | The Revelation. 3s. |
| St. John. 3s. 6d. | Colossians, Thessalonians, and Timothy. 3s. | An Introduction to the New Testament. 2s. 6d. |
| The Acts of the Apostles. 3s. 6d. | | |

**Commentary, Old Testament.** Edited by Bishop ELLICOTT. Handy Volume Edition. Suitable for School and general use.

| | | |
|---|---|---|
| Genesis. 3s. 6d. | Leviticus. 3s. | Deuteronomy. 2s. 6d. |
| Exodus. 3s. | Numbers. 2s. 6d. | |

**Copy-Books, Cassell's Graduated.** Complete in 18 Books. 2d. each.

**Copy-Books. The Modern School.** Complete in 12 Books. 2d. each.

**Drawing Books for Young Artists.** 4 Books. 6d. each

**Drawing Books, Superior.** 3 Books. Printed in Fac-simile by Lithography, price 5s. each.

**Drawing Copies, Cassell's Modern School Freehand.** First Grade, 1s.; Second Grade, 2s.

**Drawing Copies, Cassell's Standard.** In 7 Books. Price 2d. each.

**Energy and Motion.: A Text-Book of Elementary Mechanics.** By WILLIAM PAICE, M.A. Illustrated. 1s. 6d.

**English Literature, A First Sketch of,** from the Earliest Period to the Present Time. By Prof. HENRY MORLEY. 7s. 6d.

**Euclid, Cassell's.** Edited by Prof. WALLACE, A.M. 1s.

**Euclid, The First Four Books of.** In paper, 6d.; cloth, 9d.

**French, Cassell's Lessons in.** *New and Revised Edition.* Parts I. and II., each 2s. 6d.; complete, 4s. 6d. Key, 1s. 6d.

**French-English and English-French Dictionary.** *Entirely New and Enlarged Edition.* 1,150 pages, 8vo, cloth, 3s. 6d.

**Galbraith and Haughton's Scientific Manuals.** By the Rev. Prof. GALBRAITH, M.A., and the Rev. Prof. HAUGHTON, M.D., D.C.L.

| | |
|---|---|
| Arithmetic. 3s. 6d. | Optics. 2s. 6d. |
| Plane Trigonometry. 2s. 6d. | Hydrostatics. 3s. 6d. |
| Euclid. Books I., II., III. 2s. 6d. Books IV., V., VI. 2s. 6d. | Astronomy. 5s. |
| Mathematical Tables. 3s. 6d. | Steam Engine. 3s. 6d. |
| Mechanics. 3s. 6d. | Algebra. Part I., cloth, 2s. 6d. Complete, 7s. 6d. |
| | Tides and Tidal Currents, with Tidal Cards, 3s. |

**German-English and English German Dictionary.** 3s. 6d.

**German Reading, First Lessons in.** By A. JAGST. Illustrated. 1s.

**German Reading, Modern.** By Prof. HEINEMANN. 1s. 6d.

**Handbook of New Code of Regulations.** By JOHN F. MOSS. 1s.; cloth, 2s.

**Historical Course for Schools, Cassell's.** Illustrated throughout. I.—Stories from English History, 1s. II.—The Simple Outline of English History, 1s. 3d. III.—The Class History of England, 2s. 6d.

**Latin-English and English-Latin Dictionary.** By J. R. BEARD, D.D., and C. BEARD, B.A. Crown 8vo, 914 pp., 3s. 6d.

*Selections from Cassell & Company's Publications.*

**Little Folks' History of England.** By ISA CRAIG-KNOX. With 30 Illustrations. 1s. 6d.

**Making of the Home, The** : A Book of Domestic Economy for School and Home Use. By Mrs. SAMUEL A. BARNETT. 1s. 6d.

**Marlborough Books.**

| Arithmetic Examples. 3s. | French Exercises. 3s. 6d. |
| Arithmetic Rules. 1s. 6d. | French Grammar. 2s. 6d. |
| German Grammar. 3s. 6d. | |

**Music, An Elementary Manual of.** By HENRY LESLIE. 1s.

**Natural Philosophy.** By Rev. Prof. HAUGHTON, F.R.S. Illustrated. 3s. 6d.

**Popular Educator, Cassell's.** *New and Thoroughly Revised Edition.* Illustrated throughout. Complete in Six Vols., 5s. each; or in Three Vols., half calf, 42s. the set.

**Physical Science, Intermediate Text-Book of.** By F. H. BOWMAN, D.Sc., F.R.A.S., F.L.S. Illustrated. 3s. 6d.

**Readers, Cassell's Readable.** Carefully graduated, extremely interesting, and illustrated throughout. *(List on application.)*

**Readers, Cassell's Historical.** Illustrated throughout, printed on superior paper, and strongly bound in cloth. *(List on application.)*

**Reader, The Citizen.** With Preface by the Rt. Hon. W. E. FORSTER, M.P. 1s. 6d.

**Readers, The Modern Geographical.** Illustrated throughout, and strongly bound in cloth. *(List on application.)*

**Readers, The Modern School.** Illustrated. *(List on application.)*

**Reading and Spelling Book, Cassell's Illustrated.** 1s.

**Right Lines ; or, Form and Colour.** With Illustrations. 1s.

**School Manager's Manual.** By F. C. MILLS, M.A. 1s.

**Shakspere Reading Book, The.** By H. COURTHOPE BOWEN, M.A. Illustrated. 3s. 6d. Also issued in Three Books, 1s. each.

**Shakspere's Plays for School Use.** 5 Books. Illustrated. 6d. each.

**Spelling, A Complete Manual of.** By J. D. MORELL, LL.D. 1s.

**Technical Manuals, Cassell's.** Illustrated throughout :—

| Handrailing and Staircasing. 3s. 6d. | Machinists and Engineers, Drawing for. 4s. 6d. |
| Bricklayers, Drawing for. 3s. | Metal-Plate Workers, Drawing for. 3s. |
| Building Construction. 2s. | Model Drawing. 3s. |
| Cabinet-Makers, Drawing for. 3s. | Orthographical and Isometrical Projection. 2s. |
| Carpenters and Joiners, Drawing for. 3s. 6d. | Practical Perspective. 3s. |
| Gothic Stonework. 3s. | Stonemasons, Drawing for. 3s. |
| Linear Drawing and Practical Geometry. 2s. | Applied Mechanics. By Prof. R. S. Ball, LL.D. 2s. |
| Linear Drawing and Projection. The Two Vols. in One, 3s. 6d. | Systematic Drawing and Shading. By Charles Ryan. 2s. |

**Technical Educator, Cassell's.** Illustrated throughout. Popular Edition, in Four Vols., 5s. each.

**Technology, Manuals of.** Edited by Prof. AYRTON, F.R.S., and RICHARD WORMELL, D.Sc., M.A. Illustrated throughout.

| The Dyeing of Textile Fabrics. By Prof. Hummel. 5s. | Spinning Woollen and Worsted. By W. S. Bright McLaren. 4s. 6d. |
| Watch and Clock Making. By D. Glasgow. 4s. 6d. | Design in Textile Fabrics. By T. R. Ashenhurst. 4s. 6d. |
| Steel and Iron. By W. H. Greenwood, F.C.S., Assoc. M.I.C.E., &c. 5s. | Practical Mechanics. By Prof. Perry, M.E. 3s. 6d. |
| Cutting Tools Worked by Hand and Machine. By Prof. Smith. 3s. 6d. | |

*Other Volumes in preparation. A Prospectus sent post free on application.*

*Selections from Cassell & Company's Publications.*

## Books for Young People.

"**Little Folks**" **Half-Yearly Volume.** With 200 Illustrations, 3s. 6d.; or cloth gilt, 5s.

**Bo-Peep.** A Book for the Little Ones. With Original Stories and Verses. Illustrated throughout. Boards, 2s. 6d.; cloth gilt, 3s. 6d

**The "Proverbs" Series.** Consisting of a New and Original Series of Stories by Popular Authors, founded on and illustrating well-known Proverbs. With Four Illustrations in each Book, printed on a tint. Crown 8vo, 160 pages, cloth, 1s. 6d. each.

Fritters; or, "It's a Long Lane that has no Turning." By Sarah Pitt.
Trixy; or, "Those who Live in Glass Houses shouldn't throw Stones." By Maggie Symington.
The Two Hardcastles; or, "A Friend in Need is a Friend Indeed." By Madeline Bonavia Hunt.
Major Monk's Motto; or, "Look Before you Leap." By the Rev. F. Langbridge.
Tim Thomson's Trial; or, "All is not Gold that Glitters." By George Weatherly.
Ursula's Stumbling-Block; or, "Pride comes before a Fall." By Julia Goddard.
Ruth's Life-Work; or, "No Pains, no Gains." By the Rev. Joseph Johnson.

**The "Cross and Crown" Series.** Consisting of Stories founded on incidents which occurred during Religious Persecutions of Past Days. With Four Illustrations in each Book, printed on a tint. Crown 8vo, 256 pages, 2s. 6d. each.

By Fire and Sword: a Story of the Huguenots. By Thomas Archer.
Adam Hepburn's Vow: a Tale of Kirk and Covenant. By Annie S. Swan.
No. XIII.; or, The Story of the Lost Vestal. A Tale of Early Christian Days. By Emma Marshall.

**The World's Workers.** A Series of New and Original Volumes by Popular Authors. With Portraits printed on a tint as Frontispiece. 1s. each.

General Gordon. By the Rev. S. A. Swaine.
Charles Dickens. By his Eldest Daughter.
Sir Titus Salt and George Moore. By J. Burnley.
Florence Nightingale, Catherine Marsh, Frances Ridley Havergal, Mrs. Ranyard ("L. N. R.") By Lizzie Alldridge.
Dr. Guthrie, Father Mathew, Elihu Burritt, Joseph Livesey. By the Rev. J. W Kirton.
Sir Henry Havelock and Colin Campbell, Lord Clyde. By E. C. Phillips.
Abraham Lincoln. By Ernest Foster.
David Livingstone. By Robert Smiles.
George Müller and Andrew Reed. By E. R. Pitman.
Richard Cobden. By R. Gowing.
Benjamin Franklin. By E. M. Tomkinson.
Handel. By Eliza Clarke.
Turner, the Artist. By the Rev. S. A. Swaine.
George and Robert Stephenson By C. L. Mateaux.

**The "Chimes" Series.** Each containing 64 pages, with Illustrations on every page, and handsomely bound in cloth, 1s.

Bible Chimes. Contains Bible Verses for Every Day in the Month.
Daily Chimes. Verses from the Poets for Every Day in the Month.
Holy Chimes. Verses for Every Sunday in the Year.
Old World Chimes. Verses from old writers for Every Day in the Month.

**New Five Shilling Books for Boys.** With Original Illustrations, printed on a tint. Cloth gilt, 5s. each.

"Follow my Leader;" or, the Boys of Templeton. By Talbot Baines Reed.
For Fortune and Glory; a Story of the Soudan War. By Lewis Hough.
The Champion of Odin; or, Viking Life in the Days of Old. By J. Fred. Hodgetts
Bound by a Spell; or, the Hunted Witch of the Forest. By the Hon. Mrs. Greene.

**New Three and Sixpenny Books for Boys.** With Original Illustrations, printed on a tint. Cloth gilt, 3s. 6d. each

On Board the "Esmeralda;" or, Martin Leigh's Log. By John C. Hutcheson.
For Queen and King; or, the Loyal 'Prentice. By Henry Frith.
In Quest of Gold; or, Under the Whanga Falls. By Alfred St. Johnston.

**The "Boy Pioneer" Series.** By EDWARD S. ELLIS. With Four Full-page Illustrations in each Book. Crown 8vo, cloth, 2s. 6d. each.

Ned in the Woods. A Tale of Early Days in the West.
Ned on the River. A Tale of Indian River Warfare.
Ned in the Block House. A Story of Pioneer Life in Kentucky.

*Selections from Cassell & Company's Publications.*

**The "Log Cabin" Series.** By EDWARD S. ELLIS. With Four Full-page Illustrations in each. Crown 8vo, cloth, 2s. 6d. each.

The Lost Trail. | Camp-Fire and Wigwam.

**Sixpenny Story Books.** All Illustrated, and containing Interesting Stories by well-known Writers.

Little Content.
The Smuggler's Cave.
Little Lizzie.
Little Bird.
The Boot on the Wrong Foot.
Luke Barnicott.

Little Pickles.
The Boat Club.
The Elchester College Boys.
My First Cruise.
The Little Peacemaker.
The Delft Jug.

Helpful Nellie; and other Stories.

**The "Baby's Album" Series.** Four Books, each containing about 50 Illustrations. Price 6d. each; or cloth gilt, 1s. each.

Baby's Album.
Dolly's Album.

Fairy's Album.
Pussy's Album

**Illustrated Books for the Little Ones.** Containing interesting Stories. All Illustrated. 1s. each.

Indoors and Out.
Some Farm Friends.
Those Golden Sands.
Little Mothers and their Children.

Our Pretty Pets.
Our Schoolday Hours.
Creatures Tame.
Creatures Wild.

**Shilling Story Books.** All Illustrated, and containing Interesting Stories.

Thorns and Tangles.
The Cuckoo in the Robin's Nest.
John's Mistake.
Pearl's Fairy Flower.
The History of Five Little Pitchers.
Diamonds in the Sand.
Surly Bob.
The Giant's Cradle.

Shag and Doll.
Aunt Lucia's Locket.
The Magic Mirror.
The Cost of Revenge.
Clever Frank.
Among the Redskins.
The Ferryman of Brill.
Harry Maxwell.

A Banished Monarch.

**Cassell's Children's Treasuries.** Each Volume contains Stories or Poetry, and is profusely Illustrated. Cloth, 1s. each.

Cock Robin, and other Nursery Rhymes.
The Queen of Hearts.
Old Mother Hubbard.
Simple Rhymes for Happy Times.
Tuneful Lays for Merry Days.
Cheerful Songs for Young Folks.
Pretty Poems for Young People.

The Children's Joy.
Pretty Pictures and Pleasant Stories.
Our Picture Book.
Tales for the Little Ones.
My Sunday Book of Pictures.
Sunday Garland of Pictures and Stories.
Sunday Readings for Little Folks.

**"Little Folks" Painting Books.** With Text, and Outline Illustrations for Water-Colour Painting. 1s. each.

Fruits and Blossoms for "Little Folks" to Paint.
The "Little Folks" Proverb Painting Book.

The "Little Folks" Illuminating Book.
Pictures to Paint.
"Little Folks" Painting Book.
"Little Folks" Nature Painting Book.

Another "Little Folks" Painting Book.

**Eighteenpenny Story Books.** All Illustrated throughout.

Three Wee Ulster Lassies.
Little Queen Mab.
Up the Ladder.
Dick's Hero; and other Stories.
The Chip Boy.
Raggles, Baggles, and the Emperor.
Roses from Thorns.
Faith's Father.

By Land and Sea.
The Young Berringtons.
Jeff and Leff.
Tom Morris's Error.
Worth more than Gold.
"Through Flood—Through Fire;" and other Stories.
The Girl with the Golden Locks.

Stories of the Olden Time.

*Selections from Cassell & Company's Publications.*

### The "Cosy Corner" Series.
Story Books for Children. Each containing nearly ONE HUNDRED PICTURES. 1s. 6d. each.

- See-Saw Stories.
- Little Chimes for All Times.
- Wee Willie Winkie.
- Pet's Posy of Pictures and Stories.
- Dot's Story Book.
- Story Flowers for Rainy Hours.
- Little Talks with Little People.
- Bright Rays for Dull Days.
- Chats for Small Chatterers.
- Pictures for Happy Hours.
- Ups and Downs of a Donkey's Life.

### The "World in Pictures" Series.
Illustrated throughout. 2s. 6d. each.

- A Ramble Round France.
- All the Russias.
- Chats about Germany.
- The Land of the Pyramids (Egypt).
- Peeps into China.
- The Eastern Wonderland (Japan).
- Glimpses of South America.
- Round Africa.
- The Land of Temples (India).
- The Isles of the Pacific.

### Two-Shilling Story Books.
All Illustrated.

- Stories of the Tower.
- Mr. Burke's Nieces.
- May Cunningham's Trial.
- The Top of the Ladder: How to Reach it.
- Little Flotsam.
- Madge and her Friends.
- The Children of the Court.
- A Moonbeam Tangle.
- Maid Marjory.
- The Four Cats of the Tippertons.
- Marion's Two Homes.
- Little Folks' Sunday Book.
- Two Fourpenny Bits.
- Poor Nelly.
- Tom Heriot.
- Through Peril to Fortune.
- Aunt Tabitha's Waifs.

In Mischief Again.

### Half-Crown Books.

- Little Hinges.
- Margaret's Enemy.
- Pen's Perplexities.
- Notable Shipwrecks.
- Golden Days.
- Wonders of Common Things.
- Little Empress Joan.
- Truth will Out.
- Pictures of School Life and Boyhood.
- The Young Man in the Battle of Life. the Rev. Dr. Landels.
- The True Glory of Woman. By the Rev. Dr. Landels.
- The Wise Woman. By George Macdonald.

Soldier and Patriot (George Washington).

### Picture Teaching Series.
Each book Illustrated throughout. Fcap. 4to, gilt, coloured edges, 2s. 6d. each.

- Through Picture-Land.
- Picture Teaching for Young and Old.
- Picture Natural History.
- Scraps of Knowledge for the Little Ones.
- Great Lessons from Little Things.
- Woodland Romances.
- Stories of Girlhood.
- Frisk and his Flock.
- Pussy Tip-Toes' Family.
- The Boy Joiner and Model Maker.
- The Children of Holy Scripture.

### Library of Wonders.
Illustrated Gift-books for Boys. 2s. 6d. each.

- Wonderful Adventures.
- Wonders of Animal Instinct.
- Wonders of Architecture.
- Wonders of Acoustics.
- Wonders of Water.
- Wonderful Escapes.
- Bodily Strength and Skill.
- Wonderful Balloon Ascents.

### Gift Books for Children.
With Coloured Illustrations. 2s. 6d. each.

- The Story of Robin Hood.
- Sandford and Merton.
- Playing Trades.
- True Robinson Crusoes. (Plain Illustrations.)
- Reynard the Fox.

The Pilgrim's Progress.

### Three and Sixpenny Library of Standard Tales, &c.
All Illustrated and bound in cloth gilt. Crown 8vo. 3s. 6d. each.

- Jane Austen and her Works.
- Better than Good.
- Mission Life in Greece and Palestine.
- The Dingy House at Kensington.
- The Romance of Trade.
- The Three Homes.
- My Guardian.
- School Girls.
- Deepdale Vicarage.
- In Duty Bound.
- The Half Sisters.
- Peggy Oglivie's Inheritance.
- The Family Honour.
- Esther West.
- Working to Win.
- Krilof and his Fables. By W. R. S. Ralston, M.A.
- Fairy Tales. By Prof. Morley.

*Selections from Cassell & Company's Publications.*

The "**Home Chat**" **Series.** All Illustrated throughout. Fcap. 4to. Boards, 3s. 6d. each ; cloth, gilt edges, 5s. each.

Home Chat.
Sunday Chats with Our Young Folks.
Peeps Abroad for Folks at Home.
Around and About Old England.

Half-Hours with Early Explorers.
Stories about Animals.
Stories about Birds.
Paws and Claws.

### Books for the Little Ones.

**The Little Doings of some Little Folks.** By Chatty Cheerful. Illustrated. 5s.
**The Sunday Scrap Book.** With One Thousand Scripture Pictures. Boards, 5s. ; cloth, 7s. 6d.
**Daisy Dimple's Scrap Book.** Containing about 1,000 Pictures. Boards, 5s. ; cloth gilt, 7s. 6d.
**Leslie's Songs for Little Folks.** Illustrated. 1s. 6d.
**The Little Folk's Out and About Book.** By Chatty Cheerful. Illustrated. 5s.
**Myself and my Friends.** By Olive Patch. With numerous Illustrations. Crown 4to. 5s.
**A Parcel of Children.** By Olive Patch. With numerous Illustrations. Crown 4to. 5s.
**Little Folks' Picture Album.** With 168 Large Pictures. 5s.

**Little Folks' Picture Gallery.** With 15 Illustrations. 5s.
**The Old Fairy Tales.** With Original Illustrations. Boards, 1s. ; cloth, 1s. 6d.
**My Diary.** With Twelve Coloured Plates and 366 Woodcuts. 1s.
**Three Wise Old Couples.** With 16 Coloured Plates. 5s.
**Old Proverbs with New Pictures.** With 64 Fac-simile Coloured Plates by Lizzie Lawson. The Text by C. L. Matéaux. 6s.
**Happy Little People.** By Olive Patch. With Illustrations. 5s.
**"Little Folks" Album of Music, The.** Illustrated. 3s. 6d.
**Elfie Under the Sea.** By E. L. Pearson. W Full-page Illustrations. 3s. 6d.

### Books for Boys.

**King Solomon's Mines.** A Thrilling Story founded upon an African Legend. By H. Rider Haggard. 5s.
**The Sea Fathers.** By Clements Markham. Illustrated. 2s. 6d.
**Treasure Island.** By R. L. Stevenson. With Full-page Illustrations. 5s.
**Half-Hours with Early Explorers.** By T. Frost. Illustrated. Cloth gilt, 5s.
**Modern Explorers.** By Thomas Frost. Illustrated. 5s.

**Cruise in Chinese Waters.** By Capt. Lindley. Illustrated. 5s.
**Wild Adventures in Wild Places.** By Dr. Gordon Stables, M.D., R.N. Illustrated. 5s.
**Jungle, Peak, and Plain.** By Dr. Gordon Stables, R.N. Illustrated. 5s.
**O'er Many Lands, on Many Seas.** By Gordon Stables, M.D., R.N. Illustrated. 5s.

### Books for all Children.

**Cassell's Robinson Crusoe.** A handsome Quarto Edition, with 100 striking Illustrations. Cloth, 3s. 6d. ; gilt edges, 5s.
**Cassell's Swiss Family Robinson.** Illustrated. Cloth, 3s. 6d. ; gilt edges, 5s.
**Sunny Spain: Its People and Places,** with Glimpses of its History. By Olive Patch. With Full-page Illustrations. 5s. 6d.
**Rambles Round London Town.** By C. L. Matéaux. Illustrated. 5s.
**Favorite Album of Fun and Fancy, The.** Illustrated. 3s. 6d.
**Familiar Friends.** By Olive Patch. Illustrated. Cloth gilt, 5s.
**Odd Folks at Home.** By C. L. Matéaux. With nearly 150 Illustrations. Extra fcap. 4to, cloth gilt, gilt edges, 5s.
**The Children's Own Paper.** With numerous Illustrations. Volumes, 3s. each.
**Field Friends and Forest Foes.** By Olive Patch. Profusely Illustrated. Extra fcap. 4to, cloth gilt, gilt edges, 5s.
**Silver Wings and Golden Scales.** By One of the Authors of "Poems Written for a Child." Illustrated. Cloth, gilt edges, 5s.

**The Wonderland of Work.** By C. L. Matéaux. With numerous Original Illustrations. Extra crown 4to, cloth gilt, 7s. 6d.
**Little Folks' Holiday Album.** A charming collection of Stories, &c. Illustrated throughout. Crown 4to, cloth, 3s. 6d.
**Tiny Houses and their Builders.** By the Author of "Poems Written for a Child." Illustrated. Fcap. 4to, cloth gilt, gilt edges, 5s.
**Children of all Nations.** Their Homes, their Schools, their Playgrounds. With Original Illustrations. 5s.
**Tim Trumble's "Little Mother."** A New Story for Young Folks. By C. L. Matéaux. With 18 Illustrations by Giacomelli. 5s.
**The World's Lumber Room.** By Selina Gaye. Illustrated. 5s.
**The Wonderland of Work.** By C. L. Matéaux. With numerous Original Illustrations. Extra crown 4to, cloth gilt, 7s. 6d.
**A Moonbeam Tangle.** Original Fairy Tales. By Sydney Shadbolt. With numerous Illustrations. Cloth, gilt edges, 3s. 6d.
**The Children's Album.** Containing nearly 200 Engravings, with Short Stories by Uncle John. Cloth gilt, 3s. 6d.

---

CASSELL & COMPANY, Limited, Ludgate Hill, London, Paris, New York and Melbourne.

www.ingramcontent.com/pod-product-compliance
Lightning Source LLC
Chambersburg PA
CBHW022107300426
44117CB00007B/623